Lecture Notes in Computer Science 2451

Edited by G. Goos, J. Hartmanis, and J. van Leeuwen

T0134785

Springer
Berlin
Heidelberg
New York
Barcelona
Hong Kong
London
Milan
Paris
Tokyo

Lecture Notes in Computer Science 2451

Springer
Berlin
Heidelberg
New York
Barcelona
Hong Kong
London
Milan
Paris
Tokyo

Bertrand Hochet   Antonio J. Acosta
Manuel J. Bellido (Eds.)

Integrated
Circuit Design

Power and Timing Modeling, Optimization and Simulation

12th International Workshop, PATMOS 2002
Seville, Spain, September 11-13, 2002
Proceedings

Springer

Series Editors

Gerhard Goos, Karlsruhe University, Germany
Juris Hartmanis, Cornell University, NY, USA
Jan van Leeuwen, Utrecht University, The Netherlands

Volume Editors

Bertrand Hochet
Swiss University of Applied Science
Ecole d'Ingénieurs du Canton de Vaud
Microelectronics and Systems Institute, Digital Communications Team
Rue de Galilée 15, 1400 Yverdon, Switzerland
E-mail: bertrand.hochet@eivd.ch

Antonio J. Acosta
Manuel J. Bellido
Universidad de Sevilla
Instituto de Microelectrónica de Sevilla-CNM-CSIC
Departamento do Diseño Digital
Departamento de Electrónica y Electromagnetismo
Departamento de Tecnología Electrónica
Avda. Reina Mercedes, s/n. Edif. CICA
E-mail: {acojim, bellido}@imse.cnm.es

Cataloging-in-Publication Data applied for

Die Deutsche Bibliothek - CIP-Einheitsaufnahme

Integrated circuit design : power and timing modeling, optimization and
simulation ; 12th international workshop ; proceedings / PATMOS 2002,
Seville, Spain, September 11 - 13, 2002. Bertrand Hochet ... (ed.). - Berlin ;
Heidelberg ; New York ; Barcelona ; Hong Kong ; London ; Milan ; Paris ;
Tokyo : Springer, 2002
    (Lecture notes in computer science ; Vol. 2451)
    ISBN 3-540-44143-3

CR Subject Classification (1998): B.7, B.8, C.1, C.4, B.2, B.6, J.6

ISSN 0302-9743
ISBN 3-540-44143-3 Springer-Verlag Berlin Heidelberg New York

Springer-Verlag Berlin Heidelberg New York,
a member of BertelsmannSpringer Science+Business Media GmbH

http://www.springer.de

© Springer-Verlag Berlin Heidelberg 2002
Printed in Germany

Typesetting: Camera-ready by author, data conversion by PTP-Berlin, Stefan Sossna e.K.
Printed on acid-free paper   SPIN: 10871110   06/3142   5 4 3 2 1 0

Preface

The International Workshop on Power and Timing Modeling, Optimization, and Simulation PATMOS 2002, was the 12th in a series of international workshops previously held in several places in Europe.[1] PATMOS has over the years evolved into a well-established and outstanding series of open European events on power and timing aspects of integrated circuit design. The increased interest, especially in low-power design, has added further momentum to the interest in this workshop. Despite its growth, the workshop can still be considered as a very focused conference, featuring high-level scientific presentations together with open discussions in a free and easy environment. This year, the workshop has been opened to both regular papers and poster presentations. The increasing number of worldwide high-quality submissions is a measure of the global interest of the international scientific community in the topics covered by PATMOS.

The objective of this workshop is to provide a forum to discuss and investigate the emerging problems in the design methodologies and CAD-tools for the new generation of IC technologies. A major emphasis of the technical program is on speed and low-power aspects with particular regard to modeling, characterization, design, and architectures. The technical program of PATMOS 2002 included nine sessions dedicated to most important and current topics on power and timing modeling, optimization, and simulation. The three invited talks try to give a global overview of the issues in low-power and/or high-performance circuit design. The first one presents a history of low-power clock watches, while the other ones deal with high performance clocking and the state of the art in low-voltage memories.

In 2002, the venue was Seville, Spain. The workshop was organized by the Microelectronics Spanish Center (IMSE-CNM) and the University of Seville. Seville is the major city in the southwest of Spain and one of its richest historic, cultural, and artistic spots. Seville is the capital of the Autonomous Community of Andalusia and has a population of over 700,000. Many very different cultures have figured in the history of Sevilla. The legacy of these cultures has grown over the centuries into the cultural, monumental, and artistic heritage that can be admired in the city's streets and museums today.

September 2002

Bertrand Hochet
Antonio J. Acosta
Jorge Juan-Chico

---

[1] Visit the web site: http://www.patmos-conf.org

Organization

Organization Committee

General Co-chairs:

Prof. Antonio J. Acosta
Prof. Jorge Juan-Chico
(Inst. de Microelectrónica de Sevilla, CNM;
Universidad de Sevilla, Spain)

Program Chair:

Dr. Bertrand Hochet
(Inst. of Microelectronics and Systems,
University of Applied Sciences,
Yverdon-les-Bains, Switzerland)

Local Committee

Finance Chair:
Prof. Manuel Valencia

Local Arrangements:
Prof. Carmen Baena
Prof. Pilar Parra

Publication and Web:
Prof. Manuel Bellido
Prof. Alejandro Millán
Prof. Paulino Ruiz de Clavijo
(Inst. de Microelectrónica de Sevilla, CNM;
Universidad de Sevilla, Spain)

Program Committee

D. Auvergne (U. Montpellier, France)
J. Bormans (IMEC, Belgium)
J. Figueras (U. Catalunya, Spain)
C.E. Goutis (U. Patras, Greece)
A. Guyot (INPG Grenoble, France)
R. Hartenstein (U. Kaiserslautern, Germany)
S. Jones (U. Loughborough, UK)
P. Larsson-Edefors (U. Linköping, Sweden)
E. Macii (Politecnico di Torino, Italy)
V. Moshnyaga (U. Fukuoka, Japan)
W. Nebel (U. Oldenburg, Germany)
J.A. Nossek (T.U. Munich, Germany)
A. Núñez (U. Las Palmas, Spain)
M. Papaefthymiou (U. Michigan, USA)
H. Pfleiderer (U. Ulm, Germany)
C. Piguet (CSEM, Switzerland)

R. Reis (U. Porto Alegre, Brazil)
M. Robert (U. Montpellier, France)
A. Rubio (U. Catalunya, Spain)
J. Sparsø (T.U. Denmark)
A. Stauffer (Swiss Fed. Inst. of Tech. Lausanne)
A. Stempkowsky (Acad. of Sciences, Russia)
T. Stouraitis (U. Patras, Greece)
A.-M. Trullemans-Anckaert (U. Louvain)
R. Zafalon (STMicroelectronics, Italy)

PATMOS Steering Committee

Daniel Auvergne (U. Montpellier, France)
Reiner Hartenstein (U. Kaiserslautern, Germany)
Wolfgang Nebel (U. Oldenburg, Germany)
Christian Piguet (CSEM, Switzerland)
Antonio Rubio (U. Catalunya, Spain)
Joan Figueras (U. Catalunya, Spain)
Bruno Ricco (U. Bologna, Italy)
Dimitrious Soudris (U. Trace, Greece)
Jean Sparsø (T.U. Denmark)
Anne Marie Trullemans-Anckaert (U. Louvain, Belgium)
Peter Pirsch (U. Hannover, Germany)
Bertrand Hochet (EIVd, Switzerland)
Antonio J. Acosta (U. Sevilla/IMSE-CNM, Spain)
Jorge Juan (U. Sevilla/IMSE-CNM, Spain)
Enrico Macii (Politecnico di Torino, Italy)

Executive Subcommittee for 2002

President: Christian Piguet
                        (CSEM, Switzerland)
Vice-president: Joan Figueras
                        (U. Catalunya, Spain)
Secretary: Reiner Hartenstein
                        (U. Kaiserslautern, Germany)

Sponsoring Institutions

- IEEE Circuits and Systems Society
- IEEE Circuits and Systems Society Spanish Chapter
- European Commission, 5th Framework Programme, Grant IST2001-92074
- Consejo Superior de Investigaciones Científicas (CSIC), Centro Nacional de Microelectrónica

- Universidad de Sevilla, Vicerrectorados de Investigación y de Extensión Universitaria
- Ministerio de Ciencia y Tecnología, Plan Nacional de Investigación Científica, Desarrollo e Innovación Tecnológica, Acción especial TIC2001-4774-E
- Junta de Andalucía. III Plan Andaluz de Investigación

Table of Contents

CAD Tools and Algorithms

Timing

Gate-Level Modeling

Memory Optimization

High-Level Modeling and Design

Communications Modeling and Activity Reduction

Posters

The First Quartz Electronic Watch

Christian Piguet

CSEM Centre Suisse d'Electronique et de Microtechnique SA
2000 Neuchâtel, Switzerland
LAP-EPFL, Lausanne, Switzerland
christian.piguet@csem.ch

Abstract. The goal of this paper is to present the history of the quartz electronic watch. The first quartz watch was a Swiss wristwatch presented in 1967. It was exactly 20 years after the invention of the transistor. Today Switzerland and Japan are dominant on the market of electronic watches. In Switzerland, the competencies in low-power electronics come directly from the watch industry.

> "The watch, almost more than the steam engine, was the real protagonist of the Industrial Revolution". Lewis Mumford, American social philosopher.

1 Introduction

In December 2002, we will celebrate the 35th anniversary of the first electronic watch, a Swiss quartz watch named Beta [1], developed by the Swiss Horological Center or Centre Electronique Horloger (CEH). This paper presents the history of the quartz electronic watch [1, 2, 3, pp. 66-88]. It shows how in Switzerland private research was performed at that time without any public support but pushed by some visionary people. Hopefully, this research was quite successful, producing the first quartz electronic watch. However, commercialization was quite difficult, due to the structure of the Swiss watch industry. The impact of this research has gone beyond the watch industry, as it opened the way to very low-power integrated circuits and microprocessors developed by the CEH and from 1984 by CSEM. A recent book in French [1] presents this interesting story.

2 Before the Quartz

Digital watches, employing ICs and analog or solid-state displays, trace their beginning to 1960 [4]. In that year Bulova introduced its Accutron tuning-fork watch [2] designed by a Swiss engineer, Max Hetzel. The frequency of the tuning fork was

B. Hochet et al. (Eds.): PATMOS 2002, LNCS 2451, pp. 1–15, 2002.
© Springer-Verlag Berlin Heidelberg 2002

360 Hz. The precision was one minute per month. The discrete-component watch established that electronic accuracy in timekeeping products was possible.

3 The Centre Electronique Horloger (CEH)

In the late fifties, the Swiss watch industry was very successful in the production of mechanical watches. However, the President of the Swiss Horological Federation, Gérard Bauer, a visionary and powerful President, was under the feeling that electronic could be a source of trouble for the watch industry. He was not an engineer, has been Swiss ambassador in Paris, so he was not able to motivate his feeling with technical advantages, but his vision was that electronic or microelectronic was a very dynamic discipline, producing many inventions and new products. So he was under the impression that the watch industry could either completely ignore this new technology, and many Swiss watchmakers wrote him letters supporting this attitude, or that the Swiss watch industry has to take into account very seriously the microelectronic revolution. A second strong reason was that he was conscious that Japanese and US watchmakers would not ask the permission to Swiss watchmakers to design electronic watches. Despite strong opposition, he decided to facilitate the creation of a new laboratory financed by all the Swiss watchmakers. It was the first time that a research laboratory has to be financed by several different companies traditionally totally reluctant to disclose any information to other companies. But Gérard Bauer was so convincing that in September 1960, it was decided to hire the future director of this Centre Electronique Horloger, Roger Wellinger, a Swiss citizen coming from General Electric, USA.

The main work of Roger Wellinger was to hire engineers and to prepare the official creation of the CEH. In the two years 1962-63, 12 engineers joined the CEH, 7 coming back from US and 5 from EPFL (Swiss Federal Institute, Lausanne, Switzerland). The CEH was officially created in Neuchâtel on January 30, 1962 with the mission "to develop an electronic wrist-watch with at least one advantage over existing watches". It was clear that not so many watchmakers believed than electronic watches were a so big threat! It was decided that the CEH scientific results should be kept secret to prevent some company to use them before the other companies.

4 The CEH between 1962 and 1967 [1]

The research plan of the CEH in 1962 was based on the following question: does the CEH have to focus on a long term research about a 100% electronic watch, or does the CEH have also to take into account short term research on some electromechanical watch like the Accutron? In fact, both approaches were followed, with code name Beta for the electronic watch and Alpha for the Accutron-like.

The CEH was organized with a technology department, in charge to set up a microelectronic technology (bipolar, then MOS) and with a circuit department in charge of designing low-power watch circuits. In 1963, Max Hetzel, the inventor of the Bulova Accutron, was hired by another laboratory in Neuchâtel working on mechanical watches. He was attached to the CEH in 1966, working on an improved version of the Accutron, but the success of the CEH in quartz watches was a killer for the project of Max Hetzel that was sold to Omega.

There were 12 engineers in 1963, there were 24 persons in the CEH in 1965, and the shareholders were waiting on results. As it was decided to keep secret the preliminary results, shareholders complained that the CEH did not show interesting results nor in 1966, neither in 1967. A meeting in May 1967 was organized by the shareholders to try to have more information, but their attitude was mainly to encourage CEH engineers not to be demotivated by the lack of results. They did not know that these engineers were pretty close to present the first quartz electronic watch! Furthermore, shareholders were secretly hoping that the CEH results would not threat the mechanical watches they were producing at that time!

However, the result of this meeting was that confidence between shareholders and the CEH director was broken. An external audit was performed in late 1967, interpreted as a lack of confidence over the CEH director. In May 1968, Roger Wellinger resigned and Max Forrer was chosen as the new CEH director.

5   Research Projects

To execute the research plan of the CEH in 1962, several projects were defined. For the "Alpha" line, mechanical symmetrical resonators were proposed to avoid position errors, as it was the case in the Accutron. Max Hetzel also improved the Accutron resonator at 480 then 720 Hz for the Swissonic watch. But these projects were based on patents from Bulova, so their use by the Swiss watch industry was not easy. In 1967, these projects were stopped due to the success of the "Beta" line. Although Max Hetzel, with his scientific prestige, has predicted at that time: "for many reasons a practical quartz controlled watch will probably remain a theoretical dream"!

For the "Beta" line, a silicon process was clearly required. Kurt Hübner, coming from the Shockley Semiconductor (Shockley is one of the inventors of the transistor), was hired to set up a microelectronic technology as Head of the semiconductor department. In 1963, the bipolar technology was more reliable than the MOS technology, and the first CEH circuits were bipolar circuits. It was the first silicon process installed in Switzerland. The goal was to have 1.0 to 1.3 Volt and 10 μA of power consumption for the complete circuit. The CMOS technology was chosen by the CEH only few years later, 10 μm in 1964 and 6 μm in 1966, largely before the main semiconductor companies that switched to CMOS only around 1984. Figure 1 shows a MOS transistor in this 6 μm technology.

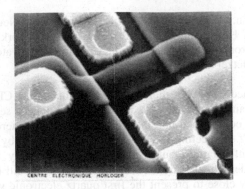

Fig. 1. MOS transistor in the CEH CMOS 6μ technology

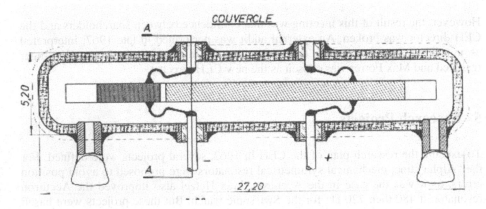

Fig. 2. Quartz at 8'192 Hz

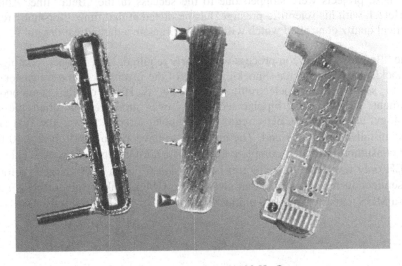

Fig. 3. Photographs of the 8'192 Hz Quartz

Always for the "Beta" line, under the direction of Max Forrer, Head of the circuit department, several projects were also defined, such as, for instance, a time base using statistics, i.e. the disintegration of a radio-active piece of plutonium! More serious projects about a time base were the development of a quartz resonator, but it was considered as a very risky project.

The main problem was the miniaturization of such a resonator at 8 kHz, work achieved by Amin Frei and Jean Hermann. The 8 kHz frequency was chosen to avoid a too large power consumption of the divider chain, while providing a not too big quartz. Figure 2 shows this 24 mm long quartz and Figure 3 a photograph.

Projects were also defined in the design of low-power circuits based on transistors, with a collaboration of Prof. Dessoulavy from EPFL. Eric Vittoz was hired from EPFL to work on quartz oscillators and frequency dividers in bipolar technology consuming microwatts at 1.3 Volt in bipolar technology [6]. Figure 4 shows the schematic of this first divider.

The electronic wristwatch was supposed to have handles and not a solid-state display. So it was required to develop a micromotor. Two different electronic watches were considered, i.e. Beta 1 with a step by step motor at 0.5 Hz and Beta 2 with a vibrating motor at 256 Hz. So from 8192 Hz, only five frequency dividers by two were necessary, and not 14 for the Beta 1 that will consume more than the target power consumption (Figure 5). But the two types of micromotors were developed by Henri Oguey, assuming that future electronic watches could have 14 dividers in an improved silicon process.

It was decided not to develop a new battery cell, but to collaborate with existing companies (Union Carbide, Mallory, Renata) to have battery cells adapted to watches.

Another project was to develop a radio receiver to regulate an electronic clock, with the idea to have an excellent precision without any quartz, using emitters at 75 kHz that distribute second and minute signals. This project was started in 1966 by Jean Fellrath and 1000 clocks were constructed in 1969. But the development of the quartz wristwatch was so fast that this project was stopped (although today Junghans, Germany, is a leader in radio controlled wristwatches).

These research projects were dealing with all the basic components of a quartz electronic wristwatch, but several problems remain, mainly their assembly in the watch case. One can notice that all these research projects were focused on a common goal, i.e. a quartz electronic wristwatch. This period was very challenging and enthusiastic for all the CEH employees coming from various disciplines, as they were all contributing to this unique common goal.

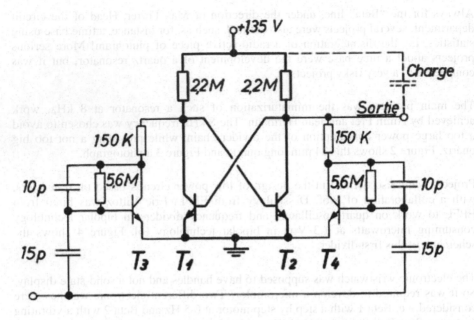

Fig. 4. Frequency Divider by 2 in bipolar technology

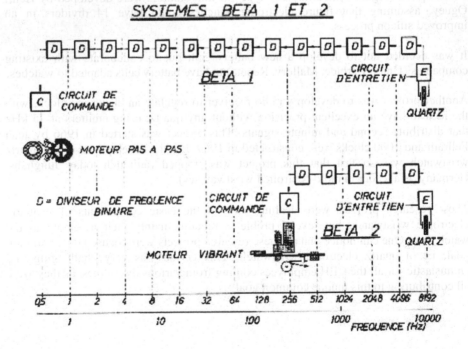

Fig. 5. The two systems Beta 1 and Beta 2

6 The First Quartz Wristwatch [1, 5]

The two Beta 1 and Beta 2 systems (Fig. 5) were presented to the members of the CEH Board in August 1967. As all the research projects have been executed secretly, it was a major shock for the Swiss watchmakers. A seminar was organized for CEH shareholders in December 1967, and it was the beginning of the very strong reputation of the CEH in low power.

The Beta 2 wristwatch was the only capable of reaching a one-year duration with the chosen battery. There were several bipolar chips, one being the 8192 Hz quartz oscillator and the other circuits being the five frequency dividers to provide a 256 Hz signal used for the mechanical motor and the handles [5, 6]. The technology was a 10 μm bipolar process. Each divider by 2 consumes approximately 1 μA. The frequency divider was a flip-flop designed with 4 NPN transistors and some integrated resistors and capacitors. Two frequency dividers (12 elements) were integrated on the same die of 2.1 mm^2. The motor control circuit was also a digital circuit producing the right motor pulses. The total chip power consumption was 15 to 30 μA at 1.3 V. and the chosen battery cell was supposed to deliver 18 μA during one year. So the lifetime of one year was satisfied. Figure 6 shows the Beta 2, with the electromechanical part at left and the printed circuit with the IC and the quartz on the right of the Figure.

Ten Beta 2 watches were presented to the "Observatoire de Neuchâtel" in December 1967. Figure 7 shows this Beta 2 watch, Figures 8 and 9 the watch module. The result was ten CEH watches at the first ten places, only a few tenths of a second per day, followed by four other quartz watches from Seiko, Japan. This competition was suspended the next year. The improvement was of a factor 10 in precision compared to the mechanical wristwatches presented one year before.

Visiting Hewlett Packard in California in 1968, waiting at the reception, Max Forrer, the CEH director, was looking at the clock reception, controlled by an atomic frequency device. Looking at his Beta 2 wristwatch, he noticed that his watch was wrong of three seconds. He asked if the atomic controlled clock was delivering the exact time. Naive question. But after, it was confirmed that this atomic controlled clock was wrong of 2 seconds, and that it has been discovered by using a Swiss quartz wristwatch. The story was immediately known from all the Hewlett Packard employees, and Max Forrer was invited by the famous President of HP, R. Hewlett, that visited the CEH in 1969.

It was decided to move towards a commercial product renamed Beta 21. A single bipolar integrated circuit (called ODC-04, feature size 6 μm, Fig. 10), containing about 110 components, managed all the electronic functions of the watch, including quartz crystal excitation, frequency division, and motor drive. The fabrication was performed by the CEH itself (Faselec Zurich, created in 1967 by the Swiss Horological Federation and Philips, was not able to produce about 6'600 bipolar circuits in the required delays). EM Microelectronic Marin will be created only in 1975 by Ebauches/Asuag (today Swatch Group).

Fig. 6. Beta 2: a) on the left: electromechanical part b) on the right the printed circuit with the IC and the quartz

Fig. 7. The Beta 2 Watch

An industrial consortium of Swiss watch manufacturers was created in 1968 in order to mass produce this Beta 21. Its members were shareholders of CEH, and several of them were responsible for a specific component of the watch. CEH designed the watch module and produced the integrated circuit. Ebauches SA. manufactured the mechanical parts and the quartz crystal resonator. Omega produced the micromotor. The watches were assembled by three separate shops that produced the final products according to the design requests of the Swiss watch companies that placed orders. Sixteen Swiss watch companies began selling the quartz watches under their own brand names in 1970.

It was a very expensive watch of several thousands of Swiss francs. This new technology was not seen as a thread for the mechanical Swiss watches. Everybody in watch industry was thinking that quartz watches would only be expensive watches, reaching a very small part of the world market. In April 1970, the first 1000 Beta 21 pieces were announced on the market.

On the other hand, the Swiss watch industry was a growing industry, only based on mechanical products. Nobody was motivated to introduce a new technology for a so interesting market: the Swiss mechanical watch. It explains why so little research and development have been performed in Switzerland in the late sixties.

Only the CEH was motivated to perform further research and development. Fortunately, the Swiss industry continued to fund this laboratory with 100 people. Its main work was the development of a CMOS technology, recognized as the best technology for electronic watches. A pilot line was created, as well as a design department that worked on quartz oscillators, frequency dividers, and watch chips in CMOS technology [9].

7 Electronic Watches in Japan [2]

What about Japan, well-known to produce multifunctional electronic watches? In the development of a quartz wristwatch, Seiko was close to CEH, but behind. Seiko has started to look at quartz timekeeping in 1958 with the development of a quartz crystal clock. In 1959, Seiko was developing a quartz watch. Obviously, they had to reduce the size of a quartz-based clock to that of a quartz wristwatch. The result of this project was the world's first analog quartz watch to reach the market, the Seiko 35SQ Astron, introduced on Christmas, 1969. The next Seiko 36SQC was introduced in 1970 and was the first quartz watch to use a CMOS chip.

One can see that Seiko and other Japanese watch companies like Citizen and Casio quickly and successfully switched to electronics. As a result, Japan took the lead in worldwide watch production in 1978. They worked at perfecting new technologies and have successfully develop watches that do more than tell time.

Fig. 8. The Beta 2 Watch Module (back face)

Fig. 9. The Beta 2 Watch Module (upper face)

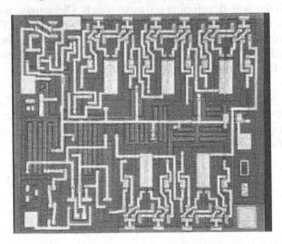

Fig. 10. The bipolar circuit of the Beta 21 (called ODC-04)

8   Electronic Watches in USA [3, pp. 157–158]

The first US electronic watch was announced in 1970 and introduced on the market by Hamilton in the fall of 1971: it was a digital one called Pulsar. A button was necessary to show the time through light-emitting diodes. The company bought the chip from RCA that was the first US company to produce CMOS chips.

But other US companies were thinking to develop watch CMOS chips. For instance, Motorola offered the first integrated electronic watch kit to manufacturers in early 1972. The CMOS circuit, the quartz and a miniature microwatt motor were offered old for only $15, starting a move toward very cheap electronic watches.

In 1974, the "Electronics" magazine was asking: "Is 1974 the year of the digital watch?" CMOS and display suppliers thought it was, but watch companies observed that semiconductor manufacturers did not understand styling and watch market. It was true.

In 1974 National Semiconductor introduced six watches. American Microsystems came out with a digital-watch module that was smaller and less power consuming. By February 1975, about 40 companies were offering electronic watches with digital displays. Everybody was thinking that solid-state digital watches will soon take a large part if not the total watch market.

The digital watch boom, begun in 1975, followed a trajectory similar to the four-function calculator: attracting numerous competitors (most of them were also calculator makers, plus Intel with its Microna line), producing hundreds of millions of dollars in revenues in a couple of years, then suffering a massive shakeout [10].

In 1976, Texas Instruments introduced its plastic-cased, five-function LED watch that sold for $19.95. As a result, National Semiconductor reduced its watch prices. Six months later watchmakers were predicting a $9.95 digital watch would be on the market by Christmas. As LCD prices dropped, TI introduced a watch with liquid-crystal hands in August 1976.

By 1977, the price of digital watches had fallen from more than 100$ to less than $10 in just two years. Profits evaporated. Once again as with calculators, in 1977 there were only three real survivors: again, two Japanese competitors, Casio and Seiko, and Texas Instruments [10]. Twenty years later, Intel chairman Gordon Moore still wore his ancient Microna watch ("my $30 million watch", he called it) to remind him of that lesson.

In 1978, Hewlett-Packard produced a calculator watch with LED display containing a 38'000 transistors microprocessor [11]. One has to use a pencil to touch the tiny keys. The watch was able to give the day of a given date for a 200-year period or the number of days between two dates. It performed some sophisticated calculations and provided a timer, a chronograph and memories for phone numbers. It was heavy (170 gr), thick (16 mm) and expensive (650 $).

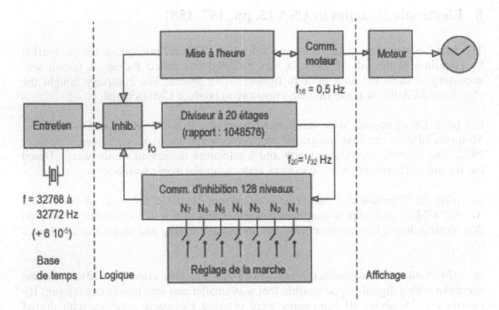

Fig. 11. The Beta 3 inhibition system: switches manually activated allow to remove the required number of clock pulses

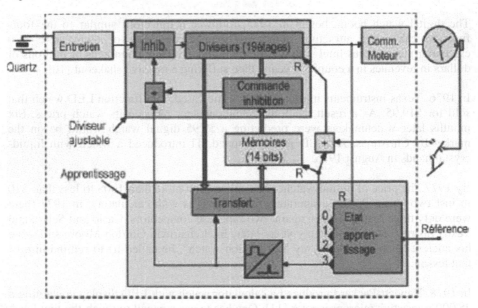

Fig. 12. The Beta 4 inhibition system: control bits are generated automatically by comparing the watch time base to a reference

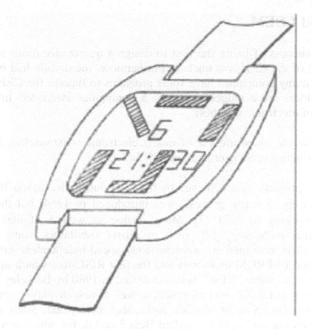

Fig. 13. To write handwritten symbols on the watch glass

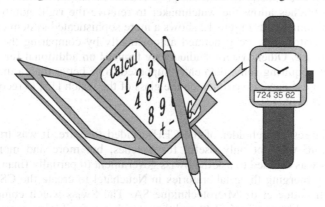

Fig. 14. Watch a pencil to send paper agenda data into the watch

The US watch market rise then fall was largely due to a brilliant but dangerous strategy of Texas Instruments. By reducing so dramatically the TI electronic watch prices, they succeed to eliminate all their US competitors. But, ultimately, TI found itself with watches whose prices had been driven so low that they do not make any profit. Worse, the Japanese Seiko and Casio, with relatively low labor costs at that time, were fierce competitors. Those firms soon did to TI what TI had done to its American competitors [10].

9   CEH and CSEM

After the big success of being the first to design a quartz electronic wristwatch, the future mission of the CEH was unclear. Furthermore, the decade had been born with recession and many companies have some problems to finance the CEH. In 1972, the CEH shareholders were reduced to only 5 companies interested in research and development of electronic watches.

In 1975, worldwide, there are 4% of quartz electronic wristwatches. In 1980, there 20% of such watches in Switzerland but 80% in Japan.

Many research projects were achieved by the CEH during the period 1972-1984. The CMOS technology, a 6 µm process, was introduced in 1964, but the first CMOS divider was working in 1971 [7]. Many analog as well as digital circuits were designed in this technology [9], such as quartz oscillators, static and dynamic frequency dividers designed as asynchronous speed-independent circuits [8, 12], ROM, RAM and EEPROM memories and the first RISC-like watch microprocessors [13] (before that the name "RISC" was introduced in 1980 by Berkeley and Stanford). New quartz, such as the ZT, was proposed, as well as new displays composed of LED for analog displays. System studies were also performed, such as the digital adjustment of the quartz oscillator, called Beta 3 and 4, for which some pulses were removed to achieve exactly 32'768 Hz [14]. Figure 11 shows the Beta 3 system in which some switches allow the watchmaker to remove the right number of pulses using an inhibition circuit. Figure 12 shows a more sophisticated system called Beta 4 in which the control bits are generated automatically by comparing the watch time base to a reference. Other system studies were focused on additional functions to the classical watch requiring some data entry supposed to be better than a tiny keyboard. The idea was to use capacitive touch on the glass of the watch [15] to recognize hand-written symbols (Fig. 13).

In 1982, the largest shareholder of the CEH decided to retire. It was true that low-power electronic was not only useful for watches, but more and more for other domains. So it was decided to ask the Swiss government to partially finance the CEH. It was done by merging three laboratories in Neuchâtel to create the CSEM, Centre Suisse d'Electronique et de Microtechnique SA. The Swiss watch companies were important shareholders of the CSEM and they decided to fund many research projects in low-power design. These projects were mainly the development of low-power microprocessors [16], digital standard cell libraries, memories, analog circuits for watches such as oscillators and end of life, simulation tools and multifunctional watches [17]. An interesting idea was to use a pencil to read the appointment data in a paper agenda to send them in the watch that rings the bell at the right time (Fig. 14).

10   Conclusion

Today, the CSEM is well known in the design of very low-power IC including analog, digital and RF circuits. As it is now clear, these low-power competencies come from the early research about the first quartz electronic watch.

References

1. M. Forrer, R. Le Coultre, A. Beyner, H. Oguey, «L'aventure de la montre à quartz, Mutation technologique initiée par le Centre Electronique Horloger, Neuchâtel», O. Attinger Editor, Switzerland, 2002, ISBN 2-88380-016-2.
2. Museum Smithsonian http://www.si.edu/lemelson/Quartz/index.html
3. "An Age of Innovation, The World of Electronics 1930-2000, by the Editors of Electronics, McGraw-Hill, 1981
4. "200 Years of Progress", Electronic Design 4, February 16, 1976, pp. 60–133.
5. M. Forrer et al. «Montre-bracelet électronique à quartz», Colloque international de chronométrie, Paris, September 16-19, 1969, papers B 242 to B 246.
6. C. Fonjallaz, E. Vittoz, «Circuits électroniques pour montre bracelet à quartz», Colloque international de chronométrie, Paris, September 16–19, 1969, paper B 244.
7. F. Leuenberger, E. Vittoz, «Complementary-MOS Low-Power Low-Voltage Integrated Binary Counter», Proc. of the IEEE, Vol. 57, No 9, September 1969.
8. E. Vittoz, B. Gerber, F. Leuenberger, "Silicon-Gate CMOS Frequency Divider for the Electronic Wrist Watch", IEEE JSSC, Vol. SC-7, No 2, April 1972.
9. Vittoz, Dijkstra, Shiels Editors., "Low Power Design : A Collection of CSEM Papers", Electronic Design Books, A Division of Penton Publishing, 1995.
10. M. S. Malone, "The Microprocessor. A Biography", 1995, Springer-Verlag, New-York, Inc., published by Telos, Santa-Clara, CA, p. 133.
11. R. Zaks, "Les microprocesseurs: une intelligence programmée", La Recherche No 85, Janvier 1978, p. 21
12. C. Piguet, "Logic Synthesis of Race-Free Asynchronous CMOS Circuits" IEEE JSSC-26, No 3, March 1991, pp. 271–380.
13. C. Piguet, J. F. Perotto, N. Péguiron, J.J. Monbaron, E. Sanchez, R. St-Girons, "Microcomputer Design for Digital Watches", EUROMICRO 81, September 8-10, 1981, pp. 431–442, Paris, France.
14. E. Vittoz, W. Hammer, H. Oguey, "Montre électronique à autoréglage instantané", SSC, Congrès International Chronométrie CIC, 1974, Stuttgart, Comm. No 9, p.625
15. C. Piguet, J-F. Perotto, J. Fellrath, "Entrée de données pour montre multifonctionelle", Proc. Journées d'Electronique et de Microtechnique 1976, EPFL, Lausanne, Suisse, 1976, Conf. E1.
16. J-F. Perotto, C. Lamothe, C. Arm, C. Piguet, E. Dijkstra, S. Fink, E. Sanchez, J-P.Wattenhofer, M. Cecchini, "An 8-bit Multitask Micropower RISC Core", IEEE JSSC, Vol. 29, No 8, August 1994.
17. C. Piguet, "Défis de la microélectronique horlogère", Congrès Européen de Chronométrie, Genève, 28–29 Septembre 2000

An Improved Power Macro-Model for Arithmetic Datapath Components

D. Helms, E. Schmidt, A. Schulz, A. Stammermann, and W. Nebel

OFFIS Research Institute
D - 26121 Oldenburg, Germany
Domenik.Helms@OFFIS.de,
http://www.lowpower.de

Abstract. We propose an improved power macro-model for arithmetic datapath components, which is based on spatio-temporal correlations of two consecutive input vectors and the output vector. Based on the enhanced Hamming-distance model [3], we introduce an additional spatial distance for the input vector and the Hamming-distance of the output vector to improve model accuracy significantly. Experimental results show that the models standard deviation is reduced by 3 % for small components and up to 23 % for complex components. Because of its fast and accurate power prediction, this model can be used for fast high-level power analysis.

1 Introduction

A high-level power estimation and optimization tool, like our ORINOCO®needs fast and precise macro models for high-level power estimation. There are several different approaches reported. Analytical models are very precise but they lack of generality because they have to be adopted to every different component. Empirical models are nearly exact, but they have to store every combination of two consecutive input vectors at every input. Such a full characterization for a component with i inputs and a input bit-width of n has to precalculate and to store 2^{2in} different power values, e.g. $4.3 \cdot 10^9$ values for a simple 8 bit adder. Thus we decided to use macro models, abstracting to input data statistics in order to limit the number of stored power values while keeping the model precision as high as possible.

Such a high-level power estimation model [5,7] can not use knowledge about the components internal (gate-level) structure because it can't be adopted to each different component. Therefore these 'black box models' have to abstract to statistical data of the input stream. There are several approaches to do so. Macii et al. [7] gives a good overview of different recent techniques.

1.1 Description of the Signal Statistics

Trying to predict the power consumption of a component, the normalized Hamming-distance of the inputs is often used in the area of high-level synthesis as

D. Hochet et al. (Eds.): PATMOS 2002, LNCS 2451, pp. 16–24, 2002.

reported in [8,9]. The Hamming-distance is equal to the number of transitions between two consecutive input vectors.

$$Hd = \frac{\sum_{i \neq j} t_{i \to j}}{\sum_{i,j} t_{i \to j}}, \tag{1}$$

where $t_{i \to j}$ is the number of transitions from i to j within two consecutive input bit-vectors.

In our previous model [2,3], we added the input's normalized signal-distance, which is equivalent to the number of input bits that are fixed to logic one in two consecutive input vectors.

$$S = \frac{t_{1 \to 1}}{\sum_{i,j} t_{i \to j}}, \tag{2}$$

In CMOS designs, switching activity is nearly proportional to the power consumption [10]. A single bit that is set to logic one or logic zero over two consecutive input vectors does not directly contribute to the switching activity. But the number of logic ones increases the components saturation and therefore the probability that the switching activity of the inputs is propagated through the component. Thus the number of fixed ones also contributes to the component's power consumption. Knowing the Hamming-distance and the signal-distance, it's easy to also obtain the number of fixed zeros $\bar{S}$ of the input. Note that components using inverted logic (i.e. the number of fixed zeros increases the component's saturation) can be characterized the same way using the fact that $\bar{S}$ is determined by Hd and S.

$$\bar{S} = \frac{t_{0 \to 0}}{\sum_{i,j} t_{i \to j}} = 1 - Hd - S. \tag{3}$$

1.2   Equivalence of Switched Capacitance, Power, and Energy

To be independent of operation voltage and frequency, the switched capacitance is stored instead of the power consumption, taking advantage of the fact that power consumption, switched capacitance and even energy conversion are proportional when voltage and frequency are kept constant as given by

$$E = \frac{1}{2} C_{sw} V^2, \ P = \frac{1}{2} f C_{sw} V^2, \tag{4}$$

where E is the component's energy conversion and P the power consumption for one operation, C_{sw} is the switched capacitance, V the operating voltage and f the inverse of the time, the component needs for one operation. This equation is valid if one halve of the energy for charging a capacitance and afterwards discharging it is assigned to the charging and the other halve to the discharging [10].

1.3   Separating the Components Bit-Width

To eliminate the influence of the component's input bit width, it is shown in [3] that the input bit width dependency of the power can be separated (also rf. [1, 4]). This can be done within a model precision of $10\,\%$ when using normalized Hamming-distance and signal-distance and if the two halves of the input vector, pertaining to the two inputs of the component, are treated separately.

The switched capacitance of a component can now be modelled by the following separation equation

$$C_{sw} = f(bw_i) \cdot g(Hd_A, S_A, Hd_B, S_B),\qquad(5)$$

where bw_i is the components input bit width, Hd_A and Hd_B are the two Hamming-distances at both inputs of the component and S_A and S_B are the corresponding signal-distances.

The bit width dependency function $f(bw_i)$ is obtained by a regression over the power consumption of a component with different bit widths and $Hd_A = Hd_B = 1$ and $S_A = S_B = 0$, (rf. [11]). The data statistic function $g(Hd_A, S_A, Hd_B, S_B)$ is sampled at a fixed bit width for special combinations of Hd_A, S_A, Hd_B and S_B where each value varies between 0 and 1 in n equidistant steps.

1.4   The Size of the Model

Using the fact, that $Hd_A + S_A + \bar{S}_A = 1$ and $Hd_B + S_B + \bar{S}_B = 1$, it's easy to obtain the number M of different possible sets of Hd_A, S_A, Hd_B and S_B as follows

$$M = (1 + 2 + \dots + n)^2 = 1/4(n^4 + 2n^3 + n^2).\qquad(6)$$

For example, if n is 5 then $Hd_A, S_A, \bar{S}_A, Hd_B, S_B, \bar{S}_B \in [0; 0.25; 0.5; 0.75; 1]$ and the power consumption for $M = 225$ different parameter combinations have to be generated.

The model we proposed ('standard' model) is tested for different components at different bit widths and the results are given in [3]. Further examination shows that this model accuracy can be increased modifying the set of parameters.

The rest of the paper is organized as follows: In section 2 a new set of input data statistic parameters is introduced leading to a much smaller and slightly better model. In section 3 the model is augmented by a new parameter, describing the output data Hamming-distance that again increases model accuracy. Section 4 presents experimental data and evaluates the different presented models. In Section 5 we conclude this paper.

2   Improving the Input-Parameters

Because the propagation of the internal carry-in differs for every arithmetic component, a 'black-box model' can only handle the power consumption with respect to the carry-in in a statistical way. As mentioned in section 1.2, this can be done by storing the number of switching input bits, which can generate a switching carry signal and the number of input bits stuck at one, which can propagate the carry signal.

2.1   Analysis of a Full Adder

Therefore we begin by investigating the different states of a full adder – the integral part of nearly every arithmetic component. Ignoring the carry-in, four different input transitions can be identified: No bit is switching at the input (0-case), just one bit is switching (1-case), both bits are switching the same way (parallel) or one bit is switching up and one bit is switching down (anti-parallel). At this point we can identify a principal problem of the model proposed in [3]. Even though the power consumption of the anti-parallel case is close to the 0-case, the two switching bits contribute to the total Hamming-distance so that high power parallel and low power anti-parallel states are treated the same way in our previous proposed model. Note that this holds for every state of the carry-in.

2.2   The Improved Model

As a result, the Hamming-distance of our model is split into up-switching distance (Up) and down-switching distance (Do).

$$Up = \frac{t_{0 \to 1}}{\sum_{i,j} t_{i \to j}} \tag{7}$$

$$Do = \frac{t_{1 \to 0}}{\sum_{i,j} t_{i \to j}} \tag{8}$$

The distribution of parallel and anti-parallel states can now be modelled ('improved' model). The experimental results in section 4 show that the model accuracy is drastically improved. The cost of doing so are much more entries in the lookup table of the g-function. Consequentially

$$M = \frac{1}{36}(n^3 + 3n^2 + 2n)^2 \tag{9}$$

lookup-values for n steps are needed. That means $M = 1225$ if $n = 5$ (every parameter is in $[0; 0.25; 0.5; 0.75; 1]$).[1]

2.3   The Reduced Model

Now, the focus is on reducing the size of M by identifying different symmetries within the new set of parameters. First we notice that the separation of the parameters for both inputs is important for simple, bit-linear components like the ripple-adder. But it is less significant, when the structure of the component becomes more complex.

---

[1] Note that models of components, modifying the input vector's data statistics (e.g. the booth-coded multiplier) can't be significantly improved. After being booth-coded, the input reaching the full adders has totally different data statistics. Neither Up and Do nor Hd are further correlated to input transition probability of the full adders.

Therefore we omitted separating the bit-vector into two peaces (The 'reduced1' model has just one signal-distance but two Hamming-distances, the 'reduced2' model has just one signal-distance and one Hamming-distance). In order to keep the models comparable, the width of the steps of the unnormalized parameters is kept constant. Thus the new model's parameters vary within $n = 9$ steps instead of $n = 5$ steps. The results is, that the model accuracy is slightly reduced but the size of M is drastically decreased (rf. section 4). With the 'reduced2' model the number of lookup values is $M = 1/6(n^3 + 3n^2 + 2n)$, e.g. $M = 165$ for $n = 9$ instead of $M = 1225$ with the previous approach.

In a final step M is nearly cut in halve without reducing model accuracy by using a symmetry within the parameters: If the energy consumption for charging a capacitance equals the energy consumption for uncharging it, power consumption and therefore switched capacitance should remain constant when every up-switching bit is replaced by a down-switching bit and vice versa. By storing Hd and the smaller one of the parameters Up and Do we take this symmetry into account. Input transitions with $Up = x$ and $Do = y$ will now be mapped to the same lookup-table value as $Up = y$ and $Do = x$. According to this we finally choose the input parameters given in table 1 for our new model. The number of M is of order $1/12n^3 + 1/4n^2$ which means in our example $M = 95$ for $n = 9$.

Table 1. Parameters of the proposed model

S	the signal-distance of the full input bit vector
Hd	the Hamming-distance of the full input vector
$\min(Up, Do)$	the minimum of up-switching and down-switching distance

3   Output Statistics Dependency

The extreme small size of lookup-table values M in the 'reduced2' model of the previous section implies that this model can now be expanded by an additional parameter. Even with an exhaustive search over a large set of additional parameters, we found no possibility to increase model accuracy without drastically increasing the size of M, which should be limited to a few hundreds for our application.

3.1   Model Enhancement by Additional Parameters

We tried to add every parameter Pa, that can be described as

$$Pa = \sum_i^N \beta_4(a_i^0, b_i^0, a_i^1, b_i^1)/N \tag{10}$$

or

$$Pa = \sum_i^N (\beta_2(a_i^0, a_i^1) + \beta_2(b_i^0, b_i^1))/2N, \tag{11}$$

where N is the width of one input, a_i^0 and b_i^0 are the first bit vectors at the inputs and a_i^1 and b_i^1 are the consecutive bit vectors. β_2 and β_4 are arbitrary boolean functions. β_j is implemented the following way: A lookup table of 2^j entries is generated and every combination of the j input bits is assigned to one value of the table. Every combination in this 2^j-size table is generated and the resulting parameter is added to the model. The product of standard deviation and M is build to classify the improvement caused by the parameter. But for every choice of Pa, this product is significantly bigger than without a further parameter.

A possible explanation is, that the power consumption of a component not only depends on the statistical distances of the input data, but also on some internal values of the component like the carry signal of an adder or a multiplier. These internal values depend on the input bit vector in a deterministic way, but may have totally different statistical properties compared to the input itself. So going deeper into the component, the internal data loses the correlation with the input data which means that the model accuracy is getting worse. Furthermore the internal data statistic is dependent of the input data statistic in a different way for every component. Refer to [6] for a detailed correlation analysis of a multiplier.

3.2   The Output Model

There are two different solutions to this problem. One could find a different set of input parameters for every component exactly describing the internal correlations (analytical model), which would lead to a set of very accurate and small models. With this approach the model's universality would get lost, because the model has to relate to the component's internal structure. The other possibility is to include data statistics of the component's output. While the internal data is going through the component, losing correlation to the input data statistic, it gains correlation to the output data statistic. This enables a 'deeper look' into the component (rf. [7]).

One disadvantage is that getting characterization data with specific output data statistics isn't as easy as for the input values. It isn't possible to generate every combination of input and output distances, because the output distances normally depend on the input distances.

Accordingly the model is limited to just store three different output Hamming-distances for every set of input parameters.[2] When the model is calculating it's estimation, it does a linear interpolation of these three values. This method has a positive impact to our model in three ways. First, we take the output's Hamming-distance into account. Secondly we increase the accuracy of the input parameters, because the over-determined linear interpolation over three sampling points also averages these values. The effect of dissenting sampling points is

---

[2] Note that it's not easy and sometimes impossible to achieve different Hamming-distances so that this cases have to treated separately.

reduced. Finally this method just increases the size of M by factor 3, which is acceptable for our application.

In section 4 the new model ('output' model) is compared to the different models of section 2.

Table 2. Calculated values

$\langle C_i \rangle$	the component's average switched capacitance
$\langle |\langle C_i \rangle - C_i| \rangle$	the average absolute error of the model
abs [%]	the ratio between avg. abs. error and avg. C_{sw}
$\sigma_C = \sqrt{\langle (\langle C_i \rangle - C_i)^2 \rangle}$	the standard deviation σ_C of the model
std. dev. [%]	the ratio between σ_C and avg. C_{sw}
M	the total number of lookup-table values needed

4   Experimental Data

We tested our model with three components, a 4 bit carry-ripple subtracter with carry-in stuck at zero, a 4 bit carry-look-ahead adder with carry-in stuck at zero and a 4 bit standard-multiplier (carry-save adder multiplier) for unsigned values with 8 bit output. The evaluation data are obtained in the following way: The inspected component is synthesized by Synopsys®Design Compiler to gate level, then simulated with MentorGraphics®ModelSim simulator. The switching activity data is then back-annotated into Synopsys®PowerCompiler, which computes the power figures. As input for the gate level simulation we used every permutation of two consecutive 8 bit vectors (full characterization) which took about 4 days an a 550 MHz workstation. We choose components that small because of the ability to generate a full characterization. The full characterization prevents additional statistical errors when evaluating the models. Nevertheless, the results of the 4 bit components can be compared to larger components, because components with different bit-width but identical data statistics have proportional power consumption within an error of 10 % (rf. [3]). For every model mentioned in this paper we calculated the values described in table 3.

Table 3. Used parameters and number of steps for this parameters for the different models

standard	$S_A(5), S_B(5), Hd_A(5), Hd_B(5)$
improved	$S_A(5), S_B(5), Up_A(5), Up_B(5), Do_A(5), Do_B(5)$
reduced1	$S(9), Up_A(5), Up_B(5), Do_A(5), Do_B(5)$
reduced2	$S(9), Hd(9), min(Up(9), Do(9))$
output	$S(9), Hd(9), min(Up(9), Do(9)), Hd_Z(3)$

The parameters used for the different models are shown in table 3 where Up is the up switching-distance and Do the down switching-distance. Values without indices relate to the total input bit vector and Hd_Z is the output's Hamming distance. The number n of different parameter values is given in brackets for every parameter. The results are shown in table 4 for the rpl-subtracter, in table 5 for the cla-adder and in table 6 for the csa-multiplier.

Table 4. Evaluation results for the rpl subtracter component

SubRpl	$\langle C_i \rangle [pCb]$	$\langle |\langle C_i \rangle - C_i| \rangle [pCb]$	abs [%]	$\sigma_C [pCb]$	std. dev. [%]	M
standard	0.580	0.116	20.0	0.143	24.7	225
improved	0.580	0.110	19.0	0.137	23.6	1225
reduced1	0.580	0.115	19.8	0.142	24.5	825
reduced2	0.580	0.125	21.6	0.152	26.2	95
output	0.580	0.112	19.3	0.139	24.0	285

Table 5. Evaluation results for the cla adder component

AddCla	$\langle C_i \rangle [pCb]$	$\langle |\langle C_i \rangle - C_i| \rangle [pCb]$	abs [%]	$\sigma_C [pCb]$	std. dev. [%]	M
standard	1.203	0.221	18.4	0.275	22.9	225
improved	1.203	0.211	17.5	0.261	21.7	1225
reduced1	1.203	0.211	17.5	0.262	21.8	825
reduced2	1.203	0.215	17.9	0.267	22.2	95
output	1.203	0.188	15.6	0.234	19.5	285

Table 6. Evaluation results for the csa multiplier component

MultCsa	$\langle C_i \rangle [pCb]$	$\langle |\langle C_i \rangle - C_i| \rangle [pCb]$	abs [%]	$\sigma_C [pCb]$	std. dev. [%]	M
standard	3.951	0.707	17.9	0.901	22.8	225
improved	3.951	0.513	13.0	0.656	16.6	1225
reduced1	3.951	0.556	14.1	0.708	17.9	825
reduced2	3.951	0.618	15.6	0.779	19.7	95
output	3.951	0.551	13.9	0.698	17.7	285

The gain in accuracy for the new models is bigger for the more complex csa multiplier than for the far simpler rpl subtracter. The error for the 'reduced2' is even worse than the error of the 'standard' model. For linear components, the separation of the two input's seems to be of more importance than for the complex ones. Nevertheless, even the 'reduced2' model is an improvement because it is much smaller and it increases accuracy for high power components while slightly decreasing the accuracy for low power components. Thus, for a whole design build up of these components, the total error should be decreased by 10% to 15%.

5   Conclusion

In this paper, we have shown that the accuracy of power-macro models for arithmetic datapath components can be increased using a different set of parameters. We proposed three new sets of parameters. The most expensive model, called 'improved' is decreasing the average absolute error by 5 % up to 27 % and the standard deviation by 4 % up to 27 %.[3] A very fast model was found, slightly increasing model precision but using just 42 % of the values which are necessary to generate the well known 'standard' model. A satisfying trade-off between model size and model precision was found, including the output signal as a parameter. This decreases the average absolute error and standard deviation by 3 % up to 23 %.

Further work has to be done enabling this model to characterize components, modifying the input vectors before calculation (e.g. booth-coding). One way to handle this is to characterize the coding part and the calculation part of such components separately.

Even though the 'reduced2' model is much smaller than all other models, it slightly improves model precision, compared to the 'standard' model. Further work has to be done to find a better fourth parameter to this model which seems to be very promising. It seems to be possible to improve the estimation error while keeping the number of look-up table values needed at several hundreds.

References

[1]   A. Bogliolo, R. Corgnati, E. Macii, M. Poncino: Parameterized RTL Power Models for Combinational Soft Macros. *Proc. of the IEEE-ACM International Conference on Computer Aided Design*, pp. 284–287, 1999.

[2]   G. Jochens, L. Kruse, E. Schmidt, W. Nebel: A New Parameterizable Power Macro-Model for Datapath Components. *Proc. of Design, Automation and Test in Europe (DATE)*, 1999.

[3]   G. Jochens, L. Kruse, E. Schmidt, A. Stammermann, W. Nebel: Power Macro-Modelling for Firm-Macros. *Int. Works. Power and Timing Modelling of Integrated Circuits (PATMOS)*, 2000.

[4]   P. Landman, J. Rabaey: Architectural Power Analysis: The Dual Bit Type Method. *IEEE Transactions on VLSI Systems, Vol. 3, No. 2*, 1995.

[5]   P. Landman: High-Level Power Estimation. *IEEE Proc. of ISLPED*, 1996.

[6]   R.P. Llopis: An Analytical Power Model of Multipliers and its Impact on FIR Filters. *Internal Report. Phillips Research Laboratories, Eindhoven.*, 1998.

[7]   E. Macii, M. Pedram, F. Somenzi: High level power modelling, estimation and optimization. *Trans. on Design Automation of Electronic Systems*, 1998.

[8]   E. Mussol, J.Cortadella: Scheduling and resource binding for low power. *IEEE Int. Symposium on System Synthesis*, pp.104–109, 1995.

[9]   E. Mussol, J.Cortadella: High-level synthesis techniques for reducing the activity of functional units. *IEEE Int. Symposium on Low Power Design*, 1995.

[10] J.-M. Rabaey, M. Pedram: Low Power Design Methodologies. *Kluwer Academic Publishers*, 1996.

[11] E. Schmidt, G. Jochens, L. Kruse, A. Stammermann, W. Nebel: Automatic Non-linear Memory Power Modelling. *Proc. of Design, Automation and Test in Europe (DATE)*, 2001.

---

[3] The 'standard' model represents the state-of the art macro modeling. Thus the improvement for every new model is given as the percentage, the model's error is smaller than the 'standard' model's error.

Performance Comparison of VLSI Adders Using Logical Effort[1]

Hoang Q. Dao and Vojin G. Oklobdzija

Advanced Computer System Engineering Laboratory
Department of Electrical and Computer Engineering
University of California, Davis, CA 95616
http://www.ece.ucdavis.edu/acsel
{hqdao,vojin}@ece.ucdavis.edu

Abstract. Application of logical effort on transistor-level analysis of different 64-bit adder topologies is presented. Logical effort method is used to estimate delay and impact of different adder topologies and to evaluate the validity of the results obtained using logical effort methodology. The tested adder topologies were Carry-Select, Han-Carlson, Kogge-Stone, Ling, and Carry-Lookahead adder. The quality of the obtained estimates was validated by circuit simulation using H-SPICE for 1.8V, 0.18μm Fujitsu technology.

1 Introduction

Delay estimation is critical in development of efficient VLSI algorithms [2]. Unfortunately, delay estimates used are usually presented either in terms of gate delays or in terms of logic levels. Neither of these estimates allows us to properly evaluate different VLSI topologies. One such component, VLSI adder, is critical in the design of high-performance processors. Using gate delay is no longer adequate because gate delays are dependent on gate types, the number of inputs (fan-in), output load (fan-out), and particular implementation. Further, a particular VLSI implementation can use static or dynamic CMOS where logic function is usually packed into a complex logic blocks. Thus, the notion of logic gate and associated gate delay becomes artificial and misleading. In this analysis, we are evaluating the use of the logical effort method not only for the purpose of better delay estimation but also for evaluation of different adder topologies and their impact on design of VLSI adders.

The logical effort (LE) analysis [1] models the gate delay using gate characteristics and its loading and compares the gate delay to τ, the delay of a Fan-Out of 1 (FO1) inverter. This latter delay is normally known for a given technology and can serve to estimate the speed. When a gate is loaded, its delay varies linearly with the output load expressed in terms of fan-outs. LE also accounts for the effect of circuit

---

[1] This work has been supported by SRC Research Grant No. 931.001, Fujitsu Laboratories of America and California MICRO 01-063

B. Hochet et al. (Eds.): PATMOS 2002, LNCS 2451, pp. 25–34, 2002.

topology, by including path branching in the model. The delay estimation using LE method is quick and sufficiently accurate.

In order to evaluate the efficiency and usefulness of LE we have chosen several diverse adder topologies and compared the estimated delay with the one obtained via simulation. The adders chosen for this analysis were: multiplexer-based adder (MXA) which is implemented as a static radix-2 4-bit adder with conditional-sum in the final stage [3]; Han-Carlson consisting of static and dynamic radix-2 adders [5][6]; Kogge-Stone, static and dynamic radix-2, and dynamic radix-4 adder [7][8]; Naffziger's implementation of Ling's adder [9] in a dynamic radix-4 topology [10]; and a Carry-Look-ahead (CLA) adder implemented in dynamic radix-4 topology [4].

The multiplexer-based adder (MXA) takes advantage of its simplicity and speed of transmission-gate multiplexer implementation [3]. The sums are generated conditionally in groups of 4 bits. The carries to these groups are formed using radix-2 propagates and generates. The generate path is critical, passing through 9 stages including the total of 7 multiplexers. Thus, the transmission-gate multiplexer speed is a dominant factor determining the speed of this adder.

The Han-Carlson and Kogge-Stone adders use similar radix-2 structure as MXA. However, they combine the carries with the half-sum signals in order to obtain the final results. Direct CMOS implementation of generate and propagate logic had been used, allowing usage of both static and dynamic gates. The Han-Carlson adder differs from the Kogge-Stone adder by not creating all the carries from the radix-2 structure. Instead, only even carries are created and odd carries are generated from even carries. Therefore, in terms of logic stages, Han-Carlson uses one extra stage while Kogge-Stone adder is equivalent in the number of stages to MXA.

Ling's adder obtains high performance by exploiting wired-OR gate property of emitter-coupled logic. With CMOS implementation, such advantage is lost. However, it was shown in [10] that high performance could be realized using radix-4 propagates and generates for carries and conditional sum.

The CLA adder allows fast implementation, especially the dynamic radix-4 type [4]. CLA is a textbook example and it is most commonly used. However, with dynamic radix-4 implementation, its large transistor stack and many stages made it appear slow compared to other adders.

Using logical effort method for quick optimization, these adders were evaluated and compared in [12] and extended next with the inclusion of radix-2 Han-Carlson and radix-2 Kogge-Stone adders. Section 2 outlined the optimization conditions for the adders. The delay of adders using logical effort method was discussed in section 3. The results were compared with H-SPICE simulation in section 4. The conclusion of the work was given in section 5.

2 Optimization Conditions

All adders were optimized under the following conditions: maximum input size of 20μm, maximal allowable transistor size of 20μm and an equivalent load of 30μm-inverter. These conditions were set to get reasonable transistor sizes and loads to an adder.

The wiring capacitance was included. It was computed using the unit-length wiring capacitance and the 1-bit cell width. This width was determined from the preliminary layout of the most congested bit cell. The wire length was determined from the number of bits it spanned and the number of wires running in parallel.

Using logical effort method, the adders were optimized according to the critical paths that were estimated from the adder topology. Delay effort in other paths was computed from the critical one. The optimization process was applied recursively to update the branch factors along the critical path. It finished after all transistor sizes converged and the final result recorded the adder delay.

3 Delay Effort of Adders

The logical effort of gates was obtained from simulation. This adjustment was necessary for two reasons: first, pMOS and nMOS driving capability vary with technology, and secondly, better average per-stage delay can be achieved using the p-n ratio in the range of 1.4-1.6. Thus, we needed to repeat the gate delay simulation in order to accurately model the delay; the drain and source areas of transistors were

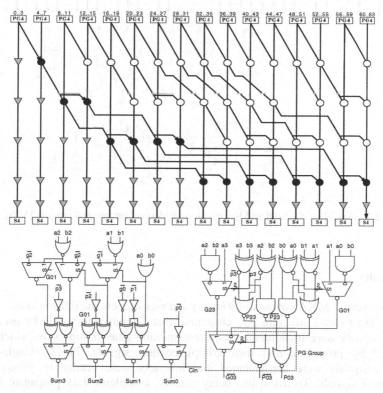

Fig. 1. Multiplexer-based carry select adder: diagram and circuits [3]

included to match better with real layout. We used p-n ratio of 1.5 for the performance reason. Nonetheless, all gates continued to show linear delay with fan-out. In addition, to accurately model the delay, the domino gates were broken into dynamic and static gates. First, the latter have different driving capability and needed to size differently. Second, domino gates can be very complex (for example, in CLA and Ling adder, group generates and group carries drive multiple inverters at different locations on its NMOS stack). Without such separation, it is very difficult to model its delay accurately.

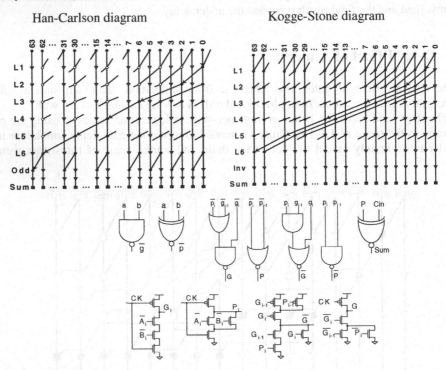

Fig. 2. Radix-2 Han-Carlson and Kogge-Stone adders: diagrams and circuits [5][6][7]

3.1 Results

The static radix-2 MXA consists of 9 stages and was implemented using static CMOS (Fig. 1). The radix-2 structure was chosen so that 2-input gates could be used. The generate signals were implemented with transmission-gate multiplexers, which were controlled by propagate and their complementary signals. In [3], single-ended propagate signals were implemented and inverters were needed to generate the complement signals. To avoid this delay penalty, complementary propagate signals were generated directly. The critical path was from bit-1 propagate through generate paths to the MSB sum. Along this path, the fan-out was slightly larger than 2. The logical effort optimization achieved the total delay of 55.8τ (11.4FO$_4$).

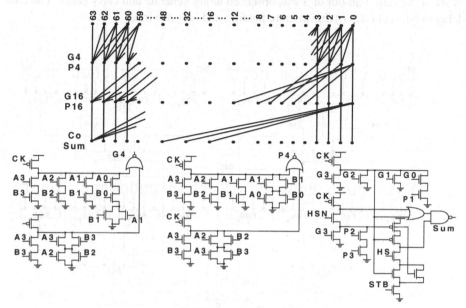

Fig. 3. Radix-4 Kogge-Stone adder: diagrams and circuits [7]

The radix-2 Han-Carlson adder (Fig. 2) realizes even carries with propagate and generate signals of even bits. The odd-bit carries are generated at the end using even carries. The critical path goes from bit 1 through the generate path to the MSB sum, traversing 10 stages. The propagate paths had the equal number of stages but they were loaded less heavily than the most critical generate path. The fan-out along the critical path was less 2. The total delay was 62.5τ (12.8FO$_4$) and 55.8τ (11.4 FO$_4$) for static and dynamic implementation.

The radix-2 Kogge-Stone adder is similar in architecture to the Han-Carlson. The difference is that propagate and generate signals of all bits are created in Kogge-Stone adder (Fig. 2). This results in 9 stages, one less as compared to Han-Carlson adder. The cost, however, was in twice as many gates for propagate and generate signals and doubling of the number of wires. The critical path went through the generate signals, traversing 9 stages. The fanout was also less than 2. The total delay after optimization was 57.6τ (11.8FO$_4$) and 42.6τ (8.7FO$_4$) for static and dynamic implementation. The delay is better compared to Han-Carlson adder.

The dynamic radix-4 Kogge-Stone adder was implemented in only 6 stages, by using redundant logic in propagate and generate stages and strobe signals for final sum (Fig. 3). The cost was very high input and internal loading and large amount of wiring between stages. In addition, dynamic stages that followed were slow NOR gates. The critical path went through the generate path from bit 0 to the MSB sum. The total delay is 30.1τ (6.2FO$_4$). This is the best delay seen – showing the advantage of using fewer stages over its complexity.

The dynamic radix-4 CLA was realized in 16 stages or 8 domino gates (Fig. 4). The critical path was from bit 0 through the generate path and higher-bit carries to the MSB of the sum. Fan-out of 3 was observed along generate and carry paths. The total delay is 54.3τ ($11.1FO_4$) due to more loading and longer wires.

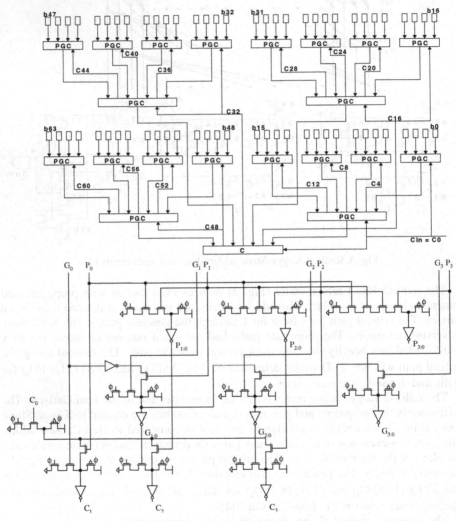

Fig. 4. Radix-4 CLA adder: diagrams and circuits [4]

Naffziger's implementation of modified Ling's adder [10] utilizes Ling pseudo-carries and propagate signals [9] in order to generate long carries and the conditional-sum adder for local carries (Fig. 5). The critical path was chosen through the long carry to the MSB Sum and it was realized in 9 stages, due to larger gate and wire loading. Local carry and sum paths have more stages than the critical path. They were implemented with faster gates to avoid becoming critical. The total delay is 43.9τ ($9.0FO_4$).

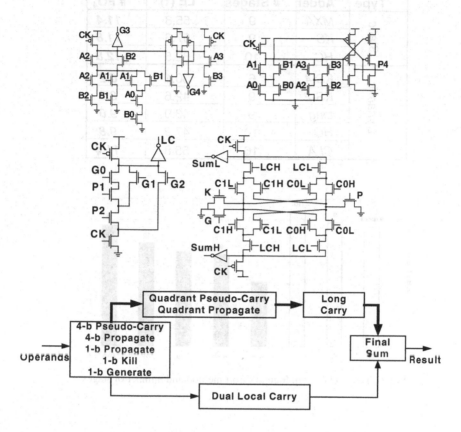

Fig. 5. Radix-4 modified Ling adder: diagrams and circuits [7]

3.2 Comparison

Table 1 summarized the delay of adders using logical effort analysis. The delays are expressed in terms of inverter delay τ and FO_4. The adders with fewer stages are consistently faster. Figure 6 shows the total delay and number of stages. The delay was found to be linearly proportional to the number of stages in the critical path. It was capitalized into $1.2FO_4$ and $0.6FO_4$ per stage, respectively, for static and dynamic implementation.

Table 1. Adder delays using logical effort method

Type	Adder	# Stages	LE (τ)	# FO$_4$
Static	*MXA*	9	55.8	*11.4*
	KS	9	57.6	*11.8*
	HC	10	62.5	*12.8*
Dynamic	*KS-4*	6	30.1	*6.2*
	KS-2	9	42.6	*8.7*
	Ling	9	43.9	*9.0*
	HC	10	47.9	*9.8*
	CLA	16	55.8	*11.4*

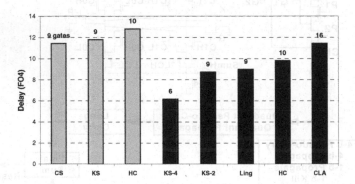

Fig. 6. Total delay from logical effort method and number of stages

4 Simulation Results

The worst-case delay of each adder's critical path was simulated with H-SPICE using the 0.18µm, 1.8V CMOS at 27°C temperature. The results obtained were presented in Table 2.

The results obtained using H-SPICE simulations are fairly consistent with the logical effort analysis in term of relative performance among adders. That is a good indicator and it confirms our belief that LE estimates should replace number of stages or gate counts as delay estimates when developing VLSI algorithms. Figure 7 showed the delays obtained using H-SPICE and a relative difference with logical effort results. The delay of adders remained dependent on the number of stages. In addition, the per-stage delay difference was degraded to 1.4FO$_4$ and 0.8FO$_4$ for static and dynamic implementation, respectively.

Some inconsistency was observed between logical effort result and H-SPICE for MXA, which had larger errors compared to Kogge-Stone and Han-Carlson. The main error came from larger delay in the multiplexers than modeled. Because pMOS-to-nMOS ratio of 1.5 was used, the rising signal was faster than the falling signal. So, multiplexer did not fully switch until the rising control to the multiplexer. Therefore, the multiplexer delay was always determined by the slow rising signal. It corresponded to the worst-case delay, not the average. Large errors were also seen in radix-4 dynamic adders. They used high-stack nMOS and had many branches. Therefore were harder to model accurately, especially on parasitic delay.

Table 2. Logical effort and simulation delay results

Type	Adder	# Stages	LE (FO4)	HSPICE (FO4)	HSPICE (ps)	Diff. (%)
Static	KS	9	11.8	10.9	853.0	-8.04
	MXA	9	11.4	12.8	1003	10.99
	HC	10	12.8	13.3	1036	3.49
Dynamic	KS-4	6	6.2	7.4	581	17.11
	KS-2	9	8.7	9.2	717	4.82
	Ling	9	9.0	9.5	742	5.34
	HC	10	9.8	9.9	772	0.85
	CLA	16	11.4	14.2	1107	19.3

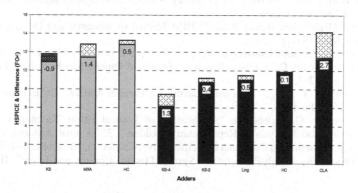

Fig. 7. Total delay with H-SPICE and delay difference

Nonetheless, the relative performance among adders did not vary significantly. It was realized that having less stages in critical path helped to improve delay. Although less stage meant more complex gates that translated into worse per-stage delay, such delay degradation was offset by more delay reduction due to fewer stages.

5   Conclusion

Use of Logical Effort method for performance comparison of different adder topologies was presented with wire capacitance included. Obtained results were consistent with simulation and are encouraging. They show that incorporating Logical Effort into the analysis of VLSI adders can help find better adder topologies.

References

1. I. Sutherland, B. Sproull, D. Harris, "Logical Effort: Designing Fast CMOS Circuits," Morgan Kaufmann Publisher, 1999.
2. V. G. Oklobdzija, E. R. Barnes, "Some Optimal Schemes for ALU Implementation in VLSI Technology", Proceedings of 7th Symposium on Computer Arithmetic, June 4-6, 1985, University of Illinois, Urbana, Illinois.
3. A. Farooqui, V. G. Oklobdzija, "Multiplexer Based Adder for Media Signal Processing," 1998 Symposium on Circuits and Systems.
4. A. Naini, D. Bearden, W. Anderson, "A 4.5nS 96-b CMOS Adder Design", in Proc. CICC, Feb. 1992, pp. 25.5.1–25.5.4.
5. S. K. Mathew et al., "Sub-500ps 64-b ALUs in 0.18μm SOI/Bulk CMOS: Design and Scaling Trends," Journal of Solid-State Circuits, Nov. 2001.
6. T. Han, D. A. Carlson, "Fast Area-Efficient VLSI Adders," 8th IEEE Symposium on Computer Arithmetic, Como, Italy, pp. 49–56, May 1987.
7. P. M. Kogge, H. S. Stone, "A Parallel Algorithms for the Efficient Solution of a General Class of Recurrence Equations", IEEE Transactions on Computers, Vol. C-22, No 8, Aug. 1973. p. 786–93.
8. J. Park et al., "470ps 64-Bit Parallel Binary Adder," 2000 Symposium on VLSI Circuits Digest of Technical Papers.
9. H. Ling, "High Speed Binary Adder", IBM Journal of Research and Development, Vol. 25, No 3, May 1981, p. 156.
10. Naffziger, S., "A Sub-Nanosecond 0.5 um 64 b Adder Design", 1996 IEEE International Solid-State Circuits Conference, Digest of Technical Papers, San Francisco, February 8-10, 1996. p. 362–3.
11. R. P. Brent, H. T. Kung, "A Regular Layout for Parallel Adders," IEEE Trans., C-31(3), pp. 260–264, Mar 1982.
12. H. Q. Dao, V. G. Oklobdzija, "Application of Logical Effort Techniques for Speed Optimization and Analysis of Representative Adders," 35th Annual Asilomar Conference on Signals, Systems and Computers, Pacific Grove, California, November 4–7, 2001.
13. V. G. Oklobdzija, "High-Performance System Design: Circuits and Logic", IEEE Press, 1999.

MDSP: A High-Performance Low-Power DSP Architecture

F. Pessolano, J. Kessels, and A. Peeters

Philips Research, Prof. Holstlaan 4, 5656 AA Eindhoven, The Netherlands
{francesco.pessolano, joep.kessels, ad.peters }@philips.com

Abstract. The Multi-process DSP architecture (MDSP) is presented and evaluated for high-performance low-power embedded processors. The proposed architecture extends the standard control-flow DSP architecture with simple data-flow primitives. Such primitives are used to generate concurrent processes at run-time, which independently generate and consume data without accessing the instruction flow. We have evaluated the MDSP proposal by designing an asynchronous DSP core, since previous studies showed it to be better suited as an implementation technique. The experiment showed interesting improvements in overall performance and external device management compared to the current commercially available DSP cores. It also showed good scalability and compiler-friendliness with respect to alternative approaches.

1 Introduction

The increasing demand for low-power high-performance embedded DSP cores has been the impetus behind the experiments with new architectures [1]. A system should be economical in its energy requirements, it should be EMC compliant, and it must have a short time-to-market. No performance trade-off is acceptable. The adoption of sub-micron technologies is raising questions about which design methodology to use, however, which also affects our choices at system level. A recent approach in DSP design is the adoption of VLIW-like architectures [2, 3], which moves the burden of code analysis from the hardware to the software. VLIW-like DSP can provide high performance with a relatively short time-to-market. It does present the following challenging drawbacks, however: complex compiler design, reduction of software/hardware flexibility, increase in energy demand and compiler-sensitive performance.

A possible solution is to provide scalar cores with LIW capabilities, which are to be used in critical code sections [4]. The instruction flow is then composed of standard scalar instructions and "customisable" ones, which are dynamically translated into LIW instructions. This hybrid approach gives a good balance between the hardware and software complexity. However, the compiler cannot easily generate customisable instructions. Another disadvantage is that only the customisable instructions can access all of the available resources; in the standard running mode, instructions are executed by a resource sub-set.

In this paper, we describe the novel *Multi-process DSP architecture (MDSP)*, which extends the basic scalar approach with dynamic instruction scheduling and simple

B. Hochet et al. (Eds.): PATMOS 2002, LNCS 2451, pp. 35–44, 2002.

data-flow primitives. Dynamic scheduling is introduced to use all available resources without the need for customisable instructions. Instructions are therefore assigned and executed by the first available compatible resource. The proposed architecture can easily be extended so that it can receive more complex instruction flows (similar to VLIW-like DSP's, for example) with partially dynamic scheduling. Data-flow primitives are introduced to generate and execute concurrent processes, which independently produce and consume data. Processes are intrinsic in the executing resources themselves, and therefore do not require explicit instructions. Communication between processes is carried out by means of special (first-in first-out or FIFO) registers, thus achieving a high-degree of de-coupling between the resources.

This paper is organised as follows. Section II describes the main features of the proposed DSP architecture. The expected advantages of the proposed DSP architecture are examined in section III. Our first design experiment is analysed in section IV, together with a general evaluation of the proposed solution. Some conclusions are drawn in section V.

2 Multi-process DSP Architecture

The MDSP architecture is a hybrid control-driven/data-flow core, which supports concurrent execution of both independent instructions and processes. It is built on the standard scalar DSP architecture by adding three main features: resource clustering, register classes and intrinsic processes.

2.1 Resource Clustering

Multi-process DSP architectures are based on the idea of clusters: register files, dispatch queues and functional units are distributed across multiple heterogeneous clusters. A generic MDSP core is composed of five main module types (Fig.1):

- The *x-Memory* holds the coefficient data and instructions. A more complex memory hierarchy can be used, but it is seen as a single x-Memory module.
- The *y-Memory* holds general-purpose data. A more complex memory hierarchy can be used, but it is seen as a single y-Memory module.
- The *Fetch and Dispatch Unit* (*FDU*) reads instructions from memory and dispatches them to the proper cluster. No assumption on the dispatching strategy is made at this point.
- The *Heterogeneous Clusters* are responsible for instruction execution apart from directive used for instruction scheduling (if any). There are five cluster classes: *ALU, AXU, ACU, GP-RF* and *IMU*. The first three clusters are equivalent to the functional units used in the previous DSP generation: arithmetic and logic clusters (*ALU*), application specific clusters (*AXU*) and memory pointer clusters (*ACU*). The *GP-RF* clusters represent a global locking register file, which every cluster can freely access. The fetch unit is responsible for its correct operation.

The *Instruction Memory Unit* (*IMU*) is responsible for handling the instruction pointer during both sequential operation and branching.

• The *Communication Registers* are used to allow full connectivity between clusters. They are implemented as FIFO's and are used as normal registers without requiring dedicated operations. They can be either global (i.e. all clusters can access a register) or point-to-point connections (i.e. the sender and receiver are fixed beforehand).

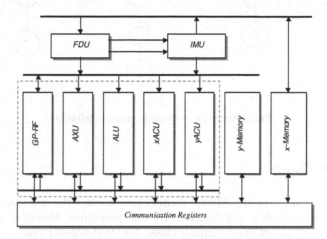

Fig. 1. Architectural view of Multi-process DSP cores.

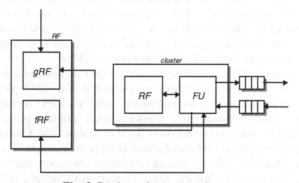

Fig. 2. Register classes overview

A general MDSP core fetches instructions through the FDU/IMU pair. Instructions are then dispatched to the proper cluster together with operand information (sent to the GP-RF), if required. All clusters are effectively de-coupled by either their pipelined nature or circular buffers in the FDU. De-coupling is also achieved through communication registers, which are used to exchange data between clusters.

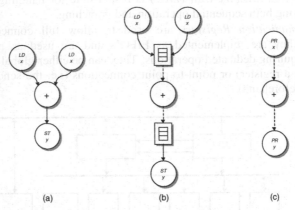

Fig. 3. Example of intrinsic process definition.

2.2 Register Classes

The clustered nature of the MDSP architecture implies the existence of different register classes with different functionality and properties. However, all operations can theoretically accept every register class. The first register class is composed of local registers, visible only to one cluster (RF in Fig.2). The second class implements shared registers that all clusters can access, such as the *GP-RF* cluster (*gRF* in Fig.2). The last class is composed of *FIFO registers*, which are used for inter and intra-clusters (*fRF* in Fig.2). Shared registers are used to hold computational data; cluster-visible registers are generally used for special information (e.g. looping data in the *IMU*); FIFO registers are used to store temporary variables.

While the first two classes can also be found in other architectural models [1], FIFO registers are a new feature. Each register is composed of a FIFO queue, and they are used either for communication or as temporary variables. FIFO registers can be either local or global resources: local FIFO registers allow point-to-point communication between two clusters. Such a connection is fixed at design time. Global FIFO registers receive data from one cluster and make it available to all clusters. Their most obvious use is to enforce data-dependency among operations; however, they can be also used to forward the last result within the same cluster. An example is described in the graph of Fig.3(a/b): since ACUs handle memory operations (*LoaD* to *xACU* and *STore* to *yACU[1]*), data is exchanged through communication registers (Fig.3b). In this way, we also reduce register file pressure (only two communication registers are required instead of three). If the code were the body of a loop, communication registers would also introduce elasticity. In figure 3b, for example, elasticity would allow the load operation to be executed before the result is required.

---

[1] The destination is specified by the x or y at the end of each operation in the node of the graph.

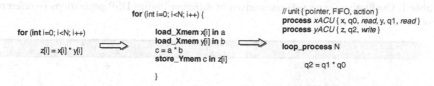

Fig. 4. Example of intrinsic process definition and use. The process construct is used here to specify the process in terms of the name (xACU/yACU) and list of pointer, FIFO register and action groups (e.g. {x,q0,read,..}). The construct loop_process is a simple loop construct which starts previously defined processes.

2.3 Intrinsic Processes

Each cluster has a well-defined basic functionality. For example, the basic functionality of an ACU is to handle pointers. Each ACU generally either reads or writes data through a pointer, and updates it by a constant step. If the pointer properties were constant during computation, we could just provide the ACU with the range and the usage rate. The ACU would read all data and put it into FIFO registers. Even if this assumption does not hold for the entire application, it generally holds in loops (especially with stream-like operations). In this case, we could try providing each ACU with the number of loop steps and the range of each pointer required in the loop. When the loop starts, data is already being read from memory, and memory is ready to be updated. This is the basic description of what we define as an *intrinsic process*[2] that is a process executed by a single cluster, which requires basic information and performs a basic functionality and does not receive any operations. If we consider the example in Fig.3(a), the loads can be seen as a process for the *xACU* (*PRx*, Fig.3c), and the store as a process for the *yACU* (*PRy*, Fig.4c).

An interesting example where processes can be used is the vector product (Fig.4). We can think of the standard *C* loop (Fig.4a) as being composed of three basic operations (Fig.4b): reading the *i-th* vector values, multiplying them and storing the result. Supposing data is read from *x-memory* and written to *y-memory*[3] we could create then one process reading vectors and one writing the result. The *xACU* cluster would execute the reading process, while *yACU* would execute the writing one. Both processes could be defined (*process* directive - Fig.4c) by the pointer names (*x*, *y* and *z* - Fig.4c), the communication registers (*q0*, *q1* and *q2* - Fig.4c) and the operation (*read* and *write*, Fig.4c). Processes are created by the loop initialisation (*loop_process* operation) that explicitly uses previously-defined processes.

3 Expected Benefits

In the previous sections, we described the main features of Multi-process DSP architectures. Such features lead to an efficient architecture with an expected general improvement with respect to the customisable LIW and VLIW DSP's (Tab.1):

---

[2] For sake of simplicity, we will generally refer to it simply as process.
[3] This is only required to keep the example simple.

Table 1. Qualitative (expected) comparison of different Philips DSP generations (~ reference, - worse, + better)

	Customisable LIW	MDSP	VLIW
COMPILER	~	+	~
PARALLEISM	~	+	+
PROCESS PARALLEISM	*na*	+	*na*
SCALABILITY	~	+/++	+
POWER CONSUMPTION	~	+	-
CODE SIZE	~	~/+	-/--
REGISTER PRESSURE	~	+	-

- *Compiler complexity*: no dedicated instruction scheduling strategy underlies MDSP architectures, instead, the examples we describe are based on a standard sequential model. In this way, we can evaluate the sensitivity of MDSP performance with respect to the compiling process. We do expect reduced sensibility in comparison with the other two approaches.
- *Exploitation of parallelism*: The availability of all resources combined with a simple dynamic instruction scheduling strategy is expected to allow the proposed solution to easily exploit parallelism in both arithmetic code and control code.
- *Process parallelism*: Processes are a more complex component with respect to instructions. The adopted process strategy easily allows overlapped execution of at least one process per cluster; performance is expected to improve significantly when processes can be used. In addition, processes only use the allocated resources; all the other resources are free to receive and execute instructions that do not depend on the running processes.
- *Scalability*: Due to its inherent clustered nature and partially dynamic dispatching (through the FDU), a MDSP core can easily be extended without significant design or compiler complexity.
- *Energy requirements*: Processes reduce the number of instruction memory accesses as well as decoding and dispatching. In fact, process operations are implicit in their definition
- *Code size*: Processes and communication registers may provide a way of reducing the number of overall instructions with respect to similar instruction sets, while reducing the number of operations executed dynamically (process).

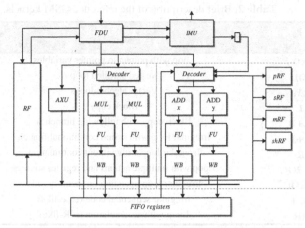

Fig. 5. MDSP OW* core architecture.

- *Register-file pressure.* The use of FIFO registers to hold temporary variables reduces the number of required global registers, thus allowing more parallelism to be exploited.

4 Experimental Results

4.1 Experiment Methodology and Motivations

The distributed nature of the proposed architecture seems better suited to an asynchronous implementation, where no global clock is used to synchronise clusters [5]. For this reason, we evaluated the MDSP approach through a design experiment in asynchronous logic. A simple asynchronous MDSP core, called *OW**, was designed using the proprietary Philips Tangram CAD tool [6]. The *OW** processor's overall organisation (Fig.5) resembles a customisable LIW Philips RD16xxx DSP core, which is used here as a term of comparison. It was provided with the same functionality as the commercial core. Even if the adopted CAD tools aim at extreme low power and reduced speed, the OW* MDSP core allowed testing of the overall architecture, thus giving us a first impression of its performance.

In the OW* MDSP (Fig.5), instructions are fetched one at a time. Each fetched instruction is partially decoded (FDU) and sent to the proper cluster together with its operands (if any). Here, the instruction decoding is completed and the operation is executed. In the case of the integer cluster (DCU), two sub-units (FU) are available and they receive operations on a first-available-first-served basis. However, the ACU decoder steers the operation to the competent sub-unit (either x-memory or y-memory). The OW* core can therefore execute two integer and two memory processes at the same time (excluding processes executed by the AXU cluster). The OW* is provided with only global FIFO registers: each cluster is free to use all of them.

Table 2. Brief description of the reference GSM kernels.

Kernel Name	Features	Static Size
IND	Loop with pointer to pointer variables	8
COL	Nested unrolled loop	15
R;W	Vector algebra loops	6
LL	Block copy	7
LL+B	Block copy with following branches	7
IL+J	Irregular loop with nested conditional branch	20
IR	Irregular loop with/without continuation	29
CXPROC	Loop with multiple actions and register sources	12
CXPROC2	Irregular loop operating on memory only	8
CASE	Irregular sequential code (case-like)	14
FIR	Loop with accumulation (FIR-like)	6

4.2 A Real-World Application: The GSM Code

In order to evaluate the proposed solution, we extracted some kernels from a real-world application: the GSM speech codec used in Philips mobile phones. Kernels were selected on the basis of a typical run trace (Tab.2) for their frequency and characteristics. They were chosen in order to gain more insight into the MDSP architecture, rather than into the OW* core itself. We already gained information about these aspects from the assembler code: complex processes can be generated but they may require more resources and code size is not generally increased. The RISC-like instruction set penalties are well balanced by reduced loops unrolling, since it is inherently provided by the process mechanism.

A very positive effect has been found in instruction memory accesses (Fig.6): we saw a ~16% reduction (resulting in an ~10% overall energy saving[4]) and no access to internal tables as for the translation of customisable LIW instructions found in current DSP cores (resulting in an ~20% overall energy saving[4]). These results, together with more frequent repeat-loops (which do not require decoding and fetching) and simpler decoding modules lead to relevant energy savings. This result is only due to the use of intrinsic processes that allow the replacement of a group of instructions with a process definition.

We also evaluated the kernel run-time. The results showed that the OW* is generally faster (Fig.7) when processes are simple but otherwise comparable. These results were greatly influenced by the pipeline organisation which we considered, and thus they are therefore of limited interest and accuracy.

---

[4] Estimated on a typical use of a proprietary mobile telecommunication handset.

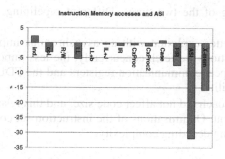

Fig. 6. Instruction memory and customisable LIW instruction table access (ASI) variation for the OW* core relative to the RD16xxx (reference value 1).

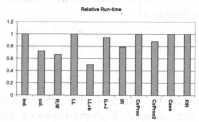

Fig. 7. GSM kernel run-time of the OW* DSP relative to the RD16xxx (reference value 1).

4.3   Final Remarks on Results

The results we have obtained so far show that the MDSP approach leads to relevant energy savings (at system level) , no loss in performance and code size, and a reduction in compiler complexity. However, it is worth analysing the results in order to understand which advantages are independent of the design experiment itself. Let us consider the possible effects derived from a MDSP approach starting from the expected advantages list (Section III):

- Reduced compiler complexity is, because we have considered a scalar RISC core without VLIW or customisable instructions.
- Exploitation of parallelism is enhanced by processes; in fact, we can express parallelism so to concurrently execute processes instead of simple operations. With a wide issue MDSP core, more processes could be generated and manipulated, resulting in more parallelism (if available).
- The OW* showed an impressive energy consumption reduction due to the MDSP architecture alone. This saving is at "system level", since it is achieved by reducing access to external resources such as memories. We can safely assume that it is a basic characteristic of an MDSP core.
- The results on kernel run-time are a delicate point. The OW* MDSP was evaluated under assumptions that force it to work as a synchronous core. Improvement was achieved for process parallelism, only due to the MDSP approach. However, in some kernels we have comparable run-time (e.g. FIR) due

to the pipelining of the two cores. Different pipelining could lead to different results.

- Scalability is better exploited with an asynchronous implementation. However, the more instructions we fetch concurrently, the less important the asynchronous factor is. We expect communication registers and the FDU to become the critical factor for scalability.

- The MDSP approach did not affect code size, and this result is independent of the design experiment. Optimisations of the instruction set could be used to improve it while increasing decoding complexity.

Besides the above described advantages, we should point out that the MDSP does introduce new elements, however, which may prove critical for performance. For example, FIFO latency affects performance and incorrect software can deadlock the system because of wrong FIFO utilisation. Further work is required to analyse these components more comprehensively.

5   Conclusions

In this paper, we introduced the Multi-process DSP architecture (MDSP) solution. The proposed architecture aims at high-performance low-power DSP cores by extending the basic DSP architecture with simple data-flow primitives. The data-flow behaviour is introduced to generate and execute concurrent processes, which can independently produce and consume data. These processes are simple to handle, since they are intrinsic in the resources themselves. We evaluated the MDSP proposal by designing an asynchronous DSP core, since previous studies showed it to be better suited as an implementation technique. However, most evaluation depends on the instruction set architecture rather than the implementation technique. The DSP core was evaluated using small kernels extracted from the GSM speech codec application. The proposed solution shows interesting improvements in energy consumption (at system level) and do no introduce performance or code size penalties.

References

1. Faraboschi P., et al., *The latest word in digital and media processing*, IEEE Signal Processing Magazine, 59–85, March 1998
2. Van Eijndhoven J.T.J., et al., *TriMedia CPU64 Architecture*, Proc. Of the Intl. Conference on Computer Design, Austin, Oct. 1999
3. Scott J., et al., *Designing the M*CORE M3 CPU Architecture*, Proc. Of the Intl. Conference on Computer Design, Austin, Oct. 1999
4. Lambers E., et al., *R.E.A.L. DSP: Reconfigurable Embedded DSP Architecture for Low-Power/Low-Cost Telecom Baseband Processing*, 1st IEEE Workshop on Circuit and Systems for Wireless Communication, Lucerne, 1998
5. Pessolano, F. et al., *Towards a high-performance asynchronous-friendly DSP architectural template*, 4th AciD Workshop on Asynchronous Circuits and Systems, France, 2000
6. Van Gageldonk H., *An Asynchronous Low-power 80C51 Microcontroller*, Ph.D. Thesis, Technical University of Eindhoven, The Netherlands, 1999

Impact of Technology in Power-Grid-Induced Noise

Juan-Antonio Carballo and Sani R. Nassif

IBM Austin Research Laboratory

Abstract. Due to technology scaling, the trend for integrated circuits is towards higher power dissipation, higher frequency and lower supply voltages. As a result, the power supply current delivered through the on-chip power grid is increasing dramatically, which is recognized in the International Technology Roadmap for Semiconductors as a difficult challenge. Early power grid design and the addition of decoupling capacitance have become crucially important to control power-grid-induced noise. We show analytical relationships and simulation results that highlight key relationships between noise and technology parameters. The results underline trends in noise based on current roadmap predictions and reinforce the importance of early planning of global power grids.

1 Introduction

Noise margins have been greatly reduced in modern designs due to lower supply voltages, lower threshold voltages, and the presence of numerous noise generators that eat into the noise margins of a design. The power grid, which provides the power (denoted by V_{dd}) and ground signals in a chip, is also a major source of signal integrity issues. Supply voltage variations can lead not only to delay variations, which can cause malfunction, but also to spurious transitions when dynamic logic is used [1]. Thus it is critical to model and predict power supply performance from a noise perspective.

Chip power supply models can be very large. The analysis of such large systems is challenging in itself [4]. In this paper, however, we focus on simple equivalent-circuit power grid models that are amenable to analytical modeling. We then use these models to understand the impact of power grid noise on the power delivery system. We also study the trends in the values of the model component based on the ITRS [5] and other documents, and translate those into trends for noise. Finally, we present simulation data that reveals the relative importance of key parameters absent in the model.

2 Canonical Power Grid Circuit

Consider the circuit shown in Fig. 1, which can be thought of as a canonical model of a power grid and loading circuit. We make the following assumptions:

- V_{dd} and Ground are largely symmetric nets and so it suffices to consider V_{dd} alone.
- The package is predominantly inductive, and is modeled by the series inductance L (about 4nH for a C4 solder-ball in a high performance package). This assumption is well suited for realistic current packages and appears to be on track to remain so.

B. Hochet et al. (Eds.): PATMOS 2002, LNCS 2451, pp. 45–54, 2002.
© Springer-Verlag Berlin Heidelberg 2002

– The power grid is predominantly resistive, and is modeled by the resistance R_g. This modeling decision is possible because (1) although power grid inductance is growing, it is still much smaller than the package inductance for GHz-range frequencies[2], and (2) grid capacitance is negligible compared to active device capacitance, given by Angstrom-range thicknesses.

– The circuit using the power grid is modeled by two components. First, switching circuits are represented by a time-varying current source (I_{load}) modeled using a triangular waveform with period T:

$$I_{load} = \begin{cases} 0 & : \quad t < 0 \\ \mu t & : \quad t < t_p \\ \mu(2t_p - t) & : \quad t < 2t_p \\ 0 & : \quad 2t_p < t < T \end{cases} \tag{1}$$

Second, non-switching circuits are represented by C_d. The power grid components *between* the switching and non-switching circuits are modeled by R_d.

– Any additional decoupling capacitance added is modeled by C_d and R_d as well.

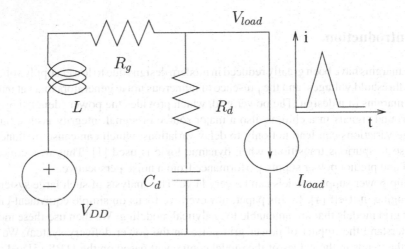

Fig. 1. A canonical representation of a power network.

3   Analytical Model for Power Grid Noise

The circuit in Fig. 1 allows to solve for the analytical form of the voltage delivered to the active circuitry, denoted by V_{load}, then determine the minimum such voltage and therefore the maximum power grid noise. Two cases are examined. First, the inductance L is set to zero, thereby yielding a far simpler result from which insight is easier to gain. Second, the inductance L is included, but the resulting expressions are complicated and the insight is gained by making simplifying assumptions.

3.1   Noise Model When $L = 0$

We observe that V_{load} over the time interval from $t = 0$ to $t = t_p$ can be expressed as:

$$V_{load} = V_{dd} - \mu R_g \left(t - C_d R_g (1 - e^{-t/\tau}) \right) \tag{2}$$

where

$$\tau = (R_g + R_d) C_d \tag{3}$$

The minimum V_{load}, or maximum normalized power-supply-induced noise occurs at $t = t_p$ and the magnitude of the noise is:

$$V_{max} = \mu R_g \left(t_p - C_d R_g (1 - e^{-t_p/\tau}) \right) \tag{4}$$

It is interesting to take the Taylor expansion of Eq. 4:

$$V_{max} = \mu R_g \left(t_p - \frac{t_p R_g}{R_g + R_d} + \frac{t_p^{\ 2} R_g}{(R_g + R_d)^2 C_d} - \cdots \right) \tag{5}$$

We claim that terms in the Taylor expansion of order greater than one can be ignored, an approximation made increasingly more valid as C_d is increased. Thus for large $C_d >> t_p$, the behavior of V_{max} is dominated by the term:

$$\mu R_g \left(t_p - \frac{t_p R_g}{R_q + R_d} \right) \tag{6}$$

Increasing the decoupling capacitance C_d has little effect on the noise in the system. Instead, one ought to decrease the resistance associated with the decoupling capacitance R_d, pointing to the need to carefully control the *placement* of this capacitance.

3.2   Noise Model with $L \neq 0$

When the inductance is included, the solution for V_{load} for the time interval from $t = 0$ to $t = t_p$ can be expressed as:

$$V_{load} = V_{dd} - \mu(L + R_g t - C_d R_g^2) + \psi_1 + \psi_2 \tag{7}$$

where

$$\psi_1 = (e_1 + e_2) \frac{\mu(L - C_d R_g^2)}{2} \tag{8}$$

$$\psi_2 = (e_1 - e_2) \frac{\mu C_d}{2\beta} (\tau R_g^2 - L(3R_g - R_d)) \tag{9}$$

$$e_1 = \exp \frac{-(\tau + \beta)t}{2C_d L} \tag{10}$$

$$e_2 = \exp \frac{-(\tau - \beta)t}{2C_d L} \tag{11}$$

$$\beta = \sqrt{\tau^2 - 4LC_d} \tag{12}$$

If the system is over-damped (i.e. β is *real*), the minimum V_{load} occurs at $t = t_p$ and the magnitude of the noise is:

$$V_{max}^L = \mu(L + R_g t_p - C_d R_g^2) - \psi_1 - \psi_2 \tag{13}$$

If the system is not over-damped, overshoot and oscillations represent a greater problem than the power grid noise we are tackling here, and treating the grid in the lumped-element manner implied by Fig. 1 may not be appropriate.

Due to the complexity of the equation, we first start by calculating the maximum noise of a full factorial experiment over the main variables R_g, R_d, C_d, and L. Figure 2 shows scatter plots of the experiment results. The variable ranges are shown in Table 1.

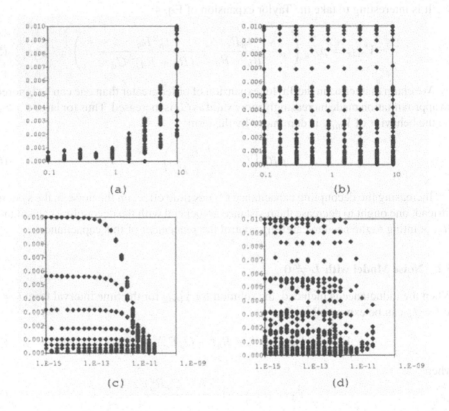

Fig. 2. Maximum noise from Eq. 13 vs. (a) grid resistance R_g in Ohms, (b) decap resistance R_d in Ohms (c) decap capacitance C_d in Farads, and (d) package inductance L in Henrys.

We plot the maximum noise evaluated from Eq. 13 vs. each variable. The plots indicate the relative importance of the various variables. For example, the major factor which determines the noise is the grid resistance R_g because the plot (Figure 2(a)) shows

Table 1. Range of values over which Eq. 13 was evaluated.

Variable	min	max	unit
R_g	0.1	10	Ohm
R_d	0.1	10	Ohm
C_d	10^{-15}	10^{-10}	Farad
L	10^{-15}	10^{-10}	Henry

a very strong dependence of noise on R_g. This implies that the noise performance of the grid is limited by *IR* noise rather than (for example) *Ldi/dt* noise. The plots also show that the second most important factor is the amount of decoupling capacitance C_d, but it is only effective in reducing noise when its value is relatively large (Figure 2(c)).

A good simplified approximation to Eq. 13 can be derived based on Eq. 4:

$$V_{max}^L \approx V_{max} + \mu L \tag{14}$$

Fig. 3 shows a plot of the results of evaluating Eq. 13 and 14 for the same range of values of R_d, R_g, C_d and L as in Table 1 for Figure 2. We see that Eq. 14 is within a few percent of Eq. 13, especially when the noise magnitude is high. We also note that the estimate provided by Eq. 14 is always pessimistic, making it safe to use as a design guideline in performing worst-case analysis.

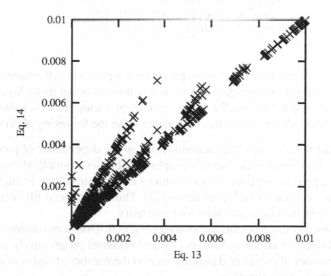

Fig. 3. Comparison of Eq. 13 and an approximation to it Eq. 14

4 Technology Trends

We now use data from the International Technology Roadmap for Semiconductors [5], summarized in Table 2, to predict the dependence of the load voltage V_{load} on circuit parameters, thereby predicting trends in power-grid-induced noise with technology scaling. The table shows the projected yearly trends for transistor effective length, circuit frequency, supply voltage, chip size, power dissipation and power dissipation density.

In analyzing the roadmap data, we make the following first order approximations. First, the characteristic time for the load current source $t_p \propto f^{-1}$. This assumes that the same circuit family is pushed to a higher frequency, resulting in a general compression of relevant voltage and current waveforms proportional to operating frequency. Second, The power per unit area or power density, $P_D \propto V_{dd}\mu t_p$ since the maximum current is μt_p. Thus the slope of the current $\mu \propto P_D f / V_{dd}$.

Thus t_p decreases by about 0.6X across the table, while μ increases by about 3.25X.

Table 2. Trends in IC technology parameters.

Year	L_{eff} nm	f MHz	V_{dd} V	Size mm^2	Power W	Density W/mm^2
1999	140	1200	1.8	450	90	0.2
2000	120	1321	1.8	450	100	0.22
2001	100	1454	1.5	450	115	0.26
2002	85	1600	1.5	509	130	0.26
2003	80	1724	1.5	567	140	0.25
2004	70	1857	1.2	595	150	0.25
2005	65	2000	1.2	622	160	0.26

A harder problem is trend analysis in power supply parameters themselves since such trends are a function of design parameters such as the number of metal layers, pitch and width of power grid wires, and the relative among of wiring resources allocated to the power grid. From existing work, it is reasonable make the following observations:

- Power grid wires are rarely minimum width, so the dependence of power grid performance on overall back-end lithography capabilities is small. However, the planarization process used for the interconnect induces variability in metal thickness dependent on local metal layout density [3]. Thus using metal fill becomes crucial to controlling thickness, and therefore resistivity.
- Once a power grid spans more than 4 layers of metal, it has been empirically observed that the benefit of additional layers of metal becomes progressively smaller. Thus the dependence of power grid performance on the number of layers of metal should be a relatively weak function.
- With no change in the relative amount of area allocated to decoupling capacitance, the ratio of C_d to the maximum current μt_p should remain fairly constant since -to first order- the maximum device current can be expressed as:

$$I_{ds} = C_{ox}K\left(V_{dd} - V_T\right)^{\alpha} \tag{15}$$

where C_{ox} is the gate oxide capacitance which is directly proportional to the decoupling capacitance.
- The package inductance L is a strong function of package *cost*. It is unlikely that current packaging technology will result in significantly reduced package inductance without cost increases. Thus it is safe to assume that L stays the same.

Keeping the comments above in mind, we re-examine the equation for maximum noise relative to the power supply V_{dd}:

$$V_{noise} = \frac{\mu}{V_{dd}} \left(L + R_g t_p - C_d R_g^2 (1 - e^{-t_p/\tau}) \right) \qquad (16)$$

Based on the trends in Table 2 we observe that over the interval from 1999 to 2005, V_{dd} decreases by about 0.6X, t_p decreases by about 0.6X, μ increases by about 3.25X, C_d increases by about 1.95X, and L, R_g and R_d stay approximately constant.

We observe that the major source of concern is a 5.4X increase in μ/V_{dd}. L is approximately constant and t_p is changing relatively modestly, thus we need to decrease the grid resistance R_g and increase the decoupling capacitance C_d. This reinforces the insights gained from evaluating Eq. 13 in Figure 2.

5 Impact of Routing Resources

Further insight can be gained by examining the power grid routing layer assumptions in more detail. Power grid performance depends on the number of routing layers and the percentage of routing resources in those layers that is associated with the power grid. Let n be the number of routing layers, and $\{\alpha(i), i = 1, ..., n\}$ be the percentages of routing assigned to the power supply for each layer, referred to as their *power grid wiring density*. Assuming a resistive load for simplicity, the V_{dd} drop is reduced to $V_{max} = R_g I_{load}$. In this case, it is reasonable to expect the following:

- If $\alpha(i)$ is constant, since upper layers must carry larger currents (because current flows in a tree-like fashion towards lower layers), the IR drop will have a high dependence on the power grid wiring density for upper layers.
- If $\alpha(i)$ is monotonic increasing (i.e., lower layers have fewer routing resources), the strong dependence of maximum noise on the wiring density for higher layers may be reduced.

Figure 4 shows maximum noise v. wiring density results from an experiment run using a power grid simulator [4].

Simulations were run on a seven-metal-layer grid. The experiment consisted of a full factorial set on $\{\alpha(i), i = 1, ..., 6\}$, with each $\alpha(i)$ covering a (5%, 20%) interval. The seventh routing level was kept constant, since it defined where contact with the C4 grid balls were. We observe that the power grid wiring density for the second, fourth, and sixth layers are more significant factors in maximum Vdd noise than the other densities. We also observe that the density for the sixth layer is the most significant density factor. To confirm these results, Table 3 shows for each layer the routing direction, wire thickness in microns, and a normalized coefficient, based on a linear regression fit.

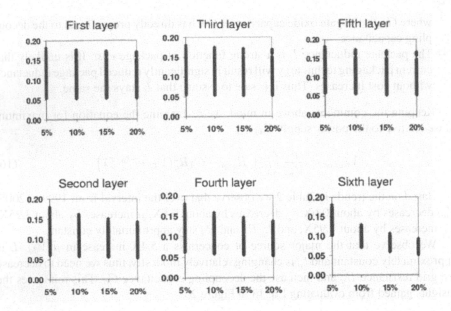

Fig. 4. Maximum power supply drop or V_{dd} noise v.s. power grid density.

Table 3. Routing direction, wire thickness, and normalized coefficients for routing layers.

Layer	Routing direction	Wire thickness	Normalized coefficient
1	Vertical	80	0.03
2	Horizontal	80	0.12
3	Vertical	160	0.03
4	Horizontal	160	0.15
5	Vertical	320	0.06
6	Horizontal	320	0.47

Figure 5 shows maximum noise data versus average wiring density (left) and a measure of its monotonicity or skew given by $3(\alpha(6) - \alpha(1)) + 2(\alpha(5) - \alpha(2)) + (\alpha(4) - \alpha(3))$ (right). We observe that raising the average wiring density reduces the maximum noise, as expected. We also observe that when the wiring density is most monotonic increasing the noise reduction is the most significant.

Based on this data, we can make the following observations:

– The most important factor is the wiring density for the highest (sixth) layer, because of the larger currents that flow through this layer.
– The densities for the second and fourth layers are also significant factors, because they are routed in the same (horizontal) direction as the sixth layer. The other three layers are routed in the same (vertical) direction as the seventh layer, which presents very low effective resistance (thanks to direct connections to C4s).
– Raising wiring density helps reduce maximum power grid noise. Making wiring density a very monotonic increasing function may help as well.

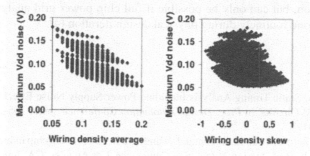

Fig. 5. Maximum noise data versus average wiring density (left) and its skew (right).

This data suggests that it is critical to plan global grid routing early in the design process, especially the highest routing layers. Power grid routing resources should be carefully allocated for these layers. Second, the relationships between the various routing layers, their routing parameters, and packaging technology have an influence in the amount of associated power grid noise. Specifically, certain groups of layers (e.g., layers routed in opposite direction to the highest layer) have more impact on voltage drop than others. Thus these layers should be either avoided for power routing when possible or supplied with abundant routing resources.

6   Conclusions

This paper has presented an analytical formulation for power grid noise as a function of key technology and design parameters. The results show that recent trends to lower power supply voltages, faster frequencies and increased power dissipation result in dramatically increased power grid noise which can be controlled by careful decoupling capacitance placement and sizing. This points to the need to do early power estimation, power grid design, and decoupling capacitance sizing in order to avoid costly re-design. It also reveals the importance of accounting early for routing constraints derived from manufacturing technology factors, such as packaging technology, and the number and distribution of routing resources and layers.

As an example of techniques that can be applied to improve this situation, paper [6] shows an approach to maximize the effect of the decoupling capacitance. However, managing the power grid resistance R_g demands more radical improvement of current power grid planning methodologies. Power grids are usually designed early in an integrated circuit's design cycle, and is difficult to adapt once designed. Unfortunately, our ability to predict power dissipation in general, and its spatial distribution in particular, mean that it is not possible to estimate power grid performance until very late in the design cycle. This has engendered a design methodology in which power grids are significantly over-designed. The continued scaling of technology and the rapid increase in supply currents is making this increasingly expensive in routing resource. One possible solution is to develop algorithms to sparsify power grids as the physical design phase nears completion, but can only be possible if full chip power grid analysis is efficient enough to be done routinely during physical design iteration [4].

References

1. G. Bai and B. Static Timing Analysis Including Power Supply Noise Effect on Propagation Delay in VLSI Circuits. In *Proc. Design Automation Conference*, pages 295–300, Las Vegas, NV, June 2001.
2. K. Gala, V. Zolotov, R. Panda, B. Young, J. Wang, and D. Blaauw. On chip inductance modeling and analysis. In *Proc. Design Automation Conference*, Los Angeles, CA, June 2000.
3. V. Mehrotra, S. Nassif, D. Boning, and J. Chung. Modeling the effects of manufacturing variations on high-speed microprocessor interconnect performance. In *Proceedings of IEDM*, 1998.
4. S. R. Nassif and J. N. Kozhaya. Fast Power Grid Simulation. In *Proc. Design Automation Conference*, pages 156–161, Los Angeles, CA, June 2000.
5. Semiconductor Industry Association, *http://public.itrs.net/Files/1999_SIA_Roadmap. The International Technology Roadmap for Semiconductors*, 1999.
6. H. Su, S. Sapatnaker, and S. Nassif. An algorithm for optimal decoupling capacitor sizing and placement. In *Proc. Internations Symposim on Physical Design*, 2002.

Exploiting Metal Layer Characteristics for Low-Power Routing

Armin Windschiegl, Paul Zuber, and Walter Stechele

Institute for Integrated Circuits, University of Technology Munich, D-80290 Munich,
GERMANY
Armin.Windschiegl@ei.tum.de

Abstract. Wire load has become an important variable for power and timing optimization. As standard cell geometries are shrinking and average wirelength increases due to increasing design complexities wire capacitance has become dominant over gate capacitance. However the wire load of a net not only depends on wirelength but also on which metal layer a net is routed. In this paper we investigate the characteristics of metal layers and propose a power driven routing scheme, which exploits the different metal layer properties in deep sub-micron semicustom design flows. Layer assignment for final routing will be done according to the switching activity of a net and the layer characteristics. In section 3 we describe the investigation of the characteristics of routing layers. A parameter for the validation of metal layers for use in routing for low-power is derived. In sections 4 and 5 an objective function for power driven routing and the layer assignment methodology is described.

1 Introduction

Limited standby and operational time of mobile applications and increasing thermal problems of chips lead to an increasing importance of low-power design. Low-power design techniques should be applied on all levels of abstraction. Methodologies on algorithmic and register transfer level are seen to have the highest potential for power reduction. However techniques on layout level are attractive, if power savings can be achieved without additional working time by applying automatic design flow steps with CAD tools. As the layout has to be routed anyway, the application of a power driven routing algorithm will hardly affect tape-out time.

The power consumed by digital CMOS circuits can be decomposed into two basic classes: static and dynamic power. Static power is due to the leakage current, which occurs at reverse bias diodes and due to sub-threshold currents. Static power can become significant, if the device spends much more time in standby mode than in operational mode. As for a lot of designs operating time is much longer than standby time, the dynamic component of power is the dominant part of the total power, consumed by a chip.

B. Hochet et al. (Eds.): PATMOS 2002, LNCS 2451, pp. 55–64, 2002.

The two sources of dynamic power dissipation in digital CMOS circuits are short circuit and switched capacitance power. Short circuit power is due to the direct-path short circuit current, drawn from V_{dd} to ground, when both the NMOS and the PMOS transistors of a gate are simultaneously active. In [1] and [2] is reported that the percentage of average switching current over total average current depends on the circuit/ application, technology and the logic family. Short circuit current can be minimized by careful design of transition edges in order to speed up the transistors switching. This is possible in pure logic designs but cannot be influenced from the designer e.g. within memory blocks, where long nets between the sense amplifier and the memory cells are never fully charged to Vdd or discharged to 0V due to longer transition time compared with the clock cycle. From [1] we can derive that switched capacitance power is the dominant factor of power dissipation in static CMOS circuits in applications with high switching rate (implies low static power relative to dynamic) and typically low short circuit current (pure logic).

$$P_{switching} = \alpha_{0 \to 1} \cdot V_{dd}^2 \cdot f_{clk} \cdot (C_{pin} + C_{wire}) \tag{1}$$

The switched capacitance power in CMOS circuits depends on the node transition activity factor $\alpha_{0 \to 1}$, the supply voltage V_{dd}^2, the clock frequency f_{clk} and the load capacitance. The load capacitance can be split into two components. Where the pin capacitance C_{pin} of the gates on a driven node decreases due to shrinking geometries, the wire capacitance C_{wire} increases with each technology generation. CAD tools steadily enable handling of more complex designs, but due to conflicting demands on a placement optimizer that is attempting to optimally place the cells directly connected to a net, less optimal placement will be achieved. As a result the average wirelength of the nets in a design increases with increasing gate count [3]. Industrial experience and independent sources have shown that in deep submicron (DSM) technologies 80% or more of the delay of critical paths are caused by interconnects [4]. Where wire delay has its origins in resistance and capacitance, power dissipation due to interconnects has its origins mainly in the wire capacitance. As wire load has an emerging impact on delay and power, it can become a significant factor in future low-power designs.

2 Related Work

Saxena and Liu proposed a performance driven routing approach by taking advantage of the different layer characteristics [5]. They minimize the peak tree delays of global interconnects by layer assignment during routing. Since the parasitic coupling of different layers varies over a large range, assignment of the edges in interconnect trees to specific routing layers has a large impact on the interconnect delay. Kahng and Stroobandt propose wiring layer assignment of local interconnects in order to achieve timing closure within macro blocks. The method describe in [6] serves as a guide for the router of local interconnects as well as for placement tools. Especially the impact of vias on delay is considered during layer assignment. While the approaches described in [5] and [6] deal with timing driven routing approaches, the methodology, described in this paper is focussed on power driven routing. Chandrakasan/ Brodersen and Bellaouar/ Elmasry make a very general proposal that high switching activity wires should be kept short and for signals with lower switching activity can be allowed progressively longer wires [7] [8]. A methodology for low-power driven cell placement has been published by Vaishnav and Pedram [9]. They propose to consider switching activities in order to place cells closely together, which are connected by high switching nets. Guidlines for power driven routing in consideration of metal layer properties are advised by Laurent and Briet [10]. They propose to route nets with high node activity on the higher levels of metal as they have lower area capacitances and lower resistivities.

3 Characteristics of Metal Layers

The wiring capacitance of a metal line on ground plane can be modeled as given in [11]. The total capacitance of two adjacent lines can be interpreted as sum of several components of capacitance:

$$C_{total} = C_{area} + C_{fringe\_0} + C_{fringe\_1} + C_{coupling} \tag{2}$$

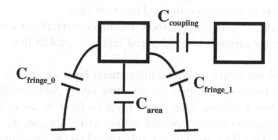

Fig. 1. Components of capacitance of two adjacent lines on ground plane. The area capacitance C_{area} is due to the bottom plane of the wire. Fringing capacitance in case of coupling to a line in the neighbourhood (right side: C_{fringe_1}) has to be distinguished from single fringing capacitance (left side: C_{fringe_0}). Coupling capacitance $C_{coupling}$ between adjacent lines on the same metal layer is the dominant component of the total capacitance.

The area capacitance C_{area} is due to the bottom plane of the wire. Fringing capacitance appears at the side wall of the wire. If there is an adjacent line in the neighborhood of the wire, fringing capacitance C_{fringe_1} is lesser than it appears without a neighboring wire C_{fringe_0}. The reduction in the fringing capacitance is due to shielding effects of the adjacent line, which interferes the electric field of C_{fringe_1}. Coupling capacitance appears between two adjacent lines.

Deep submicron technologies use up to seven or more metal layers for wiring. Problems arise in the alignment of masks in later semiconductor process steps. A lot of manufacturer lessen the problem by using bigger width and spacing for wires on the upper layers [12]. The effect is a big difference in the characteristics of wires on the lower and the upper layers and wire capacitance and resistance depends on which layer a net is routed.

There are several variables, which affect wire capacitance and resistance. As resistance of wires more affects delay than power, we limit the discussion on the influence of wire capacitance. Additionally we only will consider design rules for semicuston routing. The following technology, design rule and layout characteristics have been investigated:

1. Wires on the lower and the upper routing layers can have different aspect ratios of wirewidth to wireheight. Due to lesser wirewidth, in DSM technologies wires become taller than wide on the lower layers. Side wall area becomes larger than the surface area of the bottom and top side of a metal line and hence coupling capacitances become larger than area capacitances.
2. Minimal spacing between adjacent lines on the same layer depends on which metal layer the lines are routed. Spacing on the upper layers is wider than on lower layers. Thus intralayer coupling capacitance on upper layers is lesser than on lower layers.
3. Real spacing of routed wires depends on how congested are the routing layers, that means if a line is directly surrounded by other lines.
4. Wires on the middle routing layers show the highest interlayer coupling capacitance as the average distance to neighbored layers is smaller than on bottom and top layers.
5. The insulator tier between two consecutive metal layers is thicker than minimal spacing between two adjacent lines on the same layer. Additionally, in semicustom designs routing direction alters from layer to layer of about 90 degree and thus capacitive coupling mainly appears between two lines on the same layer as they run parallel. Two lines are more closely routed on the same layer than on two consecutive layers. Thus intralayer coupling is dominant.
6. Different materials for conductors and insulators are used on the different layers. Typically metal 1 is made of tungsten due to high thermal stability, other layers are made of aluminium or copper. The insulator between two consecutive metal layers and between adjacent lines on the same layer may have different relative dielectrics in technologies that use low-k materials.

We found that for the 0.25-micron technology, which is used by us, the dominant component of the wire capacitance is the coupling capacitance which occurs between adjacent lines on the same metal layer.

As a result of the items 1 – 6, the relationships of the components of the total capacitance like given in equation (2) are summarized in table 1 for the 0.25-micron technology, which is used by us. The basic data for this calculations have been provided by the semiconductor company. This is only an example for a deep submicron technology with different layer characteristics, but it shows the dominance of the coupling capacitance on each metal layer. Considering the data which are given in [4] and [12], we can derive that this relations will be qualitative valid for next generations of technologies, too. In table 1 we can see that coupling capacitance in average is about 8 times the area capacitance, 2 times the fringing capacitance without neighbored line and about 18 times the fringing capacitance with neighbored line.

Table 1. Relationship of components of the total capacitance like given in equation (2) for all metal layers of a 0.25-micron technology.

layer	$C_{coupling} / C_{area}$	$C_{coupling} / C_{fringe\_0}$	$C_{coupling} / C_{fringe\_1}$
Metal 1	10.7	2.3	19.8
Metal 2	8.4	2.5	22.5
Metal 3	9.6	2.6	24.0
Metal 4	10.1	1.6	19.4
Metal 5	4.4	1.7	9.2
Metal 6	6.6	1.9	14.5
Average	8.3	2.1	18.2

Due to the dominance of the coupling capacitance we have to pay attention to this parameter. A simple estimation, if layer assignment is suitable in order to reduce the capacitance of switched wires can be derived from the parallel plate capacitor formula:

$$C = \varepsilon \cdot \frac{l \cdot t}{s} \qquad (3)$$

The coupling capacitance C of a wire segment on a layer depends on the dielectric ε of the insulator material, the length l of the wire segment, the thickness t of the wire and the space s between adjacent lines on the same routing layer.

For our investigations, let us consider two adjacent lines on layer i and two adjacent lines on layer i+1 of length l, respectively. In this model coupling capacitance is defined by the parallel side wall plates of two adjacent lines and hence given by wirelength and the thickness of the conductor. As in both cases the dielectric and the length are the same, only the ratio of thickness to wire spacing t/s determines the coupling capacitance between two wires on the same routing layer. Like described above, conductor thickness and spacing differs for several layers due to semicustom design rules and technology parameters. If the ratio of t/s on several layers is significantly different, layer assignment is a promising technique for low-power routing. Table 2 shows the ratios with the parameters wire thickness, width and spacing of a 0.25-micron technology. The most important parameter, the ratio of thickness to spacing t/ s, has lowest values on the upper layers. Wire width increases on the upper layers but wire spacing increases, too. Coupling capacitance and not area capacitance is the most dominant component in equation 2 (see also table 1). Thus the upper layers have least capacitance per unit length.

Table 2. Ratios conductor thickness/ spacing, thickness/width and width/spacing for all metal layers of a 0.25-micron technology.

layer	thickness/ spacing	thickness/ width	width/ spacing
Metal 1	1.5	1.5	1
Metal 2	1.8	1.2	1.5
Metal 3	1.8	1.2	1.5
Metal 4	1.8	1.2	1.5
Metal 5	1.02	1.02	1
Metal 6	0.57	1.02	0.56

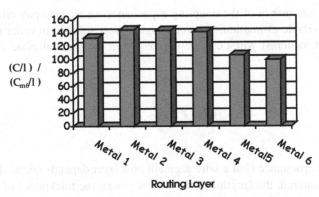

Fig. 2. Capacitance per unit length of all routing layers of a 0.25-micron technology. C/l is given in % and has been normalized on C/l of metal 6 as this layer shows lowest capacitive coupling.

In figure 2 we can see that capacitance per unit length differs up to 42% for different metal layers. Due to larger space between adjacent lines, wires on metal 5 and metal 6 have lowest capacitive coupling and lowest capacitance per unit length. We can conclude that layers with minimal ratio of t/s should be prefered for the routing of nets with high switching activity.

4   Switching Activity Driven Layer Assignment

The objective function for routing of large cell-based designs is the total wirelength or the minimization of vias. An objective function for power driven routing additionally includes the switching rate of each net. Switching rate is used for evaluating the potential of power dissipation. Nets with high switching activity should be carefully treated.

The basic set of parameters for the characterization of a cell based design is given by

$$
\begin{aligned}
I &= \{i_1, i_2, .. i_I\} \\
O &= \{o_1, o_2, .. o_O\} \\
G &= \{g_1, g_2, .. g_G\} \\
W &= \{w_{n1}, w_{n2}, .. w_{nW}\},
\end{aligned}
\tag{4}
$$

where I denotes the number of inputs, O the number of outputs, G the number of gates and W the number of nets in the design. Hence the total number of nets W is given by I + G. Each node n can be associated with the switching activity α at the output of gate g, the arrival time t_a at the output of gate g, the required time t_r at the output of gate g, the input capacitance C_{gates} of the driven gates and the wire capacitance C_{wire} due to net w. The switching activities α at the outputs of all gates are computed in consideration of functionality of logic gates and zero wire delay. The inputs of the design are assumed to be switching independently at probability of 0.5.

The total power consumption for the circuit is given by

$$
P = 0.5 \cdot V_{dd}^2 \cdot f_{clk} \sum_n \alpha_n \cdot (C_{wire\_n} + C_{gates\_n}).
\tag{5}
$$

As the capacitance of the driven gates is independent of the routing, the objective function can be reduced by the expression $\sum_n \alpha_n \cdot C_{gates\_n}$. The term $0.5 \cdot V_{dd}^2 \cdot f_{clk}$ is constant and can be dropped, too. Wire capacitance C_{wire} depends on the characteristics of the layer on which a net is routed (equation 3). Therefore, the objective function for power driven routing becomes

$$C = \sum_{n} \alpha_n \cdot (c_i' \cdot l_{wire}), \tag{6}$$

where l_{wire_n} is the length of wire n and c_i' is the layer-specific capacitance per unit length on layer i. There is a trade-off between wirelength and layer assignment, if wirelength increases due to detours enforced by assignment of nets to locally congested routing layers. Hence assignment of nets to specific layers is only performed, if the total capacitance of the net is lesser than the capacitance when routing the net in the shortest possible way using all available layers.

5   Layer Assignment Methodology

Nets with tight delay constraints are routed first as all resources are available for treating these nets with highest priority. In this case the minimization of vias is as important as the reduction of wire load.

Routing for low-power must be performed by weighting the nets based on switching activity. We will use a greedy algorithm for routing, i.e. nets are treated one at a time. The nets are sorted by their switching activity in order to get knowledge about the 'priority-to-route'. Then the total wirelength of the nets is calculated. Wirelength is estimated by half-perimeter of placed standard cells. The goal is to form groups of nets where each group has almost the same total wirelength (step 3). It should be noted that each group does not include the same number of nets but the sum of the wirelengths of all nets in each of the three groups is almost the same. This way we guarantee a balanced distibution of nets on all available layers and via blockages are avoided. The balanced distribution has a second effect: Not all available tracks on a routing layer are occupied, the average space between wires increases and thus coupling capacitance decreases. In a common routing flow usually the bottom layers are more congested than the upper layers as the tool minimizes the number of vias. In this case we can observe respectable coupling between lines in neighborhood.

Nets within one group are assigned to two specific routing layers. In semicustom designs routing direction alters from layer to layer and every cell pin can be reached by using at least two layers. As we use a technology with six metal layers, we formed three groups of nets.

The groups include nets according to their switching activity:

Group 1: Nets with high switching activity
Group 2: Nets with medium switching activity
Group 3: Nets with low switching activity

Nets in group 1 are constrained to metal 5 and metal 6. In the technology which is used by us, those layers are best suitable for power driven routing as they have the best ratio thickness/ spacing (see equatation 3 and table 2). Nets in group 2 are constrained to metal 3 and metal 4. Nets with lowest priority (group 1) are assigned to metal 1 – metal 3. Metal 1 is made of tungsten and is only suitable for short interconnects due to its high resistance. In order to offer additional routing resources for long wires in the direction of metal 1, we set the top layer limit to metal 3.

Figure 3 shows the algorithm for layer assignment. Capacitance per unit length for each layer has been calculated in consideration of basic data of the technology. In order to reduce computation time the searching of tracks on the constrained layers could be limited to a set of wires with highest switching activity.

```
function layer_assignment( N: input netlist,
                           P: placement information,
                           S: switching activity,
                           L: technology library )
   foreach group of nets
      foreach net w in group
         find track for net w on constrained layers;
                                     // = solution 1
         calculate wire_load_1;
         find track for net w without layer constraints;
                                     // = solution 2
         calculate wire_load_2;
         if wire_load_1 < wire_load_2
            route net w in subject to solution 1;
         else
            route net w in subject to solution 2;
end function;
```

Fig. 3. Algorithm for layer assignment. In this example the objective function for minimizing the wire load is applied to all nets in the desing. In order to reduce computation time it also could be reduced on the set of nets with highest switching rate.

6 Application Example

It has to be noted that the routing scheme described above is not yet completely implemented. However, this is one of the next steps that are planned.

The proposed methodology will be used for intra block routing (not for global interconnects). Nets are assigned to specific layers according to their switching activity. Wiring will be optimized with respect to wire load, which is characterized by wirelength and the properties of the layers on which the net is routed.

7 Conclusions and Future Work

The influence of wiring on power and delay is a key factor for meeting the design goals when using deep submicron technologies. Therefore it is essential to have efficient methodologies for optimizing the circuit on all stages in the design flow. On layout level many researchers have been studying the impact of vias on timing and the impact of wirelength on power and timing.

This paper gave a first overview of metal layer properties in semicustom designs and its impact on wire capacitance. A parameter for the validation of metal layers for use in routing for low-power was derived. The objective function for power driven routing and the layer assignment methodology will be implemented in a CAD tool, which considers layer characteristics during routing.

References

[1] A.-C. Deng, "Power Analysis For CMOS/BiCMOS Circuits", International Workshop on Low Power Design, p. 3–8, 1994

[2] M. A. Ortega, J. Figueras, "Short Circuit Power Modeling in Submicron CMOS", Proceedings of PATMOS'96, p. 147–166, 1996

[3] M. Pedram, B. T. Preas, "Interconnection Analysis for Standard Cell Layouts", in IEEE Transactions On Computer-Aided Design Of Integrated Circuits And Systems, Vol. 18, No. 10, 1999.

[4] Semiconductor Industry Association, *National Technology Roadmap for Semiconductors*, p. 101, 1997.

[5] P. Saxena, C. L. Liu, "Optimization of the Maximum Delay of Global Interconnects During Layer Assignment", IEEE Transactions On Computer-Aided Design Of Integrated Circuits And Systems, Vol. 20, No. 4, p. 503–515, 2001.

[6] A. B. Kahng, D. Stroobandt, "Wiring Layer Assignment with Consistent Stage Delays", Proceedings of the Int. Workshop on System-Level Interconnect Prediction, p. 115–122, Kluwer Academic Publishers, 2000.

[7] A. P. Chandrakasan, R. W. Brodersen, *Low Power Digital CMOS Design*, p. 256, Kluwer Academic Publishers, 1995.

[8] A. Bellaouar, M. I. Elmasry, *Low-Power Digital VLSI Design*, p. 490, Kluwer Academic Publishers, 1995.

[9] H. Vaishnav, M. Pedram, "PCUBE: A Performance Driven Placement Algorithm for Low Power Designs", Proceedings of the EURO-DAC, p. 72–77, 1993.

[10] M. Laurent, M. Briet, "Chapter 3: Low Power Design Flow And Libraries", *Low Power Design in Deep Submicron Electronics*, edited by W. Nebel and J. Mermet, p. 65, Kluwer Academic Publishers, 1997.

[11] T. Sakurai, K. Tamaru, "Simple Formulas for Two- and Three-Dimensional Capacitances", IEEE Transactions on Electron Devices, Vol. ED-30, No. 2, p. 183–185, 1983.

[12] Integrated Circuit Engineering Corporation, *Status 1999 – A Report On The Integrated Circuit Industry*, p. 8-18, 1999.

Crosstalk Measurement Technique for CMOS ICs

F. Picot[1], P. Coll[1], and D. Auvergne[2]

[1] ATMEL Rousset, Zone Industrielle, 13106 Rousset, France
[2] LIRMM, UMR CNRS/Université de Montpellier II, (C5506),
161, rue Ada, 34392 Montpellier, France

Abstract. Signal integrity is of primary concern for designs in submicron processes. Based on the characterization of an industrial driver library in terms of crosstalk-induced noise possibility [1], we present a specific test structure to measure crosstalk signal on interconnect lines. An original implementation is proposed for direct amplitude and pulse width measurement of the crosstalk-induced parasitic signal. A validation is given with an HSPICE simulation of the extracted layout of the structure implemented in a $0.25\mu m$ process.

1   Introduction

With the increasing level of integration in VLSI the problem of signal propagation in interconnect lines is of great concern in designing high-performance ICs. Particularly, the robustness to crosstalk has been widely investigated. Crosstalk glitches appear on metal layers when the cross-coupling capacitance between metal lines has a greater value than a certain threshold, determination of which is of fundamental importance to guarantee the signal integrity. These glitches, resulting from crosstalk coupling effects may pollute the propagation of signals through logic cells.

Many studies have been conducted to model [2]-[4] metal lines and crosstalk effects but few are concerned with a measurement system. The main difficulties in directly measuring parasitic crosstalk effects are the very high rapidity of the induced signal and its very low energy. Accordingly, this parasitic signal cannot be observed on an output pad. Several methods [5] can be used for measuring the crosstalk effect but no technique has been set up to directly obtain both the amplitude and the timing characteristics of the parasitic signal. If some systems give both items of information, they are based on sampling techniques [6], which can be difficult to tune.

The method presented here is able to evaluate the voltage amplitude and the timing characteristics of the crosstalk-induced parasitic signal with a high level of accuracy. The novelty of this method is to facilitate the accurate measurement of short-duration glitch widths.

In part 2 we describe the principle of the measurement to be performed. The technique used to measure the amplitude and the pulse width is described in parts 3 and 4 respectively. The experimental validation of the proposed structure is given in part 5, and we conclude in part 6.

B. Hochet et al. (Eds.): PATMOS 2002, LNCS 2451, pp. 65–70, 2002.

2   Description of the Measurement Technique

In the Fig. 1 we illustrate the configuration used for the measurement. The structure is composed of two parts: a threshold-based detection of overshoots and a time-duration measurement of these overshoots. The test structure is composed of two aggressor lines surrounding the victim line. The inputs of the lines are controlled by tri-state buffers which allow the imposition of different drive strengths to control the state of the lines.

The worst-case configuration defined in the Atmel (0.25μm) process is considered: the victim driver imposes a low level at the input of the line and the aggressor drivers generate a positive pulse.

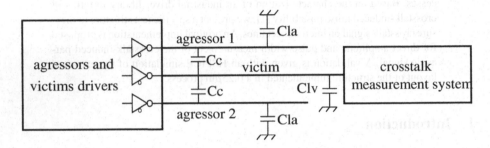

Fig. 1. Crosstalk measurement arrangement.

The output of the victim line is fed to the amplitude and pulse width measurement system.

3   Amplitude Measurement

As the crosstalk signal may exhibit very fast edges and does not have sufficient energy, it cannot be directly measured. In Fig. 2 we describe the voltage amplitude measurement of the crosstalk signal induced on the victim line. It consists of an elementary analog-to-digital flash converter. The array of inverters connected to the output of the victim line has been designed with different threshold voltages. As illustrated in the figure, the filtering action of this array allows one to observe only a stable level at the output of the inverter with a threshold voltage value immediately superior to the amplitude value of the crosstalk signal.

As can easily be understood, the accuracy of this system depends on the number of inverters and is directly determined by the step between the different threshold values.

4   Pulse Width Measurement

The crosstalk signal is generally too fast to be directly observable at the inverter outputs. Considering that the pulses at the inverter output have the width of the parasitic signal

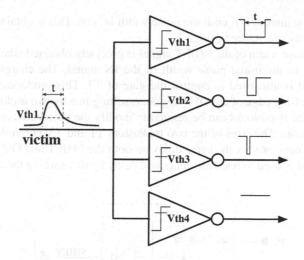

Fig. 2. Crosstalk peak amplitude detection.

at the threshold voltage, we have designed a specific system to multiply the pulse width in order to measure the parasitic signal width, with a good accuracy

The principle of the proposed structure is to design a specific cell to multiply the pulse width by two. Then cascading these cells results in the targeted multiplication factor. The by two multiplication is obtained by generating a delay equal to the pulse width of the input signal. The general schematic of the structure is represented in Fig. 3. The input pulse generates the SHIFT and OUT signals which are out of phase. The falling edge of SHIFT is delayed until its width reduces to zero. At this time the generated delay is equal to the initial pulse width. At the same time the falling edge of the OUT signal is delayed by the same amount and its resulting width is then two times wider.

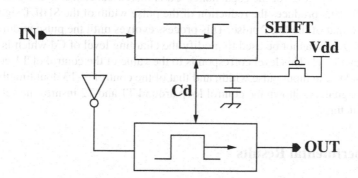

Fig. 3. Schematic diagram of the double width generator.

The detail of the structure is represented in Fig. 4. The delay on the falling edge of SHIFT signal is controlled by the charging level of the capacitance Cd, that controls the gate of the transistor T1, until the pulse width is zero. This is obtained by regularly charging the Cd capacitance through T2.

The zero pulse width of the SHIFT signal is precisely obtained when the generated delay is equal to the initial pulse width of the IN signal. The charging level of the capacitance Cd is also used to control the gate of T3. This produces an equal delay action on the falling edge of the OUT signal, resulting in a by two multiplication of the pulse width. The two blocks can be resized to modify the ratio between the IN and the OUT pulse widths. The sizes of the two transistors T1 and T3, controlled by Cd, must be matched to have exactly the same delay on both the SHIFT and OUT falling edges. The RST signal is used to reset the generated delay by discharging the capacitor Cd.

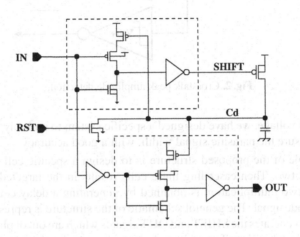

Fig. 4. Electrical structure of the double width generator.

The illustration of the signal wave shape is given in Fig. 5. As shown, at each input pulse the loading level of the capacitance Cd is increased, reducing the gate control of T1 and T3. This produces the reduction of the pulse width of the SHIFT signal and the increase of that of the OUT pulse. This process evolves until the pulse width of SHIFT, reduced to zero, cannot be used to modify the charging level of Cd which is then fixed at a constant level. This level corresponds to the value of the control of T1 generating a delay equal to the input pulse width, and that of the control of T3 doubling the width of the OUT signal. As shown the control loop around T1 and T2 insures the self-converge of this structure.

5   Experimental Results

Validation is obtained with respect to HSPICE simulation of the structure implemented in a 0.25μm process. We compare in Fig. 6, for an input pulse width varying from 100ps to 3ns, the simulated pulse width of the OUT signal to two times the value of the width of the input signal. As shown, this double width generator exhibits a satisfactory response for pulses as short as a few hundred pico-seconds.

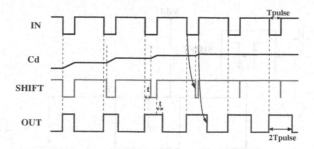

Fig. 5. Double width waveform generator.

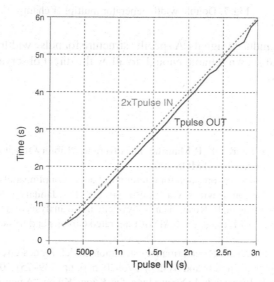

Fig. 6. Comparison of the output pulse width to the expected (2xTpulse IN) value.

This cell can then be used for 2^N pulse width multiplication, allowing the direct observation of the crosstalk-induced signal at the output pad. A drawback of this arrangement is the significant area necessary to implement the 2^N Cd capacitance. An interesting solution is proposed in Fig. 7, using only one double width generator and a cascade of several delay blocks (half-double width generators). In this configuration only one Cd common capacitor is necessary.

6   Conclusion

The crosstalk characterization defining the interconnect line implementation strategy is of great concern for robust design of high-performance ICs. We have shown that using a simple arrangement of output drivers it is possible to measure both the amplitude and time

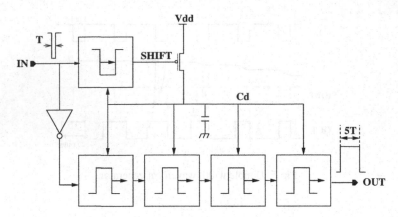

Fig. 7. Double width generator multiplier chain.

width of crosstalk induced signals. A specific structure for pulse width multiplication has been proposed and shown reliable enough to allow the direct observation of the signals at the output pads.

References

1. F. Picot, P. Coll, A. Landrault, P. Maurine, D. Auvergne, "Library sensitivity characterization to interconnect crosstalk", PATMOS 2001
2. T. Sakurai, "Closed-form expression for interconnect delay, coupling and crosstalk in VLSI's", IEEE Trans. on Electron. Devices, vol 40, n°1, pp. 118–124, January 1993
3. K. Shepard, V. Narayanan, R. Rose, "Harmony: static noise analysis of deep submicron digital integrated circuits" IEEE trans. on CAD of Integrated Circuits and Systems, vol. 18, n°8, pp. 1132–1150, 1999.
4. J. Cong, Z. Pan, "Interconnect performance estimation models for design planning" IEEE trans. on CAD of Integrated Circuits and Systems, vol.20, n°6, pp. 739–751,2001.
5. Yungseon Eo, W. R. Eisenstadt, Ju Young Jeon, Oh-Kyong Kwon, "A new on-chip interconnect crosstalk model and experimental verification for CMOS VLSI circuit design" IEEE transactions on Electron Devices, vol.47, n°1, pp. 129–140, 2000.
6. F. Caignet, S. Delmas-Ben Dhia, E. Sicard, "On the measurement of Crosstalk in Integrated Circuits", IEEE Trans. VLSI systems, vol. 8, n°5, pp. 606–609, October 2000.

Instrumentation Set-up for Instruction Level Power Modeling*

S. Nikolaidis, N. Kavvadias, P. Neofotistos, K. Kosmatopoulos,
T. Laopoulos, and L. Bisdounis[1]

Section of Electronics and Computers, Department of Physics,
Aristotle University of Thessaloniki, 54124 Thessaloniki, Greece
[1]INTRACOM S.A., Development Programmes Department
19.5 Km Markopoulo Ave., GR-19002 Peania, Greece
snikolaid@physics.auth.gr

Abstract. Energy constraints form an important part of the design specification for processors running embedded applications. For estimating energy dissipation early at the design cycle, accurate power consumption models characterized for the processor are essential. A methodology and the corresponding instrumentation setup for taking current measurements to create high quality instruction level power models, are discussed in this paper. The instantaneous current drawn by the processor is monitored at each clock cycle. A high performance instrumentation setup has been established for the accurate measurement of the processor current, which is based on a current sensing circuit, instead of the conventional solution of a series resistor.

1 Introduction

Embedded computer systems are characterized by the presence of a dedicated processor which executes application specific software. A large number of embedded computing applications are power or energy critical, that is power constraints form an important part of the design specification [1]. Early work on processor analysis had focused on performance improvement without determining the power-performance tradeoffs. Recently, significant research in low power design and power estimation and analysis has been developed. The determination of a method for the accurate estimation of the power consumption in processors and its dependencies on data and the architectural characteristics of the processor are required for the creation of high quality models.

Power modeling techniques existing in literature, are distinguished into two main categories: a) physical *measurement-based* and b) *simulation-based* ones. In *simulation-based* methods [2][3], energy consumed by software is estimated by calculating the energy consumption of various components in the target processor through simulations, which can be performed at different levels of abstraction. The

* This work was supported by EASY project, IST-2000-30093, funded by the European Union

B. Hochet et al. (Eds.): PATMOS 2002, LNCS 2451, pp. 71–80, 2002.
© Springer-Verlag Berlin Heidelberg 2002

common case is evaluating the power consumption figures from a mix of gate-level and RT-level descriptions. The main drawback of these simulation-based techniques is the need of information about the circuit level design of the processor which is not usually available. In addition to that, these techniques do not provide a mechanism to relate the energy consumption of software with executing the instruction sequence.

In *measurement-based* approaches [1], [4]–[7], the energy consumption of software is characterized by data obtained from real hardware. Power metrics extracted by monitoring the execution of software on the target processor involve either current measurements or direct energy measurements. The advantage of measurement-based approaches is that the resulting energy model proves close to the actual energy behavior of the processor.

In measurement techniques, the usual concept is to associate instructions running on the processor with their corresponding energy cost. These techniques are evaluated by analyzing the power consumption of the processor via decomposing its workload into sequentially executing assembly-level instructions. A profound advantage of measurement-based methodologies towards simulation-based methods is that knowledge of micro- architectural details of the processor under study is not necessary.

2 Related Work

Power analysis techniques for embedded processors that employ physical measurements were firstly suggested in mid 90's. Significant effort on software optimization for minimizing power dissipation is found in [1],[8]-[10], where a technique based on physical measurements is developed. Power characterization is done with the extraction of cost factors for the average current drawn by the processor as it repeatedly executes short instruction sequences. The base cost for an instruction is determined by constructing a loop with several instances of the same instruction. Inter-instruction effect induced when executing different adjacent instructions, is measured by replacing the one-instruction loops used in the measurement of base costs, with loops consisting of appropriate instruction pairs. The sum of the power costs of each instruction executed in a program, refined by the power cost of the inter-instruction effects, are considered to provide the power cost of the program. This method has been validated for commercial targets based on embedded core processors.

The majority of work published on the field of measurement-based techniques, refers to the Tiwari method as a base point. By Tiwari method only average power estimates can be utilized for modeling task, since the measurements are taken with a standard digital ammeter. Direct application of the Tiwari technique is found in [11] where an extensive study of the ARM7TDMI processor core is reported. In order to confine the set of all instruction variations, the ARM instructions are organized according to addressing mode. In [5], physical measurements for the processor current are also obtained by a precise amperemeter. However, power modeling effort is more sophisticated, as architectural-level model parameters are introduced and integrated within the power model. These consist of the weight of instruction fields or data words, the Hamming-distance between adjacent ones, and basic costs for accessing the CPU, external memory and activating/deactivating functional units.

Instantaneous current is firstly measured in [4], where a digitizing oscilloscope is used for reading the voltage difference over a precision resistor that is inserted between the power supply and the core supply pin of the processor. Instantaneous power is then calculated directly from the voltage waveform from which average figures are extracted to guide instruction power modeling. A similar measurement methodology is described in [12], where a high bandwidth differential probe is utilized for reading instantaneous power on a resistor, consumed by an ARM7 processor. Resistor-based methodologies suffer from the supply voltage fluctuations over the measurement resistor and noise induced in supply current path, phenomena which inherently reduce the accuracy of the method.

All the above techniques acquire the current drawn by the processor on instruction execution. A complex circuit topology for cycle-accurate energy measurement is proposed in [6,7], which is based on instrumenting charge transfer using switched capacitors. The switches repeat on/off actions alternately. A switch pair is charged with the power supply voltage during a clock cycle and is discharged during the next cycle powering the processor. The change in the voltage level across the capacitors is proportional to the square of the consumed energy and this value is used for the calculation of energy in a clock cycle. However, this method can not provide detail information for the shape of the current waveform, which may be significantly useful in many applications and also in case high quality power models including the architectural characteristics of the processor are required. In order to measure the energy variations, various (*ref*, *test*) instruction pairs are formed, where *ref* notes a reference instruction of choice and *test* the instruction to be characterized. This setup combined with the above modeling concept are then utilized to obtain an energy consumption model for the ARM7 processor. With this circuitry, measures are gathered for each pipeline stage energy consumption.

Power analysis methods that utilize simulation tools for constructing energy estimation models for software are also accounted. In these methodologies, the target system is first synthesized from an RTL description and gate-level power estimators are used to construct a power cost database for all instructions or instruction pairs [2]. In a different approach, system-level components such as the processor and the cache are considered on either active or idle state, and for each state a factor analogous to the power consumption is assigned [13].

3   Instrumentation Setup for Measuring the Instantaneous Current

As it is mentioned in the previous section different techniques have been proposed for the estimation of the impact of software to the overall power consumption of a given processor. The common methodology is based on the derivation of instruction level power models by measuring the average current of each executed instruction of the processor. The average power P consumed by a microprocessor while running a program is given by: $P=I_{DD}V_{DD}$, where I_{DD} is the average current and V_{DD} is the supply voltage. The energy E consumed by a program is further given by: $E=PN\cdot$, where N is the number of clock cycles taken by the program, and $\cdot$ is the clock period. Thus, the ability to measure the current drawn by the CPU during the execution of the program is essential for measuring its power/energy cost.

The proposed method is based on the measurement of the instantaneous current drawn by the processor during the execution of the instructions. The measurement of the instantaneous current gives the opportunity for higher quality models to be developed since the behavior of the processor can be observed on a finer level exploiting the knowledge of its architectural characteristics. A current measurement method that used a digitizing oscilloscope to measure instantaneous power was proposed in [4] to develop a power model for the JF and HD implementations of the i960 family. However, the current was measured as the voltage drop on a resistor set directly in the current supply network. This configuration is inherently influencing the actual level of the voltage applied to the chip and thus creating an offset noise on the current values, and consequently reducing the accuracy of the method. Using small values of resistors reduces the resolution of the measurements.

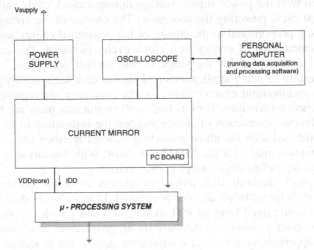

Fig. 1. Measurement setup

The proposed current measurement approach which is based on a current mirroring configuration with Bipolar Junction Transistors (for high frequency operation and limited power-supply voltage fluctuation), aims to overcoming the insufficiencies of the previous methods. The instruction level power models may now be derived on the basis of the information about the "instantaneous" current variations monitored and measured continuously by a high-accuracy and high-speed automated measurement and data acquisition setup. The main task is performed by a high performance current mirroring circuit which is capable of providing a precise copy of the instantaneous current drawn by the processor core. The output (copy) current is then monitored by a precision Digital Storage Oscilloscope (DSO) and transferred to a PC for further calculations by the appropriate software (in Labview environment). The measurement setup is shown in Fig. 1. This measurement approach is similar to the Built-In Self Test (BIST) techniques [14] used for testing high frequency analog circuits (monitoring the current, drawn by the circuit under test, results to evaluation of the different operating conditions). By applying proper timing and signature analysis techniques to these measurements, the power consumption of each instruction sequence used in the software can be estimated.

A simple 4-transistor configuration, shown in Figure 2, is used, which has been proven by extensive tests to offer a quite remarkable performance in terms of copying accuracy and time (frequency) response. The first of these characteristics is obviously important for the accurate measurement of the instantaneous current value. The second one is also important for this case, since the current variations in each clock, are short pulse-like shape waveforms. A key point in this measurement problem is to monitor accurately the shape of the current variations since this characteristic affects strongly the energy consumption value. The energy is calculated by integrating the current in the clock period and multiplying by the supply voltage. In Figure 2, *Rbias* is used for biasing purposes. Current measurements are taken on *Rmeas* (actually the voltage is measured) at the output branch which reflects the current through the processor (DUT). An offset DC current value due to the *Rbias* has to be subtracted from the measured value.

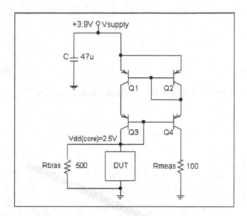

Fig. 2. Current mirror. DUT corresponds to the processor.

The experimental circuit of the current mirror and other components for the proper operation of the system are placed in a specially designed printed circuit board. Then, multiple experimental tests were performed to ensure the proper operation of this setup for the specific application. These tests include current copying accuracy, operation range (min-max values of the current), frequency response, phase difference measurements (between input and output current waveforms), etc. Note that the current drawn by the processor (and during these tests this current was controlled by different high performance generators) is considered as input current waveform while as output current, the current copy generated by the current mirror at its output is considered. The set of instruments used for the experimental test of the current sensing configuration includes an HP-3325B – 20MHz Function Generator, an IFN-2025 – 2GHz Sinusoidal Signal Generator, an HP-3575 – Gain/Phase Meter, and power supply units from Delta Electronika and Kikusui. The monitoring instrument is an HP54601B – 100MHz Digital Storage Oscilloscope connected to a PC computer by a GPIB local network. The experiments presented here have been done using BC212 transistors. The main characteristics of them are given in Table 1.

76   S. Nikolaidis et al.

Table 1. Transistor Characteristics

Collector Current – Continuous	I_c	-100 mA,DC
DC Current Gain (I_c = -100mAdc)	h_{FE}	120 (typ)
Current Gain – Bandwidth Product	f_T	280 MHz

In Figure 3 a comparison between the processor supply voltage fluctuation when the proposed circuit, (a), and a resistor, (b), for the same resolution, are used as the current sensing circuit. The supply voltage fluctuation at the processor is significantly less (more than 7 times) compared to a simple resistor configuration.

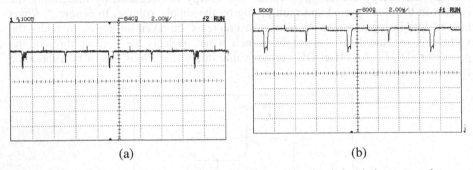

(a)   (b)

Fig. 3. Supply voltage fluctuation when the current sensing circuit is (a) the proposed current mirror and (b) a resistor.

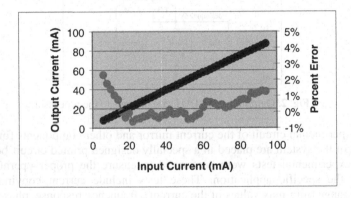

Fig. 4. Experimental results for the input-output characteristic of the current sensing configuration and for the error of output current values (current copying)

The following experimental diagrams present the performance characteristics of this instrumentation system in terms of the different specifications considered above. These diagrams present typical cases from the multiple measurement tests, which were repeatedly performed in the lab. As it is shown in Figure 4 the main specification characteristics of the measurement system include an operation range of 2-100mA, with an error less than 2.5% and less than 1% in an operating range large enough to monitor different variations. Note that in case a lower current value needs to be monitored, then a constant current value may be added changing *Rbias*, which will therefore shift the operating range within the useful range.

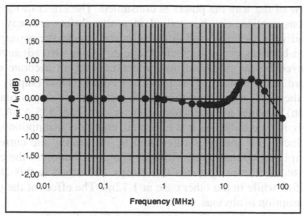

Fig. 5. Experimental results for the frequency response of the current sensing configuration

Current copying capability (gain) is practically maintained constant to equal input-output current values up to 100Mhz. As it is shown in Figure 5, the gain fluctuation is less than 0.5dB all over the frequency range up to 100MHz. For higher frequencies a fixing coefficient has to be considered for the reproduction of the actual input current value. In addition to the above mentioned performance, a series of oscilloscope recordings were transferred to the PC and are shown here, to illustrate the efficiency of the proposed solution for accurate monitoring and measurement of the current waveforms. These examples present the accurate monitoring of a 10MHz square wave (Figure 6), showing also the details of the comparison between input and output.

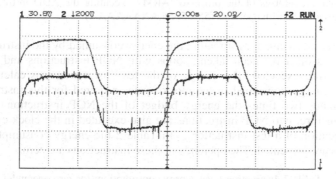

Fig. 6. A 10MHz test current-signal recorded by the instrumentation setup. Upper trace corresponds to the input current and lower trace to the output current of the mirroring circuit

4   Results

Using the proposed instrumentation set up, accurate instruction level power models can be derived based on the measurement of the instantaneous current drawn during the execution of the instruction. By monitoring current at each clock cycle we can have a clear view of the way the power is consumed. The effect of the factors, which affect power consumption, can be studied in a straightforward way. For example, different operand values can be used in the instructions and the corresponding power consumption can be measured at each clock cycle. Taking also advantage of the high resolution achieved by the proposed instrumentation set up accurate models can be created. In Figure 7, the current of the processor ARM7TDMI (supply voltage 2.5V) when executes the ADD instruction (while the rest pipeline stages run phases of the NOP instruction) having as operands (0,0) and (55555555,AAAAAAAA) and running at 6MHz is monitored. As it is expected, energy consumption appears in both phases of the clock. The energy is calculated by integrating the current for a clock period and multiplying by the supply voltage. The contribution of the current through *Rbias* is subtracted. For zero operands the energy consumed in this clock cycle was estimated at 0.95nJ while in the other case at 1.12nJ. The effect of the operand values on energy consumption is obvious.

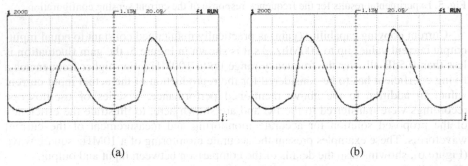

(a)   (b)

Fig. 7. Current waveforms of the processor ARM7 executing the ADD instruction (a) with operands (0,0) and (b) with operands (55555555,AAAAAAAA)

Many measurements for estimating the energy consumed by the instructions of the ARM7 processors have been taken. Loops with NOP instructions and the one test instruction were executed. The energy of the test instruction was calculated as the sum of the energy consumed in the clock cycles required for this instruction to be executed minus two times the energy budget of the NOP instruction. (Due to the pipeline structure, two NOP instructions are also executed in the clock cycles needed for the execution of a test instruction). In Table 2 the energy consumption of some instructions are presented. The operands in these instructions are zero.

Table 2. Instruction-level energy consumption (for zero operands)

Instruction	E (nJ)	Instruction	E (nJ)
ADD R2,R0,R1	0.910	LDR R2,[R1,R3]	2.774
AND R2,R0,R1	0.856	STR R2, [R1,R3]	1.961
ORR R2,R0,R1	0.907	MUL R2, R0, R1, R10	2.768
ORRS R2,R0,R1	0.967	MLA R2, R0, R1, R10	3.748
MOV R2,R1	0.935	CMP R0,R1	0.751
MOV R0,R0	0.903	SWP R2,R0,[R1]	3.917
Instruction	E (nJ)	Instruction	E (nJ)
ADD R2,R0,R1, ASR R3	2.137	MRS R2, CPSR	0.977
B label	3.095	MSR CPSR_f, R2	1.143

With the proposed measuring environment models including the dependencies of the energy consumption on the operand values and their addressing values can be easily created. Although we have not yet completed the measurements, we have observed that there is a dependence of the energy on the number of 1s in the values of the operands and their addresses, which is close to be linear. This is illustrated for the ADD and the LDR instructions in Figure 8. The dependence of the energy on the position of the 1s in the operand words doesn't have significant effect on energy.

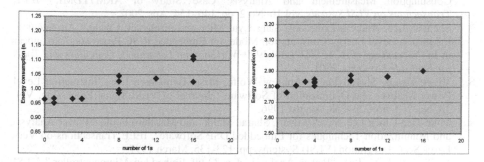

Fig. 8. Energy consumption of (a) ADD and (b) LDR as a function of the number of 1s in the operand values

5 Conclusions

More accurate instruction level power models can be derived by measuring the instantaneous current drawn by the processor at each clock cycle. The instrumentation setup for monitoring the instantaneous current and for calculating the corresponding energy is presented. A current mirror is used as a current sensing circuit to minimize the supply voltage fluctuation and to increase the resolution of our measurements. Exhaustive experiments were done to ensure the propriety of the proposed circuit examining in terms of accuracy and frequency response. Some results from measurements of the current drawn by the ARM7TDMI processor are presented.

References

1. Vivek Tiwari, Sharad Malik and Andrew Wolfe, "Power Analysis of Embedded software: A First Step Towards Software Power Minimization", *IEEE Transactions on Very Large Scale Integration (VLSI) Systems*, Vol. 2, No. 4, pp. 437–445, December 1994
2. Chaitali Chakrabarti, Dinesh Gaitonde, "Instruction Level Power Model of Microcontrollers", Proceedings of the IEEE International Symposium on Circuits and Systems, pp. 176–179, 1999.
3. Tony D. Givargis, Frank Vahid, Jorg Henkel, "Instruction-based System-level Power Evaluation of System-on-a-chip Peripheral Cores," in Proc. of IEEE/ACM International Symposium on System Synthesis (ISSS '00), pp. 163–169, September 2000.

4. J. T. Russell and M. F. Jacome, "Software Power Estimation and Optimization for High Performance, 32-bit Embedded Processors, In Proceedings of the International Conference on Computer Design (ICCD '98), October 1998.
5. S. Steinke, M. Knauer, L. Wehmeyer, P. Marwedel, "An Accurate and Fine Grain Instruction-Level Energy Model supporting Software Optimizations," in Proc. of the International Workshop on Power and Timing Modeling, Optimization and Simulation, Yverdon-les-bains, Switzerland (PATMOS '01), September 2001.
6. Naehyuck Chang, Kwanho Kim, and Hyun Gyu Lee, "Cycle-Accurate Energy Consumption Measurement and Analysis: Case Study of ARM7TDMI," *IEEE Transactions on VLSI Systems*, vol 10, No 2, pp. 146–154, Apr. 2002.
7. Sheayun Lee, Andreas Ermedahl, Sang Lyul Min, and Naehyuck Chang, "An Accurate Instruction-Level Energy Consumption Model for Embedded RISC Processors," to appear in In *Proceedings of ACM SIGPLAN 1999 Workshop on Languages, Compilers and Tools for Embedded Systems, 2001.*
8. Vivek Tiwari, Sharad Malik, Andrew Wolfe, Mike Tien-Chien Lee, "Instruction Level Power Analysis and Optimization of Software", Journal of VLSI Signal Processing, Vol. 13, No. 2–3, pp. 223–238, August 1996.
9. Mike Tien-Chien Lee, Vivek Tiwari, Sharad Malik, and Masahiro Fujita, "Power Analysis and Minimization Techniques for Embedded DSP Software", IEEE Transactions on Very Large Scale Integration (VLSI) Systems, pp. 123-135, March 1997.
10. V. Tiwari, T. C. Lee, "Power Analysis of a 32-bit Embedded Microcontroller," VLSI Design Journal, Vol. 7, No. 3, 1998.
11. SOFLOPO, Low Power Development for Embedded Applications, Esprit project, Deliverable 2.2: Physical measurements, by Thanos Stouraitis, University of Patras, December 1998.
12. Xavier Amela, Joan Figueras, Salvador Manich, Josep Rius, Rosa Rodriguez, Antonio Rubio, "ARM Instruction Set Energy Models and Power Simulation Tools (ARM7TDMI)," UPC Internal Report for the IST 10425 VIP (Versatile Integrated Payphone) Project, March 2001.
13. Tajana Simunic, Luca Benini and Giovanni De Micheli, "Cycle-Accurate Simulation of Energy Consumption in Embedded Systems," In Proceedings of the Design Automation Conference (DAC '99), 1999.
14. A. Hatzopoulos, S. Siskos and Th. Laopoulos, "Current conveyor based test structures for mixed-signal circuits", IEE Proceedings – Circuits, Devices and Systems, V.144, N.4, 1997

Low-Power Asynchronous A/D Conversion

Emmanuel Allier, Laurent Fesquet, Marc Renaudin, and Gilles Sicard

TIMA Laboratory, Concurrent Integrated Systems Group,
46 Avenue Felix Viallet,
38031 Grenoble, France
{Emmanuel.Allier, Laurent.Fesquet, Marc.Renaudin,
Gilles.Sicard}@imag.fr
http://tima.imag.fr/cis

Abstract. This paper presents a new architecture of Analog-to-Digital Converter (ADC) for low-power applications. The converter is a tracking circuit without any global clock, based on an asynchronous design. Samples conversion is only triggered by the analog input signal amplitude variations, hence an irregular sampling of it. System simulations demonstrate that a significative reduction of the circuit activity can be achieved with it. Moreover, such a converter has been designed with 6-bit resolution, using a 0.18-μm, 1.8-V standard CMOS technology from ST-Microelectronics. Electrical simulations show that, the asynchronous converter has an average power dissipation of only 1.9mW in the worst case, with a sample conversion time of 37.9ns, and an important noise reduction is achieved, compared to its synchronous counterparts.

1 Introduction

Classical Analog-to-Digital Converter (ADC) architectures are synchronous i.e. driven by a global clock, and belong to one of these two families: the Nyquist rate converters, or the over-sampled one. The choice of the architecture and the design of the circuit depend on the characteristics to privilege: resolution, speed, area, power consumption,... signal-to-noise ratio (SNR) [1], [2]. Like any electronic circuit, ADC design follows the current trend that is to reduce power consumption, especially when it is used in embedded systems or SoC's (System on Chip) powered by batteries or remotely powered.

Most of the systems using ADC bring signals with interesting properties into operation, but common synchronous signal processing architectures do not take advantage of them. Actually, these signals are almost constant but may vary a lot during brief moments, such as temperature sensors, pressure sensors, electro-cardiograms, speech signals, ... In this way, classical synchronous converting systems, presented above, are highly constrained, due to the Shannon theory, which is to ensure for the sampling frequency to be at least twice the input signal frequency bandwidth. Therefore, in the time domain, this condition can be translated as a wide number of useless samples: as soon as the input signal is constant during a time superior to the sampling period. This

B. Hochet et al. (Eds.): PATMOS 2002, LNCS 2451, pp. 81–91, 2002.

effect implies a useless increase of activity of the circuit compared to the supplied output digital information relevance, and so a useless increase of the power dissipation.

It is well known that asynchronous designs exhibit interesting properties such as low energy dissipation, immunity to metastable behaviour, low electromagnetic interference generation, ...[3], [4]. In this way, D. Kinniment et al. have described a new kind of ADC named "asynchronous" [5] in order to eliminate conversion errors due to metastability when conversion time is bounded. This ADC architecture looks like the successive approximation converter one, but the comparator has been replaced with an asynchronous block generating a local signal driving the digital part. This converter is locally asynchronous, but the conversion principle still remains identical to synchronous analog-to-digital conversion. They have also studied a micropipelined flash ADC [6] and proved metastability removal, as well as noise and power consumption reduction, with asynchronous micropipelining techniques. Although the ADC is locally asynchronous, the general conversion principle always remains a well known synchronous one. Conti et al. described an analog current memory using 5-bit asynchronous successive approximation ADC's [7]. The running of the converters is asynchronous, but an external clock triggers samples conversion: the sampling is always made in a regular way. Nevertheless, a comparison with the corresponding synchronous ADC architecture underlines a removal of the conversion errors due to metastability in the case of the "asynchronous" converter.

The concept of conversion proposed in this paper consists in building a completely asynchronous system, with an irregular sampling of the analog signal to process. This principle has been first described in [8]. N. Sayiner et al. studied its advantages in [9], in order to use it in a classical synchronous signal processing system. The principle of conversion is the following one: the converter supplies a digital sample only if a perceptible variation of amplitude is detected in the analog input signal. This converter is not controlled by any global clock, but only by the analog input signal: it is a tracking system, enslaved on the signal. Moreover, this design is using a classical CMOS technology, which allows the use of that kind of ADC in low-cost mixed-signal applications.

The paper is organized as follows: Section 2 describes the architecture and the principle of this new ADC. Design and electrical simulations are given in Section 3. Section 4 provides the low-power aspect of the converter. Lastly, Section 5 concludes the paper.

2 Architecture Design

2.1 Principle of Asynchronous Conversion

The block diagram of the conversion circuit is shown in Fig.1, it is composed of a difference quantificator, a state variable modelling the inner state (an up/down binary counter), and a Digital-to-Analog Converter (DAC) processing this digital signal to make it compatible with the input voltage amplitude. The converter resolution N and the input analog voltage amplitude dynamic range ΔV_{in} are known. They set the quantification step q (or LSB) according to (1).

$$q=\frac{\Delta V_{in}}{2^N-1} \tag{1}$$

The output digital value V_{num} is converted to V_r by the DAC, compatible with V_{in}, and compared to it. If the difference between them is superior to $\frac{1}{2}.q$, the state variable is incremented ($inc='1'$), if it is inferior to $-\frac{1}{2}.q$, it is decremented ($dec='1'$). In all other cases, nothing is done ($inc=dec='0'$): the converter output signal V_{num} remains constant. This conversion is not made at a constant frequency like in clocked converters, but at instants that only depend on the input signal amplitude variations. The output signal is so composed of couples (b_i, Dt_i) where b_i is the digital value of the sample and Dt_i the time spent since the previous converted sample b_{i-1} given by the timer (with $V_{num}=\{b_i\}_{i\in N}$). We have here an irregular sampling, hence the designation "asynchronous". We use now the term A-ADC for Asynchronous Analog-to-Digital Converter. As for any asynchronous digital circuit, the key point of this converter is that information transfer is locally managed with a bi-directional control signalling. Each "data" signal is associated with two "control" signals: a request ($Req.$ in Fig.1) and an acknowledge ($Acq.$ in Fig.1). A first stage sends a request ($Req.='1'$) to a second stage when data are ready to be computed. The second stage sends an acknowledge ($Acq.='1'$) to the first stage when data are processed which indicates that it is ready to treat another data.

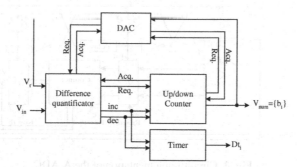

Fig. 1. Block diagram of the asynchronous A/D converter.

There are two running modes for the A-ADC: tracking and convergence. If the input signal V_{in} slope is too high, some digital samples b_i may not correspond to any right analog value of the input: that is the convergence mode. Otherwise all converted samples fit the analog signal: the A-ADC is in the tracking mode. A condition on the slew-rate of the input signal insures that the converter still remains in tracking is:

$$\left|\frac{dV_{in}(t)}{dt}\right|\leq\frac{q}{\delta} \tag{2}$$

where q is the LSB quantum (or quantification step) and δ the total delay of the conversion loop. The limitation given by (2) may be exceeded by adding a very simple circuit, which detects the validity of each digital sample of the output.

Moreover, a little architecture modification of the A-ADC can be achieved, by adding a simple digital part, to limit the A-ADC activity, if the exceeding of condition (2) is allowed. In the worst case, for an N-bit A-ADC, the $(2^N\text{-}1)$ output codes are swept between two valid digital samples. If an analog input V_{in} of maximum frequency f_{max} must be computed, the free running of the conversion loop may be restrained by inhibiting it, between two valid samples, during a time T given by (3). In this case, f_{max} must verify (4). Thus, in the worst case, the number of valid digital samples is equal to the number of samples if the signal were sampled in a regular way with a sampling period respecting the Shannon theorem. This architecture is named "constrained A-ADC".

$$T=-\left(2^N-1\right)\delta+\frac{1}{2.f_{max}} \tag{3}$$

$$f_{max}\leq\frac{1}{2.\left(2^N-1\right)\delta} \tag{4}$$

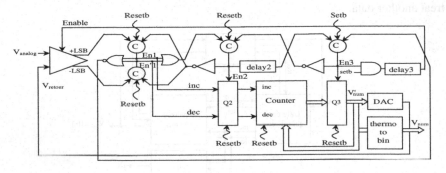

Fig. 2. Circuit implementation of the A-ADC.

2.2 Architecture

The A-ADC structure presented in Fig.1 may be implemented in different ways, a full-handshake protocol, three levels micro-pipelined architecture [10] has been chosen. It is constituted of two distinct parts: a delay-insensitive four-phase-protocol control part composed of delays, Muller gates (or C-elements [11]) and standard gates, and a data part composed in our case of digital or analog blocks (cf. Fig.2). The asymmetric delay element *delay2* (resp. *delay3*) is matching the counter critical path (resp. DAC critical path). They only delay the rising edge of the input signal and not the falling edge. The time generator is not represented on the circuit. The bits *Setb* and *Resetb* on the circuit are controlled by the same initialisation signal *RESET*.

When *RESET='0', En1=En'1= En2='0', En3='1', Enable='0'* and so +*LSB* and -*LSB* are forced to *'0'*. The positive-edge-triggered-clock flip-flop *Q2* is reset and the positive-edge-triggered-clock flip-flop *Q3* is set to the numerical value corresponding to half the input voltage dynamic range. When *RESET='1'*, the AND gate in the third half-buffer sends an initialisation *'1'* in the control part, then *Enable='1'*, the input comparators are now active. If the input voltage varies noticeably, for example: $V_{in}-V_r>\frac{1}{2}.q$ (resp. $V_{in}-V_r<-\frac{1}{2}.q$), we have +*LSB='1'* (resp. -*LSB='1'*), a request (*'1'*) is sent to one of the two Muller gate of the first half-buffer, then *En1='1'* (resp. *En'1='1'*), an acknowledge (*'0'*) is sent to the third half-buffer, then *En3='0', Q3* is memorising its last value, and *Enable='0'*: the comparator outputs +*LSB* and -*LSB* are reset to *'0'*. The sample can now be converted. A request (*'1'*) is sent to the second half-buffer: data are ready in front of *Q2*, then *En2='1'* so *Q2* samples the two bits on the positive edge: the counter incremented the value V'_{num} of *Q3* as *En='1'* (resp. decremented if *En'1='1'*) and an acknowledge (*'0'*) is sent to the first half-buffer (so *En1=En'1='0'*). After *delay2* nanoseconds, data are ready in front of *Q3*, a request arrives to the third half-buffer, so *En3='1'*, an acknowledge is sent to the second half-buffer so *En2='0', Q3* samples the data, the output digital value is now converted by the DAC. After *delay3* nanoseconds, the conversion is done, V_r is ready in front of the comparator, *Enable='1'*: the A-ADC is now in a stable situation until another conversion sample is needed.

2.3 Validation

This asynchronous converter has been compared to a similar synchronous one: the successive approximation converter. This ADC nearly presents the same components as the A-ADC: a comparator, a latch, a complex digital part, and a DAC. The two principles of conversion have been simulated in behavioral level in VHDL-AMS with Mentor ADV-MS tool. The two loop delays δ have been set equal to have valid comparisons between the two modes of running. The ADC always needs *N* cycles to convert a sample, thus the sampling frequency of the sample-and-hold (S/H) of the synchronous converter is given by:

$$f_{sample}=\frac{1}{N.\delta} \qquad (5)$$

An example of conversion is studied in Fig.3, with an input voltage (cf. Fig.3a) respecting (2). There are 150 cycles (cf. Fig.3b) for the ADC (25 converted samples – 6 cycles per sample) vs. 50 (cf. Fig.3d) for the A-ADC (50 converted samples – 1 cycle per sample), which represents here a reduction of 66% of the number of cycles for asynchronous conversion. As it is shown in Fig.3c and Fig.3e, conversion cycle triggering for the A-ADC only depends on the analog signal amplitude variations, contrary to the ADC where conversion is ordered by a clock. Conversion cycles are sparse for the A-ADC compared to the ADC. This irregular sampling, and low circuit activity of the asynchronous converter allow power consumption savings and electromagnetic emissions reduction as it is demonstrated in Section 4.

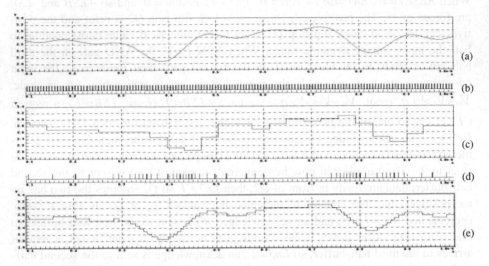

Fig. 3. Comparison between synchronous and asynchronous conversion - (a) Analog signal to be converted, (b) Conversion cycles of the successive approximations converter, (c) DAC output of the ADC, (d) Conversion cycles of the A-ADC, (e) DAC output of the A-ADC

3 Circuit Design

The A-ADC has been designed, in a standard 1.8-V supply, 5-metal-layers, 2-poly, 0.18-μm CMOS process from ST Microelectronics [12] and simulated with Cadence Spectre. Digital parts of the converter, as well as the timer, have been described in VHDL and synthetised. A current mode design has been chosen to implement all the analog parts of the A-ADC. Four distinct analog circuits have been designed for the A-ADC: a transconductor (to translate the analog input voltage V_{in} to convert in a current), a current comparator, a latch and a DAC.

The resolution of the converter is 6-bit and the chosen current quantum is $q=0.8\mu A$. Delay elements are implemented with standard cells of the targeted technology, and their values are chosen to match, with a margin of 10%, the delay of the corresponding parts. The maximum power dissipation is obtained when the input voltage dynamic ΔV_{in} is swept by a voltage ramp, with the non-respect of (2): the A-ADC is also running at its maximum speed: $1/\delta$ cycles per second. The measured dissipated power is $P_{diss}=1.95mW$, with maximum consumed current pick amplitude of $6mA$. The average dissipated power for one cycle of conversion depends on the value of the current digital sample V_{num} stored in Q3. It varies from $1.3mW$ (when the first step of the output dynamic V_{out} is made) to $2.54mW$ (when the last step of the output dynamic V_{out} is made) with an average value of $1.87mW$ (half of the dynamic). When the converter is inactive, analog parts, especially the DAC, consume current. They involve a static consumption, which also depends on the value of the current digital sample V_{num} stored in $Q3$. It varies from *0.8mW* to *2.16mW* with an average value of *1.51mW*. The A-ADC parameters are summarized in Table 1.

Table 1. Performances summary of the A-ADC.

Resolution	6-bit
Technology	5-metal, 2-poly 0.18-μm CMOS
Power supply	1.8 V
Quantum	0.8 μA
Loop delay	37.96 ns
INL	0.45 LSB
DNL	0.35 LSB
Static power consumption	Average : 1.51 mW Max : 2.16 mW Min : 0.8 mW
Dynamic power consumption	Average : 1.87 mW Max : 2.54 mW Min : 1.3 mW

4 Power Saving Aspects

4.1 Transistor Level Aspect

The synchronous ADC has been designed in the targeted technology too, however the sample and hold (S/H) has just been described in behavioral level, in Verilog-A. The ADC uses the same analog blocks as the A-ADC (transconductor, comparator, and DAC), and the sampling clock frequency of the S/H has been set according to (5). The consumed current spectrum of the two converters are given in Fig.4 for a sinusoidal input signal V_{in} respecting (2).

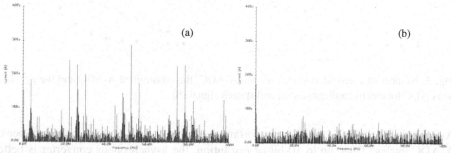

(a)   (b)

Fig. 4. Consumed current spectrum for the synchronous ADC (a), and the A-ADC (b), with a pure sine wave input.

The DC components of the two current spectrums are the same ($I_{DC}=400\mu A$) because of the presence of the same analog parts, which consume the same static current. Nevertheless, for the A-ADC current spectrum (Fig.4b), current peaks are lower and sparser than in the synchronous case (Fig.4a), because of the absence of any global clock driving the entire digital part. These spectrum properties underline a lower dynamic consumption and lower electromagnetic emissions for the A-ADC. This gap in performances between the two types of conversion would be widened if the S/H were added in transistor level. The spreading of the harmonic peaks in the asynchronous case leads to a very significant noise reduction, which will improve the dynamic parameters of the converter, more particularly the signal-to-noise ratio (SNR). This aspect will be investigated in future works.

4.2 System Level Aspect

Although it is difficult to find a general relation between a signal and the number of conversion cycles for the A-ADC, statistical studies enabled us to estimate it for different kind of signals: speech, electrocardiograms (ECG's). System level simulations in Matlab of the A-ADC, the constrained A-ADC and the ADC are given in Fig.5a for a standard ECG taken from the database [13], and Fig.5b for a speech signal. The amplitude of each input signal has been adapted to match 5% to 95% of the converters dynamic. Moreover, ADC's models characteristics used in Section 2.3 have been set according to their values measured in electrical simulations (cf. Section 3). The number of conversion cycle in the synchronous case is proportional to $N.f_s$ with $f_s=360Hz$ for ECG's and $f_s=22.05kHz$ for speech.

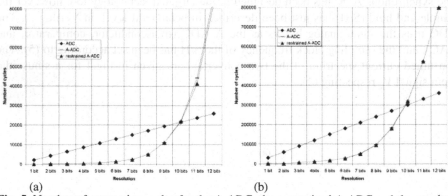

(a)   (b)

Fig. 5. Number of conversion cycles for the A-ADC, the constrained A-ADC and the synchronous ADC for electrocardiograms (a) and speech signal (b).

As it is visible in Fig.5, the more important the resolution N, the higher the number of A-ADC cycles. Thus, up to 10-bit of resolution, the asynchronous converter is better than the synchronous one. The gain in the number of conversion cycles is 97% for ECG and 91% for speech when $N=4$-bit, 90% for ECG and 84% for speech when

$N=6$-bit, and 70% for ECG and 59% for speech when $N=8$-bit. Over 10-bit of resolution, the number of conversion cycles can be reduced with the "constrained A-ADC", but only with a gain of 4%, compared to the A-ADC, in the best case. Should we insist that here, curves are only valid for the considered signals. Generally speaking, such previous studies must be done before converting another type of signal with the A-ADC. Yet, to conclude, the A-ADC will be interesting in low-resolution applications, or in systems bringing up slow signals in action. Another architectures of the A-ADC will be investigated in future works, such as multi-resolution, to improve widely the number of conversion cycle gain for resolution higher than 10-bit.

To sum up, the A-ADC activity and power consumption are reduced thanks to three distinct aspects:

- the asynchronous running of the converter reduces activity because samples are processed only if it is useful.
- if condition (2) is true, conversion only needs one cycle for the A-ADC whatever the number of bit of the converter may be, but needs N cycles for the synchronous ADC (if N is the resolution).
- the fact that a sample and hold circuit is useless in the asynchronous case.

5 Conclusion

A new asynchronous ADC architecture, for low-power, low-resolution applications, has been proposed. It provides an irregular sampling of the analog input voltage to convert, according to its amplitude variations. System simulations show that the activity of the circuit is significantly reduced when resolution is lower than 10-bit. Such a converter has been designed, with 6-bit resolution, as well as its synchronous counterpart: the successive approximation converter. It is shown that, in the worst case (in the case of identity in (2)), the A-ADC has an average consumption similar to the synchronous converter, to which the sample and hold power consumption must be added. Hence, as it is shown in Section 4, the asynchronous converter allows significant power reduction.

This kind of converter fits the new microelectronics topicality which is to integrate complex systems on one chip (SoC's or System on Chip): sensors (analog), signal processing (digital), and data transmission (radio-frequency). It brings an efficient alternative to restrain power consumption and electromagnetic emissions.

Firstly, the A-ADC could be used in a classical signal processing system as a low-power conversion front-end. In this way, signal processing studies presented in [9] can be applied. The A-ADC can also be used in a totally asynchronous system with a specific asynchronous microcontroller [14] or microprocessor [15], and a specific signal processing theory [16].

Moreover, redundant sample conversion absence allows lossless data compression, which is interesting for signal storage and transmission. This characteristic may be used for instance in medical domain for analog-to-digital conversion and compression of ECG's, in order to use new irregular sampled signal techniques like those described in [17].

Future works will be focused on improving analog and digital parts of the architecture. For example the four-phase-handshake protocol will be replaced by a two-phase-handshake one. Multi-resolution schemes will also be investigated to improve the converter performances. Further investigation will be carried out concerning the other benefits of the asynchronous conversion: low electromagnetic emissions and metastability error removal.

References

1. Jespers, P. G. A.: "Integrated Converters, D to A and A to D Architectures, Analysis and Simulation", *Oxford University Press*, 2001.
2. Walden, R. H.: "Analog-to-Digital Converter Survey and Analysis", *IEEE Journal on Selected Areas in Communications*, Vol. 17, n° 4, pp. 539–550, April 1999.
3. Renaudin, M.: "Asynchronous Circuits and Systems: a Promising Design Alternative", *Journal of Microelectronic Engineering*, Vol. 54, pp. 133–149, 2000.
4. Piguet, C., Renaudin, M., Omnes, T.: "Special Session on Low-power Systems on Chips", Design Automation and Test in Europe (DATE), Munich, Germany, March 13-16, 2001, pp. 488–494.
5. Kinniment, D., Yakovlev A., Gao B.: "Synchronous and Asynchronous A-D Conversion", *IEEE Transactions on VLSI Systems*, Vol. 8, n° 2, pp. 217–220, April 2000.
6. Kinniment, D.J., Yakovlev, A.V.: "Low Power, Low Noise Micropipelined Flash A-D Converter", *IEE Proceedings on Circuits Devices Systems*, Vol. 146, n° 5, pp. 263–267, October 1999.
7. Conti, M., Orcioni, S., Turchetti, C., Biagetti, G.: "A Current Mode Multistable Memory Using Asynchronous Successive Approximation A/D Converter", *IEEE International Conference on Electronics, Circuits and Systems*, Cyprus, September 1999.
8. Mark, J.W., Todd, T.D.: "A Nonuniform Sampling Approach to Data Compression", *IEEE Transactions on Communications*, Vol. COM-29, n° 4, pp. 24-32, January 1981.
9. Sayiner, N., Sorensen, H.V., Viswanathan, T.R.: "A Level-Crossing Sampling Scheme for A/D Conversion", *IEEE Transactions on Circuits and Systems II*, Vol. 43, n° 4, pp. 335–339, April 1996.
10. Sutherland, I. E.: "Micropipelines", *Communications of the ACM*, Vol. 32, pp. 720–738, June 1989.
11. Shams, M., Ebergen, J.C., Elmasry, M.I.: "Optimizing CMOS Implementations of the C-Element", *International Conference on Computer Design 1997*, pp. 700–705.
12. Allier, E., Fesquet, L., Renaudin, M., Sicard, G.: "A 6-bit Low-Power Asynchronous Analog-to-Digital Converter", *Internal Report*, ISNR: TIMA-RR–02/03-06–FR, February 2002.
13. MIT, Physionet Electrocardiograms Database: "http://www.physionet.org/physiobank/database/#ecg"
14. Abrial, A., Bouvier, J., Renaudin, M., Senn, P., Vivet, P.: "A New Contactless Smart Card IC using On-Chip Antenna and Asynchronous Microcontroller", *Journal of Solid-State Circuits*, Vol. 36, 2001, pp. 1101–1107.

15. Renaudin, M., Vivet, P., Robin F.: "ASPRO: an Asynchronous 16-Bit RISC Microprocessor with DSP Capabilities", *ESSCIRC'99*, Duisburg, September 21–23, 1999.
16. Publications of NUHAG (The Numerical Harmonic Analysis Group), University of Vienna, Austria: http://tyche.mat.univie.ac.at/papers/index.html
17. Fontaine, L., Granjon, Y., Ragot, J., Aliot, E.: "Events Detection in Irregularly Sampled Electrocardiograms", *Proceedings of the World Congress on Medical Physics and Biomedical Engineering at Nice – France*, September 14–19 1997, *Medical & Biological Engineering & Computing*, 1997, vol. 35, supplement part I, p. 436.

Optimal Two-Level Delay – Insensitive Implementation of Logic Functions

Igor Lemberski and Mark Josephs

South Bank University, SCISM, 103 Borough Road, London SE1 0AA,UK
{lemberi,josephmb}@sbu.ac.uk

Abstract. We proposed an approach to multi-output optimal two-level delay-insensitive (DI) implementation of logic functions. It bases on the procedure of logic minimization. We formulated and proved constraints the minimized logic implementation remains delay-insensitive for. Also, we pointed out an existing tool that produces result under constraints formulated. Using this tool we processed several examples and compared implementation complexity with one obtained using known approach. We achieved more than 4 times improvement.

1 Introduction

DI implementation is becoming more and more attractive for designers because correct behaviour of such a scheme doesn't depend on 1) logic elements and wiring delays, 2) timing interval the input signals switch their values, 3) number of inputs switching their values. As a result, the DI scheme is robust and reliable. Within synthesis process, designers don't need to perform timing verification.

A DI model was introduced in [4,6] where its behavioural an environmental constraints were specified. Different implementation approaches were described in [1,3,4,5,7,8]. DI two – level (C + OR elements) implementation was considered in [1,5]. However, it is very expensive (in terms of logic gates) because for any function of n inputs, 2^n terms of length n each should be implemented using C-elements. In [4], PLA – like approach is offered. Again, it bases on implementation of 2^n terms of length n each. The method [3] generates less expensive structure for both two-level and multi – level implementation. However, instead of input-output mode, fundamental mode of operation is supposed.

In synchronous logic, two - level implementation is attractive due to its high performance, regularity and good starting point for multi - level design. Also, there are a lot of powerful tools oriented on two-level representation. In DI logic, we expect the same benefits and therefore, intend to develop an approach to DI optimal design targeting two-level implementation.

The methods [1,3,4,5] are based on the model offered in [4,6] where input – output sequence satisfies so called strong or weak constraints. Under the strong constraints, EACH output changes its state only when ALL inputs change their states. Under the weak constraints, it is permitted for SOME outputs to change the states when SOME

B. Hochet et al. (Eds.): PATMOS 2002, LNCS 2451, pp. 92–100, 2002.
© Springer-Verlag Berlin Heidelberg 2002

inputs change their states. In both cases, it is supposed that ALL outputs depend on values of ALL inputs. However, sometimes not ALL inputs may affect outputs value. Suppose one output structure described by the function: $f = x_1x_2x_3 \lor x_1 \bar{x}_2x_3 \lor x_1x_2\bar{x}_3$. Although this function depends on 3 inputs one can see that within particular switching where inputs x_1, x_2 accept value 1, function f accepts value 1 independently on input x_3 value. In this case, input x_3 is treated as redundant, the first and third terms can be merged and function f representation may be simplified: $f = x_1x_2 \lor x_1\bar{x}_2x_3$. The procedure of detecting and merging terms to avoid redundant variables is called function minimization[2]. It is widely used to reduce the implementation complexity. It is pointed out in [5] the exterior similarity between two-level (AND-OR) synchronous and DI (C-OR) implementations where C-elements replace AND elements. However, no minimization procedure for DI logic is offered. Furthermore, it is stated that generally, minimization for DI implementation is not allowed. As a result, known DI two-level implementation [1] is very expensive. In our paper, we offer DI logic minimization procedure. To take the fact of possible input redundancy into account we consider a structure [9] implemented as a composition of two schemes: multi – output two - level (C-OR) function scheme and two-level (OR-C) signalling one (in contrast to [1] where implementation is completed as a single scheme). For optimal two – level implementation of the function scheme, we formulate conditions the minimized functions should satisfy. We point out an existing minimization tool that produces the result under these conditions. Finally, we compute several examples, evaluate their complexity and compare with the known implementation [1].

2 DI Implementation

Let $F = \{f_1, f_2, ..., f_q\}$ be the set of functions of n inputs: $x_1, x_2, ... , x_n$. We also call F as a multi-output function of n inputs.

In DI logic, it is supposed that each input and output (function) may be in three states: state 1, 0 or undefined (u). To implement three – state input x_i, $i = 1, 2, ... , n$, two signals $x_i(1)$ and $x_i(0)$ are introduced, where $x_i(1) = 1$ and $x_i(0) = 0$, if x_i is in the state 1, $x_i(1) = 0$ and $x_i(0) = 1$ if x_i is in a state 0, $x_i(1) = x_i(0) = 0$ if x_i is „u" state (combination $x_i(1) = 1$ and $x_i(0) = 1$ is not allowed). Similarly, to implement three - state output, function f_c, $c = 1, 2, ..., q$, should be represented in both right $f_c(1)$ and inverse $f_c(0)$ form . If $f_c(1) = 1$, $f_c(0) = 0$ than function f_c is in the state 1, if $f_c(1) = 0$, $f_c(0) = 1$ - function f_c is in the state 0, and if $f_c(1) = f_c(0) = 0$ -function f_c is in „u" state (combination $f_c(1) = f_c(0) = 1$ is not allowed). We say that function F is in defined / undefined state if all functions from the set F are in defined / undefined states.

Let t_i be the product term the function $f_c(1)$ ($f_c(0)$) accepts value 1 for:
$t_i = x_{i1}(j1) \, x_{i2}(j2) ... x_{ip}(jp)$, where $j1, j2 , ... , jp \in \{0,1\}$, $p \leq n$. We call it as a 1- (0-) product term. Note that some product terms may contain less than n variables. Suppose, functions $f_c(1)$, $f_c(0)$ are represented as the sum of 1- and 0- product terms respectively: $f_c(1) = t_1 \lor t_2 \lor ... \lor t_k$, $f_c(0) = t_{k+1} \lor t_{k+2} ... \lor t_m$, $m \leq 2^n$. Denote: $S(t_i)$ – set of term t_i variables: $S(t_i) = \{x_{i1}(j1) , x_{i2}(j2) ,..., x_{ip}(jp)\}$.

Previously, we announced our approach to DI implementation. It bases on the composition of multi-output function and one-output signalling schemes where the function scheme (fs) features from two types of constraints (further, fs – constraints):
– its behaviour cycle is as follows:
1. If all inputs are undefined then all outputs remain undefined.
2. If some inputs are defined then all outputs become defined.
3. If all inputs are defined then all outputs remain defined.
4. If some inputs are undefined then all outputs are undefined.
5. Go to 1.

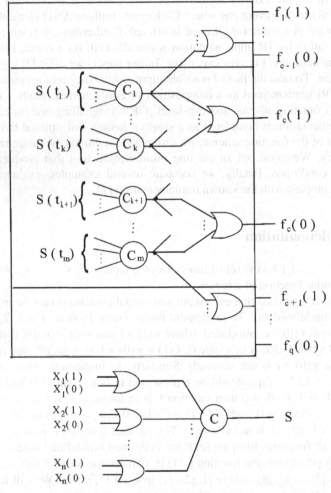

Fig. 1. DI structure

– input-output operation mode is supposed (in contrast to fundamental input-output mode [3]). It means that the inputs may change its state (to defined /undefined state) when not only all outputs are in a stable (undefined or defined) state but also signals inside circuit are stable and can't change their value until, at least, one input changes its state.

The structure is given in fig.1. Within the multi-output function scheme, each function f_c, c=1,2,...,q, right (f_c (1)) and inverse (f_c (1)) forms are implemented as a two-level (C-OR) circuits and C – elements are supplied by the part of (not all) inputs described by the sets: $S(t_1)$, $S(t_2)$,..., $S(t_m)$. Several functions may share the same C –element.

From the behaviour cycle description (see fs - constraints) it is clear that changing function F state (for defined / undefined one) may occur if only PART of (not all) inputs change their states for defined / undefined ones (steps 2 and 4 respectively). Therefore, the structure includes one – output signalling scheme implemented as an OR-C circuit. It notifies (by signal S rising / falling) environment that ALL inputs became defined / undefined (steps 1,3). Once function F changes its state from undefined to defined one and signal S rises the environment may change inputs from undefined states to defined ones. Similarly, once function F changes its state to undefined one and signal S falls the environment changes inputs for defined states. Note, that the structure (fig.1) satisfies DI model under weak constraints[9].

3 Minimization of Logic Functions

Let T^1, T^0 be a set of function f_c 1- and 0- terms respectively, $f_c \in F$. Two terms t_i, $t_j \in T^1 \cup T^0$ are orthogonal (define: t_i ort t_j) if there is at least, one input x_v, such that $x_v(0) \in S(t_i)$, $x_v(1) \in S(t_j)$ or $x_v(0) \in S(t_j)$, $x_v(1) \in S(t_i)$. Otherwise, terms t_i, t_j are non – orthogonal (t_i nort t_j). For non-orthogonal terms, joint set $S(t_{ij})$ is defined: $S(t_{ij}) = S(t_i) \cup S(t_j)$. Let X = { $x_1(v1)$, $x_2(v2)$,..., $x_n(vn)$} be the set of all inputs accepting value 1 within given cycle: $x_1(v1) = x_2(v2) =... = x_n(vn) = 1$, v1,v2, ... ,vn $\in$ {0,1}.

Note, if for given switching, all inputs from the set $S(t_{ij})$ accept value 1 than both term t_i and t_j accept value 1.

Example. Consider function $f = x_1x_2 \lor x_1\bar{x}_2x_3$. Its right form representation is as follows: $f(1) = x_1(1)x_2(1) \lor x_1(1)x_2(0)x_3(1)$. Terms $t_1 = x_1(1)x_2(1)$ and $t_2 = x_1(1)x_2(0)x_3(1)$ are orthogonal because there is a variable x_2 such that $x_2(1) \in S(t_1)$, $x_2(0) \in S(t_2)$. Further function f minimization results in its representation as follows: $f = x_1x_2 \lor x_1x_3$ and its right form f(1) description as follows: $f(1) = x_1(1)x_2(1) \lor x_2(1)x_3(1)$. Terms $t_1=x_1(1)x_2(1)$, $t_2= x_2(1)x_3(1)$ obtained are non-orthogonal. Therefore, we can create set $S(t_{1,2}) = x_1(1)x_2(1)x_3(1)$. Suppose, that within given cycle, input signals $x_1(1)x_1(0)$ $x_2(1)x_2(0)$ $x_3(1)x_3(0)$ switch their values from undefined state 00 00 00 to defined one as follows: 10 10 10. In this case, X = { $x_1(1)$, $x_2(1)$, $x_3(1)$} and $S(t_{12}) \subseteq X$. Therefore, terms t_1 and t_2 accept value 1.

Our goal is to develop logic function minimisation rules provided that the multi-output function scheme behaviour satisfies fs - constraints.

To ensure it, we have to guarantee that within multi-output function scheme (fig.1), behaviour of each function f_c , c = 1,2,...,q, satisfies (or more exactly, doesn't violate) fs - constraints.

Theorem 1. Function f_c implementation (fig.1) satisfies fs - constraints if and only if: t_i ort t_j, for $\forall(t_i,\ t_j) : t_i,\ t_j \in T^1$, and $t_i,\ t_j \in T^0$.

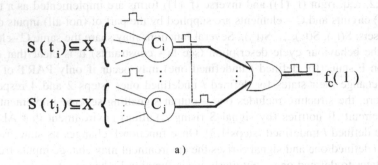

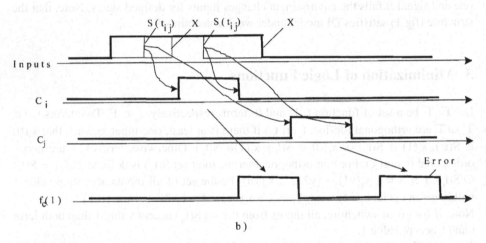

Fig. 2. Necessity conditions

Proof. Necessity. Consider the scheme fragment containing C_i, C_j – elements connected to OR element (fig.2, a). Suppose: t_i nort $t_j : t_i,\ t_j \in T^1$. Let X be the set of all input signals accepting value 1 within given cycle. Suppose: $S(t_{ij}) \subseteq X$, where $S(t_{ij}) = S(t_i) \cup S(t_j)$. Accepting value 1 by all signals from the set $S(t_{ij})$ may be treated as „some inputs become defined". It results in rising both C_i- and C_j- elements outputs (fig.2,b). The signal from the scheme inputs to its output propagates through two paths (fig.2, a, bold lines). However, if the path C_i - OR delay is shorter than C_j – element delay, signal $f_c(1)$ accepts value 1 and keeps it after rising of all signals from the set X. However, C_j-element may still generate (unstable) signal 0 on its output. This fact violates input-output mode conditions. Furthermore, falling of all input signals from the set $S(t_{ij})$ („some inputs become undefined") causes falling of C_i – and OR elements outputs (fig.2, a, b) while C_j may still generate (due to its large delay) signal 0 on its output. As a result, OR element output may have erroneous pulse (caused by Cj-

element output signal rising and falling) AFTER changing inputs for undefined states. It violates a fs - constraint describing function scheme behaviour cycle.
The above conclusion is valid for function $f_c(0)$, if t_i, $t_j \in T^0$.
Sufficiency. Again, consider the same scheme fragment (fig.3, a). Suppose: t_i ort t_j, t_i, $t_j \in T^1$ Let X be set of ALL input signals accepting value 1 within given cycle. Suppose, $S(t_i) \subseteq X$. Rising of all signals from the set $S(t_i)$ may be treated as „some inputs become defined". However, in contrast to previous case, there is only one path (fig.3,a, bold line) the input signals switching propagates through the scheme and results in rising of signal $f_c(1)$. Indeed, $S(t_j) \not\subset X$, $\forall tj \in T^1 \setminus t_i^1$ due to orthogonality of t_i, t_i.

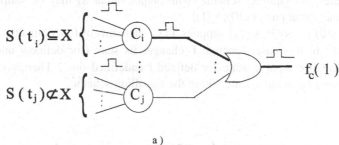

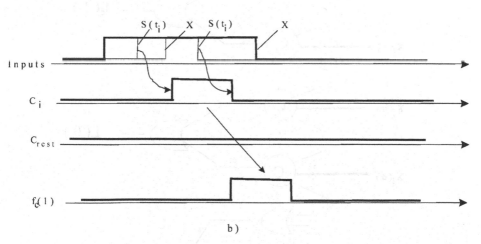

a)

b)

Fig. 3. Sufficiency conditions

It means that even after rising of all signals from the set X , C_i –element still generates value 1 on its output but the rest of C-elements (C_{rest}) generate value 0 on their outputs (fig.3,b). Therefore, signal f_c (1) accepts value 1 and keeps it when all input signals become defined („if all inputs are defined then all outputs remain defined"). Similarly, after falling of all signals from the set $S(t_j)$ („some inputs are undefined") C_i-element output falls. It results in accepting value 0 by the signal $f_c(1)$ („if some inputs are undefined then outputs become undefined"). Also, signal f_c (1) keeps value 0 when ALL signals from the set X fall („if all inputs are undefined then all outputs remain undefined"). Therefore, the scheme behaviour satisfies fs - constraints. The above conclusion is valid for function $f_c(0)$ if we suppose: $t_i, t_j \in T^0$. ∎

Orthogonal form implementation was offered in [8] and called as a standard one for DI logic. However, no proof has been done that this form satisfies necessary and sufficient conditions for minimized two-level implementation.

4 Signalling Scheme Optimization

In some cases, a signalling scheme (with output signal S) may be simplified by excluding some signal pairs $(x_i(0), x_i(1))$.

Let signal $x_i(d) \in \{x_i(0), x_i(1)\}$ supplies all C –elements of functions $f_c(0)$ and $f_c(1)$, $f_c \in F$ (fig.4). In this case, function f_c changes its state (for defined/ undefined one) AFTER changing input x_i state for defined / undefined one. Therefore, no need to indicate changing signal x_i state within the signalling scheme.

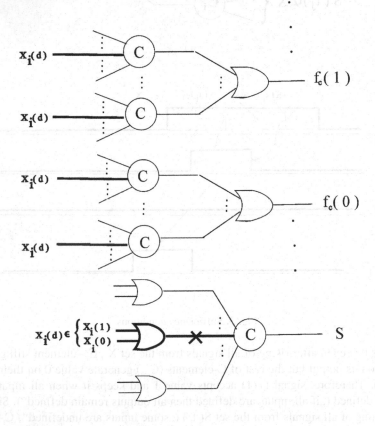

Fig. 4. Removing redundant inputs from a signalling scheme

5 Example

Consider DI implementation of two input function AND described as follows: $f(1)=x_1(1)x_2(1)$, $f(0)=x_1(1)x_2(0)$ v $x_1(0)x_2(1)$ v $x_1(0)x_2(0)$. Function $f(0)$ minimised representation that satisfies theorem 1 conditions is : $f(0)= x_2(0)$ v $x_1(0)x_2(1)$. Implementation is given in fig.5. Note, that all terms in expressions for $f(1)$ and $f(0)$ contain either signal $x_2(1)$ or signal $x_2(0)$. Therefore, signals $x_2(1)$, $x_2(0)$ may be excluded form the signalling scheme.

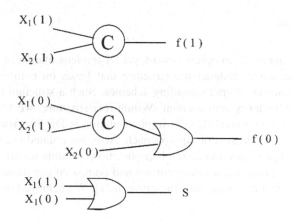

Fig. 5. Two-input function AND DI implementation

6 Experimental Results

We processed 3 examples: function AND of 2, 3 and 4 variables (further: AND(2), AND(3), AND(4)). For minimization, espresso tool [2] with the option – d1merge (espresso –d1merge) is used. It merges terms that differ in one variable and therefore produces minimised solution that satisfies theorem 1 condition. We compared complexity (expressed as the number of 2 input gates) of our implementation and known one [1] using methodics from [3]. Within this approach, the complexity of logic elements is measured as follows: $C(n) = (n-1)C(2)$, $C(2)=4G(2)$, $G(n)=(n-1)G(2)$, where $C(n)$ – C-element of n inputs, $G(n)$ – n-input gate. One can see (tab.1) that improvement is more than 4 times for function AND(4) and we believe that it is even higher for function AND of more variables.

Table 1. Function AND implementation complexity

Examples	Anantharaman's implementation	Our implementation	Improvement
AND(2)	18 x G(2)	10 x G(2) (fig.5)	1,8 times
AND (3)	102 x G(2)	28 x G(2)	3,64 times
AND (4)	206 x G(2)	50 x G(2)	4,12 times

7   Conclusion

We proposed an approach to optimal two-level DI implementation of logic functions. For this purpose, we considered the structure that bases on composition of multi-output function and one-output signalling schemes. Such a structure takes the fact of possible input redundancy into account. Within this structure, we formulated conditions the minimized two-level (C-OR) implementation is DI. We described an example to demonstrate application of our approach. Also, we pointed out the existing tool (espresso [2]) that produces minimized logic functions suitable for DI implementation. With this tool, we processed a few examples and compared our implementation complexity with one obtained using known approach [1]. We achieved more than 4 times improvement.

References

1. T.S. Anantharaman, A Delay Insensitive Expression Recognizer, IEE VLSI Tech.Bull, Sept, 1986
2. R.K.Brayton, et al, Logic Minimization Algorithm for VLSI Synthesis, Norwell, MA: Kluwer Academic, 1984
3. I.David, R.Ginosar, M.Yoeli, An Efficient Implementation of Boolean Functions as Self-timed Circuits, IEEE Trans. Computers, Vol.41,No 1, 1992, pp. 2–11
4. C.L Seitz, System Timing , In: Introduction to VLSI Systems, C. Mead, L. Conway, Addison—Welsey Publishing Company,1980, pp. 218–262
5. J.Sparsø, J.Staunstrup, M. Dantzer-Sørensen, Design of Delay Insensitive Circuits Using Multi-Ring Structures, pp. 15–20
6. W.J.Dally, J.W.Poulton, Digital Systems Engineering, Cambridge University Press, 1998, 663 p.
7. M. Saarepera, T. Yoneda, A Self-Timed Implementation of Boolean Functions, ASYNC'99, pp. 243–250
8. V.Varshavsky, Self-Timed Control of Concurrent Processes. Kluwer Academic Publisher, 1990
9. C.D. Nielsen, Evaluation of Function Blocks for Asynchronous Design, Euro-DAC'94, pp. 454–459

Resonant Multistage Charging of Dominant Capacitances

Christoph Saas[1] and Josef A. Nossek[1]

Munich University of Technology,
Institute for Circuit Theory and Signal Processing,
Arcisstr. 16,
80290 Munich, Germany
chsa@nws.ei.tum.de
http://www.nws.ei.tum.de

Abstract. It has been shown [2] that adiabatic switching can significantly reduce the dynamic power dissipation in an integrated circuit. Due to the overhead in the realization of adiabatic logic blocks [3] the best results are achieved when it is used only for charging dominant loads in an integrated circuit [7]. It has been demonstrated [4] that a multi stage driver is needed for minimal power dissipation. In this article a complete three stage driver including the generation of oscillating supply is described. To obtain a minimal power dissipation during synchronization the resonant frequency has to be constant. Therefore the waveforms for the logic states of the signal and the realization of a single stage differ from those presented in [4]. In the H-SPICE simulations losses of the inductor are taken into account. This allows to estimate the power reduction that is achievable in a real system.

1   Introduction

Adiabatic switching [2] is a method to reduce the dynamic dissipation of a circuit by charging the capacitances with a time-variant source. The minimal energy dissipation is achieved for a constant charging current.

$$E_{diss} = \frac{RC_L}{T} C_L V_{dd}^2 \tag{1}$$

It has been shown in [3] that the realization of general complex logic blocks based on pass transistor logic utilizing adiabatic switching leads to a considerable overhead for maintaining true adiabatic behavior.

Therefore, to avoid this overhead it is worthwhile to consider the energy dissipation associated with dominant load capacitances only, without the inclusion of complex logic. Keeping in mind that usually a large part of the power is dissipated in the I/O-cells of complex chips [5], a significant reduction of this dissipation can be expected. Since the driver is only a single cell, which can be controlled by standard CMOS logic, it can easily be included into a standard design flow.

B. Hochet et al. (Eds.): PATMOS 2002, LNCS 2451, pp. 101–107, 2002.
© Springer-Verlag Berlin Heidelberg 2002

The standard pad driver has to charge the pad as quickly as possible. In particular this time has to be shorter than the clock period T_{clk}. For this reason the $\frac{W}{L}$ of the driving inverters has to be adjusted to keep the channel resistance R_{ch} sufficiently low. The dissipated energy $E_{diss} = \frac{1}{2}C_L V_{dd}^2$ is independent of the channel resistance R_{ch}. Therefore the properties of the transistor do not influence the energy dissipated but only the switching speed of the driver.

To fulfill the speed requirements in most cases an inverter chain is used. The dissipated energy is increased by the additional gate capacitances of the driving inverters in a multiple stage CMOS driver.

The basic idea of an adiabatic driver is very simple. The capacitor is charged through a transmission gate by the phase Φ. The gate is controlled by standard CMOS gates. Both, p and n channel transistors are used to obtain a full voltage swing at the output. Of course the $\frac{W}{L}$ has to be reasonably large as the dissipated energy is dependent on the channel resistance R of the transmission gate (1). Despite the simplicity of the circuit there is a significant saving of energy. The biggest part of the remaining power loss is dissipated in the controlling CMOS gates which charge the large gate capacitances of the transmission gate non-adiabatically. To minimize the losses in the controlling gates, the adiabatic stages can be cascaded. By this, only transistors in the first stage are charged non-adiabatically.

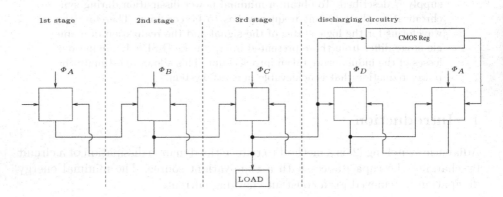

Fig. 1. Schematics for the multi stage driver

2   The Multistage Driver

For the present examination a three stage design has been chosen. To find the optimum number of stages is an open problem.

A diagram for a three stage driver is shown in Fig.1. About half of the energy is dissipated during the discharging process. Therefore, also the discharging has to be done adiabatically. For the discharging process the initial state in the controlling gate has to be present for the whole discharging process [6]. So, it is necessary to store this state. Of course, the discharging has to be done with appropriately sized transistors. For discharging the same cascade as for charging is obtained. For the last stage a CMOS register is used to store the state needed for adiabatic discharging. Only the minimal transistors in the register are discharged non-adiabatically, but this energy is negligible.

Four clock phases are used. Each stage is connected to a clock phase which is delayed by a quarter of the clock period to the phase of the predecessor (see Fig. 5). By this the output of the preceding stage is valid during the rising edge. Therefore, the output is well suited as a control signal for the charging of the succeeding stage. On the other hand, the succeeding stage is valid during the falling edge at the preceding stage. Therefore, its output can be used to discharge the output of the preceding stage adiabatically. For the charging of the first stage standard CMOS gates are used. They have to be valid for three quarters of the clock period. The output of a standard CMOS register is used to discharge the last stage. Its state has been set in one of the preceding stages. Using this timing a three stage driver generates a latency of $\frac{3}{4}T_{clk}$.

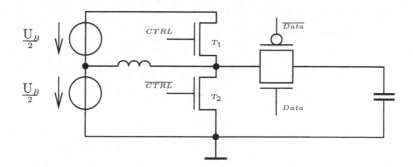

Fig. 2. Oscillator

3   The Oscillator

As the generation of ideal ramps is not possible with a high efficiency, they are approximated by a sinus. To generate the sinusoidal power supply, an oscillator is needed. The oscillator (Fig. 2) is a resonant LC oscillator working in class E operation. It is based on [8] and inherits some advantages when compared to the widely known blip circuit which is often proposed as the source for adiabatic circuits [1]. It needs only a single inductor for each phase and generates a full sinusoidal voltage oscillation at the capacitor.

The transistors T_1 and T_2 are closed for a short period at the maximum, respectively the minimum, of the output voltage to compensate for losses during one cycle. In addition this keeps the oscillation synchronous to the CMOS clock. Although some work has been published [8], the controlling signals for the T_1 and T_2 are generated manually.

4   Different Waveforms for the Logic States

The idea of the adiabatic driver is to charge the output with the oscillating power supply in case of a logical 1 and to cut it off in case of a logical 0. A logic 1 will result in one period of the oscillation at the output. It starts and ends in the minimum of the sinus. The DC offset as well as the amplitude of the oscillation is $\frac{V_{dd}}{2}$.

As already mentioned above, dual rail encoding is required. There are two ways to invert the signal, which will be named "truly inverted" and "logically inverted". They are depicted in Fig. 3

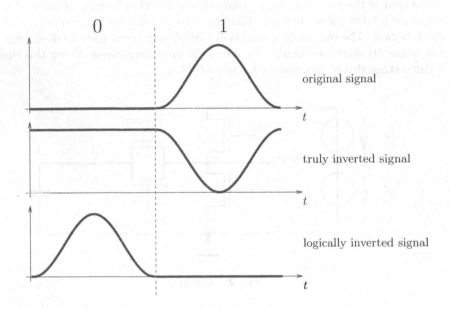

Fig. 3. Waveforms and inverted waveforms

If the original waveform is inverted, a signal which is at V_{dd} for a logical 0 and sinusoidal for a logical 1 is obtained. This signal would be the perfect one to control the p-MOS transistor of the transmission gate in the previous and the subsequent stage. The drawback of this solution is, that one can not combine the original and the inverted signal to a DC and a sinusoidal one just by switching between them. This is needed to present a constant load to the oscillator. Since the load capacitance is part of the resonator and therefore has a direct influence on the operating frequency it is mandatory to keep it constant for all logic states of the circuit. To achieve this, there has to be a constant sinusoidal oscillation on a constant number of load capacitances. Therefore, an "logically inverted" output is used. During a logic 1 the output will oscillate and the inverted one will stay at 0 Volt. For the logic 0 it is vice versa. If only a logically inverted signal is created at the output of each stage, there is no signal available which is suitable for controlling the p-channel transistors in the other stages. Therefore, the individual stages of the driver have to be redesigned.

5   A Single Stage

It is well known, that a p-channel transistor is needed to obtain a full charging up to V_{dd}. If only a n-channel transistor is used, it is only conducting until V_{GS} is larger than V_{th}. Therefore, the source voltage can only reach $V_{dd} - V_{th}$. If one has a look at the waveform of a logical "1", one can see, that the p-channel transistor is only needed during a rather short period. The basic evaluation of the input of a single stage is done by the n-channel transistors. It seems to be a good idea to abandon the signal to control the p-channel transistors and use some internal signal which already reached its final level due to the n-channel transistors instead.

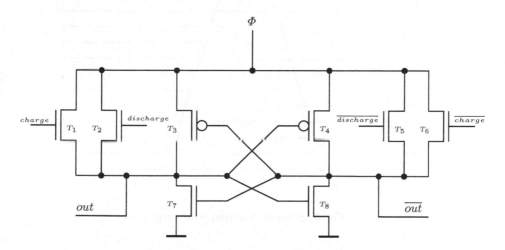

Fig. 4. Circuit of a single stage

Such a solution is proposed in Fig. 4. It consists of 2 n-channel transistors per output signal (T_1, T_2 and T_5, T_6). One is used to charge the output, and the other one for discharging. The charging signal is "delayed" by $-\frac{T}{4}$ whereas the discharging signal is delayed by $\frac{T}{4}$.

There is only one p-channel transistor per output signal (T_3, T_4). If the output is logical 1 it has to be conducting during the maximum of Φ. On the other hand, it has to be non-conducting for $U_\Phi > V_{th}$ in the case of a logical 0.

This behavior can be achieved by controlling the p-channel transistor with the inverted output. The n-channel transistors T_7 and T_8 ensure the output signal to be 0 V for a logical "0". They are not present in stage 3, as the load capacitor is large enough to achieve a stable output.

6   Simulation Results

The adiabatic multistage driver has been simulated using a $0.25\,\mu$m process and H-SPICE. As the inductor is intended to be an external one, a Q value of 100 seems to be realistic.

A driver for 4 off-chip connections has been simulated. Each of the connections represents a load of 28 pF. The adiabatic drivers are implemented to load one output at a time. Thus a group of 4 drivers represents a constant data independant load to the oscillators for all times.

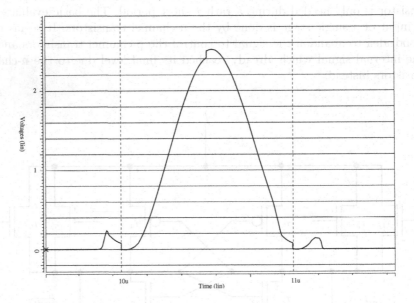

Fig. 5. Simulated output waveform

$\frac{1}{2}CU^2$	$350\ \frac{pJ}{Bit}$
CMOS	$351.52\ \frac{pJ}{Bit}$
adiabatic driver at 1 MHz	$52.7\ \frac{pJ}{Bit}$

The simulations results show that the circuit is working adiabatically. The driver dissipates much less energy than a conventional CMOS driver at useful operating frequencies. It has to be noted, that these results summarize the whole energy that is dissipated including the generation of the sinusoidal power supply.

7   Summary and Outlook

In this paper a multi stage adiabatic driver has been presented. An oscillator has been chosen to generate the sinusoidal power supply. To ensure minimal losses during the synchronization to the CMOS clock the oscillator has to work on a constant load. The single stages of the driver have been designed to fit the requirements for constant load. To avoid the need for truly inverted signals the p-channel transistors which are needed to reach V_{dd} are realized as clamping devices. The simulation results show a pretty large potential for adiabatic drivers. Although a small latency of $\frac{T}{4} * Num. of stages$ is introduced the energy savings still make this concept interesting for a number of applications. Of course high speed applications are not the aim for adiabatic circuits, but the simulations have shown that reasonable operation is possible up to 100 MHz. Further research has to be done on the modeling of the load and the other off-chip connections. Most probably the results can be improved by more advanced methods for the transistor sizing.

References

1. W. Athas, N. Tzartzanis, W. Mao, L. Peterson, R. Lal, K. Chong, J-S. Moon, L. Svenson, and M. Bolotski. The Design and Implementation os a Low-Power Clock-Powered Microprocessor. *Journal od Solid State Circuits*, 35(11):1561–1570, 2000.
2. W.C. Athas, L.J. Svensson, J.G. Koller, N. Tzastzanis, and E.Y.-C. Chou. Low-Power Digital Systems Based on Adiabatic Switching Principles. *IEEE Transactions on Very Large Scale Integration (VLSI) Systems*, 2(4):398–406, 1994.
3. C. Saas, A. Schlaffer, and J.A. Nossek. An adiabatic multiplier. *PATMOS 2000*, pages 276–284, 2000.
4. C. Saas, A. Schlaffer, and J.A. Nossek. An adiabatic multi stage driver. *ECCTD*, 2001.
5. Takayasu Sakurai, Hiroshi Kawaguchi, and Tadahiro Kuroda. Low Power CMOS Design through V_{th} Control and Low-Swing Circuits. *Institute of Industrial Science, Univ. of Tokyo*, 1997.
6. A. Schlaffer. *Entwurf von adiabatischen Schaltungen*. PhD thesis, Munich University of Technology, 2000.
7. B. Voss and M. Glesner. A low power sinusoidal clock. *ISCAS*, 2001.
8. C.H. Ziesler, S. Kim, and M.C. Papaefthymiou. A Resonant Clock Generator for Single-Phase Adiabatic Systems. *ISLPED*, Aug 2001.

A New Methodology to Design Low-Power Asynchronous Circuits*

Oscar Garnica, Juan Lanchares, and Román Hermida

Dept. Computer Architecture. Universidad Complutense de Madrid.
{ogarnica,julandan,rhermida}@dacya.ucm.es

Abstract. The aim of this paper is to present a new approach to creating high performance and low-power asynchronous circuits using high level design tools. In order to achieve this, we introduce a new timing model called Pseudo Delay-Insensitive model. To prove the goodness of this model, we present the results after comparing, for a set of benchmarks, our implementation with other implementations (synchronous and asynchronous).

1 Introduction

Asynchronous circuits are increasing in popularity among designers because of their advantages. They have no problems associated with clock signal and circuit performance is the performance of the average case, due to in an asynchronous circuit the next computation can start immediately after the previous computation has completed. As a result, and particularly in data-dependent computation, asynchronous circuit performance is higher than synchronous circuit performance [9]. Furthermore, asynchronous circuits may have a better noise and EMC (Electromagnetic Compatibility) properties [2] than synchronous circuits; they are modular and parameter-variation tolerant, and they are insensitive to meta-stability problems. Finally, asynchronous circuits consume less power than synchronous [3] because asynchronous circuits only dissipate energy when and where active. In contrast, synchronous circuits are either quiescent or active entirely. In this field, there are a number of successful demonstrations [5]. In [1] are collected many examples of how asynchronous design help to increase circuit performance, increase EMC and reduce power consumption.

There are different approaches to synthesizing asynchronous circuits [11] [4]. The most general approach to synthesizing asynchronous circuits is based on the delay-insensitive model [8]. However, this timing model is both area and power consuming when it is compared with other timing models.

Our main goal has been to obtain a methodology to simplify the design of high performance low-power circuits using a timing model as unconstrained as possible. This methodology is based in three key decisions: the use of a new asynchronous timing paradigm, integration with high-level design tools (such as high

---

* This paper has been funded by Spanish Government Grant TIC 99/0474.

D. Hochet et al. (Eds.): PATMOS 2002, LNCS 2451, pp. 108–117, 2002.

level synthesis, HLS, tools) as a way to increase productivity, and the minimization of the number of switchings in the circuit. Hence, using this methodology,
we simply have to design the circuit using HLS tools, find the nearly critical
paths of the circuit, translate the structural specification generated by a HLS
tool from a standard cell library to a new library of gates that are capable of
working in the desired way, and finally include some additional gates. So, the
methodology requires a new library of gates capables of fulfilling the constraints
that asynchronous design requires. However, we have not focused in presenting
this new library in this paper.

This paper is structured as follows. In section 2 we present the delay-
insensitive circuits. In section 3 we describe our approach, the pseudo delay-
insensitive model. Section 4 introduces the behavior of the circuits built using
this timing model. Finally, we show the results we have obtained after building
some circuits, from which we draw the conclusions and propose future work.

2   Delay-Insensitive (DI) Circuits

When designing circuits, designers make several assumptions about the wire and
gate delays, the number of transitions between emitter and receptor, and the level
of integration between control and data. These assumptions define the timing
model, the protocol model, and the single-rail or a dual-rail model, respectively.

As stated above, the most general timing model is the delay-insensitive
model. This model [6] assumes unbounded wire and gate delays. In other words,
the designer does not know the delay of any element of the circuit. In [8], Martin
showed that it is not possible to make DI circuits using just one electrical signal to represent the information unit (bit). In such a circuit, there must be two
kinds of data. On the one hand there are calculation data, d, which transmit the
information involved in the computation. On the other, there are synchronizer
data, s, which are used to distinguish between two consecutive d data.

This new kind of data (s data) can be introduced in two different ways;
using either ternary logic [7] or dual-rail logic [10]. We have chosen a dual-
rail approach. In the dual-rail approach, the two kinds of data are encoded using
two wires, (w_1, w_0). Among the four possible states taken by the dual-rail signals
(w_1, w_0), the most used encoding is the four-cycle encoding. Hence, the codes
$(w_1, w_0) = 00$ or $(w_1, w_0) = 11$ represent the s data, $(w_1, w_0) = 01$ represents
logic zero and $(w_1, w_0) = 10$ represents logic one.

Finally, delay-insensitive circuits can be classified according to the communication protocol model as well. In the four-cycle protocols, every computation is
executed in two phases (and each phase involves two cycles); a calculation phase
followed by an synchronism phase. The calculation phase consists of the issue
of a request and the corresponding issue of an acknowledgment, while the synchronism phase consists of the withdrawal of the request and the corresponding
withdrawal of the acknowledgment.

The DI model has two important drawbacks, each one related to one of the
two phases of the four-cycle protocol:

Drawback 1. During the calculation phase some method is necessary to check that the circuit has finished the computation. The solution usually proposed is to use dual-rail logic to implement all the circuit wires and gates. As mentioned in the previous section, the drawback of this solution is the great amount of area required.

Drawback 2. In the synchronism phase (during this phase, the s data are produced, simultaneously, by all the circuit gates and these data propagate to the ends of the wires), a method is necessary to check when the ends of all the wires have the s data. The solution usually proposed is to use a Completion Detector circuit (CD circuit) to check the data kind and place it at the end of every circuit wire. In this way, the change from the synchronism phase to the calculation phase is produced when the ends of all circuit wires have the s data. Again, this solution requires a great amount of area.

3   Pseudo Delay-Insensitive (PDI) Model

The pseudo delay-insensitive model (PDI model), as the DI model, does not suppose any upper bound in the delay of the circuit components (gates and wires). Consequently, it is necessary to use the two kinds of data proposed by Martin: calculation data, d, and synchronizer data, s, which will be implemented using dual-rail logic. The PDI model is a variation of the DI model, in which two suppositions are included which allow a reduction in the area required to face the drawbacks of the DI model. The PDI suppositions are:

Supposition 1. The critical path/paths are always the same. In other words, the critical path does not change once the circuit is built. In the DI model this restriction does not exist, and consequently the critical path can change during the circuit life. Therefore, it is necessary to transmit the two data kinds proposed by Martin throughout all the circuit wires. This implies it is necessary to transform all the circuit wires and gates into dual-rail logic. However, in the PDI model this requirement is eliminated because it is extremely unlikely that any path can be a critical path. This implies that not all the circuit wires must be dual-rail wires. Hence, it is only necessary to transform into dual-rail logic those gates and wires within the critical path.

The weak point of this supposition is that a set of paths can exist with the possibility of being critical paths depending of the working condition of the circuit. However, this behavior does not invalidate this supposition. When a set of paths can be critical paths then two approaches exist:

1. All these paths are transformed into dual-rail logic. In this way, when the critical path changes due to the circuit conditions (aging, environmental conditions, ...), the circuit behavior will be correct due to the possibility of checking the end of the calculation phase at the output of a dual-rail path.
2. Some logic is added to one of the critical path candidates in order to convert it into the actual critical path. This solution only slows down the computation time of those computations which involve the critical path.

Supposition 2. The electrical signals propagate through the wire as waves. Hence, as a consequence, it is not necessary that the value, that we want to transmit, remains at the wire input until this value reaches the wire outputs (this supposition describes the behavior of the wires, not the gate behavior). Once the signal is propagating throughout the wire, the source which produces this signal can change its output value and the signal will not disappear from the wire. This is the ideal behavior. However, in a wire exists resistance that apart the wire behavior from the ideal behavior. These resistances reduce the amplitude of the signal as it propagates through the wire (due to Joule's effect). In this way, if the wire has a great length the electrical signal will finally disappear from the wire. The length at which the amplitude of the signal is under a certain threshold depends on the resistance value, the voltage and the duration of the signal.

We have simulated the electrical behavior of a wire of Aluminum (sheet resistance $R_S = 0.04\Omega$ and $width = 1\mu m$) as an RCL distributed network using the Spectre^{TM} tool from Cadence. From this simulation we have observed that in wires with length about microns the pulse duration must be about picoseconds in order to reduce its amplitude at 10%. This implies working frequencies about terahertz which are far away from the present working frequencies.

Due to Supposition 2, it is not necessary to check the state of all the circuit wires before changing the data kind at the circuit inputs. The CD circuit has not to be at the ends of all the circuit wires. It is sufficient to check the state of just one end of one wire. If this end has received the synchronism data, then it is certain that this data kind will arrive at the end of all wires.

The weak point of this supposition is that, a priori, it does not ensure the impossibility of mixing two calculation data, which belong to two different calculation phases, in a single computation. In other words, suppose we have a gate, G_i, with two inputs i_1 and i_2. This model does not ensure that during an interval of time, the input i_1 has calculation data of the present calculation phase (calculation data n) and the input i_2 has calculation data of the previous phase (calculation data $n - 1$). To avoid this situation, it is sufficient that the s data (synchronism data $n - 1$) are simultaneously in i_1 and i_2 during an interval of time. This condition is satisfied when the propagation delay of the fastest wire, $t_p(w_{i_1})$, plus the s data duration, $t_p(s)$, is greater than the propagation delay of the slower wire, $t_p(w_{i_2})$. Eq. (1) shows this condition.

$$|t_p(w_{i_1}) - t_p(w_{i_2})| < t_p(s) \tag{1}$$

On the other hand, the duration of s data is:

$$t_p(s) = t_p(w_{CD}) + t_p(CD) + t_p(w_{ack}) + t_p(gate) \tag{2}$$

where $t_p(w_{CD})$ is the propagation delay of the wire which reaches the CD circuit, $t_p(CD)$ is the computation delay of the CD circuit, $t_p(w_{ack})$ is the propagation delay of the ack wire (this wire transmits the signal which indicates when to change from a calculation phase to synchronism phase), and $t_p(gate)$ is the propagation delay of the gate. Substituting Eq. (2) into Eq. (1) and depreciating the gate delays (which today is not a good approximation but will be in the near future), this equation can be rewritten as:

$$t_p(w_{i_1}) - t_p(w_{i_2}) < t_p(w_{CD}) + t_p(w_{ack}) \qquad (3)$$

This equation must be true for all gates G_i. To satisfy this constraint the approach is to impose on the Place&Route tool that the placement and routing satisfy Eq. (3). Hence, if the chosen wire end (on which to check the data kind) does not satisfy Eq. (3), then either another wire can always be chosen for which Eq. (3) holds or the routing must be done again.

However, this constraint is easily satisfied because of the nature of the wires involved in this equation. (w_{ack} and w_{CD} are high fan-out wires with big delay, and on the other hand, w_{i_1} and w_{i_2} are local (short) wires with small delay).

In summary, the PDI model (due to its suppositions) has the following advantages over the DI model when it is used to build computation-oriented circuits:

1. Due to supposition 1, only a subset of all the circuit wires and gates (gates and wires within the critical path) are dual-rail. The rest of the circuit wires are single-rail wires (as usual in digital circuits). As a consequence, the logic gates not in the critical path will be Boolean gates (the kind of logic gates usual in digital circuits), but slightly modified to allow the implementation of the four-cycle communication protocol. In this way, the area needed to check the end of the calculation phase is reduced.
2. Due to supposition 2, it is only necessary to include a CD circuit at the end of just one wire. In this way, the area needed to check the end of the synchronism phase is reduced.

4   Circuit Behavior

In this section we introduce the behavior of the circuits under the PDI model. As stated above, the chosen communication protocol is the four-cycle protocol. The change from the synchronism phase to the calculation phase (or vice versa) is controlled by the value of a signal called ack signal. When $ack = 1$ then the circuit is in the synchronism phase and when $ack = 0$ then the circuit is in the calculation phase. This signal arrives at every gate in the circuit. Next, the circuit behavior in both phases is presented.

SYNCHRONISM PHASE. In this phase the circuit wires have to be initialized to one of the following data kinds:

- The dual-rail wires are initialized to s data.
- The single-rail wires are initialized to d data. The value of the d data for each wire depends on the logic function that the circuit performs and the wire position within the circuit. This value is calculated as follows:

1. Single-rail wires connected to dual-rail gates are initialized to the Boolean value that does not determine the output of the dual-rail gate. Hence, a single-rail wire connected to a dual-rail AND gate (for example wire $w6$, Fig. 1) must be initialized to the value 1, because with this value at one of the gate inputs and the s data at the other one, the output of the gate is not determined (it remains as s)

2. Single-rail wires connected to single-rail gates are initialized to d data which are consistent with the value of the single-rail wires connected to dual-rail gates. For example, the $w7$ and $w8$ wires (Fig. 1) must have value 1 in order to be consistent with the assignment of value 1 to the $w6$ wire.

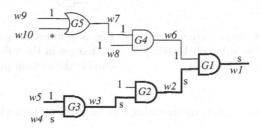

Fig. 1. A piece of an asynchronous PDI circuit in the synchronism phase. For the sake of clarity, the ack signal does not appear in this figure (it would be connected to all circuit gates). The thick line represents the critical path. s represents the s data; $w1, w2, \ldots$ represent wires; $G1, G2, \ldots$ represent gates.

3. When a wire has a fan-out greater than 1 the following problem can happen. The wire can have two different d data assigned to it. In this situation, the wire must be initialized to s data and must be transformed into a dual-rail wire.

The synchronism phase will finish when s data are detected at the w_{CD} by the CD circuit. At this moment, the ack signal changes to $ack = 0$ and the calculation phase begins.

CALCULATION PHASE. In this phase, the gates produce their outputs as a function of their inputs. The circuit wires (single-rail and dual-rail) change their values due to the new values that the environment provides to the circuit inputs. This phase finishes when the circuit outputs (which are the last stage in the critical path) change from s data to d data. At this time, the ack signal changes to $ack = 1$ and the synchronism phase starts.

Figs. 2 and 3 show two examples of the circuit behavior during the calculation phase. In Fig. 2, the value of the $w4$ wire changes from s (Fig. 1) to 1 (Fig. 2). As a consequence, the value of the $w3$ wire changes from s to 1, and similarly the $w2$ and $w1$ wires. The calculation phase finishes when the change in the $w1$ wire is detected. In this case, the response delay of the circuit has been the critical path delay.

In Fig. 3, the value of the $w8$ wire changes from 1 (Fig. 1) to 0 (Fig. 3). As a consequence, the values of the $w6$ and $w1$ wires change. As previously mentioned, the calculation phase finishes when the change in the $w1$ wire is detected. However, in this case, the response delay of the circuit is lower than the critical path delay.

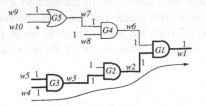

Fig. 2. Circuit output change due to changes in the value of the wires within the critical path.

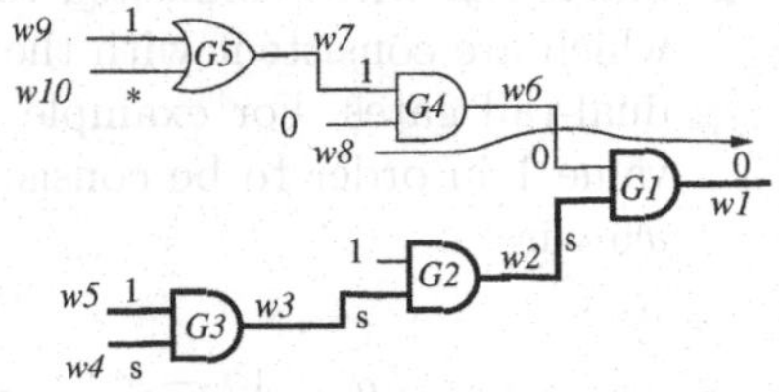

Fig. 3. Circuit output change due to changes in the value of the wires outside the critical path.

In summary, the design methodology to build circuits under the PDI model is as follows,

1. We construct the circuit using high level design tools.
2. We determine the critical path using HLD tools.
3. A precharged value is assigned to each single-rail wire using the criteria mentioned in this section. This task can be done automatically by a CAD tool.
4. All gates and wires within the critical path are substituted by their dual-rail counterparts.
5. All gates outside the critical path are substituted by Boolean gates with the capacity to pre-charge their outputs to the value calculated in bullet 3.

5   Experimental Results

In this section we present the results for the well-known benchmarks used at academia (LGSynth95). We have chosen a set of benchmarks which requires less gates to be implemented because, as yet, we have not developed an automatic tool which performs all the tasks involved in the design flow. These results have been obtained after building the circuits using the three approaches: synchronous, PDI and DI.

The design flow to build asynchronous circuits has been presented at the end of Section 4. The synchronous library used to synthesis the designs (using Design Compiler tool from Synopsys) only contained AND, OR and INV (inverter) gates. The delay of all the gates is $1ns$. The design goal is to minimize the circuit area. In the asynchronous circuits, once the circuit has been synthesized, we have eliminated all the dual-rail INV gates, because inversion is simply to interchange the dual-rail wires assigned to the logic signals (as all complementary logics). It could be possible that after transforming the circuit in this way, the critical path changes (another circuit path became the critical path). Indeed, this problem happens in 22% of the benchmarks. There are two solutions:

1. To implement the new critical path using dual-rail logic. However, the problem can appear again after this transformation.
2. To eliminate some INV gates but leave the number of INV gates which ensures that the critical path will remain as the critical path. This is the solution adopted by us.

For each experiment, the asynchronous circuit delay is estimated as the average delay for the set of chosen inputs. The synchronous circuit delay is the critical path delay and this value is calculated by the Design Compiler tool from Synopsys. Power-consumption is estimated as the number of switchings in the circuit for the set of chosen inputs. Finally, area is estimated as the number of transistors required to build the circuit. The chosen inputs are introduced at random and circuits with a number of inputs greater than 20 (apex2 and cordic) are simulated with 100.000 inputs randomly chosen.

The delay, area and power-consumption results are presented in Figs. 4, 5 and 6. In all of them *Sync* represents the results for the synchronous circuits, *PDI* the results for the PDI circuits and *DI* represents the results for delay-insensitive circuits.

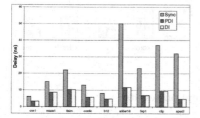

Fig. 4. Delay results.   Fig. 5. Area results.

Observing Fig. 4, the conclusion can be drawn that synchronous circuits have a lower performance (the inverse of the critical path delay) than PDI circuits. The PDI performance is between 1.71 and 6.83 times higher than synchronous performance. This is due to two reasons: the asynchronous circuit working with "average case" performance, while the synchronous circuit working with "worst case" performance, and in the asynchronous circuits, the INV gates have been eliminated from the critical path. On the other hand, comparing DI and PDI graphs in the same figure, we conclude, as expected, that PDI circuits have the same performance as DI circuits.

In Fig. 5, it can be observed that the PDI circuits use more transistors than the synchronous ones (as expected). The synchronous circuits use between 13% and 27% less transistors than PDI circuits. However, PDI circuits use less transistors than DI circuits. DI circuits use between 5% and 54% more transistors than PDI circuits.

Finally, observing Fig. 6, we conclude that that the power-consumption of the PDI circuits is similar or lower (in 67% of cases) than the power-consumption of the synchronous circuit. This is a surprising result and was not expected at the time of stating this model (we are implementing combinational logic. This results would not be surprising if we would have implemented sequential circuits in which a clock signal exists). The reasons of this behavior are that the dual-rail INV gates have been eliminated from the PDI circuit, and the PDI circuits do not have glitches and consequently there is a saving in the number of switchings in the circuit. Similarly, PDI circuits consume considerably less power than the DI circuits in all cases. DI power-consumption is 24% to 95% greater than PDI consumption. The reason for this is that the single-rail wires of the PDI circuit which have been precharged to calculation data have a low switching rate than dual-rail wires.

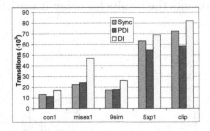 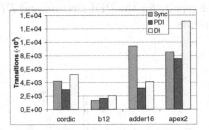

Fig. 6. Power-consumption results.

6   Conclusions and Future Work

In this paper we have presented a new approach to synthesizing low-power asynchronous circuits, based on the pseudo delay-insensitive timing model. This approach allows us to build low-power asynchronous circuits in a semi-custom way using HLS tools.

Our next step was to compare a set of benchmarks used in logic synthesis, built using the three approaches: synchronous, PDI and DI. Our approach produces circuits with higher performance in all cases and lower power-consumption in 67% of the cases when compared with synchronous implementations. Similarly, it produces circuits with the same performance and lower power-consumption in

100% of the cases when they are compared with the DI implementations. Also, this approach also produces circuits with lower area than its DI counterparts. Hence, the DI implementations use between 12% and 54% more transistors than PDI implementations. The synchronous implementations use between 13% and 27% less transistors than PDI implementations.

In future work we will develop a tool that automatically computes the initial assignment of every wire during the synchronism phase. From Figs. 4, 5 and 6, it can be deduced that performance and power-consumption improve as the circuits get bigger. So, we expect that the results would be better with bigger benchmarks.

References

1. C.H. (Kees) van Berkel, Mark B. Josephs, and Steven M. Nowick. Scanning the technology: Applications of asynchronous circuits. *Proceedings of the IEEE*, 87(2):223–233, February 1999.
2. Kees van Berkel, Ronan Burgess, Joep Kessels, Ad Peeters, Marly Roncken, and Frits Schalij. Asynchronous circuits for low power: A DCC error corrector. *IEEE Design & Test of Computers*, 11(2):22–32, Summer 1994.
3. Kees van Berkel, Ronan Burgess, Joep Kessels, Ad Peeters, Marly Roncken, and Frits Schalij. A fully-asynchronous low-power error corrector for the DCC player. In *International Solid State Circuits Conference*, pages 88–89, February 1994.
4. Tam-Anh Chu. *Synthesis of Self-Timed VLSI Circuits from Graph-Theoretic Specifications*. PhD thesis, MIT Laboratory for Computer Science, June 1987.
5. Hans van Gageldonk, Daniel Baumann, Kees van Berkel, Daniel Gloor, Ad Peeters, and Gerhard Stegmann. An asynchronous low-power 80c51 microcontroller. In *Proc. International Symposium on Advanced Research in Asynchronous Circuits and Systems*, pages 96–107, 1998.
6. Scott Hauck. Asynchronous design methodologies: An overview. *Proceedings of the IEEE*, 83(1):69–93, January 1995.
7. R. Mariani, R. Roncella, R. Saletti, and P. Terreni. On the realisation of delay-insensitive asynchronous circuits with CMOS ternary logic. In *Proc. International Symposium on Advanced Research in Asynchronous Circuits and Systems*, pages 54–62. IEEE Computer Society Press, April 1997.
8. Alain J. Martin. Asynchronous datapaths and the design of an asynchronous adder. *Formal Methods in System Design*, 1(1):119–137, July 1992.
9. Steven M. Nowick, Kenneth Y. Yun, and Peter A. Beerel. Speculative completion for the design of high-performance asynchronous dynamic adders. In *Proc. International Symposium on Advanced Research in Asynchronous Circuits and Systems*, pages 210–223. IEEE Computer Society Press, April 1997.
10. Marco Storto and Roberto Saletti. Time-multiplexed dual-rail protocol for low-power delay-insensitive asynchronous communication. In Anne-Marie Trullemans-Anckaert and Jens Sparsø, editors, *Power and Timing Modeling, Optimization and Simulation (PATMOS)*, pages 127–136, October 1998.
11. J. V. Woods, P. Day, S. B. Furber, J. D. Garside, N. C. Paver, and S. Temple. AMULET1: An asynchronous ARM processor. *IEEE Transactions on Computers*, 46(4):385–398, April 1997.

Designing Carry Look-Ahead Adders with an Adiabatic Logic Standard-Cell Library

Antonio Blotti, Maurizio Castellucci, and Roberto Saletti

Dipartimento di Ingegneria dell'Informazione:
Elettronica, Informatica, Telecomunicazioni,
University of Pisa,
Via Diotisalvi 2, I-56122 Pisa, Italy
r.saletti@iet.unipi.it

Abstract. Adiabatic circuits are usually designed with methodologies optimized for the application in which they are used. In this work we show how a conventional design-flow based on an adiabatic standard-cell library and semi-automatic tools allow the quick and easy design and verification of a complex adiabatic system, without loosing the energy reduction benefits. The methodology has been applied to the design of positive feedback adiabatic logic (PFAL) carry look-ahead adders (CLA). Post-layout simulations of the standard-cell PFAL CLAs show a 94% energy recovery as compared to a conventional static CMOS CLA at 10 MHz, and 86% at 100 MHz. The standard-cell PFAL CLAs are also more energy efficient or comparable than other custom adiabatic CLAs found in the literature.

1 Introduction

The low-power requirements of present electronic systems have challenged the scientific research towards the study of technological, architectural and circuital solutions that allow a reduction of the energy dissipated by an electronic circuit.

One of the main causes of energy dissipation in CMOS circuits is due to the charging and discharging of the node capacitances of the circuits, present both as a load and as parasitic. Such part of the total power dissipated by a circuit is called dynamic power.

In order to reduce the dynamic power, an alternative approach to the traditional techniques of power consumption reduction, named adiabatic switching [1], has been proposed in the last years. In such approach, the process of charging and discharging the node capacitances is carried out in a way so that a small amount of energy is wasted and a recovery of the energy stored on the capacitors is achieved.

Large number of research works on adiabatic circuits of various types can be found in the literature [2-3], [5], [9-17]. These circuits can be grouped in two fundamental classes: fully adiabatic circuits and partial energy recovery circuits. Even if the first class (e.g. [2], [9]) can consume, in particular working conditions, asymptotically zero energy per operation [3], the large area occupation and the high design complexity make these circuits not very competitive with respect to traditional CMOS. In the second class, we find circuits designed to recover a large portion (but not all) of the

B. Hochet et al. (Eds.): PATMOS 2002, LNCS 2451, pp. 118–127, 2002.

energy stored in the circuit node capacitances. This energy loss drawback however allows a good trade-off between circuit complexity and then area occupation.

All the works presented in the literature show the low-power feature of the adiabatic systems as compared to the opposing traditional CMOS systems and a significant energy gain is very often obtained. Therefore, it seems that adiabatic circuits can provide the optimal solution for ultra-low power consumption in some particular applications. However, the adiabatic solutions presented are usually designed as special circuits with custom methodologies, optimized for the particular application in which they are used, and thus suffer a lack of generality of application.

In order to make adiabatic logic more competitive respect to traditional CMOS logic, it is therefore necessary to develop methodologies and semi-automatic tools that allow the design and verification of complex adiabatic systems, e.g. arithmetic units, in short times, in order to enjoy the energy reduction benefits of adiabatic logic in an easy, fast and general way. The aim of our work is to demonstrate the use of a rather classical design-flow based on a standard-cell approach for the design of adiabatic logic circuits. The example given is the design of a family of adders, that are designed and verified starting from an adiabatic logic standard-cell library developed by the authors. The main result obtained is that the semi-automatic design-flow easily and quickly leads to the realization of adiabatic logic standard-cell adders characterized by a very significant energy recovery compared to conventional CMOS and other adiabatic solutions presented in the literature.

In this paper we chose the partial energy-recovery circuit structure named Positive Feedback Adiabatic Logic (PFAL) [5], since it has shown the best low-power consumption characteristics with respect to other similar families [7], [8].

2 Positive Feedback Adiabatic Logic

In an adiabatic circuit the charging and discharging of a load capacitance C through a resistive path R occur by means of a slow trapezoidal power-clock signal $\phi(t)$ that replaces the power-supply V_{DD} of the conventional logic. This process allows a recovery of the energy stored on the capacitance. A typical four-phase power-clock signal like that used in PFAL is shown in Fig. 1a: $\phi(t)$ rises from zero to V_{DD} in the *evaluate* phase (E) and supplies energy to the circuit, while $\phi(t)$ falls down from V_{DD} to zero in the *recovery* phase (R) and the energy flows back from the system to the power-clock generator; the *hold* phase (H) and the *idle* phase (I) are needed for cascade purposes.

In traditional CMOS logic, the average dynamic energy E_{CMOS} dissipated during a time interval in which the same load capacitance C is charged and discharged, is equal to CV_{DD}^2 [4] while, in the adiabatic switching, the energy dissipated on R during the evaluate and recovery phases is equal to $2(RC/T)CV_{DD}^2$ [1] and can be reduced either by reducing R or by increasing T. These two parameters are not present in E_{CMOS}, where the power is achieved by reducing C and/or V_{DD}.

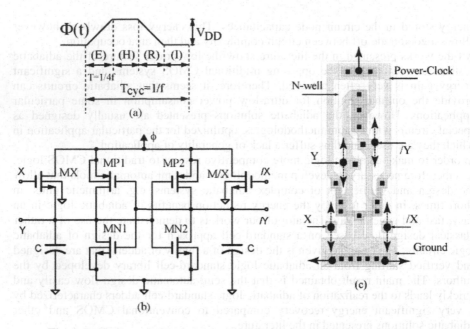

Fig. 1. Buffer/Inverter PFAL: (a) power-clock signal, (b) schematic, (c) standard-cell.

A buffer/inverter PFAL gate is shown in Fig. 1b. With reference to it, we can suppose that the input X and $/X$ are coming with the appropriate timing from another PFAL gate, the power-clock signal of which is anticipated a quarter of a period. In the evaluation phase, MX or M/X are conducting and therefore one of the outputs rises following the power-clock voltage. In the recovery phase, the output charge is recovered through the PMOS that connects that output to $\phi(t)$ [5].

The advantage of the PFAL gate, as compared to other adiabatic architectures, is the reduction of the resistive path R (that leads to a reduction of the dissipated energy) during the evaluation phase. This is due to the formation of an effective transmission gate formed by MX-MP1 or M/X-MP2, depending on the value of the inputs X and $/X$. Moreover, this logic gate can be designed as fully reversible adiabatic, as shown in [5], and the dissipated energy is further reduced, at the expense of a larger area.

3 Adiabatic Standard-Cell Library

One of the goal of this work is to allow a designer to implement an adiabatic system in short time, without getting rid of all the problems of electrical and geometrical design, as it happens in a conventional standard-cell approach. To this end, we have developed an adiabatic standard-cell library, consisting of common digital gates such as: one input gates (buffer/inverter), two inputs gates (AND2/NAND2, OR2/NOR2, XOR2/NXOR2), three inputs gates (AND3, OR3, XOR3 and their complementary gates), complex gates (2-input 1-output digital multiplexer, AO21: 2-input AND into 2-input OR, OA21: 2-input OR into 2-input AND, AO222: 2x2-input AND into 3-input OR and their complementary gates) and some special gates (1-bit half-adder, 1-

bit full-adder). The cells are designed in a 0.6 μm double-metal single-poly CMOS technology and the cell layout is designed so that the library is easily shrinkable towards more advanced technologies. The cell transistor dimensions are optimized for a typical "average" load.

Each cell is 29 μm high, whereas its width depends on the complexity of the function performed. Fig. 1c shows the layout of the buffer/inverter cell as an example. Cells can be placed adjacent to each other organized in rows, as it happens in conventional standard-cell designs. The main differences with respect to CMOS standard cells are that only the ground signal is passed from cell to cell by abutment and V_{DD} is not present. Ground signal (in metal1) runs horizontally at the bottom of the cells, whereas input and output connections for the cells are available at the top and bottom via metal2. The power-clock signal is only available at the top of the cell via metal1. Each cell has its own n-well for the PMOS devices of the adiabatic amplifier [5]; n-well is connected to the power-clock and cannot be in principle shared with other adjacent cells.

4 Design Flow

Fig. 2 shows the design flow used to design the PFAL adders presented in this work with the adiabatic standard-cell library described above. The tasks to be carried out are almost identical to those performed in a conventional design. The design starts with a *structural* description that specifies how the logic gates are interconnected to perform a given function (addition in our case). This description is a list of modules and their interconnections. The structural description is automatically generated by an adder compiler, a C++ program developed by us to perform the task of creating the adder netlist, given the number of bit of the input words. The tool also assigns the proper power-clock phase to every cell involved in the design, in order to guarantee the correct connection of the cascaded cells.

The logical functionality of the system is verified using the *functional* views of the standard cells provided with the library and the Verilog logical simulator. After this step, the design structure has been checked and in particular the correct timing between the blocks has been verified. Once the logical functionality is verified, the structural description is automatically converted in a *schematic* view. We use in this step the VerilogIn tool of the Cadence suite that imports the Verilog netlist into a schematic view.

In the next step, the schematic view is automatically converted in an *autolayout* view, in order to carry out the design floorplanning. We use the Cadence Cell Ensemble tool that allows placement and automatic routing of the standard cells of our design. This step is the most critical because the CAD tools and the silicon foundry design kits are not optimized for the dual-rail logic and the four-phase clocking strategy used in PFAL logic. With the aim of reducing the area occupation of our adders, we have verified that the best placement strategy is creating as many rows as the logic levels found in the architecture. In this case, every cell in a row belongs to the same clock phase. The main advantages of this placement strategy are: 1) since any placement region (row) has only one power-clock phase, its channel routing has minimum area; 2) the output signals of a region (located at the top of the region) are the input signals

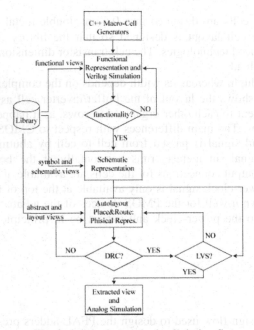

Fig. 2. Adiabatic standard-cell design flow

of the following region (located at the bottom of the region) and therefore the routing and the parasitic related to the interconnections are reduced.

Once we have obtained the *layout* view of the arithmetic units, we check that there are no geometry or physical layout violations using a Design Rule Check (DRC) program. In the next step we extract the circuit structure from the layout and we check possible connectivity or other violations by comparing the view *extracted* from layout and the original schematic view, by performing Layout Versus Schematic (LVS).

The final step of the design is the global analog simulation, performed with the Spectre circuit simulator, which allows us to calculate the power consumption of the PFAL adders as a function of the power-clock frequency.

5 Adiabatic Carry Look-Ahead Adders

PFAL circuits compute only one logic level per phase; therefore, a ripple carry adder [4] is not practical, since it needs many power-clock phases to obtain the result. We adopted a carry look-ahead (CLA) structure, by adapting to the four-phase clocking adiabatic architecture the computational model of Brent and Kung [6], which allows a reduction of carry computation (fast carry computation) and therefore a reduction of the logic depth of the adder. Let $a_N a_{N-1} \cdots a_1$ and $b_N b_{N-1} \cdots b_1$ be N-bit binary numbers. Their sum is $s_N s_{N-1} \cdots s_1$, with an input carry c_{in} and an output carry c_{out}. Let us also define a "block carry generator" G_i and a "block carry propagate" P_i:

$$(G_i, P_i) = \begin{cases} (g_1, p_1) & \text{if } i = 1, \\ (g_i, p_i) \circ (G_{i-1}, P_{i-1}) & \text{if } 2 \le i \le N, \end{cases}$$

where "$\circ$" is the following Brent and Kung operator:

$$(g_i, p_i) \circ (g_j, p_j) = (g_i + p_i \bullet g_j, p_i \bullet p_j)$$

$$g_i = a_i \bullet b_i \quad \text{and} \quad p_i = a_i \oplus b_i$$

It can be proved [6] that:

$$s_i = p_i \oplus c_{i-1} \quad \text{and} \quad c_i = G_i$$

where c_i is the carry from the i-th bit, $c_0 = c_{in}$ and $c_N = c_{out}$.

Since the operator "$\circ$" is associative, (G_i, P_i) and therefore c_i can be computed in the order defined by trees as shown in Fig. 3a for the case $N = 16$. Black circles are adiabatic gates that implement the Brent and Kung operator in one logic level. The white circles are simple adiabatic buffers inserted in the structure to maintain the correct sequence of the power-clock phases. The schematic view of an 8-bit adiabatic CLA is shown in Fig. 3b; circles and arrows show the typical tree structure for the fast carry computation. Dotted gates and dotted wires show how the schematic circuit can be extended to a 16-bit CLA. All the circuits are based on only four standard-cell types (buffer, AND2, XOR2 and AO21) from the library.

This Brent and Kung architecture is automatically generated by a C++ program (a CLA adder compiler), which receives the number N of the bit as input and gives a Verilog netlist as output: the structural representation of the N-bit CLA.

Following the methodology illustrated in the previous section, we have semi-automatically generated, verified and characterized three CLAs: 4, 8, 16-bit. The 4-bit PFAL CLA can execute one 4-bit addition per cycle and 5 stages are needed to compute the result. The 8-bit PFAL CLA and 16-bit PFAL CLA need in their turn only 6 and 7 stages respectively to compute the result.

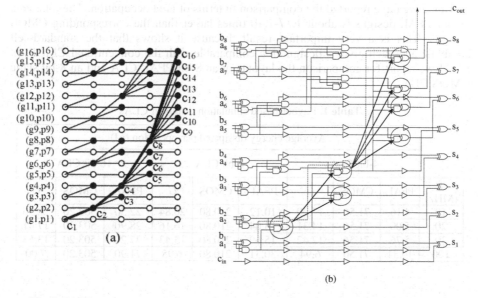

Fig. 3. (a) Fast carry computation: Brent and Kung model, (b) Schematic of an 8-bit CLA.

6 Results and Discussion

In this section we present Spectre simulation results for the adiabatic adders designed according to the previous methodology with our PFAL standard-cell library. In order to evaluate the energy efficiency achieved, we refer to purely combinational CMOS conventional adders, synthesized using the foundry standard-cell library of the same technology. For adiabatic CLAs, we have assumed no energy loss in the power-clock generator.

Both PFAL circuits and static CMOS circuits have been simulated with distributed-RC parameters extracted from layout. V_{DD} and power-clock peak voltages are 5 V. The test pattern for power calculation is a long sequence of random input values; the results give an average of the power consumption of the systems.

The *adiabatic gain* factor G is a parameter that can be useful during the comparison. The adiabatic gain factor is the ratio between the energy dissipated per operation by a conventional CMOS arithmetic unit and the energy dissipated in the same condition by the equivalent adiabatic arithmetic unit. The larger the value of G above 1, the larger the amount of energy saved by the adiabatic logic as compared to the conventional CMOS equivalent system.

In Tab. 1 we have reported the energy consumption per operation and the adiabatic gain factor of our PFAL CLAs at 10, 20, 30 and 100 MHz. Into the same table we have also reported the consumption of CMOS CLAs.

Our results show that PFAL CLAs is by far more energy efficient than static CMOS CLAs across the entire frequency range. 16-bit PFAL CLA has a gain factor equal to 22.36 at 10 MHz and 7 at 100 MHz. 95% of energy is saved over static CMOS at 10 MHz by 8-bit PFAL CLA; in this case, the energy saving is reduced to 86% at 100 MHz. 4-bit PFAL CLA dissipates only 6% of the energy of the static 4-bit CMOS CLA at 10 MHz and 14% at 100 MHz.

In Tab. 2 we have reported the comparison in terms of area occupation. The total area of our PFAL designs is about 1.17-1.40 times larger than the corresponding CMOS designs. This is a very important result, because it shows that the standard-cell adiabatic approach gives a design area comparable with the conventional CMOS. In Fig. 4 we report for comparison the layouts of an 8-bit PFAL CLA and an 8-bit static CMOS CLA.

Table 1. Layout-based simulation results of CLAs

f (MHz)	Average energy dissipated per-operation (pJ)								
	4-bit CLA			8-bit CLA			16-bit CLA		
	PFAL	CMOS	G	PFAL	CMOS	G	PFAL	CMOS	G
10	4.16	71.56	17.20	10.12	210.80	20.84	22.50	503.20	22.36
20	5.00	71.56	14.31	13.04	210.80	16.16	28.90	503.20	17.41
30	5.76	71.56	12.42	15.69	210.80	13.43	37.20	503.20	13.53
100	10.31	71.56	6.94	30.31	210.80	6.95	71.90	503.20	7.00

Table 2. Adder area occupations

	4-bit CLA	8-bit CLA	16-bit CLA
PFAL (μm^2)	33433	79465	207693
CMOS (μm^2)	28541	65551	148262
PFAL/CMOS	1.17	1.21	1.40

It is possible to compare the results obtained with other adiabatic CLA designs published in the literature. Even if technology, power-supplies and simulation methodologies widely vary, we can use the claimed adiabatic gain as a mean for a relative comparison of the various solutions. In Fig. 5 we have reported, for any adiabatic CLA, the relevant adiabatic gain (when data were available) in two interesting frequency cases: 10 MHz and 100 MHz. In [10] a CLA implemented in 0.5 µm CMOS technology (3 V) with a single-phase source-coupled adiabatic logic (SCAL), was compared to 4-bit CMOS CLA with the following results: at 10 MHz and 100 MHz adiabatic gain is equal to 9 and 6 respectively. In [14] a 4-bit 2N-2P CLA was compared with the corresponding static CMOS CLA in 1 µm technology: at 10 MHz the adiabatic gain is equal to 4. In [11] a 8-bit SCAL CMOS in 0.5 µm technology and with the smallest supply voltage that ensures the functioning shows an adiabatic gain equal to 3 and 2 at 10 MHz (1.5 V) and 100 MHz (1.8 V) respectively. In [15] an 8-bit CLA with adiabatic differential cascode voltage switch with complementary pass-transistor logic tree (ADCPL) in 0.5 µm CMOS technology (3 V) shows an adiabatic gain equal to 3.3 and 2 at 10 MHz and 100 MHz respectively. Another 8-bit CLA was found in [16], implemented by Pass-Transistor

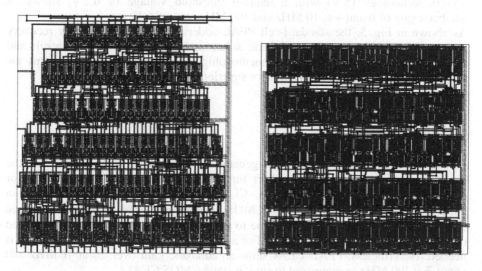

Fig. 4. Layouts of 8-bit PFAL CLA (left) and 8-bit static CMOS CLA (right)

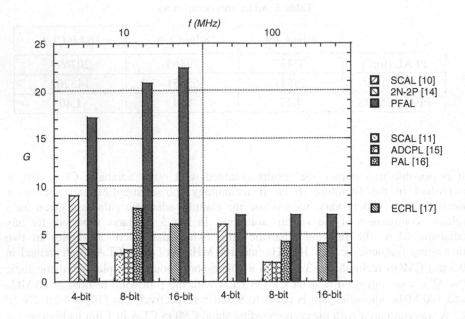

Fig. 5. Adiabatic gain comparison for different CLAs

Adiabatic Logic (PAL) in 0.6 µm CMOS technology with the smallest supply voltage that ensures the functioning; in this work, the adder show an adiabatic gain equal to 7.7 and 4.2 at 10 MHz (2.6 V) and 100 MHz (3.3 V) respectively. Finally, in [17] a 16-bit CLA implemented by Efficient Charge Recovery Logic (ECRL) in 1 µm CMOS technology (5 V) with a reduced threshold voltage of 0.2 V, shows an adiabatic gain of 6 and 4 at 10 MHz and 100 MHz respectively.

As shown in Fig. 5, the standard-cell PFAL adders show the best energy recovery performance respect to other adiabatic solutions; this is an important result and confirms the validity of the design methodology that quickly leads to designs the performance of which are comparable or superior to other custom solutions.

7   Conclusions

In this paper we have presented a general semi-automatic methodology for the designs of adiabatic systems in short time and easy way based on a library of adiabatic standard cells. Some PFAL CLAs have been designed and verified in 0.6 µm double-metal single-poly CMOS technology. The PFAL CLAs are characterized by a high throughput due to the dynamic behavior of the cells, traded off with a large input-output latency. For a 16-bit CLA, latency of 7 clock phases is needed. Nevertheless, PFAL CLAs show an adiabatic gain over 20 at 10 MHz and about 7 at 100 MHz as compared to similar static CMOS CLAs.

The validity of the standard-cell approach has also been validated by the comparison with other similar adiabatic CLAs found in the literature. The standard-cell PFAL adders have shown the best energy recovery performance in the entire frequency range (10 MHz-100 MHz) and therefore this approach is an excellent ultra low-power solution for those applications where the speed and latency are not critical parameters.

References

1. Roy, K., Prasad, S. C.: Low-Power CMOS VLSI Circuit Design. JohnWiley & Sons, New York (2000)
2. Athas, W.C., Svensson, L., Koller, J.G., Tzartzanis, N., Chou, E.Y.C.: Low Power Digital Systems Based on Adiabatic Switching Principles. IEEE Trans. on VLSI Systems. Vol. 2. No. 4. Dec. 1994, pp. 399–407
3. Younis, S.G., Knight, T.F.: Asymptotically Zero Energy Split-Level Charge Recovery Logic. Proc. Workshop Low Power Design, Napa Valley. 1994, pp. 177–182
4. Weste, N., Eshraghian, K.: Principles of CMOS VLSI Design. 2nd ed. Addison-Wesley (1993)
5. Vetuli, A., Di Pascoli, S., Reyneri, L.M.: Positive Feedback in Adiabatic Logic. Electronics Letters. Vol. 32, No. 20, Sep. 1996, pp. 1867–1869
6. Brent, R. P., Kung, H. T.: A Regular Layout for Parallel Adders. IEEE Trans. on Computers. Vol. C-31. No. 3, Mar. 1982, pp. 260-264
7. Blotti, A., Di Pascoli, S., Saletti, R.: A Comparison of Some Circuit Schemes for Semi-Reversible Adiabatic Logic. Int. J. of Electronics. Vol. 89, No. 2, Feb. 2002, pp. 147–158
8. Amirante, E., Bargagli-Stoffi, A., Fischer, J., Iannaccone, G., Schmitt-Landsiedel, D.: Variations of the Power Dissipation in Adiabatic Logic Gates. Proc. of 11th Int. Workshop on Power and Timing Modeling, Optimization and Simulation, PATMOS'01, Yverdon-les Bains, Switzerland, 2001, pp. 9.1.1–9.1.10
9. Lim, J., Kim, D.: A 16-bit Carry-Lookahead Adder using Reversible Energy Recovery Logic for Ultra Low-Energy Systems. IEEE J. of Solid State Circuits. Vol. 34, No. 6. Jun. 1999, pp. 898–903
10. Kim, S., Papaefthymiou, M.C.: Single-Phase Source-Coupled Adiabatic Logic. Proc. on Int. Symp. on Low Power Electronics and Design, Piscataway, New York (1999), pp. 97–99
11. Kim, S., Papaefthymiou, M.C.: True Single-Phase Adiabatic Circuitry. IEEE Trans. on Very Large Scale Integration (VLSI) Systems. Vol.9. No. 1. Feb. 2001, pp. 52–63
12. Ye, Y., Roy, K.: Energy Recovery Circuits Using Reversible and Partially Reversible Logic. IEEE Trans. on Circuits and Systems-I: Fundamental Theory and Applications. Vol. 43. No. 9. Sep. 1996, pp. 769–778
13. Ye, Y., Roy, K.: QSERL: Quasi-Static Energy Recovery Logic. IEEE J. of Solid-State Circuits. Vol. 36. No. 2. Feb. 2001, pp. 239–248
14. Knapp, M.C., Kindlmann, P.J., Papaefthymiou, M.C.: Design and Evaluation of Adiabatic Arithmetic Units. Analog Integrated Circuits and Signal Processing. Vol. 14. No. 1-2. Sept. 1997, pp. 71–79
15. Lo, C.K., Chan, P.C.H.: An Adiabatic Differential Logic for Low-Power Digital Systems. IEEE Trans. on Circuits and Systems-II: Analog and Digital Signal Processing. Vol. 46. No. 9. Sep. 1999, pp. 1245–1250
16. Mahmoodi-Meimand, M., Alzali-Kusha, A.: Low-Power, Low-Noise Adder Design with Pass-Transistor Adiabatic Logic. Proc. of the 21th Intern. Conf. on Microelectronics, Theran, Nov. 2000, pp. 61–64
17. Moon, Y., Jeong, D.K.: An Efficient Charge Recovery Logic Circuit. IEEE J. of Solid-State Circuits. Vol. 31. No. 4. Apr. 1996, pp. 514–521

Clocking and Clocked Storage Elements in Multi-GHz Environment

Vojin G. Oklobdzija[1], Fellow IEEE

Integration Corp. Berkeley, California, USA
http://www.integration-corp.com
[1] Department of Electrical Engineering, University of California, Davis
vojin@ece.ucdavis.edu
http://www.ece.ucdavis.edu/acsel

Abstract. An overview of clocking and design of clocked storage elements is presented. Systematic design of Flip-Flop is explained as well as "time borrowing" and absorption of clock uncertainties. We show how different clocked storage elements should be compared against each other. The issues related to power consumption and low-power designs are presented.

1 Introduction

Deciding on the clocking strategy in digital system is one of the single most important decisions. If not considered properly at the beginning of a design it can be very costly

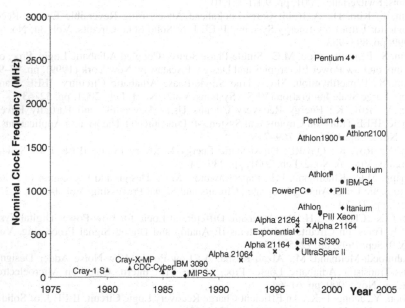

Fig. 1. Clock frequency over the years for various representative machines

B. Hochet et al. (Eds.): PATMOS 2002, LNCS 2451, pp. 128–145, 2002.
© Springer-Verlag Berlin Heidelberg 2002

afterwards. The importance of clocking is gaining momentum as the clock speed rises rapidly; doubling every three years as shown in Fig.1. At today's frequencies ability to absorb clock skew and to use faster Clocked Storage Element (CSE), results in direct performance improvement. Those improvements are

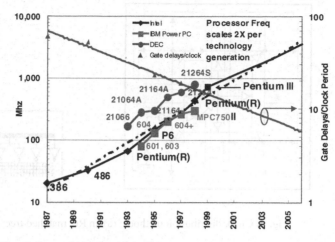

Fig. 2. Increase in the clock frequency and decrease in the number of logic levels in the pipeline (courtesy of Intel Corp.)

very difficult to obtain through architectural techniques or micro-architecture level. As the clock frequency reaches 5-10GHz traditional clocking techniques will be reaching their limit. New ideas and new ways of designing digital systems are required.

Following the speed increase, the number of logic levels in the *critical path* diminishes. In today's high-speed processors, instructions are executed in one-cycle, which is driven by a single-phase clock. In addition the pipeline depth is increasing to 15 or 20 in order to accommodate the speed increase. Today 10 levels of logic in the critical path are more common and this number is expected to be decreasing further as illustrated in Fig. 2. Thus any overhead associated with the clock system and clocking mechanism that is directly and adversely affecting the machine performance is critically important.

1.1 Clock Distribution

The two most important timing parameters affecting the clock signal are: *Clock Skew* and *Clock Jitter*:

Clock Skew is a spatial variation of the clock signal as distributed through the system. It is caused by the various RC characteristics of the clock paths to the various points in the system, as well as different loading of the clock signal at different points on the

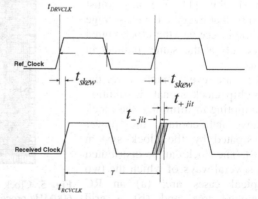

Fig. 3. Clock Parameters: Period, Width, Clock Skew and Clock Jitter

chip. Further we can distinguish *global clock skew* and *local clock skew*. Both of them are equally important in high-performance system design.

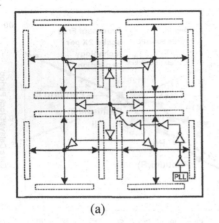
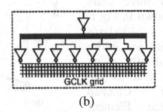

(a)   (b)

Fig. 4. Clock distribution methods: (a) an RC matched tree and (b) a grid [2]

Clock Jitter is a temporal variation of the clock signal with regard to the reference transition (reference edge) of the clock signal as illustrated in Fig. 3. Clock jitter represents edge-to-edge variation of the clock signal in time. As such clock jitter can also be classified as: long-term jitter and edge-to-edge clock jitter, which defines clock signal variation between two consecutive clock edges. In the course of high-speed logic design we are more concerned about edge-to-edge clock jitter because it is this phenomena that affects the time available to the logic.

Typically the clock signal has to be distributed to several hundreds of thousands of the clocked storage elements (also known as flip-flops and latches). levels of amplification (buffering). As a consequence, the clock system by itself can Therefore, the clock signal has the largest fan-out of any node in the design, which requires several use up to 40-50% of the power of the entire VLSI chip [1]. We also must assure that every clocked storage element receives the clock signal precisely at the same moment in time.

There are several methods for the on-chip clock signal distribution attempting to minimize the clock skew and contain the power dissipated by the clock system [18]. The clock can be distributed in several ways of which the two typical cases are: (a) an RC matched tree and (b) a grid shown in Fig. 4.

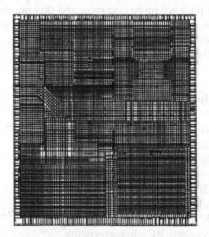

Fig. 5. Clock distribution grid used in DEC Alpha 600MHz processor [2], courtesy of IEEE.

If we had superior Computer Aided Design (CAD) tools, a perfect and uniform process and ability to route wires and balance loads with a high degree of flexibility, a matched RC delay clock distribution (a) would be preferable to grid (b). However, neither of that is true. Therefore grid is used when clock distribution on the chip has to be very precisely controlled. This is the case in high performance systems.

An example of the clock distribution grid is shown in Fig. 5 [2]. The power consumed by the clock is also the highest in cases using grid arrangement. This is not difficult to understand given that in a grid arrangement a high-capacitance plate has been driven by buffers connected at various points. Local variations in device geometry and supply voltage are important component of the clock skew. More sophisticated clock distribution than simple RC matched or grid-based schemes are thus necessary. The active schemes with adaptive digital deskewing typically reduce clock skew of the simple passive clock networks by an order of magnitude, allowing tighter control of the clock period and higher clock rates [3].

2 Clocked Storage Elements

The function of a *clocked storage element*: flip-flop or latch, is to capture the information at a particular moment in time and preserve it as long as it is needed by the digital system. It is not possible to define a storage element without defining its relationship to the *clock*.

2.1 Master-Slave Latch

In order to avoid the *transparency* feature associated with a single latch, an arrangement is made in which two latches are clocked back to back with two non-overlapping phases of the clock. In such arrangement the first latch serves as a "*Master*" by receiving the values from the Data input and passing them to the "*Slave*" latch, which simply follows the "*Master*". This is known as a Master-Slave (M-S) Latch or L1 – L2 latch (in IBM) as shown in Fig. 6. This is not to be confused with the "*Flip-Flop*", though many practitioners today do call the configuration shown in

Fig. 6. (b), a Flip-Flop (F-F). We distinguish Flip-Flop from M-S Latch. We will explain the fundamental differences between the F-F and M-S Latch in this paper.

In a Master-Slave Latch the "Slave" latch can have two or more masters acting as an internal multiplexer with storage capabilities. The first "*Master*" is used for capturing of data input while the second Master can be used for other purposes such as scan-

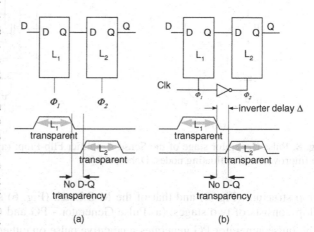

Fig. 6. Master-Slave Latch arrangement with: (a) non-overlapping clocks (b) single external clock.

input for testing purposes, and clocked with a separate clock. One such topology, utilizing two Masters, is a well-known IBM Level-Sensitive-Scan-Design [4].

2.2 Flip-Flop

Flip-Flop and Latch operate on different principles. While Latch is *"level-sensitive"* which means it is reacting on the *level* (logical value) of the clock signal, Flip-Flop is *"edge sensitive"* which means that the mechanism of capturing the data value on its input is related to the changes of the clock. Thus, the two are designed for a different set of requirements and thus consist of inherently different circuit topology. Level sensitivity implies that the latch is capturing data value during the entire period of time when clock is active (logic one) while the latch is *transparent*. The capturing process in the Flip-Flop occurs only during the transition of the clock, thus the Flip-Flop

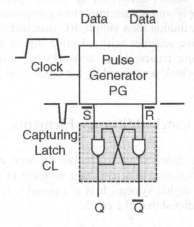

Fig. 7. General Flip-Flop structure

is *non-transparent*. However, even the Flip-Flop could have a small period of transparency associated with the narrow window during which the clock changes. A general structure of the Flip-Flop is shown in Fig. 8. The difference between a Flip-

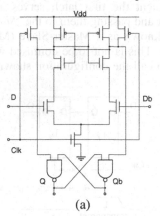

(a)

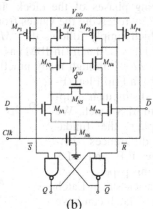

(b)

Fig. 8. Pulse Generator stage of the Sense Amplifier Flip-Flop: (a) Madden and Bowhill [5], (b) Improvement for floating nodes, Doberpuhl [9].

Flop structure (Fig. 8) and that of the M-S Latch (Fig. 6) should be noticed. A Flip-Flop consists of two stages: (a) Pulse Generator - PG and (b) Capturing Latch – CL. The pulse generator PG generates a negative pulse on either $\overline{S}$ or $\overline{R}$ lines, which are normally held at logic "one" level. This pulse is a function of Data and Clock signals and should be of a sufficient duration to be captured in the capturing latch CL. The duration of that pulse can be as long as half of the clock period or it can be as short as one inverter delay. On the contrary M-S Latch generally consists of two identical clocked latches and its non-transparency feature is achieved by non-overlapping clocks ϕ_1 and ϕ_2, clocking master latch L_1 and slave latch L_2. The relationship of S and R signals with respect to Data (D) and Clock (Clk) signal can be expressed as:

$$S_n = Clk\overline{R}(D+S) \text{ and } R_n = Clk\overline{S}(\overline{D}+R) \tag{1}$$

Those two equations (1) form a basis for derivation of a Flip-Flop structure.

Simply stated, the equation for S_n tells us that: *The next state of this Flip-Flop will be set to "1" only at the time the clock becomes "1" (raising edge of the clock), the data at the input is "1", the flip flop is in the "steady state" (both S and R are "0"). The moment Flip-Flop is set (S=1, R=0) no further change in data input can affect the Flip-Flop state: data input will be "locked" to set by (D+S)=1, and reset R_n would be disabled (by S=1).*

This assures the *"edge sensitivity"* – i.e. after the transition of the clock and setting of the S or R signal to its desired state, the Flip-Flop is "locked" for receiving a new data.

It is interesting that it took engineers several attempts to come to the right circuits topology of this Flip-Flop. The Flip-Flop used in the third generation of Digital Equipment Corp. 600MHz Alpha [1] processor used a version of the Flip-Flop introduced by Madden and Bowhill, which was based on the static memory cell design [5]. This particular Flip-Flop is known as Sense Amplifier Flip-Flop (SAFF), shown in Fig.8.a,b. Development of the Pulse Generator block

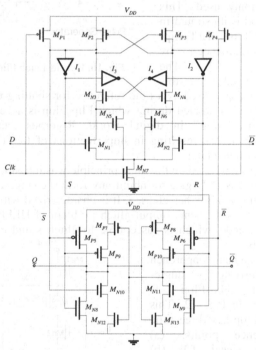

Fig. 9. Pulse Generator stage of the Sense Amplifier Flip-Flop: improvement by proper design: second stage (Stojanovic, US Patent: 6,232,810), first stage [10].

of this Flip-Flop is illustrated in Fig. 8. A substantial improvement in speed is achieved by modification of the second stage by Stojanovic (US Patent No. 6,232,810) [6].

2.3 Time Window Based Flip-Flops

Digital circuits are based on discrete time events. The time reference is a clock signal and/or finite delay through one or more logic elements. To generate a needed time reference, a pulse created by the property of *re-convergent fan-outs, with non-equal parities of inversion* is commonly used. This method is illustrated in Fig. 10. on HLFF flip-flop introduced by Partovi [7]. The trailing edge of this

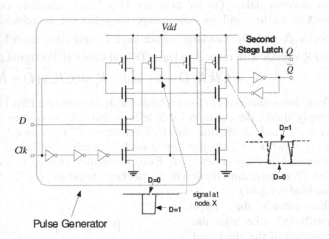

Fig. 10. Hybrid-Latch Flip-Flop introduced by Partovi [7]

short pulse is used as a time reference for shutting the Flip-Flop off. A short *"Time Window"* is created during which Flip-Flop is accepting data, which is the way of creating *"edge"* in digital world. Rigorous analysis of HLFF shows design incompleteness resulting in imperfections of the 1-1 output transition, which was demonstrated later.

A Flip-Flop based on the same principle was introduced by Klass [8], Fig. 11.
It uses a NAND gate to inhibit any further changes, and lock the existing ones after the time window has elapsed. It is characterized with one of the highest performance but suffers the same output glitch problem of HLFF. The problem is in the floating output node, which is susceptible to glitches and even slightest mismatch of clock signals.

A systematic approach in deriving a single-ended Flip-Flop is shown in Fig. 12. This Flip-Flop has three time reference points: (a) Clock signal: Clk (b) Clock signal passed through three inverters: Clk_3, (c) Clock passed through two inverters: Clk_2. The equations

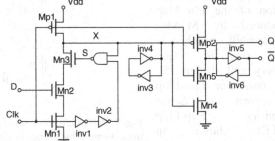

Fig. 11. Semi-Dynamic Flip-Flop: SDFF [8]

describing the pulse generator stage of this Flip-Flop is given by:

$$\bar{S} = X = \overline{(Clk + CLK_2)*(D*Clk_3 + \bar{X})}$$ (2)

The nMOS transistor section is a full realization of this equation. The pMOS section is somewhat abbreviated for performance reasons to:

$$X = \overline{(Clk + CLK_2) * (Clk_3 + \overline{X})} \qquad (3)$$

The second stage(capturing latch) is implemented as:

$$Q = \overline{X * (CLK_2 + \overline{Q})} \qquad (4)$$

This systematically derived Flip-Flop [11] does not have hazards in the output stage and is outperforming HLFF [7] and SDFF Flip-Flops [8].

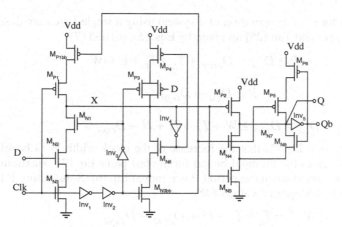

Fig. 12. Systematically derived single-ended Flip-Flop [11]

2.4 Pulsed Latches

In order to decrease the time overhead imposed by M-S Latch, or a Flip-Flop some designers resort to using Single Latch. To narrow the transparency window of the latch, they are clocked with short pulses generated locally from the global clock signal. Thus, the possibility

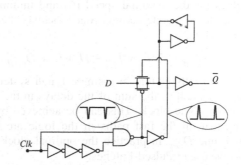

Fig. 13. Pulse Latch: Intel's Explicit Pulsed Latch [24]

of hold time violation and "races" (short paths) is not entirely eliminated, but it is traded for the convenience of a single latch and lower pipeline overhead. Given that the clock pulse is short, the hazard could be reduced by "padding" the logic, i.e. adding inverters in the fast paths so to eliminate the problem.

The clock produced by local clock generator must be wide enough to enable the Latch to capture its data. At the same time it must be sufficiently short to minimize the possibility of "critical race". Those conflicting requirements make use of such single-latch design hazardous by reducing the robustness and reliability of such design. Nevertheless, such design has been used due to the critical need to reduce cycle overhead imposed by the clocked storage elements. Intel's version of Pulsed Latch is shown in Fig. 13. Additional benefit of this design is low power consumption due to the common clock signal generator and a simple structure of the latch. This power can be traded for speed. Pulse generator used in Intel's Pulsed Latch uses the principle of re-convergent fan-out with non-equal parity of inversion in order to obtain desired short clock pulse.

Analysis of the Pulsed Latch Timing Conditions

The conditions for reliable operation of a system using a single Latch are described in the paper by Unger and Tan [25] an given by Eqs. (5), (6) and (7):

$$P_m = P \geq D_{LM} + D_{CQM} + T_L + T_T + U - W \tag{5}$$

$$P \geq D_{LM} + D_{DQM} \tag{6}$$

$$D_{Lm} > D_{LmB} \geq W + T_T + T_L + H - D_{CQm} \tag{7}$$

One can notice from Eq. (5) that the increase of the clock width W is beneficial for speed, but it increases the minimal bound for the fast paths Eq. (7). Maximum useful value for W is obtained when the period P is minimal Eq. (6). Substituting P from Eq. (6) into Eq. (5) yields optimal value of W:

$$W^{opt} = T_L + T_T + U + D_{CQM} - D_{DQM} \tag{8}$$

If we substitute the value of the optimal clock width W^{opt} into (5), then we will obtain the values for the maximal speed (6) and minimal signal delay in the logic (4.26) which has to be maintained in order to satisfy the conditions for optimal single-latch system clocking:

$$D_{LmB} = 2(T_T + T_L) + H + U + D_{CQM} - D_{CQm} - D_{DQM} \tag{9}$$

Equation (6) tells us that in a single Latch system, it is possible to make the clock period P as small as the sum of the delays in the signal path: Latch and critical path delay in the logic block. This can be achieved by adjusting the clock width W and assuring that all the fast paths in the logic are larger in their duration than some minimal time D_{LmB}. In practice the optimal clock width W^{opt} is very small and does support the use of Pulsed-Latches.

$$W^{opt} \approx 2T_{SKW} \qquad\qquad D_{LmB} = 4T_{SKW} + H - D_{CQm} \tag{10}$$

Equation (9) tells us is that under ideal conditions, if there are no clock skews and no process variations, the fastest path through the logic has to be greater than the sampling window of the Latch $(H+U)$ minus the time the signal spend traveling through the Latch. If the travel time through the Latch D_{DQM}, is equal to the sampling window, than we do not have to worry about fast paths.

3 Timing Parameters

Data and *Clock* inputs of a clocked storage element need to satisfy basic timing restrictions to ensure correct operation of the flip-flop. Fundamental timing constraints between data and clock inputs are quantified with *setup* and *hold* times, as illustrated in Fig. 14. Setup and hold times define time intervals during which input has to be stable to ensure correct flip-flop operation. The sum of setup and hold times define the *"sampling window"* of the clocked storage element.

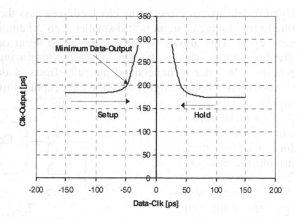

Fig. 14. Setup and Hold time behavior as a function of Clock-to-Output delay

3.1 Setup and Hold Time Properties

Failure of the clocked storage element due to the Setup and Hold time violations is not an abrupt process. This failing behavior is shown in Fig. 14. Considering how close should data be allowed to change with respect to the locking event, we encounter two opposing requirements:
– it should be kept further from the failing region for the purpose of design reliability.
– it should be as close to the clock in order to increase the time available for the logic operation.

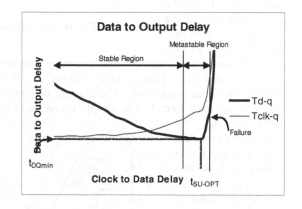

Fig. 15. Setup and Hold time behavior as a function of Data-to-Output delay

This is an obvious dilemma. In some designs an arbitrary number of 5-20% is used. Setup and Hold times are defined as points in time when the Clk-Q (t_{CQ}) delay raises for that amount. We do not find this reasoning to be valid.

A redrawn picture, Fig.15, where D-Q (t_{DQ}) delay is plotted (instead of Clk-Q), provides more information. From this graph we see that in spite of Clk-Q delay rising, we are still gaining because the time taken from the cycle is reduced.

3.2 Time Borrowing and Absorption of Clock Uncertainties

Even if data arrives close to the clock edge or pass the clock edge, the delay increase due to the storage element is still smaller than the amount of delay introduced into the next cycle. This allows for more time to be spent on useful logic operation in the previous cycle. This is known as: *"time borrowing"*, *"cycle stealing"* or *"slack passing"*. In order to understand the full effects of delayed data arrival we have to consider a pipelined design where the data captured in the first clock cycle is used as input in the next clock cycle as shown in Fig. 17.

As it can be seen in Fig. 17, the *"sampling window"* moves around the time axes. The *"sampling window"* is defined as the sum of the Setup and Hold times, i.e. the time period in which clocked storage element is *"sampling"* and data is not allowed to change. As the data arrive closer to the clock, the size of the *"sampling window"* shrinks (up to the

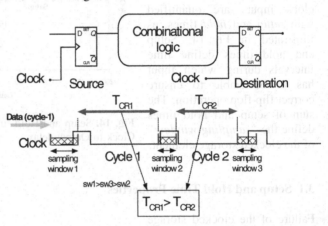

Fig. 16. *"Time Borrowing"* in a pipelined design

optimal point). Even though, the sampling window is smaller, the data in the next cycle will still arrive later compared to the case where the data in the previous cycle was ahead of the setup-time. The amount of time for which the T_{CR1} was augmented did not come for free. It was simply taken away (*"stolen"* or *"borrowed"*) from the next cycle T_{CR2}. As a result of late data arrival in the Cycle 1 there is less time available in the Cycle 2. Thus a boundary between pipeline stages is somewhat flexible. This feature not only helps accommodate a certain amount of imbalance between the critical paths in various pipeline stages, but it helps

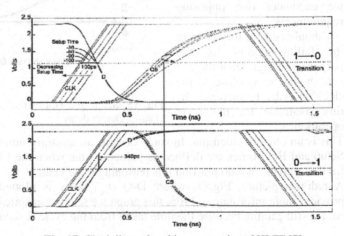

Fig. 17. Clock jitter-absorbing properties of HLFF [7].

in absorbing the clock uncertainties: *skew* and *jitter*. Thus, "*time borrowing*" is one of the most important characteristics of today's high-speed digital systems. Absorption of the clock jitter in HLFF is shown in Fig. 17 as observed by Partovi [7].

The maximal clock skew that a system can tolerate is determined by clock storage elements. If the clock-to-output delay of a clocked storage element is shorter than the hold time required and there is no logic in between two storage elements, a race condition can occur. A *minimum delay restriction* on the clock-to-output delay given by:

$$t_{CLK-Q} \geq t_{hold} + t_{skew} \tag{11}$$

If this relation is satisfied, the system is immune to hold time violations. Otherwise, it is necessary to check that all the timing paths have some minimal delay, which assures that there is no hold time violation.

4 Characterization

4.1 Power and Energy

It is important to emphasize the sources of power consumed in the Clocked Storage Element (CSE) and the correct set-up for the characterization and comparison. Power consumed by a CSE comes from various sources of which power-supply (V_{DD}) is only one of several. Using V_{DD} as a point for measuring power consumption can be misleading. Some CSE, characterized with low internal power consumption, represent a considerable load on the clock distribution network, thus taking considerable amount of power from the clock. Power can be drown from the Data input as well. Therefore the total power P_{tot} should account for all the possible power sources supplying the CSE [12].

$$P_{tot} = P_{internal} + \sum_{inputs(D,CLK)} P_{driver} \tag{12}$$

4.2 Delay

In characterizing delay it is only appropriate to take into account the amount of time taken from the cycle T due to the insertion of the CSE. This represents $D-Q$ delay (t_{DQ}) as it was discussed in III. The question is whether this delay should be $D-Q, D-\overline{Q}$ or the worse of the two? We strongly argue that it is the most appropriate to characterize the CSE with the worse of the two delays since the critical path in a design may impose that scenario. Another question is that of the output load: how large should the outputs load be?. It is only reasonable that the load on the output: $(Q, \overline{Q})$ be representative of the conditions existing in a real design. In our measurements we use 14 minimal size inverters (in the same technology) as a representative load. Finally the remaining question is: should we load only the output producing the longer delay or both? We performed our measurements by loading only the worse of the two $(Q, \overline{Q})$. This is justified by the fact that the critical path can always be improved by duplicating the CSE, and thus reducing the load to zero on the output that is not in the critical path. This is the approach that is taken by a reasonable designer and a synthesis tool as well.

4.3 Figure of Merit

It is well known that power can always be traded for speed and that superior speed can always be obtained by allowing for higher power consumption. Thus, it is hard to tell which one of the two CSE compared against each other is better. Various figures of merit have been used in the past. One commonly used and grossly misleading factor is Power-Delay-Product (PDP). It is not difficult to prove that PDP would always favor slower design, given that the energy consumed depends on the clock speed as well. It has been shown that more appropriate figure of merit is Energy-Delay-Product (EDP), [16]. However, some recent results argue that ED^2P is even more appropriate, at least in high-performance systems [19]. In our measurements we use PDP at a fixed frequency, which represents EDP.

5   Design for Low Power

The energy consumed in a clocked storage element is approximated by:

$$E_{switching} = \sum_{i=1}^{N} \alpha_{0-1}(i) \cdot C_i \cdot V_{swing}(i) \cdot V_{DD} \qquad (13)$$

where N is the number of nodes in a clocked storage element, C_i is the node capacitance, $\alpha_{0-1}(i)$ is the probability that transition occurs at node i, and V_{swing} is the voltage swing of node i. Starting from (7), several commonly used techniques applied to minimize energy consumption can be derived:

Reducing the number of active nodes and assuring that when they are switching the capacitance is minimized.
Reducing the voltage swing of the switching node
Reducing the voltage (technology scaling)
Reducing the activity of the node

The approaches listed in (a)-(d) result in several known techniques used in low-power applications. One of the most common is "clock gating" which assures that the storage elements in an inactive part of the processor are not switching. A thorough review of the common techniques for low-power can be found in [13]. In this paper we describe some recent techniques applicable to low-power design of clocked storage elements.

5.1 Conditional Capture Flip-Flop

Motivation behind Conditional Capture technique is the observation that considerable portion of power is consumed for driving internal nodes even when the value of the output is not changed (low input activity). It is possible to disable internal transitions when it is detected that they will have no effect on output. Conditional capture technique attempts to minimize unnecessary switching of the CSE. By disabling redundant internal transitions, this technique achieves power reduction at little or no delay penalties. Due to this property, it is particularly attractive from the point of view of high-performance VLSI implementations. One such

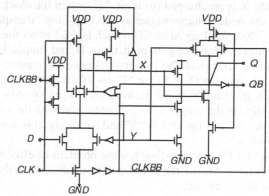

Fig. 18. Conditional Capture Flip-Flop [15]

structure is CCFF [14], which operates on the principle of J-K Flip-Flop: data can affect the Flip-Flop only if it will result in the change of the output. An improved version of CCFF is presented in [15] which reduces the overall Energy-Delay Product by up to 14% in for 50% data activity, while total power saving is more than 50% with quiet inputs (Fig. 19.). CSE equipped with conditional features have advantageous properties in low data activity conditions. In the implementation shown in Fig. 19, conditional capture is achieved by direct sampling of (inverted) input during the transparency window in single-ended CCFF. However, this approach is associated with severe drawbacks, most important of which is related to increased set-up time for sampling logic "0".

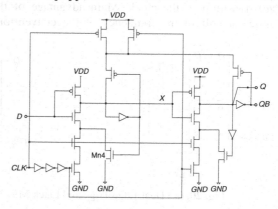

Fig. 19. Conditional Precharge Flip-Flop [15]

5.2 Conditional Precharge Flip-Flop

Conditional Precharge Flip-Flop (CPFF) [15] is shown in Fig. 19. Conditional Precharge technique is a way to save the unnecessary portion of the power in the Flip-

Flop. It eliminates power consuming precharge operation in dynamic Flip-Flops when it is not required.

Instead of gating data (in the evaluation phase), it is the precharge of the internal node that is conditioned by the state of the output. With the assumption that the internal node X is precharged (to logic "1") when the clock is in the "0" state, the evaluation of the node X happens during the Flip-Flop "transparency window". If the input D is "1", X is discharged to "0", which is used to set the output Q to "1". Node X remains at logic "0" as long as both input D and output Q are at the logic "1" level. This allows savings in the power consumed on unnecessary consecutive evaluations and precharges for D=1. Logic "1"-to-"0" transition of the output is achieved by sampling high level on X in the transparency window. Also, conditional keeping function is applied at the output to avoid contention with the output keeper - the output is kept at logic "0" as long as X is "1" and, similarly, it is kept high outside of the transparency window.

Like CCFF, this flip-flop has the problem of effectively higher set-up time for 1-to-0 transition due to the requirement to discharge the output before the transparency window is closed.

5.3 Dual Edge Triggering

One of the approaches amenable to high-performance as well as low-power application is the use of Dual-Edge Triggered (DET) clocked storage elements. Substantial power savings in the clock distribution network can be achieved by reducing the clock frequency by one half. This can be done if every clock transition is used as a time reference point, instead of using only one (*leading edge* or *trailing edge*) transition of the clock. Main advantage of this approach is that the system operates at half of the frequency of the conventional single-edge clocking design

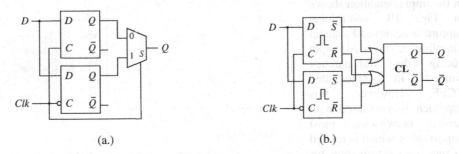

(a.)   (b.)

Fig. 20. (a.) Dual-Edge Triggered Latch-Mux (b.) Flip-Flop topology

style, while obtaining the same data throughput. Consequently, power consumption of the clock generation and distribution system is roughly halved for the same clock load. In addition, less aggressive clock subsystems can be built, which further reduces power consumption and clock uncertainties.

Dual-edge clocking requires Dual Edge-Triggered Storage Elements (DETSE), capable of capturing data on both rising and falling edge of the clock. The most critical obstacle for extensive use of dual-edge clocking strategy is the difficulty to precisely control the arrival of both clock edges. This control is essential in order to avoid large timing penalty incurred by the clock uncertainties. Even though this

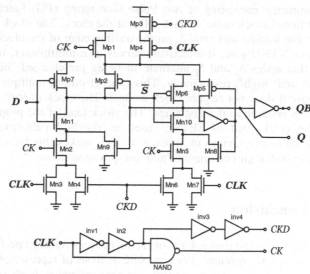

Fig. 21. Dual-Edge Conditional Pre-charge Flip-Flop, DE-CPFF [22]

requirement imposes additional complexity, it can be satisfied with reasonably low hardware overhead. In addition, the clock uncertainty due to the variation of the duty cycle can be partially absorbed by the storage element [20].

There are two fundamental ways of building dual-edge clocked storage elements: Latch-Mux and Flip-Flop as shown in Fig.21.

An example of a dual edge-triggered Flip-Flop (Dual-edge Conditional Pre-charge Flip-Flop, DE-CPFF) is shown in Fig.21 [22]. Its operation is based on creating two narrow *transparency windows* during which the logic level of the input D can be transferred to the output. This Flip-Flop is a dual-edge version of Conditional Pre-charge Flip-Flop (CPFF, [21]).

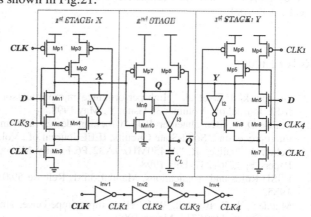

Fig. 22. Dual-Edge Triggered Flip-Flop [23]

Dual Edge Triggered Flip-Flop

An example of DET Flip-Flop design [22] is shown in Fig.22 The circuit has a narrow data transparency window and clock-less output multiplexing scheme. The first stage is symmetric consisting of two Pulse Generating (PG) Latches. It creates the data-conditioned clock pulse on each edge of the clock. The clock pulse is created at node S_x on the leading and node S_y on the trailing edge of the clock. The second stage is a 2-input NAND gate. It effectively serves as a multiplexer, implicitly relying on the fact that nodes S_x and S_y alternate in being pre-charged "high", while the clock is "low" and "high", respectively. This type of output multiplexing is very convenient because it does not require clock control. The clock energy is mainly dissipated for pulse generation in the first stage. The clock load of the proposed flip-flop is similar to the clock load of SETSEs used in the high-performance processor designs, allowing power savings of about 50%. This makes this DETSE a viable option to be used in both high-performance and low-power systems.

6   Conclusion

A review of some (but not all) of the techniques for high performance and low-power CSE design is presented. For complete analysis of representative CSE please see [27] or visit: www.ece.ucdavis.edu/acsel where extensive database of comparative results exist. In the future we expect that pipeline boundaries will start to blur and synchronous design will be possible only in limited domains on the chip.

Acknowledgement. I gratefully acknowledge contributions from my current and former students: Vladimir Stojanovic, Dejan Markovic, Nikola Nedovic, Bora Nikolic and Bart Zeydel.

References

1. Gronowski P.E, et al, "High-performance microprocessor design" Solid-State Circuits, IEEE Journal of , Volume: 33 Issue: 5 , May 1998.
2. Bailey D.W, Benschneider B.J. "Clocking design and analysis for a 600-MHz Alpha microprocessor", Solid-State Circuits, IEEE Journal of , Vol.33, No.11 , November 1998.
3. Schutz J, Wallace R. "A 450MHz IA32 P6 Family Microprocessor," ISSCC Dig. Tech. Papers, pp. 236–237, Feb. 1998.
4. LSSD Rules and Applications, Manual 3531, Release 59.0, IBM Corporation, March 29, 1985.
5. Madden, W.C, Bowhill, W.J, "High input impedance, strobed sense-amplifier," United States Patent 4,910,713, March 1990.
6. V. Stojanovic, V. G. Oklobdzija, FLIP-FLOP, US Patent No. 6,232,810, May 15, 2001.
7. Partovi, H. et al, "Flow-through latch and edge-triggered flip-flop hybrid elements", 1996 IEEE International Solid-State Circuits Conference. Digest of Technical Papers, ISSCC, San Francisco, February 8–10.
8. Klass F, "Semi-Dynamic and Dynamic Flip-Flops with Embedded Logic," Symposium on VLSI Circuits, Digest of Technical Papers, pp. 108–109, June 1998.
9. Dobberpuhl, D.W. "Circuits and technology for Digital's StrongARM and ALPHA microprocessors", Proceedings of the Seventeenth Conference on Advanced Research in VLSI, Ann Arbor, Michigan, September 15–16. p. 2–11.

10. Nikolic, B, Oklobdzija, V.G, "Design and optimization of sense-amplifier-based flip-flops", Proceedings of the 25th European Solid-State Circuits Conference, ESSCIRC'99., Duisburg, Germany, 21-23 September 21–23. p.410–13.
11. Nedovic N, Oklobdzija V.G, Dynamic flip-flop with improved power. Proceedings of the IEEE International Conference on Computer Design: VLSI in Computers and Processors. ICCD 2000, Austin, Texas, September 17-20. p.323-6.
12. V.Stojanovic and V.G. Oklobdzija, "Comaparative Analysis of Master-Slave Latches and Flip-Flops for High-Performance and Low-Power VLSI Systems," IEEE Journal of Solid-State Circuits, Vol.34, No.4, April 1999.
13. T. Kuroda and T. Sakurai "Overview of Low-Power ULSI Circuit Techniques", IEICE Trans. Electronics, E78-C, No 4, April 1995, pp. 334–344, INVITED PAPER, Special Issue on Low-Voltage Low-Power Integrated Circuits.
14. B. S. Kong, S. S. Kim, Y. H. Jun, "Conditional Capture Flip-Flop Technique for Statistical Power Reduction", Digest of Technical Papers, p. 290–291, February 2000.
15. Nikola Nedovic, Marko Aleksic and Vojin G. Oklobdzija, "Conditional Techniques for Small Power Consumption Flip-Flops", Proceedings of the 8th IEEE International Conference on Electronics, Circuits and Systems, Malta, September 2–5, 2001.
16. M. Horowitz, et al, "Low-Power Digital Design", Proceedings of the 1994 IEEE Symposium on Low-Power Electronics, 1994.
17. F. Klass et al, "A New Family of Semidynamic and Dynamic Flip-Flops with Embedded Logic for High-Performance Processors", IEEE Journal of Solid-State Circuits, vol. 34, no. 5, pp. 712–716, May 1999.
18. Friedman EG (ed.) "Clock Distribution Networks in VLSI Circuits and Systems", IEEE Press.
19. K. Nowka, P. Hofstee, private communication, IBM Research, Austin, Texas, 2000.
20. M. Saint-Laurent et al, "Optimal Sequencing Energy Allocation for CMOS Integrated Systems", Proceedings of International Symposium on Quality Electronic Design, p.94–99, March 2002.
21. N. Nedovic, V. G. Oklobdzija, "Hybrid Latch Flip-Flop with Improved Power Efficiency", Proceedings of the Symposium on Integrated Circuits and Systems Design, SBCCI2000, Manaus, Brazil, p. 211–215, September 18–22, 2000.
22. N. Nedovic, M. Aleksic, V. G. Oklobdzija, "Conditional Pre-Charge Techniques for Power-Efficient Dual-Edge Clocking", Proceedings of the International Symposium on Low Power Electronics and Design, Monterey, California, August 12–14, 2002.
23. N. Nedovic, W. W. Walker, V. G. Oklobdzija, M. Aleksic, "A Low Power Symmetrically Pulsed Dual Edge-Triggered Flip-Flop", Proceedings of the European Solid-State Circuits Conference, ESSCIRC'02., Florence, Italy, September 24–26, 2002.
24. Tschanz James, Siva Narendra, Zhanping Chen, Shekhar Borkar, Manoj Sachdev, Vivek De, "Comparative Delay and Energy of Single Edge-Triggered & Dual Edge-Triggered Pulsed Flip-Flops for High-Performance Microprocessors, Proceedings of the 2001 International Symposium on Low Power Electronics and Design, Huntington Beach, California, August 6–7, 2001.
25. Unger S.H, Tan CJ, (1986) "Clocking Schemes for High-Speed Digital Systems", IEEE Transactions on Computers, Vol. C-35, No 10, October 1986.
26. Oklobdzija V.G, (ed.) "High-Performance System Design: Circuits and Logic", Book, IEEE Press, July 1999.
27. Oklobdzija V.G, Stojanovic V, Markovic D, Nedovic N, "Digital System Clocking: High-Performance and Low-Power Aspects, J. Wiley, in press expected December 2002.

Dual Supply Voltage Scaling in a Conventional Power-Driven Logic Synthesis Environment

Torsten Mahnke[1], Walter Stechele[1], and Wolfgang Hoeld[2]

[1] Technical University of Munich, Institute for Integrated Circuits,
Arcisstrasse 21, 80290 Muenchen, Germany
{torsten.mahnke, walter.stechele}@ei.tum.de
[2] National Semiconductor GmbH, Livry-Gargan-Strasse 10,
82256 Fuerstenfeldbruck, Germany
wolfgang.hoeld@nsc.com

Abstract. Dual supply voltage scaling (DSVS) is an emerging technique in logic-level power optimization. In this paper, a novel design methodology, which enables DSVS to be carried out in a state-of-the-art environment for power-driven logic synthesis, is presented. The idea is to provide a dual supply voltage standard cell library modeled such that a typical gate sizing algorithm can be exploited for DSVS. Since this approach renders dedicated DSVS algorithms superfluous, only little modification of established design flows is required. The methodology has been applied to MCNC benchmark circuits. Compared to the results of single supply voltage power-driven logic synthesis, additional power reductions of 10% on average and 24% in the best case have been achieved.

1 Introduction

The total power dissipation of digital CMOS circuits is composed of static and dynamic components. While static power contributes significantly to the total power in certain applications that are inactive for long periods of time, it is still dominated by dynamic power in the majority of applications.

The dynamic power P_{dyn} is composed of the capacitive power P_{cap} and the short-circuit power P_{sc}. The capacitive power P_{cap} is due to currents charging or discharging the node capacitances and can be written as

$$P_{cap} = \alpha_{01} \cdot f_{clk} \cdot C_{node} \cdot V_{DD}^2 \; , \tag{1}$$

where α_{01} is the switching activity, f_{clk} is the clock frequency, C_{node} is the node capacitance, and V_{DD} is the supply voltage. Although P_{cap} usually accounts for the largest portion of the total dynamic power, P_{sc} must not be neglected. The short-circuit power is caused by currents flowing through simultaneously conducting n- and p-channel transistors. A first-order approximation of P_{sc} is

$$P_{sc} = \alpha_{01} \cdot f_{clk} \cdot (\beta/12) \cdot t_T \cdot (V_{DD} - 2V_t)^3 \; , \tag{2}$$

where β is an effective transconductance, t_T is the input signal transition time and V_t is the threshold voltage.

B. Hochet et al. (Eds.): PATMOS 2002, LNCS 2451, pp. 146–155, 2002.
© Springer-Verlag Berlin Heidelberg 2002

A very efficient means of reducing P_{dyn} is supply voltage scaling. However, since gate delay increases with decreasing V_{DD}, globally lowering V_{DD} degrades the performance. At the logic level, dual supply voltage scaling (DSVS) can be used for lowering V_{DD} only in non-timing-critical paths, thus keeping the overall performance constant [2,7,8,9,11].

In this paper, we show that cell-library-based gate sizing (GS) algorithms can be used for DSVS. This perception enabled us to carry out DSVS in a conventional logic synthesis environment, while all previously published work required proprietary tools that comprise dedicated DSVS algorithms. In our discussion of experimental results, we use the results of state-of-the-art power-driven single supply voltage (SSV) logic synthesis as reference values in order to reveal the true additional benefit of DSVS.

The remainder of the paper is structured as follows. In Sects. 2 and 3, a short overview of related work and state-of-the-art logic-level power optimization is given. In Sects. 4 and 5, we introduce the DSVS technique and we explain how DSVS can be carried out exploiting cell-library-based GS algorithms. Our novel power-driven logic synthesis methodology is described in Sect. 6. Experimental results are presented in Sect. 7. Finally, we provide concluding remarks.

2 Related Work

In this section, we review the so far most relevant publications on dual supply voltage (DSV) logic synthesis and point out the advantages of our work.

Usami et al. used a dedicated DSVS algorithm for performing clustered voltage scaling (CVS) [7,8,9]. The CVS method was applied to random logic sub-modules of real audio/video applications. Unfortunately, no information on the strictness of the timing constraints is provided. Also, it is not clear whether the SSV designs that served as references had already been optimized for power exploiting state-of-the-art logic-level techniques.

Yeh et al. developed another dedicated DSVS algorithm which is basically an improvement of the CVS method [10]. They applied the algorithm to combinational MCNC benchmark circuits subject to relaxed delay constraints. A major shortcoming of this work is the fact that the SSV reference implementations of the benchmarks had not been optimized for power and, hence, the results do not reveal the true additional benefit of DSVS.

In the most recent work by Chen et al., yet another dedicated algorithm was used for optimizing combinational MCNC benchmark circuits subject to varying delay constraints [2]. Among other things, Chen et al. compared DSVS to GS. However, in their experiments, GS lead to rather small improvements in power consumption of only 7% on average. From our experience we know that state-of-the-art power-driven logic synthesis leads to significantly larger power reduction, leaving less room for further improvement through DSVS.

It is evident from above that all known DSVS methodologies are based on dedicated algorithms and proprietary tools. This is the main reason why DSVS has not yet become an integral part of standard design flows. A major advantage of our work is that DSVS is enabled to be carried out in a conventional logic synthesis environment, thus minimizing the effort required for adopting this new optimization technique. Furthermore, due to the extensive use of state-of-the-art power-driven logic synthesis for creating the SSV reference designs, our results do not feign an unrealistically large benefit of DSVS.

3   State-of-the-Art in Power-Driven Logic Synthesis

In conventional logic synthesis methodologies, P_{dyn} can typically be minimized by means of gate sizing (GS), equivalent pin swapping and buffer insertion [6].

Down-sizing primarily aims at reducing C_{node} and, thus, P_{cap} by using smaller slower cells in non-timing-critical paths, but also reduces short-circuit currents and, hence, P_{sc} at the sized gates. On the other hand, increasing the size of a gate shortens the signal transition time t_T at its output, which in turn reduces P_{sc} at the gates driven by the sized cell. Alternatively, extra buffers can be inserted at heavily loaded nodes in order to shorten t_T. Equivalent pin swapping takes advantage of the fact that functionally equivalent input pins of logic gates often exhibit different power characteristics. With pin swapping, high activity nets are connected to power-efficient input pins with priority.

In our experiments, we made extensive use of the above mentioned techniques when we created the SSV reference designs.

4   Dual Supply Voltage Scaling (DSVS)

The purpose of DSVS is to reduce the supply voltage for gates in noncritical paths from the nominal value V_{DD} to a lower value V_{DDL} [2,7,8,9,11]. Fig. 1 illustrates a typical (DSV) circuit structure. In DSV circuits, low voltage cells must not directly drive high voltage cells. Otherwise, quiescent currents occur at the driven gates. This is the reason why gates 1 and 2 in Fig. 1 are operated at V_{DD} although they are part of a noncritical path. Level-converting cells can be inserted where transitions from V_{DDL} to V_{DD} are required [8]. However, these cells introduce additional delay and cause power and area overhead. In order to minimize this overhead, we enable level conversion only at the input and output nodes of combinational blocks as depicted in Fig. 1.

Other difficulties are the distribution of two supply voltages across the chip and the layout synthesis. One possible solution to these problems is placing low and high voltage cells in separate rows. This can be realized on the basis of conventional cell layouts but requires proprietary tools [8]. Another possibility is the use of two separate power rails for V_{DD} and V_{DDL} in each row. This requires modification of the layouts of all cells. However, low and high voltage cells can then be mixed within rows and, hence, placement and routing can be carried out using standard tools [10].

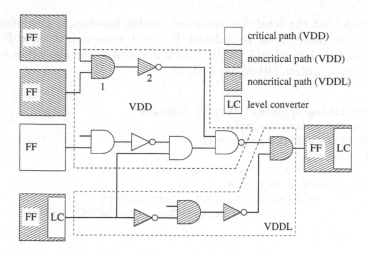

Fig. 1. A typical dual supply voltage (DSV) circuit structure

5   Algorithms and Cell Modeling for DSVS

5.1   Cell-Library-Based Gate Sizing (GS) Algorithm

At the logic level, library cells c_i can be represented by tuples of basic properties, namely the functionality F_i, delay t_{Di}, output signal transition time t_{Ti}, cell-internal dynamic power P_i, area A_i and input capacitances C_i:

$$c_i = \{F_i, t_{Di}, t_{Ti}, P_i, A_i, C_i\} \tag{3}$$

In typical cell libraries, P_i predominantly represents P_{sc}, while P_{cap} must be calculated from (1) with C_i contributing to C_{node}.

Cell-library-based GS algorithms revert to the cell properties mentioned above when picking cells that implement certain functionalities while minimizing a cost function $COST$, which evaluates the overall delay, the power and the area of a circuit [3]. In Fig. 2, a simplified GS algorithm is shown. In the case of delay-constrained power optimization, the initial solution is a timing-optimized implementation of a logic network N. Static timing analysis is used for calculating the timing slack. In each of the subsequent iterations (loops), all nodes n in the network N are visited. For each node n, the complete set $C(n)$ of library cells c_i that implement the required functionality $F(n)$ is

$$C(n) = \{c_i | F_i = F(n)\} . \tag{4}$$

The algorithm determines which cell c_{opt} in $C(n)$ must be used for replacing the cell $c(n)$ that currently implements the node under consideration, in order to maximize the cost reduction $DeltaCOST$. The substitution $c(n) = c_{opt}$ is then appended to a list of possible substitutions. Once all nodes have been visited, a subset of independent substitutions from the list off all possible substitutions is chosen, such that the total cost reduction in this iteration is maximized. Subsequently, the timing data is updated. If a cost reduction resulted from that iteration and if there is still positive slack remaining, the algorithm continues with another iteration. Otherwise it stops.

```
start from timing-optimized initial solution
perform static timing analysis
loop { possible_substitutions = {}
      foreach n in N {
            copt = c(n) ; DeltaCOSTopt = 0
            foreach c in C(n) {
                  if DeltaCOST(c(n)=c) < DeltaCOSTopt {
                        copt = c ; DeltaCOSTopt = DeltaCOST(c(n)=c)
                  }
            }
            append ''c(n) = copt" to possible_substitutions
      }
      apply max. independent subset of possible_substitutions
      update timing
      exit loop if no improvement achieved or no positive slack left }
```

Fig. 2. Gate sizing (GS) algorithm exploited for dual supply voltage scaling (DSVS)

Actual implementations of such algorithms may differ with regard to delay and power modeling, the way of updating timing data, the treatment of local minima or the way of determining maximum sets of independent substitutions. However, these aspects do not affect the basic algorithm discussed above.

5.2   Exploiting GS Algorithms for DSVS

Reducing the supply voltage for a cell affects only its timing and power characteristics. Therefore, if two different supply voltages are allowed, each library cell c_i may be represented by two low and high voltage synthesis models, c_{iLV} and c_{iHV} respectively. The two models are functionally equivalent but exhibit different timing and power characteristics:

$$c_{iLV} = \{F_i, t_{DiLV}, t_{TiLV}, P_{iLV}, A_i, C_i\} \tag{5}$$

$$c_{iHV} = \{F_i, t_{DiHV}, t_{TiHV}, P_{iHV}, A_i, C_i\} \tag{6}$$

For DSVS, however, an additional constraint is required for preventing high voltage cells from being driven by low voltage cells as described in Sect. 4. This can for instance be accomplished by means of two additional properties describing the input signal level (ISL) and the allowed fanout signal levels (FSL):

$$c_i = \{F_i, t_{Di}, t_{Ti}, P_i, A_i, C_i, ISL_i, FSL_i\} \tag{7}$$

Allowed values for ISL are LV and HV for low and high voltage signals respectively. The parameter FSL can take on one of the two values LV and DC for low voltage and don't care respectively. Functionally equivalent low voltage, high voltage and level-converting (LC) cells can then be modeled as follows:

$$c_{iLV} = \{F_i, t_{DiLV}, t_{TiLV}, P_{iLV}, A_i, C_i, LV, LV\} \tag{8}$$

$$c_{iHV} = \{F_i, t_{DiHV}, t_{TiHV}, P_{iHV}, A_i, C_i, HV, DC\} \tag{9}$$

$$c_{iLC} = \{F_i, t_{DiLC}, t_{TiLC}, P_{iLC}, A_{iLC}, C_{iLC}, LV, DC\} \tag{10}$$

If, finally, the set $C(n)$ of candidates for substituting a current implementation $c(n)$ is restricted to (4) if $ISL = LV$ for all cells driven by $c(n)$ and to

$$C(n) = \{c_i | F_i = F(n) \quad \text{and} \quad FSL_i = DC\} \tag{11}$$

otherwise, cell-library-based GS algorithms, such as the one discussed above, can be used for performing DSVS and GS simultaneously.

6 Dual Supply Voltage Logic Synthesis Methodology

In this section, we present a novel power-driven logic synthesis methodology, which builds on a conventional logic synthesis flow and incorporates DSVS without the need for any dedicated algorithm.

6.1 Design Flow and Tools

Provided that a suitably modeled DSV library exists, delay-constrained power optimization can be performed following the three-step strategy illustrated in Fig. 3. After reading the original design, delay-constrained logic synthesis is carried out (STEP 1). At this stage, low voltage (VDDL) and level-converting cells (LC) are disabled. After capturing switching activities during gate-level simulation, state-of-the-art delay-constrained power optimization comprising the techniques mentioned in Sect. 3 is carried out (STEP 2), which results in a timing- and power-optimized SSV implementation. Finally, power optimization is repeated with low voltage and level-converting cells enabled (STEP 3). This leads to a timing- and power-optimized DSV implementation.

For power-driven logic synthesis, we used Synopsys' Power Compiler (SPC). This tool is capable of minimizing power by means of a GS method similar to that discussed in Sect. 5. The cost function gives absolute priority to timing over power, i.e. the substitution of a cell is carried out only if dynamic power is reduced without sacrificing performance. Fortunately, SPC allows input and output pins of cells to be classified such that only pins of the same class will be interconnected. This feature allows us to solve the level conversion issue described in Sect. 4.

Power analysis was carried out using Synopsys' Design Power (SDP).

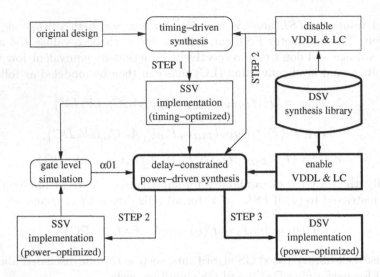

Fig. 3. Design flow comprising DSVS in addition to state-of-the-art power optimization

6.2   Dual Supply Voltage Synthesis Library

The key to DSVS exploiting GS algorithms is a suitably modeled standard cell library. We developed a DSV synthesis library from a commercial library realized in $0.25\,\mu m$ CMOS and characterized at supply voltages of $1.8\,V$ and $2.5\,V$. Note that we always used these voltage levels, which were defined by the library vendor, and forwent the costly procedure of determining an optimal voltage pair for each circuit, which was used by Usami et al. and Chen et al. [2,8].

The DSV library contains inverters, buffers, (N)ANDs, (N)ORs, X(N)ORs and D-flip-flops in up to five different sizes each. For each cell, high and low voltage synthesis models are provided and a level-converting flip-flop (DFFLC) similar to that proposed in [9] was included in order to enable level conversion at the inputs and outputs of combinational blocks as described in Sect. 4. Furthermore, we classified the input and output pins of all cells such that ouput pins of low voltage cells are not allowed to drive input pins of high voltage cells.

For SDP to properly calculate the power consumption in the presence of two supplies, we modeled P_{dyn} for each cell individually. While cell-internal look-up tables are normally used for modeling only cell-internal dynamic power [1], i.e. $P_i = P_{sci}$, we used them for modeling all the dynamic power, i.e. $P_i = P_{dyn}$. For a more detailed discussion of tool-specific DSV library modeling issues see [4].

7   Experiments, Results, and Discussion

We applied our methodology to MCNC benchmark circuits (see [5]) subject to reasonably strict delay constraints. In the following discussion, we use the results of state-of-the-art power-driven SSV logic synthesis (see Sect. 3) as reference values in order to reveal the true additional benefit of DSVS.

7.1   Delay Constraints

Achieving highest performance is a common objective in IC design. Therefore, power optimization subject to strictest delay constraints is a realistic task. However, even if the strictest delay constraint is imposed on a complex sequential design, usually the majority of combinational blocks therein are noncritical. Thus, we consider relaxed delay constraints typical of purely combinational subcircuits.

In our experiments, we determined the shortest possible delays of all circuits by timing-driven synthesis using zero-delay constraints. In the case of sequential circuits, the resulting minimum delay values were used as constraints in the power-driven synthesis. In the case of purely combinational circuits, we allowed the actual delays of the circuits to be 1.2 times the shortest possible delays.

7.2   Combinational Benchmarks

We optimized the power consumption of 15 combinational MCNC benchmark circuits, firstly, using the state-of-the-art methodology for power-driven logic synthesis (SSV optimization, STEP 1 and STEP 2 in Fig. 3) and, secondly, using our DSVS methodology (STEP 3). The results are summarized in Table 1. The numbers shown in column four confirm that SSV power optimization typically leads to significantly larger power reduction than reported by Chen et al. (see Sect. 2), leaving less room for further improvement through DSVS. The fifth column shows the advantage of our methodology over SSV power optimization. On average, the final power consumption was 10% lower if DSVS was used. In the best case, the improvement was 20%.

In order to judge the quality of our methodology in comparison with previously published DSVS algorithms, we also implemented the clustered voltage scaling (CVS) algorithm developed by Usami et al. [7]. We performed power optimization using the established SSV methodology first, followed by CVS. Column six of Table 1 shows that the additional power reduction due to CVS was only 6% on average and only 11% in the best case. This is significantly less than the additional power reduction that we achieved using our DSVS methodology.

7.3   Optimization of Sequential Benchmarks

The results of the optimization of ten sequential MCNC benchmark circuits are summarized in Table 2. Again, we include the effect of SSV power optimization in comparison with the results of timing-driven synthesis (see column five). Power optimization subject to the strictest delay constraints using our DSVS methodology lead to an average power reduction of 10% compared to the results of state-of-the-art SSV optimization as can be seen from column six. In the best case the improvement was 24%. The total cell area increased by 10% on average (see column seven). This is due to the larger area of level-converting flip-flops (DFFLCs) and can be expected to be reduced by improving the design of these cells. Note that scaling the voltage level in the clock network as proposed by Usami et al. is not possible if the strictest delay constraints apply.

Table 1. Combinational benchmarks optimized subject to relaxed delay constraints. Col. 2/3: circuit complexity. Col. 4: power reduction due to SSV power optimization. Col. 5/6: additional power reduction due to the use of a second supply voltage

	Number of		Power red.	Power red.	
	cells	I/O	SSV	DSVS	CVS
apex6	742	135/99	-26%	-10%	-8%
c432	184	36/7	-26%	-3%	±0%
c880	416	60/26	-22%	-12%	-4%
c1908	275	33/25	-30%	-7%	-6%
c3540	1077	50/22	-25%	-5%	-1%
c5315	1413	178/123	-23%	-12%	-9%
c6288	3040	32/32	-20%	-6%	-2%
c7552	1462	207/108	-17%	-9%	-5%
i10	1879	257/224	-28%	-14%	-11%
i5	373	133/66	-28%	-5%	-5%
my_adder	169	33/17	-24%	-13%	-6%
pair	1509	173/137	-26%	-9%	-8%
rot	708	135/107	-28%	-13%	-10%
x3	718	135/99	-28%	-20%	-11%
x4	368	94/71	-35%	-12%	-9%
avg.	–		-26%	-10%	-6%

Table 2. Sequential benchmarks optimized subject to the strictest delay constraints. Col. 2-4: circuit complexity. Col. 5: power reduction due to SSV power optimization. Col. 6/7: power and area after DSVS in comparison with SSV power optimization

	Number of			Power red.	Power red.	Area
	cells	FFs	I/O	SSV	DSVS	
mm4a	218	12	7/4	-25%	-6%	+8%
mm9a	507	27	12/9	-28%	-11%	+6%
mm30a	1923	90	33/30	-23%	-6%	+11%
mult32a	772	32	33/1	-2%	-24%	+8%
s713	305	19	35/23	-32%	-7%	-1%
s820	349	5	18/19	-16%	-10%	-5%
s1196	637	18	14/14	-31%	-5%	+6%
s5378	1304	163	35/49	-24%	-10%	+21%
s9234.1	1093	135	36/39	-13%	-8%	+15%
s38417	12257	1465	28/106	-12%	-17%	+28%
avg.	–			-21%	-10%	+10%

8 Conclusions

We have shown that DSVS can be carried out exploiting cell-library-based GS algorithms, provided that a suitably modeled DSV standard cell library exists. This does not necessitate special DSVS algorithms or proprietary synthesis tools. The required DSV synthesis library file can easily be created from two conventional SSV libraries. The only costly task remaining is the design of the level-converting flip-flop cells, which, of course, is required by any DSV design methodology. Consequently, if state-of-the-art power-driven SSV logic synthesis is already in use, our methodology can be adopted with a modicum of effort. This is an important step towards broad acceptance of DSVS.

The results presented in this paper prove the general feasibility of our approach. For a comparison with related work, all relevant aspects, such as the selection of circuits, the delay constraints, the technology and the library, the supply voltages and the use of state-of-the-art power optimization techniques, had to be taken into account. For this reason, we implemented a previously published DSVS algorithm and applied it to benchmark circuits within our synthesis environment. The results revealed the greater efficiency of our approach.

Currently, we are working on the optimization of an embedded microcontroller system. For this application, we will extend the voltage scaling approach to the clock network in less critical subsystems.

References

1. B. Ackalloor and D. Gaitonde, "An overview of library characterization in semi custom design," *IEEE Custom Integrated Circuits Conf.*, pp. 305-312, 1998.
2. C. Chen, A. Srivastava, and M. Sarrafzadeh, "On gate level power optimization using dual-supply voltages," *IEEE Trans. on VLSI Systems*, vol. 9, pp. 616–629, Dec. 2001.
3. O. Coudert, "Gate sizing for constrained delay/power/area optimization," *IEEE Trans. on VLSI Systems*, vol. 5, pp. 465–472, Dec. 1997.
4. T. Mahnke et al., "Power optimization through dual supply voltage scaling using Power Compiler," *Proc. European Synopsys Users Group Meeting*, 2002.
5. LGSynth93, Benchmark set used in conjunction with 1993 MCNC International Workshop on Logic Synthesis, http://www.cbl.ncsu.edu, June 2002.
6. Synopsys Inc., "Design power and power compiler technology backgrounder," Jan. 1998, http://www.synopsys.com/products/power/power_bkg.html, July 2001.
7. K. Usami and M. Horowitz, "Clustered voltage scaling technique for low-power design," *Proc. Int. Symp. on Low-Power Design*, pp. 3–8, 1995.
8. K. Usami et al., "Automated low-power technique exploiting multiple supply voltages applied to a media processor," *IEEE Journal of Solid-State Circuits*, vol. 33, pp. 463–472, March 1998.
9. K. Usami et al., "Design methodology of ultra low-power MPEG4 codec core exploiting voltage scaling techniques," *Proc. 35th DAC*, pp. 483–488, 1998.
10. C. Yeh et al., "Gate-level design exploiting dual supply voltages for power-driven applications," *Proc. 36th DAC*, pp. 68–71, 1999.
11. C. Yeh et al., "Layout techniques supporting the use of dual supply voltages for cell-based designs," *Proc. 36th DAC*, pp. 62–67, 1999.

Transistor Level Synthesis Dedicated to Fast I.P. Prototyping

A. Landrault[1,2], L. Pellier[1,2], A. Richard[1,2], C. Jay[1], M. Robert[2], and D. Auvergne[2].

[1]Infineon Technologies. 06560 Sophia Antipolis, France
[2]LIRMM, UMR 9928 CNRS, 161, rue Ada F-34392 Montpellier cedex 5 France

Abstract. Standard cell libraries have been successfully used for years, however with the emergence of new technologies and the increasing complexity of designs, this concept becomes less and less attractive. Most of the time, cells are too generic and not well suited to the block being created. As a result the final design is not well optimized in terms of timing, power and area.
This paper describes a new approach based on transistor level layout synthesis for CMOS IP cores rapid prototyping (~100k transistors).

1 Introduction

Designers are today facing two challenges of conflicting objectives. In one hand they need to design for optimum performance and in the other hand they need to design in a minimum time frame ("time to market") while using the latest up-to-date available technology. Moreover, depending on the target application, today's SoC may need to integrate the same IP block optimised using different criteria's. For example, some applications will integrate a power optimised IP as some others will prefer a timing or area optimised version of the block (Platform based design [1], [2]).

Standard cell approaches try to resolve this problem by providing the designer with multiple libraries containing cells optimized under specific electrical constraints. As a consequence, IP's generated using such a library will, most of the time, respect one specific constraint but will be over-fitted regarding other constraints.

In this paper, we describe an investigation towards the use of transistor level layout synthesis to avoid employing a standard cell library and to allow designers to very quickly prototype "application-fitted" IP blocks for any given technology.

Main advantages of the proposed methodology are in technology migration and blocks prototyping. It gives easy possibility in generating complex blocks as soon as a new technology becomes available. It avoids the costs due to developing a standard cell library and mostly allows a quick evaluation of the performances of different flavours of a same IP.

Besides the reduction in development time, this approach must allow an increased control on the trade-off between the hot design performance parameters such as speed, area and power consumption.

The paper is organized as follows: In section 2, we describe the standard cell approach and the newly emerging flows. In section 3, we present our new flow based on transistor level layout synthesis and all the currently developed tools. In section 4, we deal with the implementation. Finally in section 5, we show the first results obtained with this new flow.

B. Hochet et al. (Eds.): PATMOS 2002, LNCS 2451, pp. 156–166, 2002.

2   State of the Art

2.1   The Standard Cell Approach

The overall goal of the "Standard Cell" flow [3] is to generate the layout of the design from a behavioral description. First, it proceeds to a logical optimization and maps the Boolean equation to the target library. Then after placing and routing all the pieces of layout, representing the individual pre-characterized cells, the final layout is generated. The use of standard cells presents a well-understood trade-off. On the positive side, a standard cell library normally provides a set of pre-packaged functionalities. Each cell is speed and power characterized. Furthermore, the existing Industrial EDA tools are relatively well adapted to this approach.

However some drawbacks appear with new UDSM technologies where interconnects are of significant importance. It is more difficult to predict the cell drive and to satisfy the constraints. In order to find the right drive of each cell we have to perform back-annotation, that may induce dramatic timing inconsistency. This can seriously damage the flow convergence.

Indeed, it is well known that the quality of designs highly depend on the library that is being used [4] and of the variety of functionalities and sizes for each primitives gates [5]. Most of the designers recognize that standard cell libraries are populated with "too generic" cells [6], and that significant improvements can be achieved just by resizing some existing cells (drive continuity) or by adding a few customized cells (design dependency) in the initial library.

2.2   Emerging Cell Based Flow

In order to avoid these problems new emerging approaches are proposed by industrial companies.

1) Timing consistency problem is resolved by using tools built around a unified data model [7]: Data requirements relative to the physical design can be reached quickly by the tools thanks to this unique data structure.

2) Problem concerning "drive continuity" is partly resolved by using the "Liquid Library"[8] concept: In this approach, a static base library is used for the synthesis of the design. After that, place and route, buffer sizing or in-place optimization is done. Then needed new design dependent cells, not present in the base library, are automatically generated by a "cell Factory" and are Engineering Change Order (ECO) placed (if possible) in the design.

However, these approaches still have one main drawback: They are based on an already existing target library populated by a few hundred of logic gates specified to satisfy very specific area-performance tradeoffs. Consequently, the resulting IP block structures are forced to stay within the predetermined library precluding an optimal solution based on the total available design space of the available transistor structures.

Also, the effectiveness of the standard flow methodology is highly dependent on the library development and process migration, which is costly.

3 A Step Further: The "Layout Synthesis"

At the opposite of the methodology presented above (section 2.), we propose a standard cell independent approach, working at transistor level. Theoretically, layout synthesis offers the possibility of overcoming deficiencies of the standard cell approach. The idea is that from a structural HDL description of a circuit, a synthesis tool would generate the optimized layout for the specified technology and timing constraints.

The principal motivations for this approach are first to avoid supporting the library and standard cell generation. This enables work at the transistor level. In our methodology, each transistor can be individually sized or re-sized continuously (no discrete size limitations) at every stage of the flow.

Although our approach is not based on a standard cell library, we have to supply the logical synthesis commercial tool with a functionality set. We call this set the "virtual library". To each function, we associate a virtual cell, that can be considered as a set of connected transistors (at symbolic level, no layout generated). Virtual cells appear as an interface between synthesis, place and route and the layout generation. The vast set of functionalities in the virtual library is expected to imply less transistor count in the design.

As we don't have predefined layout as in the standard cell approach, the additional step is to deal with layout generation. So, to predict the cell size information, we should have a layout style that is highly regular. This is why the topology used is a variant of the Linear Matrix style [2]. It allows precise estimation of the width, the height and the port position of each virtual cell. The placer and the router can use this information in the additional steps. Then, each virtual cell can be flipped or merged (at symbolic level), offering the possibility to realize an optimized row-based design.

The proposed flow, illustrated in the Fig.1, is composed of the following steps:

- An input of a behavioral description of the circuit function (VHDL or Verilog),
- A high-level logic synthesis and optimization step of the circuit,
- A logic mapping (realized thanks to a commercial tool) based on the virtual library; the important point here is that the virtual library is only a collection of logic functions, no netlist or layout exist for this library,
- A transistor level layout synthesis step, at which the layout of the circuit is generated, using a set of constraints, a structural description of the design and the virtual library as inputs. This includes placement, routing, timing analysis, optimization and physical layout generation as described below.

The "transistor level layout synthesis" tool starts from a structural description of the circuit and generates the layout. This approach is constituted of three major steps. The first one consists in creating the virtual cells from their functionality, resulting in an associated transistor network. The second phase achieves a first placement, routing and timing budgeting, based on physical estimations obtained from the targeted layout style. The third step produces, step-by-step, more optimized and refined results for placement, routing and performance. Information exchanged between these tools is made through the common data structure. The physical layout of each row is then automatically generated, with full respect of the technology rules. The main problems to be addressed in constructing such a flow are as follows. It is firstly necessary to provide a technology on a transistor netlist representing an optimal number of transistors for the required functionality. Another challenge is to predict and optimize the timing and power characteristics of the circuit before the final layout availability. The last challenge consists in generating a dense layout of the netlist, based on well-specified technology rules.

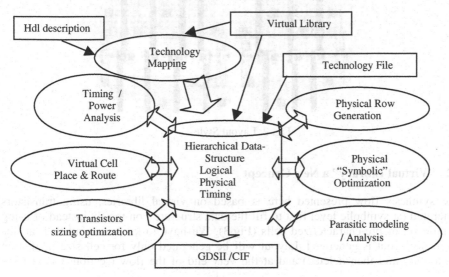

Fig. 1. Flow description.

The main goal of our approach is to deal with all these problems and to converge to the best solution. Let us now consider in more detail the different steps of this layout synthesis approach.

3.1 Layout Style

We have chosen to place all the transistors of the design in rows. The layout of the circuit is constructed by a sequence of NMOS and PMOS transistor rows.

The style chosen for the transistor-level generation is a variant of the popular "linear-matrix"[9]. It's principally characterized by the minimization of the space between the two N and P diffusion zones. Routing is realized over the diffusion zones (gain in terms of area) and avoids the use of metal 2 to save the porosity of each row and to facilitate the routing step.

This style has been chosen for its similarities with the "linear-matrix". It's perfectly adapted to software implementation (regular style with vertical placement of the poly grid)[10]. As described on Fig 2, all the ports are placed at the center between N and P diffusions. Sequential elements such as DFF are also fully generated using this layout style.

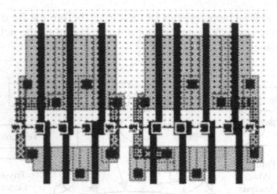

Fig. 2. Layout Style

3.2 "Virtual Library" a New Concept

The synthesis flow presented here is based on virtual libraries, using transistors generated (at symbolic level first to fill the data structure), on the fly instead of using a static set of pre-characterized cells (Fig.3). We have to keep in mind that, at this level, no layout is generated. Layout will be generated only for cell size prediction (temporary and then deleted) and at the very end of the flow (section 3.4) at row-level.

The set of cells constituting the "Virtual Library" is mostly composed of Complex Gates. As a consequence the number of cells available is virtually unlimited (in terms of logical functionality). The constraints are only fixed by the target technology and performance limitations (maximum number of serial transistors).

In addition, as the final layout will be automatically generated, the driving capability of all these "virtual cells" can be continuously adapted, at the contrary of the discrete drive possibility offered in usual standard cell design.

Logic Optimization / Technology Mapping

The library free technology-mapping problem concerns the ability to use complex gates instead of using pre-characterized libraries. The main problem with "Virtual Libraries" is that the number of cells available can be very high and we can assume that the logic synthesis commercial tools are not especially well adapted to this kind of library.

As an example, a complex gate (2, 2) will contain all the possible logical functions we can implement in CMOS technology with a maximum of 2 transistors P and 2 transistors N in series.

In Table 1 ([11]) we give an illustration of the number of different logic functions available in CMOS technology for a given limitation of N and P transistors in series.

For instance, for a maximum of four serial transistors (for each plan, N and P) we may dispose of a library with 3503 different functionalities. Moreover, we have a continuous sizing for each logical function that results in an almost infinite number of elements, considering the continuous transistor sizing facility.

Several approaches to realize technology mapping onto Virtual Library have already been proposed [11], [12], either as industrial products or in academic tools.

Table 1. Number of logic functions available for a given limitation of serial transistors.

		Number of serial PMOS transistors			
		2	3	4	5
Number of serial NMOS Transistors	2	7	18	42	90
	3	18	87	396	1677
	4	42	396	3503	28435
	5	90	1677	28435	425803

We have run a set of benchmarks on ISCAS85 circuits using the Technology Mapping Utilities available in Synopsys (Design Compiler). Several Virtual Libraries VL(x,y) were examined, where VL(x, y) represents the Virtual Library with all the possible CMOS gates with a limitation of x P-transistors and y N-transistors in series. The benchmark metric was the number of transistors used in the final generated netlist.

Table 2. Logical synthesis efficiency in term of number of transistors.

	Standard Cell Library	Lib VL(3,3)	Lib VL(4,4)
c1355	1632	1434	1424
c1908	1470	1370	1350
c2670	1942	1884	1772
c3540	3578	3050	2860
c499	1506	1420	1408
c5315	4838	4546	4384
c6288	9354	7222	7222
c7552	5916	5946	5826
c880	1156	1088	1066
Sparc IU	42400	38806	37834

Table 2 clearly shows that by increasing the available number of "logical functions" in the library gives to the Synopsys Design Compiler greater flexibility to perform the mapping. As a direct result, the number of transistors needed to realize the same circuit is decreased (Table 2) saving substantial power dissipation. Transistors count is optimized by an average 10% improvement.

Once the technology mapping on the Virtual Library cells has been completed, each logical function is transformed into the corresponding transistor network that is associated to a virtual cell. The network creation consists of generating a BDD for each virtual cell used in the design. Then an optimization of this BDD allows the generation of a transistor netlist for the circuit.

Timing & Power Modeling, Transistor Sizing

Once the transistor netlist for the circuit has been generated the essential steps of transistor sizing and timing analysis must be performed. In this section, we will briefly discuss the problem of transistor sizing in association with the inherent problem concerning the transistors structure modeling.

The efficiency of the layout synthesis flow presented in Fig.3, and the ability to obtain an optimal circuit is conditioned by the following assumptions:

- The initial phase needs to produce a layout as close to optimal as possible.
- The modifications performed during the incremental processes need to be efficient and constructive, while impacting the minimal number of transistors, to allow a quick convergence.

Consequently, two new problems need to be addressed. Firstly, as there is no characterization involved in the Layout Synthesis flow, an accurate predictive model for speed and power is required. A lot of work has been devoted to the study of CMOS models. Generally, it appears that a complete treatment of all the effects associated to the CMOS submicronic technology results in unmanageable expressions.

In [13], [14], [15], [16] it has been shown that it is possible to develop a simple analytical model of the performance of CMOS gates with a good accuracy compared to electrical simulations. An efficient implementation of this model has been integrated.

After the layout generation step, a more accurate prediction will be obtained using the parasitic data extracted from the layout (based on more accurate extracted capacitances).

As soon as we have an efficient predictive model for the transistor network, we can address the problem of transistor sizing:

It is shown in [17] that generally the delay constraint for a specific circuit can be satisfied using an optimized and regular sizing for all the transistors. This kind of approach looks very attractive as the transistor-sizing problem becomes very simple. However, regular sizing (blind sizing) involves all the transistors and results in an important increase of the circuit power dissipation. In addition, such an approach does not exploit one of the particular attractions of Layout Synthesis that allows a specific tuning for each transistor.

Meanwhile, [17] and [18] shows that for most of the circuits, a local re-sizing of only a few selected transistors in the circuit can significantly reduce the power consumption. The process of "specific transistor sizing" used here is based on a circuit path classification realized through an incremental technique. This approach reduces the number of paths to be investigated and consequently, allows a faster evaluation of different sizing options in order to select the optimal set of transistors to be resized in the circuit (even for large circuits).

3.3 Virtual Cell Place and Route

From the "Virtual Cell" structural netlist representation, Placement & Routing steps can then be performed. These operations use the predictive analytical models presented in the previous paragraph and is composed of 3 steps described below.

Partitioning & Routing

The placement is based on an iterative partitioning and routing. At each step we perform a min-cut based quadrisectioning on each partitions, then a global routing between the new created partitions is done. Additionally, during each of these steps, performance and routability is continuously improved by moving critical virtual cells between the partitions.

As a result, each partition is a set of virtual cells that respect the routing resources and the performance constraints [19]

Final Placement: Transistor Row Generation

Using the Euler trail solution [20], a virtual cell is then transformed as a set of ranged N and PMOS transistors. This constitutes the symbolic view of each virtual cell. As soon as the virtual cell detailed placement in each partition is done, the transistors rows are created. Then symbolic optimization can be performed on these rows.

1) For each Row of "Virtual Cell" a "Symbolic Row" is created
2) Then optimization algorithms are used to perform symbolic optimization, such as cell "Flip" allow "Diffusion Sharing"
3) Finally, a symbolic Maze Routing is performed to pre-route as much as possible "inner Row" interconnections (using only metal1 level)

Virtual Channel Wrouting

All the remaining connections that have not been routed inside each row and all the connections between the rows are finally established using detail routing algorithms. The routing is based on a virtual channel defined over the cells (using metal 2 to metal 5). The detail routing phase is divided in two steps. A first routing is done based on constraint graphs. Secondly the remaining nets are routed thank to a multi-layer maze router.

3.4 Layout Generation

The layout generation of each row requires as an input the symbolic view of each row (which can be seen as a set of diffusion, contact and connected transistor zones) and the technology rules for the targeted process.

From the symbolic "linear-matrix" style, a constraint graph is generated, that contains a node for every vertical slice (for X layout enhancement and horizontal slice for Y-axis) and an edge between two slices every time a constraint is required between 2 nodes [21]. Indeed, we proceed to a step that consists in forcing each node to have the smallest coordinate regarding the value of the constraints of the incoming edges. This results in a "compacted" layout inherent of the "linear-matrix" style.

The main advantage offered by this procedure is the technology independence: all the design rules are input parameters of the generator. These rules can be easily described in an user file using a rules numbering convention (fast technology migration).

4 The Implementation

As said previously, in order to be able to operate the optimization convergence most efficiently, all the developments are based on a unique data-structure.

All the development is done in C++, with plug and play facilities to ease the change of the different "engines" under development.

5 Results and Validation

A first validation has been done in a 0.18μm CMOS technology. The comparison has been done with respect to the In-House standard flow. We make a first comparison on transistor density to evaluate our tool performance. In Table 3 we compare the average transistor density obtained with a standard cell approach to that resulting from this transistor level methodology.

Table 3. Density (transistors/mm^2) Comparison between In-House Lib and Virtual LibVL (4, 4)

Circuit / Average density	Standard Cell approach	Layout Synthesis
>From 10 K to 100K transistors blocks	270000	230000

As shown in table 3, the values obtained with both methods are comparable. Transistor density comparison shows a small advantage to the standard cell implementation, but, as we need a smaller number of transistors to implement the same design (see Table 2), we can see in table 4 that the final silicon area is equivalent for the 2 approaches:

Table 4. Area (μm^2) Comparison between In-House Lib and VL(4, 4)

Circuit / Area	Standard Cell approach	Layout Synthesis
c1355	6 040	5 930
c1908	5 440	5 620
c2670	7 470	7 380
c3540	13 300	12 400
c499	5 790	5 630
c5315	17 900	19 100
c6288	34 600	31 400
c7552	22 800	24 300
c880	4 300	4 300
Sparc IU	163 000	164 000

This result is quite encouraging for these initial attempts, mostly if we consider the facility obtained in generating and migrating macro-blocks in very short time intervals (around one hour for a 100K transistor block). We hope to see significant improvements in terms of power and speed using this technique. We expect to soon have results to confirm this improvement.

On the example given in Fig.3, a benchmark circuit synthesized with our transistor-level layout generator is presented.

6 Conclusion

In this paper we have presented an original alternative to the classical standard cell based layout synthesis. In this "virtual cell library" methodology the physical generation is obtained at the transistor level, integrating the different steps of physical

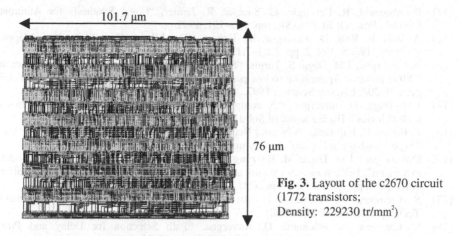

101.7 μm

76 μm

Fig. 3. Layout of the c2670 circuit (1772 transistors; Density: 229230 tr/mm^2)

layout generation, performance estimation and optimization. This may give great facilities in quickly evaluating and prototyping different flavors of IP Blocks by using the latest available technology.

We show that a first prototype of the "IP Prototyper", able to handle simple blocks (order of complexity ~10K to 100k transistors) is available. Meanwhile, much improved results are expected when every tool, currently under development, will be available. This is especially true regarding timing and power performances (currently under evaluation). More results will be presented during the conference.

References

[1] A. Sangiovanni-Vincentelli, "Platform-Based Design: A Path to Efficient Design Re-Use", First International Symposium on Quality of Electronic Design, 20–22 March, 2000, San Jose, California.

[2] M. Reinhardt, "Implementing a Migration-Based IP-Reuse Strategy", Electronic Engineering Times, May 1999.

[3] D. McMillen, M. Butts, R. Composano, D. Hill, T.W. Williams, "An Industrial View of Electronic Design Automation", IEEE Transactions on Computer, Vol. 19, No 12, December 2000, pp. 1428–1448.

[4] K. Keutzer, K. Scott, "Improving Cell Library for synthesis", Proc. Of the International Workshop on Logic Synthesis, 1993.

[5] K. Keutzer, K. Kolwicz, M. Lega, "Impact of Library Size on the Quality of Automated Synthesis", ICCAD 1987, pp. 120–123.

[6] P. de Dood, "Approach makes most of synthesis, place and route - Liquid Cell ease the flow", EETimes, September 10, 2001, Issue: 1183.

[7] O. Coudert, "Physical design closure" Monterey design system", Design Automation Conference 2000.

[8] P. de Dood, "Putting Automated Libraries into the Flow", EETimes, May 2001.

[9] A.D. Lopez, H.S. Law, "A Dense Gate-Matrix Layout for MOS VLSI", IEEE Transactions on Electron Devices, Vol. ED-27, No. 8, August 1980, pp. 1671–1675.

[10] F. Moraes, R. Reis, L. Torres, M. Robert, D. Auvergne, "Pre-Layout Performance Prediction For Automatic Macro-Cell Synthesis", IEEE-ISCAS'96, Atlanta (USA), Mai 1996, pp. 814–817.

[11] P. Abouzeid, R. Leveugle, G. Saucier, R. Jamier, "Logic Synthesis for Automatic Layout", Proc. Of EUROASIC, pp. 146–151, 1992.

[12] A. Reis, R. Reis, D. Auvergne, M. Robert, "The Library Free Technology Mapping Problem", IWLS, Vol. 2, pp. 7.1.1–7.1.5, 1997.

[13] A. Auvergne, J.M. Daga, S. Turgis, "Power and delay macro-modeling for submicronic CMOS process: Application to low power design", Microelectronic engineering, Vol.39, pp. 179–208, Elsevier Science, 1997.

[14] J.M. Daga, D. Auvergne, "A comprehensive delay macro-modeling for submicron CMOS logics", IEEE Journal of Solid State Circuits, Dec 1998.

[15] F. Brglez, H. Fujiwara, "A Neutral Netlist of 10 Combinatorial Benchmark Circuits and a Target translator in Fortran", Int. Symposium on Circuits and Systems, June 1985.

[16] D.Auvergne, J.M. Daga, M. Rezzoug, "Signal transition time effect on CMOS delay evaluation", IEEE trans. on Circuits and Systems: Fundamental theory and applications, vol.47, n°9, pp.1362–1369, Sept.2000.

[17] S. Cremoux, N. Azemard, D. Auvergne, "Path Resizing Based on Incremental Technique", Proc. of ISCAS98.

[18] S. Cremoux, N. Azemard, D. Auvergne, "Path Selection for Delay and Power Performance Optimization", Proc. of SAME98, pp. 48–53, 1998.

[19] C. M. Fiduccia, R. M. Mattheyses, "A linear-time heuristics for improving network partitions.", Proceedings of the 19th Design Automation Conference, pages 175–181, 1982.

[20] M.A. Riepe, K.A. Sakallah, "Transistor level micro-placement and routing for two dimensional digital VLSI cell synthesis", University of Michigan, Ann Arbor ISPD '99 Monterey CA USA.

[21] T.Uehara, W.van Cleemput, "Optimal Layout of CMOS Functional Arrays", IEEE Transactions on Computers, Vol. C-30, No. 5, May 1981, pp. 305–312.

Robust SAT-Based Search Algorithm
for Leakage Power Reduction

Fadi A. Aloul[1], Soha Hassoun[2], Karem A. Sakallah[1], and David Blaauw[1]

[1]Department of Electrical Engineering and Computer Science
University of Michigan - Ann Arbor
{faloul, karem, blaauw}@eecs.umich.edu

[2] Department of Electrical Engineering and Computer Science
Tufts University
soha@eecs.tufts.edu

Abstract. Leakage current promises to be a major contributor to power dissipation in future technologies. Bounding the maximum and minimum leakage current poses an important problem. Determining the maximum leakage ensures that the chip meets power dissipation constraints. Applying an input pattern that minimizes leakage allows extending battery life when the circuit is in stand-by mode. Finding such vectors can be expressed as a satisfiability problem. We apply in this paper an incremental SAT solver, PBS [1], to find the minimum or maximum leakage current. The solver is called as a post-process to a random-vector-generation approach. Our results indicate that using a such a generic SAT solver can improve on previously proposed random approaches [7].

1 Introduction

One of the challenges in designing integrated circuits is limiting energy and power dissipation. The concerns are many, including packaging and cooling costs, battery life in portable systems, and power supply grid design. As process geometries scale to achieve a 30% gate delay reduction per technology generation, the typical scaling of the supply voltage (Vdd) by 30% promises to reduce the power by 50% [4]. Such supply voltage scaling will also require scaling the threshold voltage (Vt) for MOS devices to sustain the gate delay reductions. Decreasing Vt however results in an exponential increase in the subthreshold leakage current. Thus, although the overall power is decreasing, the power dissipation due to the leakage component is increasing. It is expected that within the next 2 process generations leakage power dissipation will contribute as much as 50% of the total power dissipation for high performance designs [21]. Particularly for devices that spend a significant percentage of their operation in standby mode, such as mobile device, is leakage current a critical concern. For such devices, it is not uncommon for standby leakage power to be the dominant factor in the total battery life time.

Several circuit techniques have been proposed in recent years to minimize leakage currents. A common approach is to use a dual-Vt process where transistors are assigned either a high or a low threshold voltage, the high Vt devices having typically 30-50% more delay but as much as 30x less leakage than the low Vt device. This approach requires one additional process step and therefore entails some additional man-

B. Hochet et al. (Eds.): PATMOS 2002, LNCS 2451, pp. 167-177, 2002.

ufacturing cost. One dual-   implementation uses a high   NMOS or PMOS transistor to shut off the supply to low Vt logic [9]. This approach results in very good reduction in standby leakage current, but requires significant layout area overhead for the large power supply shutoff transistors. It also suffers from power supply integrity issues. In multi-threshold CMOS, speed critical gates are assigned a low Vt and non-speed critical gates are assigned a high Vt, thereby reducing the leakage of these gates. This approach has the advantage that it requires no additional area overhead. However, multi-threshold CMOS typically has a significant performance penalty since the majority of the gates must be assigned high Vt in order to obtain a significant savings in the standby leakage current. Recently, multi-threshold CMOS techniques that combine device sizing and Vt assignment have been proposed [17]. Another recent leakage minimization approach is based on the observation that a stack of two OFF devices has a significantly reduced leakage compared to a single OFF device [14]. A pull-down transistor is therefore replaced by two series connect transistors. However, this technique has a significant area overhead.

the key observations of gate leakage is that it strongly depends on the input state of a gate [7]. Based on this observ   and Najm modified the circuit's latches to force their outputs high or low during sleep mode without losing the state of the latch. By assigning a low leakage state to output nodes of the latches, significantly leakage current can be saved during standby mode. Forcing the output to a high or low state requires a   transistor in parallel with a shut to Vdd or Gnd and results in a slight increase in the latch delay. For flip-flop based designs, an alternate methods was recently proposed that does not increase the delay of the flip-flop noticeably [21]

In order to minimize the leakage in this approach, the circuit state with the minimum leakage state must be determined. On the other hand, the circuit state with the maximum leakage is important for designers to ensure that the circuit meets the standby power constraints which in turn impacts battery-life. Several methods for finding the minimum or maximum leakage state have been proposed. In [7], a the minimal leakage state was determined with random vectors whose number was selected to achieve a specific statistical confidence and tolerance. Bobba and Ha
estimate maximum leakage power [3]. A constraint graph is built as follows: for each gate, 2^k vertices a   of each vertex represents the leakage power when the inputs are in a particular state. Edges between the vertices represent constraints that only one of the 2^k input assignments is poss   e. Other edges are added that enforce the logic functionality between gates. The proposed heuristic uses a greedy linear algorithm to find a *maximum weight*   *ent set* to maximize the weights of the vertices under the constraint that no edges between any pair of selected vertices are selected.

Johnson et al. [8] experiment with greedy heuristics and an exact branch and bound search to find maximum and minimum leakage bounds. They propose a leakage observatory metric that reflects how a particular circuit input affects the state and thus magnitude of leakage current for all the circuit components. This metric can be calculated once before assigning any input variables, or it can be re-calculated repeatedly after each partial assignment. Their experiments show that the dynamic re-calculation of observatory metric after partial assignments improved the quality of the results found by the greedy heuristics to the point that it matched those found by the exact branch and bound algorithm.

Recently, SAT has been shown to be very successful in various applications, such as formal verification [2], FPGA routing [13], and timing analysis [16]. In this paper we propose using a SAT solver, PBS [1], to find the minimum or maximum leakage current. Finding the input vector that causes such a current corresponds to solving a SAT instance [8]. PBS was chosen because: (a) it allows incremental exploration of the solution space, (b) it handles both CNF and pseudo-Boolean constraints that can express gate leakage restrictions, and (c) it implements the latest enhancements in SAT. The obtained leakage vectors can be used by designers to determine if the circuit meets the required leakage specifications or to assign the circuit to a low leakage state during standby mode.

We begin the paper with an overview of Boolean Satisfiability. We then describe how to model the leakage problem to solve it via PBS. We conclude with an example and experimental results.

2 Boolean Satisfiability

The satisfiability problem involves finding an assignment to a set of binary variables that satisfies a given set of constraints. In general, these constraints are expressed in *conjunctive normal form* (CNF). A CNF formula φ on n binary variables $x_1, ..., x_n$ consists of the conjunction (AND) of m clauses $\omega_1, ..., \omega_m$ each of which consists of the disjunction (OR) of k literals. A literal l is an occurrence of a Boolean variable or its complement. We will refer to a CNF formula as a clause database (DB).

Most current SAT solvers [1, 10, 12, 19, 20] are based on the original Davis-Putnam backtrack search algorithm [5]. The algorithm performs a search process that traverses the space of 2^n variable assignments until a satisfying assignment is found (the formula is satisfiable), or all combinations have been exhausted (the formula is unsatisfiable). Originally, all variables are unassigned. The algorithm begins by choosing a decision assignment to an unassigned variable. A decision tree is maintained to keep track of variable assignments. After each decision, the algorithm determines the implications of the assignment on other variables. This is obtained by forcing the assignment of the variable representing an unassigned literal in an unresolved clause, whose all other literals are assigned to 0, to satisfy the clause. This is referred to as the *unit clause* rule. If no conflict is detected, the algorithm makes a new decision on a new unassigned variable. Otherwise, the backtracking process unassigns one or more recently assigned variables and the search continues in another area of the search space.

As an example, a CNF instance $f = (a + b)(\bar{b} + c)$ consists of 3 variables, 2 clauses, and 4 literals. The assignment $\{a = 0, b = 1, c = 0\}$ leads to a conflict, whereas the assignment $\{a = 0, b = 1, c = 1\}$ satisfies f.

Several powerful methods have been proposed to expedite the backtrack search algorithm. One of the best methods is known as the conflict analysis procedure [10] and has been implemented in almost all SAT solvers, such as GRASP [10], Chaff [12], PBS [1], SATIRE [19], and SATO [20]. Whenever a conflict is detected, the procedure identifies the causes of the conflict and augments the clause DB with additional clauses, known as conflict-induced clauses, to avoid regenerating the same conflict in future parts of the search process. In essence, the procedure performs a form of learning from the encountered conflicts. Significant speedups have been achieved with the addition of conflict-induced clauses, as they tend to effectively prune the search space.

Intelligent decision heuristics and random restarts [6], also played an important role in enhancing the SAT solvers performance. Chaff [12] proposed an effective decision heuristic, known as VSIDS, and implemented several other enhancements, including random restarts, which lead to dramatic performance gains on many CNF instances.

Recently, a new SAT solver, known as SATIRE [19], introduced two new enhancements, namely incremental satisfiability and handling non-CNF constraints. The former enhancement allows sets of related problems to be solved incrementally without the need to solve each problem separately. The latter enhancement, involves expressing complex constraints whose encoding in CNF is impractical. In particular, it handles Pseudo-Boolean (PB) expressions which are expressions of the form $\sum c_i l_i \leq n$, where c_i and n are constant real values and l are literals of Boolean decision variables, i.e. x_i or $\overline{x}_i$. This attracted our attention, since the ability to reason over constraints which include real-valued components is a key feature when looking at the application of SAT solvers to leakage detection.

In this paper, we will use PBS [1], a new SAT solver that handles both CNF and PB constraints. It combines the state-of-the-art search techniques implemented in Chaff [12] and SATIRE [19]. Unlike previously proposed stochastic local search solvers [18], PBS is complete (i.e. can prove both satisfiability and unsatisfiability) and has been shown to achieve several order-of-magnitude speedups when compared to SATIRE [19].

3   Bounding Leakage Using SAT Algorithms

In this section, we present our SAT-based approach for identifying the possible power leakage in a circuit. A SAT problem is created for each circuit with an objective function to minimize or maximize the possible leakage. Each problem consists of two groups of constraints: (1) a large set of CNF clauses modeling the circuit's logical behavior (2) objective constraint which specifies the amount of desired leakage.

3.1   Representing Circuits in CNF

Circuits are easily represented as a CNF formula by conjuncting the CNF formulas for each gate output. A gate output can be expressed using a set of clauses which specify the valid input-output combinations for the given gate. Hence, a CNF formula φ for a circuit is defined as the union of set of clauses φ_x for each gate with output x :

$$\varphi = \bigcup_{x \in Q} \varphi_x \tag{1}$$

where Q denotes all gate outputs and primary inputs in the circuit. Figure 1, shows

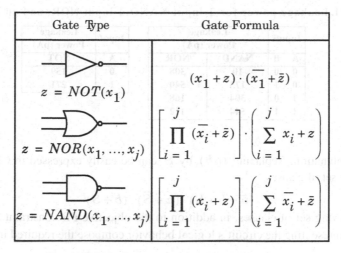

Gate Type	Gate Formula
$z = NOT(x_1)$	$(x_1 + z) \cdot (\overline{x_1} + \bar{z})$
$z = NOR(x_1, ..., x_j)$	$\left[\displaystyle\prod_{i=1}^{j} (\overline{x_i} + \bar{z}) \right] \cdot \left(\displaystyle\sum_{i=1}^{j} x_i + z \right)$
$z = NAND(x_1, ..., x_j)$	$\left[\displaystyle\prod_{i=1}^{j} (x_i + z) \right] \cdot \left(\displaystyle\sum_{i=1}^{j} \overline{x_i} + \bar{z} \right)$

Fig. 1. CNF formulas representing simple gates.

generalized CNF formulas for the NAND, NOR, and INVERTER gates.

3.2 Representing Leakage Constraints

After representing the circuit's logical behavior, we need to define an objective for the desired leakage. The objective function of the leakage estimate problem will be represented in PB, where Boolean decisions on the primary inputs of the circuit will determine a possible leakage value for the circuit. In other words, this can be viewed as a constraint representing the predicate, "There exists a leakage $< k$ ", where k is a real value. Such constraints are typically hard to express in CNF.

Given a *gate* with n inputs and leakage values c associated with each of the 2^n possible input combinations, a PB is generated as follows:

$$\sum_{i=1}^{2^n} c_i q_i \leq k \tag{2}$$

where q and k represent the input combination and maximum possible desired leakage, respectively. Consequently, a single PB constraint can be expressed for all gates in the circuit.

As an example, consider a 2-input NAND gate. Using the leakage power values listed in Table 1, the objective function will be expressed as:

$$10(\bar{a}\bar{b}) + 173(\bar{a}b) + 304(a\bar{b}) + 544(ab) \leq k \tag{3}$$

However, PBS's input format allows a single literal to be associated with each coefficient entry, i.e. the term (ab) needs to be replaced by a single literal term. Hence, for a 2-input gate, four new variables are declared, each of which represents a possible

Table 1. Leakage power for 2-input NAND, 2-input NOR, and an Inverter.

Inputs		Leakage Power (pA)		Input	Leakage Power (pA)
A	B	NAND	NOR	A	NOT
0	0	10	308	0	159
0	1	173	540	1	271
1	0	304	168		
1	1	544	112		

input combination. Replacing (ab) by S can be easily expressed in CNF using the following set of clauses:

$$(\bar{a} + \bar{b} + S) \cdot (a + \bar{S}) \cdot (b + \bar{S}) \tag{4}$$

The following set of clauses, in addition to the objective PB constraint and the CNF clauses representing the circuit's logical behavior compose the required input for PBS.

3.3 Identifying Minimum/Maximum Leakage

Initially, the objective leakage is unknown for a circuit, unless it is specified by a user. Therefore, a valid primary input assignment is derived by running the SAT solver through the CNF clauses representing the circuit's logical behavior only. Based on the leakage values in Table 1, the total leakage h for the circuit is computed using the valid primary input assignment and consequent assignments of the internal gate inputs. The total leakage is decremented by 1 to identify the minimum possible leakage in the circuit. This value denotes the new objective value. A PB constraint expressing the new objective leakage $h - 1$ is added to the problem. The problem is tested using the SAT solver. If the problem is satisfiable, the new total leakage k is computed using the new input assignment and a new PB constraint is added with the new leakage objective $k - 1$. Otherwise, the problem is unsatisfiable and the circuit is proved to have a minimum possible leakage of h. Equivalently, in order to identify the maximum possible leakage, a similar approach is used, but the objective leakage limit is incremented by 1.

The performance of the proposed approach is further improved due to the incremental satisfiability feature in PBS. Ideally, one would need to solve the problem independently from scratch every time the objective leakage is modified. However, the incremental approach provides the ability to modify previously solved problems by the addition or removal of constraints, thereby reusing decision sequences and retaining information learned by the solver from run to run.

We should note that other ways can be used to update the new leakage objective, such as using *binary search*. However, when compared with the monotonic tightening of the leakage constraint, the latter was faster, since most of the learning (i.e. conflict-induced clauses) is relinquished with binary search.

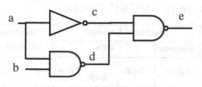

Circuit Consistency Function

$$\left[\begin{array}{c} (\bar{a} + \bar{c})(a + c) \wedge \\ (a + d)(b + d)(\bar{a} + \bar{b} + \bar{d}) \wedge \\ (c + e)(d + e)(\bar{c} + \bar{d} + \bar{e}) \end{array} \right]$$

PIs		Wires		Total Circuit
a	b	c	d	Leakage
0	0	1	1	713
0	1	1	1	876
1	0	0	1	748
1	1	0	0	825

Leakage Objective Function

$$\left[\begin{array}{c} 159\bar{a} + 271a + \\ 10A + 173B + 304C + 544D + \\ 10E + 173F + 304G + 544H \\ \leq Leakage \end{array} \right]$$

Equivalence Conditions

$$\left[\begin{array}{c} (a + b + A)(\bar{a} + \bar{A})(\bar{b} + \bar{A}) \wedge (a + \bar{b} + B)(\bar{a} + \bar{B})(b + \bar{B}) \wedge \\ (\bar{a} + b + C)(a + \bar{C})(\bar{b} + \bar{C}) \wedge (\bar{a} + \bar{b} + D)(a + \bar{D})(b + \bar{D}) \wedge \\ (c + d + E)(\bar{c} + \bar{E})(\bar{d} + \bar{E}) \wedge (c + \bar{d} + F)(\bar{c} + \bar{F})(d + \bar{F}) \wedge \\ (\bar{c} + d + G)(c + \bar{G})(\bar{d} + \bar{G}) \wedge (\bar{c} + \bar{d} + H)(c + \bar{H})(d + \bar{H}) \end{array} \right]$$

Fig. 2. Illustrative example for identifying the minimum leakage.

3.4 Example

Let's consider the circuit shown in Figure 2. The circuit consists of two 2-input NAND gates and an Inverter. There are four possible input combinations to the circuit. Clearly, the minimum leakage is obtained by setting $\{a = 0, b = 0\}$. In order to obtain the minimum leakage using PBS, the CNF clauses representing the circuit's consistency function are generated. The objective function is initialized to the maximum possible leakage of all gates, which corresponds to $544 + 544 + 271 = 1359$ and all multi-literal terms are replaced by new variables. Suppose, PBS identifies the satisfiable assignment $\{a = 1, b = 0, c = 0, d = 1, e = 1\}$., which corresponds to the total leakage of 748. A new PB constraint, with an objective leakage of 747, is added to further reduce the leakage. A satisfiable assignment is identified with values $\{a = 0, b = 0, c = 1, d = 1, e = 0\}$. This corresponds to a leakage of 713. Again, a new PB constraint is added with an objective leakage of 712. PBS fails to identify a satisfiable solution. Hence, the minimum leakage is identified at 713.

4 Experimental Results

In order to evaluate the performance of the approach, we measured the maximum and minimum leakage for the MCNC benchmarks [11]. Each benchmark was sensitized us-

Table 2. Experimental Results on the MCNC Circuits.

Circuit	# Gates	# PI	Minimum Leakage				Maximum Leakage				Max/Min Difference	
			Random	PBS			Random	PBS				
			Leak	Leak	% Imp	Time	Leak	Leak	% Imp	Time	%	Actual
x2	73	10	11937	11937	0.00	0.15	22718	22718	0.00	0.14	47.46	10781
cm152a	24	11	5536	5536	0.00	0.09	7140	7140	0.00	0.04	22.46	1604
cm151a	39	12	5679	5679	0.00	0.09	11070	11070	0.00	0.11	48.70	5391
cm162a	54	14	8172	8172	0.00	0.18	13611	13611	0.00	0.1	39.96	5439
cu	78	14	11859	11859	0.00	0.3	25293	25293	0.00	0.08	53.11	13434
cm163a	51	16	6820	6820	0.00	0.46	13276	13490	1.61	0.01	49.44	6670
cmb	62	16	8480	8480	0.00	0.58	16021	16484	2.89	0.65	48.56	8004
pm1	67	16	9444	9444	0.00	0.23	19606	19606	0.00	0.33	51.83	10162
parity	75	16	12915	12653	2.03	0.04	14608	15431	5.33	0.01	18.00	2778
tcon	41	17	7112	7112	0.00	0.04	12331	12566	1.91	0.01	43.40	5454
pcle	71	19	13511	13338	1.28	0.01	18117	18381	1.46	3.67	27.44	5043
sct	143	19	20368	19743	3.07	0.76	48958	49262	0.62	1.4	59.92	29519
cc	79	21	11820	11186	5.36	4.4	24512	24987	1.90	0.2	55.23	13801
cm150a	79	21	12855	12373	3.75	26	22687	23527	3.70	10.4	47.41	11154
mux	106	21	20135	19873	1.30	4.57	31586	31884	0.93	5.4	37.67	12011
cordic	124	23	24159	23363	3.29	8.5	30087	31687	5.32	29	26.27	8324
lal	179	26	24852	23273	6.35	3.1	57324	59213	3.30	14	60.70	35940
pcler8	104	27	15698	15303	2.52	3.39	22646	23445	3.53	682	34.73	8142
frg1	143	28	27162	24579	9.51	49.9	40383	41657	3.15	4401	41.00	17078
comp	178	32	31430	28848	8.22	526	41012	46791	14.1	2311	38.35	17943
b9	147	41	24190	21202	12.4	222	42456	46028*	8.41	5000	53.94	24826
i3	132	132	30572	28429	7.01	593	30724	36992	20.4	478	23.15	8563
ttt2	303	24	59429	57302	3.58	83.5	96223	98775	2.65	98.5	41.99	41473
C1355	552	41	143865	126413*	12.1	5000	172857	177057*	2.43	5000	28.60	50644
i6	764	138	147182	139552*	5.18	5000	240500	249871*	3.90	5000	44.15	110319
alu4	878	14	195600	195600	0.00	40	222422	222422	0.00	55	12.06	26822
x3	1174	135	246910	231567*	6.21	5000	339315	348420*	2.68	5000	33.54	116853
vda	1417	17	442553	442497	0.01	2	483430	483531	0.02	65	8.49	41034
C6288	2400	32	739329	720514*	2.54	5000	801441	816729*	1.91	5000	11.78	96215

ing *sis* [15] to a circuit consisting of only 2-input NAND, NOR and Inverter gates*. All experiments were conducted on a AMD 1.2Ghz, equipped with 512 MBytes of RAM, and running Linux. We used PBS [1] as our SAT solver with the latest enhancements, such as conflict diagnosis and random restarts, enabled. PBS implements two decision heuristics, *FIXED* and *VSIDS* [12]. In practise, both heuristics are competitive depending on the problem's structure. Therefore, we tested each circuit using both heuristics. Our table of results report the best of both runs. The runtime limit for all experiments was set to 5000 seconds.

In order to generate an initial objective goal, we generated 10K random PI vectors and identified the best leakage value among all vectors. The random approach eliminates significant part of the search space and assists in speeding up PBS.

Table 2 lists the maximum and minimum leakage results for the MCNC benchmarks. (Random) represents the best leakage value obtained using the random vector generation approach [7]. (PBS-Leak) represents the final leakage value obtained using PBS. The PBS runtime (in seconds) and the %-improvement (%-Imp) on top of the random leakage value are also reported. The random approach runtime did not exceed a few minutes in most cases. A "*" in the (PBS) leakage column indicates that PBS didn't

---

*   Table 1 was used for the gate leakage values.

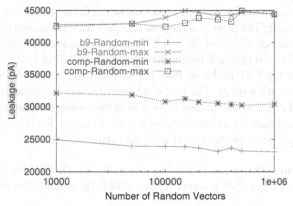

Fig. 3. Minimum and maximum leakage values obtained using random vectors.

complete the search process because it exceeded either the allowed runtime or memory limits. In such a case, the best leakage value measured is shown. All leakage values are reported in units of pA. The table also shows the percentage and actual difference between the minimum and maximum leakage found by PBS. Several observations are in order:

- in almost all reported cases, PBS was able to identify the best possible leakage value.
- in several cases the random approach was unable of identifying the optimal leakage value, especially for large circuits.
- the proposed approach was able to improve on top of the random approach by a factor of 20% in some cases. For example, PBS was able to improve on the value obtained by the random approach by a factor of 12% for the *b9* circuit. Detecting such a difference in leakage power can be very useful.
- the difference between the minimum and maximum possible leakage can be large, especially for smaller circuits. Identifying techniques to reduce the circuit leakage, as the proposed technique, is crucial.
- PBS is fast for small circuits but as the circuit size grows PBS gets slower. Perhaps, larger circuits can be partitioned to speed up the search process.
- PBS can be viewed as a checker for the random approach.

Figure 3 shows the minimum/maximum leakage values obtained using the random vector generation approach for the *b9* and *comp* circuits. In both cases, the random approach, after generating 1M random vectors, was unable of identifying any maximal leakage values greater than 45K pA, whereas PBS was successful in measuring a maximal leakage value of 46K pA for both circuits. PBS was clearly able to improve on the minimum leakage bound as well.

5   Conclusions

In this paper, we have presented a new approach for determining the minimum or maximum leakage state for CMOS combinational circuits. The maximum leakage state can be used by designers to verify that the circuit meets the required leakage constraints. On the other hand, the minimum leakage state can be useful to reduce the leakage current during standby mode. The proposed method searches for the extreme circuit state using a short random search followed by a SAT-based formulation to tighten the leakage bound. The SAT-based problem formulation used both CNF and pseudo-Boolean constraints. To solve the SAT problem, PBS [1] was used since it allows incremental exploration of the solution space. The proposed methods was implemented and tested on a extensive set of circuits. The results show that in most case, the proposed methods can determine the minimum / maximum leakage state. Also, the SAT-based approach was able to obtain significant improvement over the random approach.

Acknowledgments. This work is funded in part by the DARPA/MARCO Gigascale Silicon Research Center and an Agere Systems/SRC Research fellowship.

References

1. F. Aloul, A. Ramani, I. Markov, K. Sakallah, "PBS: A Backtrack-Search Pseudo-Boolean Solver and Optimizer," *in Symp. on the Theory and Applications of Satisfiability Testing*, 346–353, 2002.
2. A. Biere, A. Cimatti, E. Clarke, M. Fujita, and Y. Zhu, "Symbolic Model Checking using SAT procedures instead of BDDs," *in Proc. of the Design Automation Conference*, 317–320, 1999.
3. S. Bobba and I. Hajj, "Maximum Leakage Power Estimation for CMOS Circuits," *in Proc. of the IEEE Alessandro Volta Memorial Workshop on Low-Power Design*, 1999.
4. S. Borkar. "Design Challenges of Technology Scaling," *IEEE Micro*, 19(4), 23–29, 1999.
5. M. Davis, G. Logemann, and D. Loveland, "A Machine Program for Theorem Proving," *in Journal of the ACM*, (5)7, 394–397, 1962.
6. C. P. Gomes, B. Selman, and H. Kautz, "Boosting Combinatorial Search Through Randomization," *in Proc. of the National Conference on Artificial Intelligence*, 431–447, 1998.
7. J. Halter and F. Najm, "A gate-level leakage power reduction method for ultra-low-power CMOS circuits," *in Proc. of the IEEE 1997 Custom Integrated Circuits Conference*, 475–478, 1997.
8. M. Johnson, D. Somasekhar, and K. Roy, "Models and Algorithms for Bounds on Leakage in CMOS Circuits," *in IEEE Transactions on Computer-Aided Design of Integrated Circuits and Systems*, (18)6, 714–725, 1999.
9. J. Kao, A. Chandrakasan, D. Anotoniadis, "Transistor Sizing Issues and Tool for Multi-Threshold CMOS Technology," *in Proc. of the Design Automation Conference*, 409–414, 1997.
10. J. Marques-Silva and K. Sakallah, "GRASP: A Search Algorithm for Propositional Satisfiability," *in IEEE Transactions on Computers*, (48)5, 506–521, 1999.
11. MCNC Benchmarks, *http://www.cbl.ncsu.edu/CBL_Docs/Bench.html*
12. M. Moskewicz, C. Madigan, Y. Zhao, L. Zhang, and S. Malik, "Chaff: Engineering an Efficient SAT Solver," *in Proc. of the Design Automation Conference*, 530–535, 2001.
13. G. Nam, F. Aloul, K. Sakallah, and R. Rutenbar, "A Comparative Study of Two Boolean Formulations of FPGA Detailed Routing Constraints," *in the Proc. of the International Symposium on Physical Design*, 222v227, 2001.
14. S. Narendra, S. Borkar, V. De, and D. Chandrakasan, "Scaling of Stack Effect and its Application for Leakage Reduction," *in Proc. of the Int'l Symp. on Low Power Electronics and Design*, 2001.

15. E. Sentovich and K. Singh and L. Lavagno and C. Moon and R. Murgai and A. Saldanha and H. Savoj and P. Stephan and R. Brayton and A. Sangiovanni-Vincentelli, "SIS: A System for Sequential Circuit Synthesis," *University of California-Berkeley, UCB/ERL M92/41*, 1992.

16. L. Silva, J. Silva, L. Silveira and K. Sakallah, "Timing Analysis Using Propositional Satisfiability," *in Proc. of the IEEE International Conference on Electronics, Circuits and Systems*, 1998.

17. S. Sirichotiyakul, T. Edwards, C. Oh, J. Zuo, A. Dharchoudhury, R. Panda, and D. Blaauw, "Stand-by Power Minimization through Simultaneous Threshold Voltage Selection and Circuit Sizing," *in Proc. of the Design Automation Conference*, 436–441, 1999.

18. J. Walsor, "Solving Linear Pseudo-Boolean Constraint Problems with Local Search," *in Proc. of the National Conference on Artificial Intelligence*, 1997.

19. J. Whittemore, J. Kim, and K. Sakallah, "SATIRE: A New Incremental Satisfiability Engine," *in Proc. of the Design Automation Conference*, 542–545, 2001.

20. H. Zhang, "SATO: An Efficient Propositional Prover," *in Proc. of the International Conference on Automated Deduction*, 155–160, 1997.

21. A. Chandrakasan, W. Bowhill, F. Fox eds., "Design of High-Performance Microprocessor Circuits," *Piscataway, NJ: IEEE Press*, 2001.

PA-ZSA (Power-Aware Zero-Slack Algorithm): A Graph-Based Timing Analysis for Ultra-Low Power CMOS VLSI

Kyu-won Choi and Abhijit Chatterjee

School of Electrical and Computer Engineering
Georgia Institute of Technology
Atlanta, GA 30332, USA
{kwchoi,chat}@ece.gatech.edu

Abstract. This paper describes a slack budget distribution algorithm for ultra-low power CMOS logic circuits in a VLSI design environment. We introduce *Power-Aware Zero-Slack Algorithm* (**PA-ZSA**), which distributes the surplus time slacks into the most power-hungry modules. The **PA-ZSA** ensures that the total slack budget is near-maximal and the total power is minimal as a power-aware version of the well-known *zero-slack algorithm* (**ZSA**). Based on these time slacks, we have conducted the low-power optimization at gate level by using technology scaling technique. The experimental results show that our strategy reduces average 36% of the total (static and dynamic) power over the conventional slack budget distribution algorithms.

1 Introduction

Traditionally, power dissipation of VLSI chips is a neglected subject. In the past, the device density and operating frequency were low enough that it was not a constraining factor in the chips. As the scale of integration improves, more transistors, faster and smaller than their predecessors, are being packed into a chip. This leads to the steady growth of the operating frequency and processing capacity per chip, resulting in increased power dissipation. For state-of-the art systems, the trade-off solutions between the conflicting design criteria (i.e., delay, area, and power) should be considered [1,2]. In general, low-power optimizations without compromising performance are dependent on the time slack calculation and the surplus slack (slack budget) distribution. The time slack means the difference between the signal required time and the signal arrival time at the primary output of each module. The first use of the slack distribution approach is the popular *zero-slack algorithm* (**ZSA**) [3]. The **ZSA** is a greedy algorithm that assigns slack budgets to nets on long paths for VLSI layout design. It ensures that after the assignment, the net slack budget is maximal, which means that no more slack budget could be assigned to any of the nets without violating the path constraints. Most other slack-distribution algorithms are pruning versions of **ZSA** [4,5,6]. The **ZSA** and its off-springs are only for improving delay performance in layout design

B. Hochet et al. (Eds.): PATMOS 2002, LNCS 2451, pp. 178–187, 2002.

hierarchy. In this paper, we propose a low-power version of **ZSA**. The proposed **PA-ZSA** shows that our slack budget distribution strategy ensures that the power consumption is minimal without delay performance degradation for CMOS logic circuits.

2 Background

A CMOS random logic network can be represented as a directed acyclic graph $G=(V,E)$ where each node $v \in V$ represents a logic gate, and a directed edge $e_{ij} \in E$ exists if the output of gate i is an input of gate j. A primary input (PI) node has no fan-in (f_i) edges and a primary output (PO) has no fan-out (f_o) edges. A combinational circuit operates properly only within the satisfied functional and timing specification. The timing specification is given as arrival time at each primary input and required time at each primary output. The arrival time of a node is the latest time of signals to arrive at the output of any node. The arrival time, *arr(v)*, and the required time, *req(v)*, of a node v are recursively computed as:

$$arr(v) = \max_{i \in (1, f_i(v))} \{arr(u) + t_{u,i} + t_v\} \quad (1)$$

$$req(u) = \min_{j \in (1, f_o(u))} \{req(v) - t_{v,j} - t_u\} \quad (2)$$

And the slack at node v, *slack(v)*, is the difference between the require time and the arrival time at v.

$$slack(v) = req(u) - arr(v) \quad (3)$$

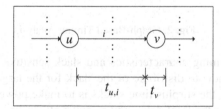

Fig. 1. Arrival Time, Required Time, and Time Slack

If an increased delay of node v is influence to the slack calculation of node u and an increased delay of node u is influence to the slack calculation of node v, we define that v is slack-sensitive to u and u is also slack-sensitive to v.

As an aid to making timing characteristics and their easier interpretation, we introduce a use of special symbols for distributing surplus slacks. The use of these symbols is shown in Figure 2. The time duration values for each node are placed in the upper quadrant of the node. The latest arrival time, *arr(i)*, is placed in the left hand quadrant of node i and the earliest require time, *req(i)*, is placed in the right hand quadrant of node i. The time slack for each node are placed in the lower quadrant of the node. The slack sensitive path is identified on its edges. Fig. 2(a) shows that an example topological structure of a random logic circuit and the delay of each node. Fig. 2(b) illustrates the result of the timing analysis. In this case, there are only three slack sensitive paths to be considered, for example, G-H, I-K-J, and D-E-F.

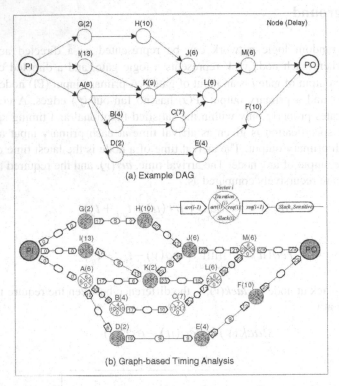

(a) Example DAG

(b) Graph-based Timing Analysis

Fig. 2. Graph-Based Timing Analysis

After obtaining the timing characteristics and slack sensitive paths as shown in Fig. 2(b), the problem is how to distribute of the slack for the target performance metrics. Our goal to distribute the surplus time slacks is to make power consumption minimal without compromising the total delay and to make all slacks of the nodes zero. In the following section, traditional slack distribution algorithms and the proposed algorithm are explained.

3   Power-Aware Zero Slack Algorithm (PA-ZSA)

Before describing our algorithm, **ZSA** (**Z**ero **S**lack **A**lgorithm) [3] and **MISA** (**M**aximum **I**ndependence **S**et **A**lgorithm) [6] are explained briefly.

ZSA: First, ZSA is to start with nodes with minimum slack and locally perform slack assignment such that the slack over the nodes get reduced to zero. This process iterated until the slack of all nodes is zero. More specifically, at each iteration, a node having the least positive slack (s_{min}) is selected. A path on which all nodes have minimum slack (s_{min}), then assign to each node an incremental delay s_{min}/N_{min}, where N_{min} is the number of nodes on the path. For example, in Fig. 3(b), if the path M1-M2 is first found as the least positive slack, each of the two nodes is assigned an incremental delay of 3/2.

MISA: As opposed to **ZSA**, the **MISA** selects a node having the most positive slack (s_{max}). Then a maximum independence set (**MIS**) of a transitive slack-sensitive graph is extracted. An incremental dealy ($S_{max}-S_{(max-1)}$) is assigned to each node in **MIS**. This iteration is continued until all slack of the nodes are zero. In Fig. 3(b), M2 and M3 is the **MIS**, respectively, and S_{max} is 3.0 and $S_{(max-1)}$ is 0. Therefore, an incremental delay for M2 is 3.0 ($S_{max}-S_{(max-1)}$) and an incremental delay for M2 is 0.

In the example of Fig. 3(a), the incremental delay of each node by **ZSA** is (M1=1.5, M2=1.5, M3=1.5) and the incremental delay of each node by **MISA** is (M1=0, M2=3.0, M3=3.0). Therefore, the **MISA**'s total slack budget (0.0+3.0+3.0=6.0) is greater than ZSA's (1.5+1.5+1.5=4.5) by 1.5. If the fan-out number is large enough, the maximum total budget by **MISA** is as large as twice the maximum total budget by **ZSA** as shown in [6]. This shows that **MISA** provides significant improvement over **ZSA** when there exist a large number of fan-outs. However, these algorithms are guaranteed that the delay performance improvement in placement, especially. If the performance metric is the power not the delay, maximum total slack budget does not guarantee low power. For example, in the Fig. 3(a), node M1 has more switching activity than M2 or M3, so, for low power, M1 must have more slack budget than M1 or M2 because power consumption can be reduced only when more surplus slack goes to more power hungry node which has more switching activity and/or capacitance. Therefore, **MISA** or **ZSA** does not ensure that their slack distribution algorithms are optimal in terms of power.

PA-ZSA: We propose a power-aware version of **ZSA** in this paper. As a pre-procedure, we perform an activity generation for each node. *Monte Carlo simulation* is performed for activity profiling of each module/sub-module as described in [2]. This approach consists of applying randomly generated input patterns at the primary inputs of the circuit and monitoring the switching activity per time interval T using a simulator. Under the assumption that the switching activity of a circuit module over any period T has a normal distribution, and for a desired percentage error in the activity estimate and a given confidence level, the number of required simulation vectors is estimated. The simulation based approach is accurate and capable of handling various device models, different circuit design styles, single and multi-phase clocking methodologies, tristate drives, etc. As shown in Fig. 3(c), after timing analysis, the optimal set of the slack distribution is chosen in proportion to the energy consumption of each node. By the following *lemma3.1*, this algorithm guarantees that the power is minimal.

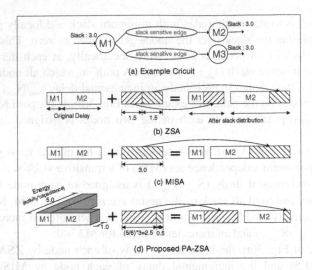

Fig. 3. Slack Distributions Overview

Lemma 3.1: EDR (Energy-Delay Ratio) Paradigm: When the ratios of the energy and the delay for each module are same, the total energy consumption is minimal. In other words, the delay for each module should be proportional to the energy of that module for minimum power consumption. Therefore the surplus time slacks should be assigned onto each module according to the cost function of $\frac{E_1}{d_1} = \frac{E_2}{d_2} = \ldots = \frac{E_n}{d_n}$ (1,..., n: module number) for low power.

Proof: Let's assume the energy consumptions of the module1 (M1) and module2 (M2) in Fig. 4 are $E_1 = K_1 V_{dd1}^2$ and $E_2 = K_2 V_{dd2}^2$ respectively. Here, the K_1 and K_2 are proportional to the switching activities and load capacitances of the Modules. The delay of the M1 can be assumed by $d_1 = L_1 / V_{dd1}$ and the delay of the M2 can be assumed by $d_2 = L_2 / V_{dd2}$, where the L_1 and L_2 are dependant on the threshold voltage of each module. And the total delay is $T = d_1 + d_2$. We want to minimize the total power P_{Total}:

$$P_{Total} = \frac{K_1 V_{dd1}^2 + K_2 V_{dd2}^2}{T}$$

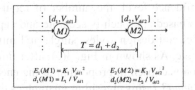

Fig. 4. Two Nodes Example

Now, $\quad d_2 = T - d_1 = (T - \dfrac{L_1}{V_{dd1}}) = \left(\dfrac{TV_{dd1} - L_1}{V_{dd1}} \right)$ and after transposing, we have

$V_{dd1} = \left(\dfrac{L_1}{T - d_2} \right)$ and $V_{dd2} = \left(\dfrac{L_2}{d_2} \right)$. Let's substitute for V_{dd1} and V_{dd2}, then the total power

P_{Total} is

$$P_{Total} = \dfrac{K_1 \left(\dfrac{L_1}{T - d_2} \right)^2 + K_2 \left(\dfrac{L_2}{d_2} \right)^2}{T}$$

$$\dfrac{\partial P_{Total}}{\partial d_2} = \left(\dfrac{K_1 L_1^2}{T} \times (-2) \times \dfrac{1}{(T - d_2)^3} \times (-1) \right) - \left(\dfrac{2K_2}{T} \cdot \dfrac{L_2^2}{d_2^3} \right)$$

At minimum power, $\dfrac{\partial P_{Total}}{\partial d_2} = 0$, therefore,

$$\left(\dfrac{K_1 L_1^2}{T} \times (-2) \times \dfrac{1}{(T - d_2)^3} \times (-1) \right) - \left(\dfrac{2K_2}{T} \cdot \dfrac{L_2^2}{d_2^3} \right) = 0$$

$$\dfrac{K_1 L_1^2}{K_2 L_2^2} = \dfrac{(T - d_2)^3}{d_2^3} \quad \text{or} \quad \dfrac{(T - d_2)}{d_2} = \left(\dfrac{K_1 L_1^2}{K_2 L_2^2} \right)^{1/3}$$

Let

$$\alpha = \left(\dfrac{K_1 L_1^2}{K_2 L_2^2} \right)^{1/3} \tag{4}$$

, then $\dfrac{T}{d_2} = \alpha + 1$. So, when P_{Total} is minimum,

$$d_2 = \dfrac{T}{\alpha + 1} \tag{5}$$

$$d_1 = T - d_2 = \dfrac{\alpha T}{\alpha + 1} \tag{6}$$

At the beginning, we recall $E_1 = K_1 V_{dd1}^2$ and $E_2 = K_2 V_{dd2}^2$, and then

$\frac{E_1}{d_1} = \frac{K_1 V_{dd1}^2}{d_1} = \frac{K_1 L_1^2}{d_1^3}$. Substituting for d_1^3 with optimal d_1 from Eqn.(3),

$$\frac{E_1}{d_1} = \frac{K_1 L_1^2}{\alpha^3 T^3}(\alpha+1)^3 \qquad (7)$$

(7)

$\frac{E_2}{d_2} = \frac{K_2 V_{dd2}^2}{d_2} = \frac{K_2 L_2^2}{d_2^3}$ and Substituting for d_2^3 with optimal d_2 from Eqn.(2),

$\frac{E_2}{d_2} = \frac{K_2 L_2^2}{T^3}(\alpha+1)^3$. But from Eqn.(1), we know that $K_2 L_2^2 = \frac{K_1 L_1^2}{\alpha^3}$, so

$$\frac{E_2}{d_2} = \frac{K_1 L_1^2}{\alpha^3 T^3}(\alpha+1)^3 \qquad (8)$$

The Eqn.(7) and Eqn.(8) are exactly same. Hence, for total minimum power,

$$\frac{E_2}{d_2} = \frac{E_1}{d_1}$$

□

Fig. 5 shows the algorithm of the **PA-ZSA**. At the initialization step, a *Monte Carlo simulation* is conducted for the activity profiling of each module/sub-module as described in [2]. In the phase I, topologic structure is identified by using breath first based search. Then, slack times are calculated by Eqn. (1-3) from the previous section 2. Finally, at the phase III, surplus time slacks are distributed iteratively according to the EDR Paradigm (*lemma 3.1*).

4 Experimental Results

Fig. 6 shows overall optimization methodology procedure.
We used an ARM core verilog description for the target system and VCS (synopsys) and design analyzer (synopsys) are used for the functional verification and logic synthesis with 0.25 micron TSMC library. A few arithmetic units (gate-level net lists) are used for the benchmark circuits after the logic synthesis. Then the **PA-ZSA** is performed for delay assignment of each module. After the maximum delays have been assigned to each module/gate in the circuit, we optimize each gate individually for minimum power. The strategy is to find iteratively, using binary search, the optimal combination of Vdd, Vth, and W for each gate that meets the maximum delay condition while achieving minimum power dissipation [7].

```
Procedure Power-Aware Zero-Slack Algorithm (PA-ZSA)

Input: directed acyclic graph G = (V,E )
Output: power-aware time slack distribution vector TSD(v )
Begin
    Phase 0: Initialization
    Initialize the class variables of all node V in G ;
    (delay, activity, early start time, early finish time, late start time,
    late finish time, slack, level=∞)

        Phase I: Labeling (Identify topological order: breath first search)
        Put a start node to FIFO Queue q;
        Initialize the level of the start node = 0;
            While (q is not 0)
            {
                Obtain a reference to the first element(x) in q;
                Remove node x from q;
                    For each fan-out node y of node x
                    {
                        If level of y = ∞ then
                        {
                            If number of fan-in node of y =1 then
                                level of y = level of x +1;
                            else if number of fan-in node of y >1 then
                                level of y =MAX( levels of fan-in nodes of y) +1;
                            Add node y into q;
                        }
                    }
            }

        Phase II: Timing Analysis and Slack Calculation
        Sort all nodes according to the topological levels;
        For each node of the sorted V
        {
            Assign early start/finish times and late start/finish time;
            Compute time slacks;
        }

        Phase III: Distributing Time Slacks
        Find slack sensitive graph Gsub ⊃ G;
        Find maximum activity set Amax  from Gsub ;
        Initialize  TSD(Vsub) = 0 ;
        While (Amax  is not empty)
        {
            Assign time slack of Amax by EDR cost function;
            Repeat Phase II
            Update  TSD(V)
        }
        Construct  TSD(V)
End
```

Fig. 5. PA-ZSA Algorithm

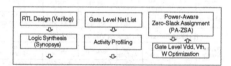

Fig. 6. Optimization Procedure

We developed the optimization simulators and the interface programs with C/C++/STL on Ultra-80 Unix machine. Table 1 demonstrates the impact of the **PA-ZSA** based optimization in terms of power and area. We can obtain over an order of magnitude of total (dynamic and static) power savings. Table 2 shows the effectiveness of the proposed slack budget distribution scheme over traditional algorithms with the same gate level optimization. Approximately 36%-39% more power reduction over conventional slack budget distribution algorithms is achieved.

5   Conclusion

We have presented a new approach to the problem of slack budget distribution for low power. The idea behind the approach has been embodied in an efficient algorithm called the power-aware zero slack algorithm. The algorithm guarantees that the distributed slack budget is optimal in terms of power and the total slack budget is near maximal. Our experiments demonstrate that the proposed slack distribution algorithm achieves results that provide significant improvement over previous approaches in power savings.

Table 1. Results of PA-ZSA Based Optimization

(a) Before Optimization (Fixed Vdd:3.3v, Vth:0.7v)

System Module	Gates/ Depth	Delay (ns)	Input Activity	ϖ, σ_ϖ	Power Dissipation			
					Static	Dynamic	Short-ckt	Total
4 - Full Adder	106/48	3.36	0.5	17.9, 24.5	2.09x10E-20	4.37x10E-11	2.15x10E-12	4.59x10E-11
			0.05	17.9, 24.5	2.09x10E-20	4.33x10E-12	2.13x10E-13	4.54x10E-12
16- Full Adder	1030/ 93	6.98	0.5	7.4, 5.6	8.60x10E-20	9.80x10E-11	8.99x10E-11	1.87x10E-10
			0.05	7.4, 5.6	8.60x10E-20	9.56x10E-11	7.51x10E-12	1.03x10E-11
16 - Look ahead (1)	1838/ 81	7.0	0.5	5.9, 6.2	1.48x10E-19	7.65x10E-10	9.33x10E-11	8.58x10E-10
			0.05	5.9, 6.2	1.48x10E-19	1.39x10E-10	9.29x10E-12	1.48x10E-10
16 - Look ahead (2)	1871/ 75	6.3	0.5	7.2, 4.9	1.88x10E-19	8.15x10E-10	9.63x10E-11	9.11x10E-10
			0.05	7.2, 4.9	1.88x10E-19	5.39x10E-10	9.59x10E-12	5.49x10E-10
16 - Look ahead (3)	1928/ 69	5.9	0.5	7.9, 5.3	1.63x10E-19	8.59x10E-10	9.91x10E-11	9.58x10E-10
			0.05	7.9, 5.3	1.63x10E-19	5.50x10E-10	9.90x10E-12	5.59x10E-10

(b) After Optimization (Vdd:0.6-3.3v, Vth:0.1-0.7v)

System Module	Gates/ Depth	Delay (ns)	Input Activity	Vdd, Vth	ϖ, σ_ϖ	Total Power	Power Reduction	Area Reduction
4 - Full Adder	106/48	3.01	0.5	0.65, 0.1	6.61, 8.41	6.52x10E-13	70.4x	63.2%
			0.05	0.625, 0.1	6.61, 8.41	7.13x10E-14	63.7x	63.2%
16- Full Adder	1030/ 93	7.14	0.5	0.625, 0.1	5.98, 5.03	8.70x10E-13	21.5x	19.1%
			0.05	0.6, 0.1	5.02, 4.03	6.30x10E-13	16.3x	32.1%
16 - Look ahead (1)	1838/ 81	7.0	0.5	0.65, 0.1	5.21, 5.70	1.80x10E-12	47.6x	11.7%
			0.05	0.625, 0.12	3.16, 2.05	3.78x10E-12	39.1x	46.4%
16 - Look ahead (2)	1871/ 75	6.3	0.5	0.625, 0.1	7.1, 6.90	5.98x10E-11	15.2x	1.3%
			0.05	0.625, 0.1	6.16, 6.05	3.08x10E-11	17.8x	14.4%
16 - Look ahead (3)	1928/ 69	5.9	0.5	0.625, 0.1	7.81, 5.01	6.19x10E-11	14.7x	1.1%
			0.05	0.625, 0.1	3.16, 2.05	7.78x10E-11	7.0x	60.0%

Table 2. Comparison with other ZSAs

(a) ZSA Based Optimization (Vdd:0.6-3.3v, Vth:0.1-0.7v)

System Module	Delay (ns)	Gates/ Depth	Input Activity	Vdd, Vth	ϖ, σ_ϖ	Total Power
64 - ALU	20.07	3417/ 226	0.5	0.625, 0.1	5.89, 4.94	1.56x10E-08
			0.05	0.625, 0.1	4.81, 3.88	1.07x10E-09

(b) MISA Based Optimization (Vdd:0.6-3.3v, Vth:0.1-0.7v)

System Module	Delay (ns)	Gates/ Depth	Input Activity	Vdd, Vth	ϖ, σ_ϖ	Total Power
64 - ALU	20.07	3417/ 226	0.5	0.625, 0.1	6.19, 5.04	1.98x10E-08
			0.05	0.625, 0.1	4.98, 4.74	1.82x10E-09

(c) PA-ZSA Based Optimization (Vdd:0.6-3.3v, Vth:0.1-0.7v)

System Module	Delay (ns)	Gates/ Depth	Input Activity	Vdd, Vth	ϖ, σ_ϖ	Total Power
64 - ALU	20.07	3417/ 226	0.5	0.625, 0.1	6.09, 9.04	9.91x10E-09
			0.05	0.625, 0.1	4.91, 6.14	6.47x10E-10

*Note: Scheme (c) is better than scheme (a) around 36% - 39% in total power reduction

References

1. A. Chandrakasan, S. Sheng, and R. Brodersen, "Low-power CMOS digital design," *IEEE Journal of Solid-State Circuits*, vol. 27, pp. 473–484, April 1992.
2. J.M. Rabaey and M. Pedram, *Low Power Design Methodologies*, Kluwer Academic Publishers, 1996, pp 21–64, 130–160.
3. R. Nair, C.L. Berman, P.S. hauge, and E.J. Yoffe, "Generation of performance constraints for layout," *IEEE Transactions on Computer-Aided Design*, pp.860–874, Aug. 1989.
4. T. Gao, P.M. Vaidya, and C.L. Liu,"A new performance driven placement algorithm," *Proc. of ICCAD*, pp. 44–47, 1991.

5. H. Youssef and E. Shragowitz, "Timing constraints for correct performance," *Proc. of ICCAD*, pp. 24–27, 1990.
6. C. Chen, X. Yang, and M. Sarrafzadeh, "Potential slack: an effective metric of combinational circuit performance," *Proc. of ICCAD*, pp. 198–201, 2000.
7. P. Pant, V. De, and A. Chatterjee, "Simultaneous power Supply, threshold voltage, and transistor size optimization for low-power operation of CMOS circuits," *IEEE Trans. On VLSI Systems*, vol. 6, no. 4, pp. 538–545, December 1998.

A New Methodology for Efficient Synchronization of RNS-Based VLSI Systems

Daniel González[1], Antonio García[1], Graham A. Jullien[2], Javier Ramírez[1], Luis Parrilla[1], and Antonio Lloris[1]

[1] Dpto. Electrónica y Tecnología de Computadores
Universidad de Granada
18071 Granada, Spain
rns@ditec.ugr.es
[2] ATIPS Laboratory, Dept. of Electrical and Computer Engineering
University of Calgary
Calgary, Alberta T2N 1N4, Canada
jullien@atips.ca

Abstract. Synchronization of VLSI systems is growing in complexity because of the increase in die size and integration levels, along with stronger requirements for integrated circuit speed and reliability. The size increase leads to delays and synchronization losses in clock distribution. Additionally, the large amount of synchronous hardware in integrated circuits requires large current spikes to be drawn from the power supply when the clock changes state. This paper presents a new approach for clock distribution in RNS-based systems, where channel independence removes clock timing restrictions. This approach generates several clock signals with non-overlapping edges from a global clock. This technique shows a significant decrease in instantaneous current requirements and a homogeneous time distribution of current supply to the chip, while keeping extra hardware to a minimum and introducing an affordable power cost, as shown through simulation.

1 Introduction

Recent advances in integrated circuit fabrication have lead to increasing integration levels and operating speeds. These factors make extraordinarily difficult the proper synchronization of integrated systems. For TSPC (True Single Phase Clock) [1], which is a special dynamic logic clocking technique and should not be used as a general example, one single clock line must be distributed all over the chip, as well as being distributed within each operating block. More complex clocking schemes may require the distribution of two or four non-overlapping clock signals [1], thus increasing the resources required for circuit synchronization. Moreover, for clock frequencies over 500 MHz, phase differences between the clock signal at different locations of the chip (skew) start presenting serious problems [2]. An added problem with increasing chip complexity and density is that the length of clock distribution lines increases along with the number of devices the clock signal has to supply, thus leading to substantial delays that limit system speed. A number of techniques exists for overriding clock skew, with the most common being RC tree analysis. This

B. Hochet et al. (Eds.): PATMOS 2002, LNCS 2451, pp. 188–197, 2002.

method represents the circuit as a tree, modeling every line through a resistor and a capacitor, and modeling every block as a terminal capacitance [3]. Thus, delay associated with distribution lines can be evaluated and elements to compensate clock skew can be subsequently added. Minimizing skew has negative sides, especially as simultaneous triggering of so many devices leads to short but large current demands. Because of this, a meticulous design of power supply lines and device sizes is required, with a large current demand resulting in area penalties. If this is not the case, parts of the chip may not receive as much energy as required for working properly. This approximation to the problems related to fully synchronous circuits and clock skew has been previously discussed [4]; this paper will present an alternative for efficiently synchronizing RNS-based circuits while keeping current demand to a minimum. The underlying idea is to generate out-of-phase clock signals, each controlling an RNS channel, thus taking advantage of the non-communicating channel structure that characterizes RNS architectures in order to reduce the clock synchronization requirements for high-performance digital signal processing systems. The synchronization strategy has been evaluated for a three-stage CIC (Cascade Integrator Comb) filter [5-6].

2   Residue Number System (RNS)

The Residue Number System (RNS) [7-8] is an integer number representation system, defined in terms of a set of relatively prime integers $\{m_1, m_2, ..., m_N\}$. The dynamic range of such an RNS system is:

$$M - \prod_{i=1}^{N} m_i \tag{1}$$

Thus, any $X \in [0, M-1]$ has a unique representation given by the N-tuple $[x_1, x_2, ..., x_N]$, where $x_i = X \bmod m_i$ $(i=1, 2, ..., N)$. The main RNS advantages lie with its arithmetic capabilities, since arithmetic is defined over the ring of integers modulo M and, for $\lozenge$ representing either addition, subtraction or multiplication and $0 \leq X, Y, Z < M$:

$$Z = (X \lozenge Y) \bmod M = [z_1, z_2, ..., z_N]$$

$$z_i = (x_i \lozenge y_i) \bmod m_i \quad (i = 1, 2, ..., N) \tag{2}$$

From (2), we see that there are no carry propagation requirements between channels, so the system throughput does not depend on system dynamic range. This fact makes RNS most suitable for digital signal processing applications [8].

3   Clock Skew

Clock skew occurs when the clock signal has different values at different nodes within the chip at the same time. It is caused by differences in the length of clock paths, as well as by active elements that are present in these paths, such as buffers.

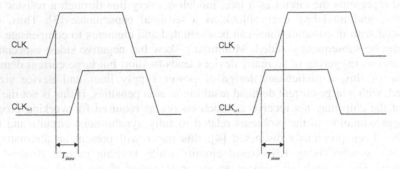

Fig. 1. Negative (left) and positive (right) skew.

Clock skew lowers system throughput compared to that obtainable from individual blocks of the system, since it is necessary to guarantee the proper function of the chip with reduced clock speed. Skew will cause clock distribution problems if the following inequality holds [9]:

$$\frac{D}{v} > \frac{k}{f_{app}} \tag{3}$$

where $k < 0.20$ (typical value) is a constant, D is the size of the system, v is the propagation speed for the clock signal and f_{app} is the applied clock frequency. Existing solutions for clock skew provide two different approaches to the problem:

- equalize the length of clock paths to processing elements using buffer and delay elements or through H-tree, mesh or X-tree topologies [3, 10-12].
- eliminate or minimize the skew caused by variations during chip fabrication [13-14].

Typically, synchronous systems consist of a chain of registers separated by combinational logic that performs data processing. The maximum clock frequency is derived from:

$$\frac{1}{f_{max}} = T_{min} \geq T_{PD} + T_{skew} \tag{4}$$

where T_{PD} is the time between the arrival of the clock signal at the i-th register and stable processed data at the output of the $(i+1)$-th register. T_{skew} is the time between the arrival of the clock signal at the i-th register and the arrival of the same signal at the $(i+1)$-th register.

Clock skew can be considered as either positive or negative, although the sign criteria is not standardized. In [12] Friedman considers the skew to be negative when the clock signal arrives at the i-th register before that to the $(i+1)$-th register, as illustrated by Fig. 1. If positive, then from equation (4), the minimum system clock period is increased, while if negative, T_{min} decreases. An excessive positive skew results in a decrease in system performance, but if the skew is negative race-related problems may arise if data processing time is lower that the skew.

4   New Synchronization Strategy

The new synchronization strategy for RNS-based systems, proposed in this paper, introduces the generation of several signal clocks from the master clock. These clocks are slightly out-of-phase, thus with non-overlapping edges. Each one of these clock signals synchronizes one of the RNS channels, while global data synchronization (mainly at the global inputs and outputs of the system) is carried out by the global master clock. Thus, each channel computes at different time instants and, consequently, the current demand is distributed over the whole clock cycle. This has the effect of reducing current spikes on the power supply lines by a factor that is approximately the number of generated clock signals. The phase difference between the generated clocks has to satisfy several specifications. First of all, the number of clock signals with overlapping active cycles has to be minimized, as well as the time two or more active cycles overlap. Moreover, clock edges must not coincide. Finally, data coherency has to be respected at both the input and output of the system. Clear advantages are obtained when these requisites are satisfied, since current spikes are reduced and power dissipation is distributed over the master clock cycle, rather than concentrated around the master clock edges. Moreover, not only absolute current values are reduced, but also temporal variations. Also, as a side effect, power supply lines may be scaled and clock distribution resources reduced, thus simplifying the chip design task.

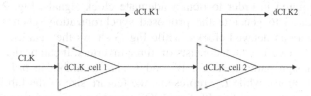

Fig. 2. dCLK_cell chain for out-of-phase clock generation

At first sight, the synchronization scheme described above may seem to be impractical because of the presence of several clock signals within the chip, with associated synchronization problems. However, the nature of RNS [7-8], with non-communicating channels, perfectly suits this clocking scheme. Thus, a generated clock signal is applied to each independent channel, while the master clock signal used to generate these other clocks can also synchronize the global input and output.

Moreover, it will be shown that the resources required for implementing this new strategy are minimal and only a few transistors, basically three inverters, are required for each channel. More specifically, the master clock signal is routed through an inverter chain, thus being delayed at every point of the chain. Meanwhile, the generated clock signals are extracted at appropriate points along the chain and conditioned to be used as clock signals for a complete RNS channel. This scheme requires the inverter chain to alternate between large and small input capacitances and low driving capabilities, so that appropriate delays can be generated. This has the effect of generating low-quality clock signals within the inverter chain, so additional

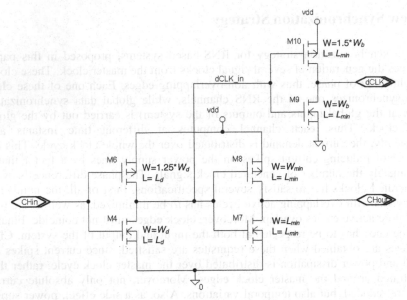

Fig. 3. dCLK_cell schematic.

buffers are required in order to obtain adequate clock signals. Fig. 2 illustrates the hardware required to generate the proposed synchronization scheme, where dCLK stands for generated delayed clocks, while Fig. 3 shows the detailed scheme for the so-called dCLK_cell, which consists of three inverters. It can be deduced from Fig. 3 that three design parameters, L_d, W_b and L_b, are available in order to obtain the system specifications, while L_{min} represents the feature size of the fabrication process and W_{min} the minimum usual width for pMOS transistors. Connecting CHout pads to CHin pads, the inverter chain described above is built, while the master global clock is used as input to this chain. Fig. 3 illustrates how large capacitance inverters are alternated with minimum-size devices. Thus, the low driving capabilities of the latter allow modeling of the required delay using the L_d parameter. Meanwhile, the generated clock signal dCLK is regenerated by a third inverter that includes the parameters W_b and L_b. These allow matching of the timing specifications for a proper clock signal for a given system, also allowing the adaptation of the cell to the overall capacitance to be driven by the generated clock. However, these three design parameters L_d, W_b and L_b are not fully independent, and their relation requires a careful study of the final system to be synchronized in order to select their optimum values. Fig. 4 shows the resulting generated clocks in a simple design example for a 300 MHz master global clock, with 0.1 pF loads for every dCLK signal. It can be noted that the requirements enumerated above with regard to non-overlapping edges and active cycles are matched, with every dCLK signal being to be used as clock for a given RNS channel.

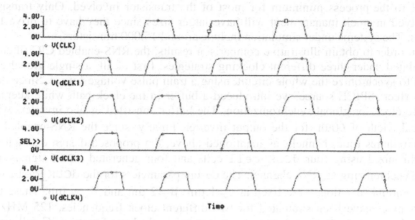

Fig. 4. Resulting generated clock signals for a design example (125 MHz).

5 Design Example and Simulation Results

A real RNS-based digital signal processing application [6] was considered for the evaluation of the proposed synchronization technique. Specifically, a three-stage decimation CIC filter with 26-bit dynamic range was designed at the transistor level simulated using PSpice for both a single global clock and the proposed technique. For these simulations, a public domain MOSIS CMOS 0.6 μm process [15] was used. This is a three-metal, one-poly, 3.3V CMOS process that is available to MOSIS costumers through Agilent Technologies. CIC filters [5] have been shown to be a useful alternative for FIR implementation for a variety of communication systems, because no multipliers are required. Although the frequency response out of the pass band may be insufficient for certain applications, low-order conventional filters may correct this effect. An S stage CIC system is defined by the transfer function:

$$H(z) = \left(\frac{1 - z^{RD}}{1 - z^{-1}}\right)^S = \left(\sum_{k=0}^{RD-1} z^{-k}\right)^S \tag{5}$$

Fig. 5 shows the structure of a three-stage CIC filter, illustrating the meaning of R and D parameters. The selected design example is a fully pipelined three-stage CIC decimation filter with $R=32$, $D=2$ and 26-bit dynamic range [6]. This RNS-enabled system requires four channels with moduli {256, 63, 61, 59}. Since the system is composed just of adders and registers, the well-known two-stage modulo adder [6] was used, while registers were implemented using negative edge triggered D flip-flops (nETDFF) based on TSPC logic [1]. TSPC was selected because it only requires a single-phase clock, thus minimizing synchronization resources and simplifying the implementation of the proposed alternative. Because of the great connection locality for the example system, load driving is kept to a minimum and device sizes can be fixed to the process minimum for most of the transistor involved. Only transistors involved in clock management will have larger sizes since they have to drive large loads. The systems under simulation include around 15.000 transistors.

In order to obtain illustrative comparison results, the RNS-enabled CIC filter was simulated under three different clocking strategies: first of all, a single global clock used to synchronize the whole circuit, using a train pulse voltage source. Since this is a "perfect" SPICE source, we introduced a buffer in the clock path while keeping a single clock for global synchronization. The buffer consisted of two inverters with a typical width of 60μm for the output inverter. Finally, since the RNS-enabled CIC filter requires four channels, as mentioned above, the proposed design example was synchronized using four dCLK_cell cells and four generated dCLK signals, each one synchronizing an RNS channel. The design parameters for the dCLK_cell cells, after careful selection, were fixed at L_d=1 μm, W_d=2 μm and W_b=9 μm. These three alternatives have been simulated for two different clock frequencies, 125 MHz and 300 MHz. Fig. 6 shows the current on power supply lines for the CIC full system working with a 125 MHz clock for both a single global clock and the proposed synchronization strategy. Fig. 7 shows the corresponding currents when a 300 MHz clock is applied. Clearly evident is the considerable decrease in the magnitude of the current spikes. In this way, current supply to the chip is distributed over time when the new strategy is considered, while for a global clock current spikes are around four times larger. This indicates that the expected benefits derived from the proposed synchronization scheme are confirmed through simulation. Table 1 summarizes the results obtained for the different simulations and both clock frequencies. We note that the maximum current spike is clearly reduced when this new clocking strategy is considered, as well as the maximum value of the current change rate (di/dt). This happens for the ideal clock and for the buffered one, thus indicating the validity of the proposed strategy.

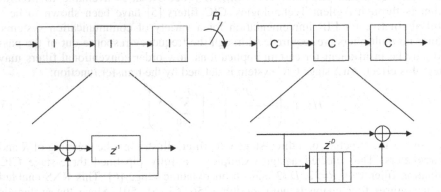

Fig. 5. Three-stage CIC decimation filter.

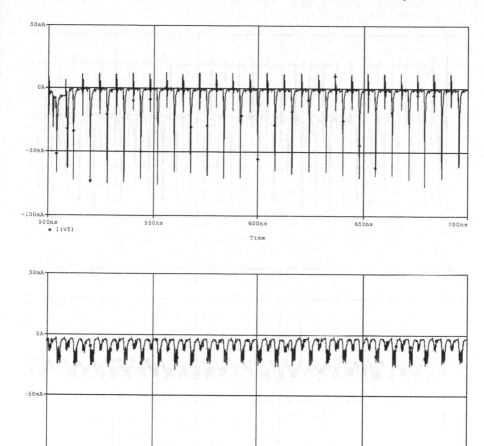

Fig. 6. Current from power supply line for a single clock (above) and the proposed alternative (below) for a 125 MHz frequency.

Table 1. Summary of simulation results for an RNS-enabled three-stage CIC filter using different synchronization approaches.

	Single clock		Single buffered clock		Proposed strategy	
	125 MHz	**300 MHz**	**125 MHz**	**300 MHz**	**125 MHz**	**300 MHz**
Max spike	78.0 mA	96.5 mA	64.4 mA	93.8 mA	23.4 mA	32.8 mA
Max di/dt	150 A/ns	254 A/ns	16.3 A/ns	36.6 A/ns	2.2 A/ns	11.9 A/ns
Power	17.0 mW	33.7 mW	19.5 mW	35.8 mW	23.5 mW	57.3 mW

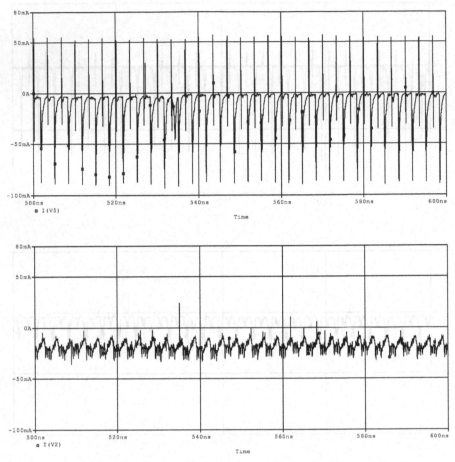

Fig. 7. Current from power supply line for a single clock (above) and the proposed alternative (below) for a 300 MHz frequency.

6   Conclusion

This paper has presented a new alternative for synchronizing RNS-based systems and reducing current demand. The proposed strategy was tested using an RNS-enabled three-stage CIC filter consisting of around 15.000 transistors. Simulation results demonstrate the effectiveness of this new clocking strategy in reducing the maximum current spike as well as reducing the maximum time derivative of the current spike. The use of this synchronization scheme may lead to reduced skew-related problems as well as to reducing chip area through the reduction of the size of power supply lines, caused by the reduction in current and current change rate requirements.

Acknowledgments. Daniel González, Antonio García, Javier Ramírez, Luis Parrilla and Antonio Lloris were supported by the *Comisión Interministerial de Ciencia y Tecnología* (CICYT, Spain) under project PB98-1354. Graham A. Jullien acknowledges financial support from Micronet R&D, iCORE and NSERC.

References

1. J. Yuan and C. Svensson, "High-speed CMOS Circuit Technique", *IEEE Journal of Solid State Circuits*, vol 24, no. 1, pp. 62–70, Jan. 1989.
2. D. W. Bailey and B. J. Benchsneider, "Clocking Design and Analysis for a 600-MHz Alpha Microprocessor", *IEEE Journal of Solid State Circuits*, vol 33, pp.1627–1633, Dec 1998.
3. P. Ramanathan, A. J. Dupont and K. G. Shin, "Clock Distribution in General VLSI Circuits." *IEEE Transactions on Circuits and Systems I: Fundamental Theory and Applications*, vol 41, no. 5, pp. 395–404, May 1994.
4. J. Yoo, G. Gopalakrishnan and K. F. Smith, "Timing Constraints for High-speed Counterflow-clocked Pipelining", *IEEE Transactions on VLSI Systems,* vol. 7, no. 2, pp. 167–173, Jun. 1999.
5. E. B. Hogenauer, "An Economical Class of Digital Filters for Decimation and Interpolation", *IEEE Transactions on Acoustics, Speech and Signal Processing*, vol. 29, no. 2, pp. 155–162, Feb. 1981.
6. U. Meyer-Bäse, A. Garcia and F. J. Taylor, "Implementation of a Communications Channelizer using FPGAs and RNS Arithmetic", *Journal of VLSI Signal Processing*, vol. 28, no. 1/2, pp. 115--128, May 2001.
7. N. S. Szabo and R. I. Tanaka, *Residue Arithmetic and Its Applications to Computer Technology*, McGraw-Hill, NY, 1967.
8. M. A. Soderstrand, W. K. Jenkins, G. A. Jullien and F. J. Taylor, *Residue Number System Arithmetic: Modern Applications in Digital Signal Processing*, IEEE Press, 1986.
9. W. D. Grover, "A New Method for Clock Distribution." *IEEE Transactions on Circuits and Systems I: Fundamental Theory and Applications*, vol 41, no. 2, pp. 149–160, Feb. 1994.
10. M.A.B. Jackson, A. Srinivasan and E.S. Kuh, "Clock Routing for High Performance IC's" *27th ACM/IEEE Design Automation Conference*, 1990.
11. D. F. Wann and N. A. Franklin, "Asynchronous and Clocked Control Structures for VLSI Based Interconnect Networks", *IEEE Transactions on Computers*, vol 32, no.5, pp. 284–293, May 1983.
12. E. G. Friedman, "Clock Distribution Networks in Synchronous Digital Integrated Circuits". *Proceedings of the IEEE,* vol. 89, no. 5, pp. 665–692, May 2001.
13. E. G. Friedman and S. Powell, "Design and Analysis of an Hierarchical Clock Distribution System for Synchronous Cell/macrocell VLSI", *IEEE Journal of Solid State Circuits*, vol. 21, no. 2, pp., 240–246, Apr. 1986.
14. M. Shoji, "Elimination of Process-dependent Clock skew in CMOS VLSI". *IEEE Journal of Solid State Circuits,* vol. 21, pp. 869–880, Oct. 1986.
15. MOSIS Process Information, "Hewlett Packard AMOS14TB", http://www.mosis.org/technical/processes/proc-hp-amos14tb.html

Clock Distribution Network Optimization under Self-Heating and Timing Constraints

M.R. Casu, M. Graziano, G. Masera, G. Piccinini,
M.M. Prono, and M. Zamboni

Electronics Dept., Politecnico di Torino, I10129, Italy

Abstract. Clock distribution networks appear to be affected by combination of thermally and electrically related issues. A new methodology is presented in this paper that produces optimized sizes for critical wires, combining thermal and electrical analysis. In particular, its application to a clock network is reported, to show alleviation strategies to undesired Deep Submicron (DSM) effects.

1   Introduction

Significant challenges in high performance interconnect design will be faced in both near and long-terms, as explicitly highlighted in ITRS. Roadmap predictions will be hardly respected with standard integration strategies and materials, such that the actual processes will not guarantee the improvements expected in the following years [2]. In particular, aggressive integration strategies are leading to higher power dissipation and increasing die and interconnect temperature: in this scenario management of thermally related issues (lifetime, performance and integrity) is becoming one of the most challenging objective. Main sources of temperature increase in a chip are the switching activities of macro-blocks over the substrate and the interconnect Joule heating, due to the current flowing through wires. In a high performance core, substrate temperature is expected to reach up to 120 °C and Joule heating can further contribute to higher interconnect temperature. This happens because current in interconnect does not scale as wire cross section does, hence current density increases and, therefore, metal temperature suffers an even higher heating contribution (see section 2). Furthermore, thermally and electrically related issues are strictly connected to each other and cannot be anymore modelled and evaluated separately: electrical issues, affecting current flow, could further increase wire temperature, while on the other side non-uniform thermal profile causes concern in performance and signal integrity of the system, that are typical electrical issues. However, thermally hot lines should be distinguished on the basis of the RMS current density flowing in each wire: power and ground busses, as well as clock distribution networks, do carry a high amount of current, due to high and continuous activity performed over them. Differently normal signal lines are supposed to carry a non significant amount of current to be severely affected by electromigration and thermal issues.

B. Hochet et al. (Eds.): PATMOS 2002, LNCS 2451, pp. 198–208, 2002.

To predict and control those effects, modelling should take into account all of them together. Currently, CAD tools are only available for separately analyzing thermal (i.e. THUNDER [14] or ITEM [15]) and electrical issues (i.e. FastHenry [9], FastCap [12] or Star-HSpice [3]), but rarely design methodologies are actually combining those two aspects, when performing timing or reliability analysis of VLSI cores.

This work aims at combining adequate tools and models for thermal and electrical analysis, that, under designer's specifications, could provide optimal interconnect width to fulfill required performances. From this point of view the first step is to identify which interconnects could suffer thermal and electrical issues. At this point, temperature of those interconnects is set to substrate temperature. A circuit timing simulation is performed to extract current densities, the sources for Joule heating; afterwards a thermal model is applied to estimate wire heating profile. If such profile does not satisfy lifetime constraints, wire width is enlarged and the operation described above is cycled up to complete specification fulfillment. Once thermal issues have been reduced, electrically related ones can be modelled and controlled: a feedback is then applyed to thermal issues if current density is varied, possibly causing significant changes in interconnect thermal profile. Optimization in this way provides a combined control on both critical interconnect issues, resulting in optimal wire width (smallest allowable according to design constraints).

This paper is organized as follows: after a brief presentation of major headlines of the adopted thermal (section 2) and electrical (section 3) models, in section 4 the optimization methodology developed will be explained; results deriving from the application of this methodology to the IBM PowerPC clock network described in [4] are reported in section 5.

2 Wire Thermal Model

A 3-D thermal simulator (i.e. THUNDER [14]) would be the best tool for temperature estimate when thermal effects assume increasing impact. However, if a complex structure is analyzed, simulation complexity, in terms of computational and time resources required, exceeds practical limits. An approximated approach to the heat diffusion equation should be considered, according to relevant integration issues.

The basis for a simpler but faster thermal model are Black's equation, for lifetime prediction:

$$TTF = AJ_{avg}^{-2} \exp\left(\frac{E_a}{kT}\right) \tag{1}$$

and self-heating equation, for interconnect Joule heating characterization:

$$T_m = T_{sub} + \Delta T_{self-heating} = T_{sub} + I_{rms}^2 R\theta_{int} \tag{2}$$

where T_{sub} is the substrate temperature, R is the interconnect electrical resistance and θ_{int} is the thermal impedance of the interconnect line to the substrate;

all parameter definitions can be found in [6] and [15]. The model based only on these equations, developed by the authors in a previous work [5], has been improved considering the work reported in [15], where the entire thermal network is mapped into an equivalent electrical circuit, in order to rapidly extract temperature profile of the interconnect network, under the assumption of heat dissipation only through the substrate. In present work only the contribution to heat generation given by Joule heating is considered, being the main focus of the analysis, while an arbitrary substrate heating profile is given. The detailed substrate temperature profile analysis performed by ILLIDAS-T in [15] is here neglected.

Interconnect sections are modelled with the same thermal cells depicted by the authors in [15]; the focus here is on wires interacting with nearby conductors too. In [15] only the case of thermally hot wires (power supply or clock interconnects) on different layers connected through via contacts is considered (see fig. 1 for modelling details).

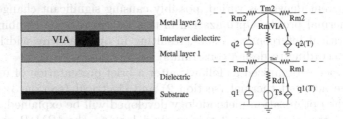

Fig. 1. Thermal model of the interconnect system near via contacts [15].

Other sources of interaction are not accounted for in temperature estimate by [15], while in [6] two main configurations are reported: one for a thermally hot wire *adjacent* to a signal wire and the other for a signal wire array *crossing* a hot wire. In the former one, temperature is not influenced by the presence of an adjacent signal wire. On the contrary the signal wire was slightly affected by the wire (temperature increases) and in [6] authors propose a simple modelling to predict possible reliability concern of that signal wire. Much more interesting is the crossing configuration (see fig. 2), implemented in the thermal model used in this work. As heat is going to be dissipated into the substrate, signal wire array acts as a further dissipation medium, because thermal impedance is lowered. It is as if power supply or clock wire is routed at a lower distance from the substrate. However, it can be simply understood that as line array spacing increases, this temperature reduction effect decreases and can be completely neglected if spacing becomes a couple of thermal diffusion lengths (around $100\mu m$, power wire behaves as an isolated interconnect).

A further improvement has been added to the used thermal model: since interconnect temperature is expected to rise ever increasingly, a better approximation of the dependency of metal electrical resistivity on metal temperature should be introduced, with respect to the linear approximation of [15]:

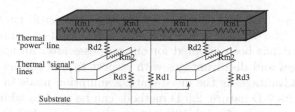

Fig. 2. Cross at 90° thermal model [6].

$$\rho = a_1 + a_2 \cdot T + a_3 \cdot T^2 \qquad (3)$$

where a_1, a_2 and a_3 are characteristics of the metal (values of Al and Cu are reported in [11]).
Details on the model implemented in this work are reported in [13].

3   Electrical Model

As operating frequency reaches to the GHz region, the RC interconnect model (Π-cell configuration) results in inaccurate timing analysis in particular for long global on-chip busses. Capacitance cannot fulfill entirely noise sources behaviour: considering crosstalk, measures show a directionality in noise spikes due to crosstalk. When inductive effect between neighboring wires becomes significant, capacitive coupling cannot model anymore crosstalk spikes, because it does not present directionality properties and inductance characterization needs to be added to the model [8]. Other two effects of increasing importance as operating frequency rises are: frequency dependent resistive losses and losses through the dielectric medium. Interconnects shorter than one-tenth of the wavelength at the highest operating frequency of interest (defined as $f_{max} \approx 0.35/t_r$) can still be properly modelled with a Π-cell RC distributed characterization. Otherwise those effects become significant and can only be predicted and controlled through a transmission line (TL) characterization.

In this work both the RC and the TL models have been implemented and compared. For the TL characterization an electrical timing simulator supporting a TL model is needed, together with a field solver, to extract, given a certain geometry, line parasitic parameters. Star-HSpice simulator has been used for timing analysis, since it provides accurate numerical simulation of multiconductor lossy frequency dependent TL (see W-element in [3]). Its main features can be briefly summarized:

- no limitation on the number of coupled conductors
- no restriction on the structures of the **RLGC** matrices
- accurate modelling of frequency-dependent losses
- very fast and low cost, in terms of memory and simulation time, without introduction of spurious numerical ringing on the propagating signal.

In both cases (RC and TL) the interconnect parasitic parameters are to be determined. Star-HSpice is provided with a 2-D built-in parasitics extractor (see [3]). Here it has been adopted for our purposes after an accurate analysis of its advantages and disadvantages with respect to 3-D solvers. In particular, one of the disadvantages is the quasi-static assumption made in the extraction algorithm and its 2-D nature. 2.5-D methods can be used to address 3-D effects if needed (see for details [7] and [10]), but in general a simple 2-D representation will be enough to model typical structures (stripline, microstripline and coplanar configurations). Comparative extractions performed on those configurations with 3-D solvers like FastHenry and FastCap, showed comfortable results. At most 10% overestimate is performed by the 2-D solver with respect to 3-D ones, with the great advantage of low cost in terms of memory and time required for the extraction (few minutes with respect to couple of hours).

4   Wire Size Optimization Methodology

Circuit timing simulation and thermal analysis is the main core of the optimization methodology developed in this work. The steps of this methodology are reported in fig. 3.

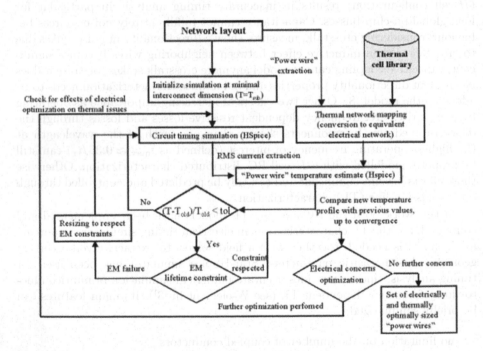

Fig. 3. Optimization methodology flow chart

The thermal model developed in this work focuses on Joule heating contribution to interconnect temperature, as the one having higher impact on wire reliability. The non-uniform substrate thermal profile has been neglected, even if the consequence of the distribution of macro-cells switching activities over the entire core can further increase thermal and electrical concerns, as shown in [1]. An apriori substrate thermal temperature (110°C) is assigned when performing wire size optimization, according to predictions for high performance cores.

Given an electrical netlist and a geometrical description of the circuit, the optimization is performed as sketched in fig. 3 using the HSPICE optimizer and simulator. Wire sizes are initialized at minimal dimension, to produce the highest possible integration density, necessary to respect roadmap requirements [2]. Timing analysis is then performed to extract RMS current densities, using both the electrical interconnect models discussed in section 3. The RMS and the average current densities are applied to the thermal model presented in section 2, such that the heating profile of those wires can be estimated. Due to the relation between metal resistance and temperature (as pointed out in equation (3)) these steps are cycled until temperature is stable. At this point EM lifetime constraint is checked (minimum Time-To-Fail is in this case 10 years). When it is not respected, wire is enlarged and a further temperature estimation is performed. When Time-To-Fail for all thermally hot wires is ensured, a complete set of wire sizes is derived and a check on timing constraints is performed. If this analysis does not meet design requirements a new interconnect optimization phase starts, using the reliability results as initial condition.

Thus the optimal solution represents a trade-off among different design requirements: thermal optimization has consequences on electrical issues and viceversa; this methodology highlights the correlation between thermal and electrical issues, producing optimal wire sizes for both effects.

This method ameliorates the one described in a previous work presented in [5], which was based on a first order approximation thermal model relying on equations (1) and (2). Comparisons of power supply interconnect configurations, reported in [5], (not reported here for brevity), show that using previous model results in a 10% overestimate, causing undesirable larger wires. This is avoided by the improved methodology described in this paper.

5   Clock Distribution Network Optimization

The proposed optimization methodology has been applied to both a power supply and a clock distribution network: this last case has been reported in this paper.

The clock scheme analyzed is derived from the one implemented on the S/390 G5 and G6 IBM microprocessors (see [4]). Structure is the one sketched in fig. 4. Operating frequency is settled at 600 MHz, for a 0.25μm technology process. The IBM technical report shows that this structure provides an extremely low skew, on the order of tens of picoseconds, even with a largely non uniform load distribution. Clock signal is driven by a central buffer into a metal 6 H-tree, that drives a second level of H-trees on metal 5, finally driving a metal 3 passive grid, guaranteeing high control on low skew demand.

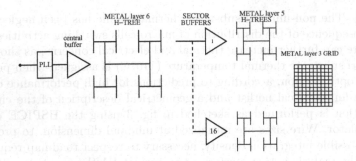

METAL layer 6
H-TREE

SECTOR
BUFFERS

METAL layer 5
H-TREES

central
buffer

METAL layer 3 GRID

PLL

I

16

Fig. 4. Clock distribution: hierarchy

In this work we are only interested to the skew component relative to thermal and electrical phenomena and not to geometrical ones, so for this reason structure has been assumed ideally symmetrical (length of interconnects having same hierarchical function are equal). Electomigration constraints and timing specifications have been met by optimizing both buffer and wire sizes using the steps described in previous section. Final constraints are a Time-To-Fail of ten year, maximum delay around 850ps and maximum skew around 50ps.

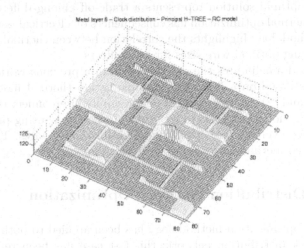

Metal layer 6 – Clock distribution – Principal H-TREE – RC model

Fig. 5. Temperature distribution in metal 6 principal H-tree. Z axis: temperature; X-Y axis: horizontal grid temperature distribution.

Figure 5 shows thermal profiles in principal H-tree: different temperature are function of load distribution. Figure 6 and 7 represent temperature profiles for metal 5 and metal 3 cases.

The two different electrical models have been applied, to show the importance of choosing the appropriate TL model when the RC one fails to correctly characterize signal behaviour (figures 5, 6 and 7 are for RC case only for brevity). The

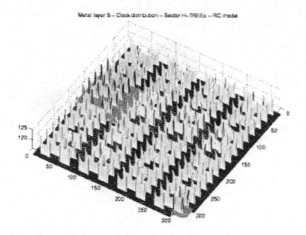

Fig. 6. Temperature distribution in metal 5 sector H-tree. Z axis: temperature; X-Y axis: horizontal grid temperature distribution.

presence of the power supply distribution network used by IBM is taken into account in parasitic extraction and timing simulation phases. Table 1 reports wire sizes and temperature in both RC and TL cases, for the three network levels. In each case, ranges represent the distance between the minimum and maximum value found for line segment in the same hierarchical level.

Table 1 Timing and EM reliability optimization of clock distribution network

	RC distributed model	Transmission line model
Principal H-tree		
Wire width [μm]	3.0 ÷ 4.0	2.0 ÷ 2.5
Thermal heating [°C]	115 ÷ 125	117 ÷ 128
Sector H-trees		
Wire width [μm]	2.2 ÷ 2.4	2.1 ÷ 2.4
Thermal heating [°C]	111 ÷ 124	114 ÷ 125
Grid		
Wire width [μm]	1.0	0.8
Thermal heating [°C]	114 ÷ 125	116 ÷ 126
Total delay	850ps	855ps
Maximal skew	50ps	52ps

Frequency dependent effects, principally inductive ones, give rise to phenomena that a RC model cannot predict. It can be seen that for almost constant skew, wire width determined with RC model is overestimated in the order of 15-60% with respect to TL prediction, leading to a slightly lower interconnect heating profile. RC characterization predicts longer lifetime with respect to TL

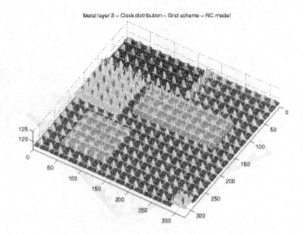

Metal layer 3 – Clock distribution – Grid scheme – RC model

Fig. 7. Temperature distribution in metal 3 grid. Z axis: temperature; X-Y axis: horizontal grid temperature distribution.

modelling (even if in both cases ten years lifetime have been achieved), but at the cost of electrical issues underestimate, like ringing and reflection, that are typical TL effects. Fig. 8, for example, shows that an RC characterization does not include those effects, that for same wire size are strongly present with the TL modelling. Thus, to reduce such degradation on propagating signal waveform, the HSPICE optimizer determined optimal wire width to be smaller.

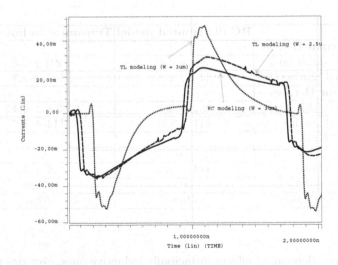

Fig. 8. Current waveforms with Π-cell RC distributed and TL modelling

This wire width reduction produces two main drawbacks from thermal issues point of view, since current density (12% higher) and line resistance are increased with respect to RC modelling, generating additional undesired contribution to Joule heating. However results show that lower TTF can be acceptable, considering that the only alternative could be the introduction of further buffering levels (causing certainly lower integration density and higher power consumption) and considering that ten years lifetime is anyway ensured.

Optimization with TL model gives rise to some other considerations. First, it is possible to see that estimated skew is only due to non uniform load distribution on the clock scheme and to non constant line resistance over equivalent wires. It can be finally noticed that temperature profile presents a low gradient as expected, since it is a necessary condition to guarantee low skew. In fact hotter and highly charged distribution sections will present wider cross section with respect to colder and lightly charged ones, to guarantee low skew.
Further improvements will consider the effect of non-uniform thermal substrate profile, as addressed in [1].

6   Conclusions

The importance of combining accurate electrical and thermal characterization of interconnects, to model undesired, but increasing Ultra Deep Submicron (UDSM) effects, has been highlighted and tested on an active, high performance clock distribution scheme. Both electromigration and timing constraints are used to optimize clock network sizes: the targets were a very low skew and a lifetime of 10 years for clock interconnects. Two different electrical models of the line have been used showing the relations between thermal and electrical phenomena in UDSM circuits and the importance of accurate modellization of high speed interconnects in order to avoid poor estimates.

The accuracy of the methodology proposed in this work relies on the simultaneous consideration of electrical and thermal aspects that are normally addressed separately in the design phases. Therefore it gives an answer to some emerging needs for high performance designs in the UDSM era.

References

1. A. H. Ajami, M. Pedram, and K. Banerjee. Effects on non-uniform substrate temperature on the clock signal integrity in high performance designs. *IEEE 2001 Custom Integrated Circuits Conference*, 2001.
2. S. I. Assoc. International Technology Roadmap for Semiconductors. Technical report, 2001.
3. Avant! Corporation. *Avant! Star-HSpice User's Manual.*
4. R. M. Averill et al. Chip integration methodology for the IBM S/390 G5 and G6 custom microprocessors. *IBM J. Res. Develop.*, 43(5/6), 1999.
5. M. Casu, M. Graziano, G. Masera, G. Piccinini, and M. Zamboni. Power supply bus sizing considering self-heating in bulk-to-soi migrated designs. *PATMOS – Power And Timing Modeling, Optimization and Simulation*, 2001.
6. D. Chen, E. Li, E. Rosenbaum, and S.-M. Kang. Interconnect thermal modeling for accurate simulation of circuit timing and reliability. *IEEE Transactions on Computer Aided Design of Integrated Circuits and Systems*, 19(2), 2000.

7. J. Cong, A. B. Kahng, D. Noice, N. Shirali, and S. H.-C. Yen. Analysis and justification of a simple, practical 2 1/2-D capacitance extraction methodology. *Design Automation Conference*, 1997.

8. Y. Ismail and E. Friedman. On-chip inductance in high speed integrated circuits. *Kluwer Academic Publishers*, Massachusetts, 2001.

9. M. Kamon, M. J. Tsuk, and J. K. White. FastHenry: A multiple accelerated 3-D inductance extraction program. *IEEE Transactions on Microwave Theory and Techniques*, 42(9), 1994.

10. W. H. Kao, C.-Y. Lo, M. Basel, and R. Singh. Parasitic extraction: Current state of the art and future trends. *Proceedings of the IEEE*, 89(5):729–739, 2001.

11. J. L. Maksiejewski. Evaluation of thermal characteristics of conductors under surge currents taking the skin effect into account. *IEE Proc.*, 137(2):85–91, 1990.

12. K. Nabors and J. White. FastCap: A multiple accelerated 3-D capacitance extraction program. *IEEE Transactions on Computer-Aided Design*, 10(11), 1991.

13. M. M. Prono. Techniques for high speed interconnect design. *Master thesis, University of Illinois at Chicago*, 2001.

14. S. D. Systems. Thunder user's manual. Technical report, 1993.

15. C.-C. Teng, Y.-K. Cheng, E. Rosenbaum, and S.-M. Kang. iTEM: A temperature-dependent electromigration reliability diagnosis tool. *IEEE Transactions on Computer Aided Design of Integrated Circuits and Systems*, 16(8), 1997.

A Technique to Generate CMOS VLSI Flip-Flops Based on Differential Latches

Raúl Jiménez[1], Pilar Parra, Pedro Sanmartín, and Antonio Acosta

Instituto de Microelectrónica de Sevilla-CNM / Universidad de Sevilla

Avda. Reina Mercedes s/n, 41012-Sevilla, SPAIN

Phone: +34-95-505-66-66; Fax: +34-95-505-66-86;acojim@imse.cnm.es

[1] also with the Universidad de Huelva, Spain

Abstract. In this comunication, a new technique to generate flip-flops based on differential structures is presented. This technique is based on the modification of size in transistors of existing differential latches. The limitations of the differential structures to apply this technique are few, so the range of application is high. The main application field is in mixed-signal analog-digital circuits, due to the low switching noise generated by these flip-flops. In this parameter, the behavior is similar in both the proposed flip-flop and the original structure, and better than existing flip-flops.

1   Introduction

The advances in integration technology allows the implementation of complex mixed-signal circuits in the same wafer. In high performance digital circuits, the use of pipe-lined architectures increases the throughput and the clock frequency, due to the parti-tioning of the circuit. In this technique, one of the most important circuit element is the flip-flop because is the building block of the pipeline stage. However, the extensive use of pipelining implies the growing importance of switching noise. The switching noise generated inside the digital part of the circuit produces deviation of the supply voltage from its ideal behavior. Traditionally, this parameter has been considered from an ana-log point of view, using layout techniques for instance [1, 2]. However, recently the study of this parameter is growing from a digital perspective, in the sense of obtaining digital families with low generation of noise [2, 3]. In this sense, two important results are that noise generation is higher in flip-flops [4, 5] and flip-flops based on differential structures generate lower level of noise [6].

The effect of switching noise can be seen as a reduction of the current supplied to the analog part, and hence, the performance is decreased. Then, an indirect possible measurement of switching noise is the peak of supply current due to the digital part. So, when this peak is higher, the current supplied to the analog part is lower and the reduc-tion of performance is higher [2, 3].

Differential structures are usually sensitive to level, and for this reason, their opti-

* This work has been sponsored by the Spanish MCYT TIC2000-1350 MODEL and TIC2001-2283 VERDI Projects

B. Hochet et al. (Eds.): PATMOS 2002, LNCS 2451, pp. 209–218, 2002.

mum use in pipeline stages is limited. So, it would be interesting to develop a technique in order to create structures sensitive to transition based on differential structures. Besides, another advantage of differential structures is the combinational block is not added as an additional block but it is integrated inside the structure.

This communication is divided as follows. First, we will do a brief introduction of level-sensitive differential structures, and following we will see the main differential structures sensitive to transition found in literature. After that, we present the new technique to generate flip-flops using differential structures sensitive to level. We will expose the simulation results. And finally, we will present the conclusions obtained.

2   Differential Structure Sensitive to Clock Level

The topology of a generic differential structure is shown in Fig 1. In these structures, we can distinguish a load block, that controls the operation and generates the -precharge- high level, and a differential NMOS tree, that generates the logic function. The operation is divided into two phases: an evaluation phase, in which the output signal acquires a valid value; and precharge phase, in which the output signal acquires a non valid value, named precharge data. Both phases are determined by the level of clock signal.

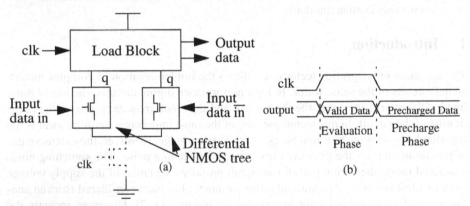

Fig. 1. (a) Scheme of a differential structure at a transistor level, and (b) its operation way.

In literature, we can found a lot of differential structures. The more known structures are Differential Cascode Voltage Switch Logic (DCVSL) [7], Differential NO RAce (DNORA) Logic [8], Enabled/disabled CMOS Differential Logic (ECDL) [9], Latched CMOS Differential Logic (LCDL) [10], Latched Differential Pass transistor Logic (LDPL) [11], Switched Output Differential Structure (SODS) [12], Sample/Set Differential Logic (SSDL) [13], EF^{2}CSL [14], etc. In all of them, the NMOS tree is the same and the change of their characteristics is due to the change of the load block and the output stages.

The main advantages of these structures when comparing to classical CMOS is their high speed, due to dynamic operation. They also implement a complex function with the delay of only one gate, because the function is implemented by the NMOS tree.

3   Differential Structure Sensitive to Clock Transition

The behavior of all early structure is sensitive to level of the clock signal. However, there exist several configurations of structure sensitive to transition. Among these configurations, the Static Single-Transistor-Clocked (SSTC) [15], the Sense Amplifier Flip-Flop (SAFF) [16] and the latch used in the K6 proccesor [17] are the most relevant.

The SSTC [15] is a master-slave configuration of differential latches, as we can see in Fig 2a. In the master structure, the input data are captured when the clock signal is low, while in the slave structure, the data are latched when the clock signal is high. In this case, the master and slave structures are different to allow static operation. In the slave structure, both inverters, neccesary to avoid the precharge data, must be weaker than the other transistors in order to allow the latching of new data.

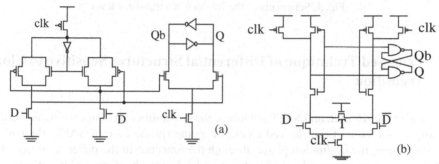

Fig. 2. Schematic of (a) SSTC and (b) SAFF at a transistor level.

The SAFF [16] is a flip-flop based on a differential latch (Fig 2b). The sensibility to transition is due to the action of transistor T. This transistor connects both outputs of NMOS tree, and so, no change in input data affect the logic value stored in the structure. In this case, the precharge data are filtered by a NAND RS latch.

The K6 latch [17] is a differential structure controlled by a skewed clock, that is, by the AND operation of the clock and its complementary signal, as we can see in Fig 3. With this operation, the active level of the equivalent control signal is reduced to the rising transition of the clock signal. In this particular case, the structure incorporates a self-reset property, and the data have not to be hold, then the precharge data are not filtered.

These solutions have not been traditionally applied to substitute more conventional flip-flops, mainly because in a master-slave configuration, the use of differential structures implies a higher number of transistors.

The SAFF solution can only be applied to a very reduced number of structures. It is necessary that the outputs of the NMOS tree are connected to the ground node of the load block because both outputs are going to have the same level in the evaluation phase. However, most of the level-sensitive differential structures do not show this property, in fact, they do not allow that the outputs of NMOS tree have the same value during the evaluation phase.

The solution of the skewed clock still presents a high number of transistors.

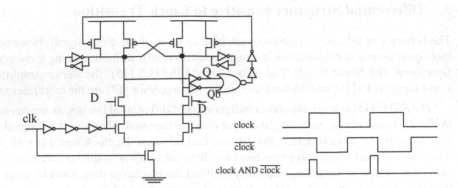

Fig. 3. Schematic of the K6 latch at a transistor level

4   Proposed Technique of Differential Structures Sensitive to Clock Transition

Analyzing the K6 latch and SAFF solutions, the sensibility to transition is achieved disabling (in the case of the skewed clock) or abling (in the case of SAFF) the path to ground during the evaluation phase, through the branches in the differential tree. The differential structure would only function as a latch at the beginning of this phase.

In order to maintain the output value, we propose the same solution used in SAFF, that is, the outputs of the differential structure are going to be connected to an output latch. Several configuration are used: a CMOS dynamic latch, as in LCDL and LDPL, an RS latch, as in SAFF, the kind of latch (NOR or NAND) will depend on the precharge data, etc. We are going to use an RS latch due to the problems of the CMOS dynamic latches [6].

The new proposed technique allows a drop of voltage in any branch of the tree. This drop would avoid the change of the output data if the input data change during the evaluation phase. The only effect in the output behavior would be a small degradation of signal, but the logic level would not be altered. To permit this drop, the differential tree will be composed of weak transistors in the sense of low W/L ratio. This technique is shown in Fig 4.

In the evaluation phase, an output node is connected to ground through the NMOS-tree and the other one is connected to supply through load block. When the input data change, the output connected to ground is isolated (there exists no connection to neither ground nor supply); while the complementary output is connected both to ground and to supply, creating a voltage divider. So, in order to guarantee a correct operation, the load block must be stronger, that is, less resistive, than the NMOS tree. In this case, there will exist a small degradation but the logic levels do not change.

From the simplified scheme, we can extract two secondary effects. Firstly, when the input data change during the evaluation phase, there exists a static path between supply and ground, and hence static power consumption. To evaluate the lost of performance

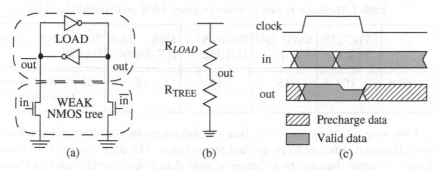

Fig. 4. (a) Simplified scheme of the new technique, (b) the equivalent electric circuit when the input data change during the evaluation phase, and (c) the waveforms.

the whole circuit must be considered including logic and pipeline blocks. Using differential structures, the combinational block can be included in them, so the increase of power consumption could be compensated. So, this parameter must be evaluated with combinational logic.

The second effect is the dynamic latching of one output of differential tree when the input data change during the evaluation phase. In this case, the output of a branch is connected to supply by the load block, and to ground by the tree; but the other output is not connected neither supply nor ground. So, the charge redistribution is problematic in this node. In order to eliminate this problem, we can add a latch that only stores the low level, that must be latched [18]. This modification is shown in Fig 5.

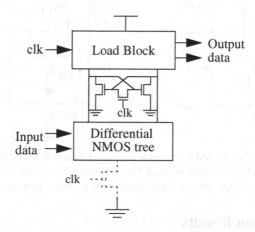

Fig. 5. Schematic of solution to avoid the problem of charge redistribution.

So, we have found a technique to build differential structures sensitive to transition that is competitive with existing flip-flops in terms of hardware resources (see table 1) The proposed SSDL flip-flop solution requiring 22 transistors uses an output NOR RS latch plus output inverters, while the 18 transistors solution uses an output NAND RS latch without output inverters.

Table 1. Hardware resources of the proposed SSDL and existing flip-flops.

	SSTC [15] (Fig 2a)	SAFF [16] (Fig 2b)	Power-PC [19] (Fig 7)	SSDL latch [13] (Fig 6a)	New SSDL flip-flop (Fig 6b)
Transistor count	16	18	18	14	22 (RS-NOR+inverters) 18 (RS-NAND)

Following, we are going to analyze the specifications and structural conditions of any differential structure to be applied this solution. The only restriction to avoid the change of output data due to a change in input data is making the load block stronger than the differential tree. So, this solution can not be applied to structures whose operation is based on the load block weaker than the tree, such DCVSL [7] where the high level is dynamically latched in precharge phase. Besides, the use of this technique does not imply great changes in the structure, basically the use of weak transistors in the tree and the change of NMOS switches by CMOS switches. Then, the range of application is greater than the range of the solution used in SAFF [16].

In order to quantify the lost of performance due to the new technique, it is going to be applied, as an example, to the SSDL structure [13]. In Fig 6, we show the schematic of the original SSDL latch and the proposed SSDL flip-flop. As the high level in the output of the tree is strong, the change only implies the use of the weak transistors in the tree, and to add an output NOR RS latch. The structure can be improved removing the output inverters and replacing the NOR RS latch by a NAND RS latch.

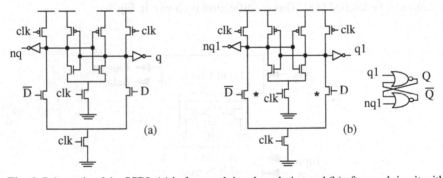

Fig. 6. Schematic of the SSDL (a) before applying the solution and (b) after applying it with output NOR latch. Transistors marked * are of weak size. Another solution saving hardware uses a NAND RS latch and removes the inverters generating q1 and nq1.

5   Simulation Results

The new technique has been evaluated through a comparison of the proposed SSDL flip-flop with previously reported flip-flops: SAFF, SSTC and the one used in the Power-PC proccesor [19] (Fig 7), as samples of conventional flip-flops with well-known good behavior. The flip-flop used in K6 is not considered in the comparison because its self-reset property.

The simulations have been done with HSPICE in a standard CMOS 0.35 µm tech-

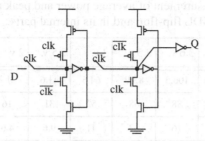

Fig. 7. Schematic of the flip-flop used in Power-PC proccesor at a transistor level.

nology. In Fig 8, we show the circuit simulated and the input pattern. In this pattern, the input data change in both levels of clock signal, in order to show the correct behavior of the flip-flops. In the comparison, the most important digital parameters considered are propagation delay, power consumption and power-delay product. The switching noise has been measured through the peak of supply current.

Firstly, we are going to study the difference of performance with the original latch due to the change of functionality. In the case of propagation delay, the increase of the new SSDL is due to the NOR RS latch. This parameter can be improved removing the output inverters and replacing the NOR RS latch by a faster NAND RS latch.

In the case of power consumption, the new SSDL shows a higher power consumption due to the static power consumption in the third and forth changes of input data. For this reason, the power-delay product also is increased. However the excess of power consumption is removed if changes in D input take place only in the low level of clock signal.

In the case of the peak of supply current, as measurement of switching noise, the new SSDL shows a higher peak due to the adding of the RS latch, but this overhead is not excessive. Anyhow, we are comparing a transparent latch with an edge-triggered flip-flop, with more complicated functionality.

Following, we are going to study the internal structure of the proposed SSDL flip-flop. Three parts have been separately considered: the inverter of input data (inv), the NOR RS latch (rs) and the differential structure (dif). In table 2, we show the parameters obtained by simulation, for the whole structure and for the internal parts.

We can see a decrease of the power consumption with supply voltage, as well as for the peak of supply current. One interesting result is that the differential structure generates the same or less noise than the RS latch or the inverter; besides, the operation of the three parts is not simultaneous, then their contributions to noise are not added.

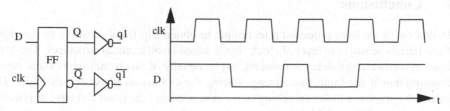

Fig. 8. Circuit under simulation and input pattern

Table 2. Simulation measurement of average power and peak noise in the proposed SSDL flip-flop and in its internal parts.

Parameter	3.6 v.	3.3 v.	3 v.	2.5 v.	2 v.	1.5 v.	1 v.
Total average power (μW.)	106.3	82.9	64.5	41.6	21.3	10.5	5.8
pow_avg_dif (μW.)	88	68	52	31	16	6.8	4.6
pow_avg_rs (μW.)	16	13	11	9.6	4.6	3.3	1.0
pow_avg_inv (μW.)	2.3	1.9	1.5	1.0	0.7	0.4	0.2
I_{VDD} peak (mA.)	0.73	0.62	0.52	0.37	0.24	0.15	0.08
I_{VDD} peak_dif (mA.)	0.55	0.49	0.42	0.37	0.24	0.15	0.08
I_{VDD} peak_rs (mA.)	0.68	0.59	0.48	0.33	0.19	0.11	0.03
I_{VDD} peak_inv (mA.)	0.64	0.57	0.50	0.37	0.24	0.14	0.07

Also, we have considered a modified SSDL with a NAND RS latch, without the output inverters. In this case, the output inverters can be eliminated because their function, isolating the differential structure, is done by the RS latch. This new cell improves in terms of speed, power and power-delay product. However, the results presented in this paper were obtained with the NOR RS latch, in order to maintain a minimum alteration with the original differential structure.

Secondly, we are going to do the comparison with the other flip-flops. In Fig 9, we show the power-delay product and the peak of supply current obtained by simulation. These parameter have been measured for several supply voltages (from 3.6 v. to 1 v). It can be seen the bad behavior of the proposed SSDL flip-flop due to static power consumption. The flip-flop with best behavior is the one used in Power-PC. However, this behavior would be taken into account as approximative. The final result must consider the combinational block inside the pipelined circuit. Also, it can be noticed a minimum value at 1.5 v., except in the proposed SSDL flip-flop.

In the case of peak of supply current, The cell with worst behavior is the Power-PC, while the cell with best behavior is the proposed SSDL. When the supply voltage goes down, the difference among different cells also decreases, but the ranking is kept. Another main result of the graphic is that flip-flops based on differential structures generate less noise.

6   Conclusions

In this paper we have proposed a technique to obtain flip-flops sensitive to transition from latches sensitive to level of clock signal, based on differential structures. The overhead in hardware and delay is reduced, but he penalty of this technique is static power consumption if the input data change during the evaluation phase (clock high). The main advantage is a lower switching noise generated and the great and easy applicability to most of differential latches.

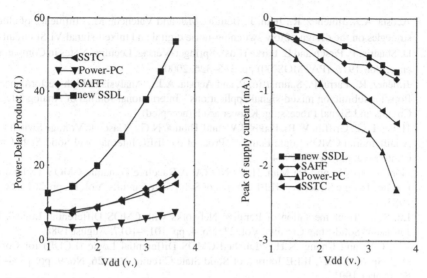

Fig. 9. Simulation results for the flip-flops: power-delay product and the peak of supply current.

The flip-flops generated with this technique produce a similar noise than the original latches, that, furthermore, is lower than the noise generated by conventional flip-flops. These results have been demonstrated via an application to a well known differential structure (SSDL), although it can be applied to most of differential latches, obtaining advantages in terms of noise in most of them. Thus, these flip-flops are recommended for mixed-signal applications.

With respect to current steering families, this technique generates a similar noise level, less power consumption and restored logic levels, making unnecessary conversion circuitry. As in the current steering logic families, this technique would be employed in the boundary with the analog blocks because the static power can be reduced and the generation of noise becomes the most limiting factor.

References

1. Tsividis, Y., "Mixed Analog-Digital VLSI Design and Technology". Ed. McGraw-Hill, 1995. ISBN: 0-07-065402-6.
2. X. Aragonès, J. L. González and A. Rubio, "Analysis and Solutions for Switching Noise Coupling in Mixed-Signal ICs". Kluwer Academic Pubs, 1999.
3. Allstot, D. J., Chee, S-H. and Shrivastawa, M.: Folded source-coupled logic vs. CMOS static logic for low-noise mixed-signal ICs. IEEE Transactions on Circuits and Systems I, vol 40, pp 553–563, Sept. 1993.
4. Jiménez, R., Acosta, A.J., Peralías, E.J. and Rueda, A., "An application of self-timed circuits to the reduction of switching noise in analog-digital circuits", in D. Soudris, P. Pirsch and E. Barke (Eds), Springer-Verlag, Lectures Notes in Computer Science, Vol. 1918, (PATMOS'00) pp. 295–305. 2000.

5.   Acosta, A.J., Jiménez, R., Juan, J., Bellido, M.J. and Valencia, M., "Influence of clocking strategies on the design of low switching-noise digital and mixed-signal VLSI circuits", in D. Soudris, P. Pirsch and E. Barke (Eds), Springer-Verlag, Lectures Notes in Computer Science, Vol. 1918, (PATMOS'00) pp. 316–326. 2000.

6.   Jiménez, R., Parra, P., Sanmartín, P. and Acosta, A.J.: "Analysis of high-performance flip-flops for submicron mixed-signal applications", International Journal of Analog Integrated Circuits and Signal Processing. Kluwer acad. (accepted)

7.   Heller, L.G., Griffin, W.R., Davis, J.W. and Thoma, N.G.: "Cascode Voltage Switch Logic: A Differential CMOS Logic Family", Proc. of the IEEE International Solid-State Circuits Conference, pp. 16–17. 1984

8.   Gonçalves, N.F. and De Man, H.J.: "NORA: A Racefree Dynamic CMOS Technique for Pipelined Logic Structure", IEEE Journal of Solid State Circuits, Vol. 18, pp. 261–266. June 1983.

9.   Lu, S.L.: "Implementation of Iterative Networks with CMOS Differential Logic", IEEE Journal of Solid-State Circuits, Vol. 23, No. 4, pp. 1013–1017. August 1988.

10. Wu, C.Y. and Cheng, K.H.: "Latched CMOS Differential Logic (LCDL) for Complex High-Speed VLSI", IEEE Journal of Solid State Circuits, Vol. 26, No. 9, pp. 1324–1328. September 1991.

11. Salomon, O. and Klar, H.: "Self-Timed Fully Pipeline Multiplier", IEEE Transaction on Computer Science and Technology, pp. 45–55, 1993.

12. Acosta, A.J., Valencia, M., Barriga, A., Bellido, M.J. and Huertas, J.L.: "SODS: A New CMOS Differential-type Structure". IEEE Journal of Solid-State Circuits, Vol. 30, No. 7, pp. 835–838. July 1995.

13. Grotjohn, T.A. and Hoefflinger, B.: "Sample-Set Differential Logic (SSDL) for Complex High-Speed VLSI", IEEE Journal of Solid State Circuits, Vol. 21, No. 2, pp. 367–369. April 1986.

14. Kundan, J. and Rezaul, S.M.: "Ehanced Folded Source-Coupled Logic Technique for Low-Voltage Mixed-Signal Integrated Circuits", IEEE Transc. on Circuits and Systems II, pp. 810–817, August 2000.

15. Yuan, J. and Svensson, C., "New single-clock CMOS latches and flip-flops with improved speed and power savings", IEEE J. Solid-States Cirtuis, vol. 32, pp. 62–69, Jan. 1997.

16. Montanaro, J., Witek, R.T., Black, A.J., Cooper, E.M., Dobberpuhl, D.W., Donahue, P.M., Eno, J., Hoeppner, G.W., Kruchemyer, D., Lee, T., Lin, P.M., Madden, L., Murray, D., Pearce, M.H., Santhanam, S., Snyder, K.J., Stephany, R. and Thierauf, S.C., "A 160-MHz, 32-b, 0.5 W CMOS RISC microprocessor", IEEE J. Solid-state Circuits, vol. 31, pp. 1703–1714, 1996.

17. Draper, D., Crowley, M., Holst, J., Favor, G., Schoy, A., Trull, J., Ben-Meir, A., Khanna, R., Wendell, D., Krishna, R., Nolan, J., Mallick, D., Partovi, H., Roberts, M., Johnson, M. and Lee, T., "Circuit Techniques in a 266-MHz MMX-enabled processor", IEEE J. Solid-State Circuits, vol. 32, pp.1650–1664, Nov. 1997.

18. Rosemberger, F.U., Molnar, C.E., Chaney, T.J. and Fang, T.P.: "Q-Modules: Internally Clocked Delay-Insensitive Modules", IEEE Trans. on Computer, vol. 37, no. 9, pp. 1005–1018, September 1988.

19. Gerosa, G., Gary, S., Dietz, C., Dac, P., Hoover, K., Alvarez, J., Sanchez, H., Ippolito, P., Tai, N., Litch, S., Eno, J., Golab, J., Vanderschaaf, N. and Kahle, J., "A 2.2 W, 80 MHz supescalar RISC microprocessor", IEEE J. Solid-State Circuits, vol. 29, pp. 1440–1452, Dec. 1994.

A Compact Charge-Based Propagation Delay Model for Submicronic CMOS Buffers

José Luis Rossello and Jaume Segura

Departament de Física, Universitat Illes Balears, 07071 Palma de Mallorca, Spain

Abstract. We provide an accurate analytical expression for the propagation delay and the output transition time of submicron CMOS buffers that takes into account the short-circuit current, the input-output coupling capacitance, and the carrier velocity saturation effects, of increasing importance in deep-submicron technologies. The model is based on the nth-power law MOSFET model and computes the propagation delay from the charge delivered to the gate. Comparison with HSPICE level 50 simulations and other previously published models for a $0.35\mu m$ and a $0.18\mu m$ process technologies show significant improvements over previously-published models.

1 Introduction

Timing analysis is one of the most critical topics in VLSI design. The nonlinear behavior of CMOS gates requires numerical procedures for accurate timing analysis at expenses of large computation times. Moreover, the impact of design parameters such as fan in, fan out or transistor sizes on the propagation delay are difficult to understand and optimize using numerical procedures.

The dynamic behavior of submicron CMOS buffers depends on several nonlinear effects like the velocity saturation of carriers due to the high electric fields in submicron technologies, the short circuit current appearing when both pMOS and nMOS transistors are conducting simultaneously [1], and the additional effect of the input-output coupling capacitance [2].

Several methods have been proposed to derive the delay of CMOS buffers [2]-[7] as a first step to describe more complex gates [8,9]. Cocchini et al. [3] obtained a piece-wise expression for the propagation delay based on the BSIM MOSFET model [10]. The model included overshooting effects (due to the input-to-output coupling capacitance) while the short-circuit current was neglected. In [2] and [4] K.O Jeppson and L. Bisdounis presented a model for the output response of CMOS buffers using a quadratic current-voltage dependence for MOSFET devices, which is not longer valid for submicron technologies. Daga et. al. [5] obtained a simple empirical expression for the propagation delay taking into account both overshooting and short-circuit currents using six fitting parameters. The relative error of this model was 19% for a $0.6\mu m$ technology as reported in [5]. Hirata et al. [6] derived a delay model based on the nth-power law MOSFET model [11] considering both short-circuit and overshooting currents and

B. Hochet et al. (Eds.): PATMOS 2002, LNCS 2451, pp. 219–228, 2002.

using numerical procedures . The model provides an accurate description for the propagation delay but the numerical procedures used increases the computation time considerably. Bisdounis et al. [7] developed a piece-wise solution with seven operation regions for the transient response of a CMOS inverter based on the α-power law MOSFET model [12] including both overshooting and short-circuit currents. In [8] T. Sakurai et. al. obtained a simple expression for the propagation delay of CMOS gates based on their nth-power law MOSFET model neglecting both short-circuit and overshooting currents.

In this work we propose a compact analytical model to accurately compute the propagation delay and the output transition time of a CMOS buffer accounting for the main effects of submicron technologies as the input-output coupling capacitance, carriers velocity saturation effects and short-circuit currents. The model is based on an accurate physically-based nth-power law MOSFET model [13] and on a power dissipation model for CMOS inverters [14]. Comparisons with HSPICE level 50 simulations and previously published models for a $0.35\mu m$ and a $0.18\mu m$ process technologies are reported showing significant improvements in terms of accuracy.

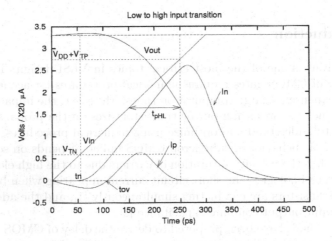

Fig. 1. In this figure we show the CMOS current and voltage switching characteristics.

This paper is organized as follows: In Section 2 the CMOS buffer switching characteristics are analyzed with detail and the MOSFET model used is presented. The delay and the output transition time models are developed in Section 3 and compared to HSPICE simulations and other previously published models for a $0.35\mu m$ and a $0.18\mu m$ process technology in section 4. Finally in section 5 we conclude the work.

2   Analysis of the CMOS Buffer Switching Characteristics

The dynamic behavior of a CMOS buffer is described by the next equation:

$$(C_L + C_M)\frac{dV_{out}}{dt} = I_p - I_n + C_M\frac{dV_{in}}{dt} \tag{1}$$

where C_L is the output capacitance, V_{out} and V_{in} are the output and input voltage respectively, while I_p and I_n are the current that crosses the pMOS and the nMOS transistor respectively. C_M is the input to output coupling capacitance which is voltage dependent. The static value of C_M when the input is low (C_M^L) is computed considering the side-wall capacitances of both transistor drains and the gate to drain capacitance of the pMOS transistor that operates in the linear region as:

$$C_M^L = C_{ox}\left(\frac{W_{p_{eff}}L_{p_{eff}}}{2} + L_{D_p}W_{p_{eff}} + L_{D_n}W_{n_{eff}}\right) \tag{2}$$

with $W_{p_{eff}}$ and $W_{n_{eff}}$ being the effective channel width of pMOS and nMOS respectively, $L_{p_{eff}}$ is the effective channel length of pMOS, while L_{D_n} and L_{D_p} are the gate-drain underdiffusion for the nMOS and pMOS transistors respectively. For a static input high the capacitance C_M^H is obtained similarly. In this work a mean value for the coupling capacitance during the transition $(C_M = 0.5\left(C_M^L + C_M^H\right))$ will be used.

Fig. 1 illustrates the input and output voltage evolution of the buffer along with the current through the nMOS and pMOS transistors for a low to high input transition. The current through the pMOS transistor (I_p in Fig. 1) has two components clearly distinguished by the sign of the current. The negative pMOS current is due to a partial discharge of the output capacitance from the output node toward the supply rail and appears when the input-output capacitance drives the output voltage beyond the supply value (V_{DD}) at the beginning of the transition [2] (this effect is known as overshooting). When the nMOS device starts to conduct, it pulls the output voltage down. Once the output voltage goes below V_{DD}, the pMOS current is positive corresponding to the short-circuit component due to the simultaneous conduction of both devices.

The propagation delay (defined as t_{pHL} for a high to low output transition) is typically defined as the time interval from the 50% V_{DD} input voltage to the 50% V_{DD} output voltage. The dependence of the propagation delay with design parameters is non-linear and difficult to model given that eq.(1) can not be solved in a closed form even using the simple Shockley MOSFET model [15]. Moreover carrier saturation effects become important with technology scaling and more complex MOSFET models accounting for such effects must be considered.

The *nth*-power law MOSFET model [11] is a widely used short-channel drain current model, and will be used in this work to derive the propagation delay and the output transition time of CMOS inverters. The drain current is expressed as:

$$I_D = \begin{cases} 0 & (V_{GS} \leqq V_{TH}) \\ (2 - \frac{V_{DS}}{V''_{D0}})\frac{V_{DS}}{V''_{D0}}I'_{D0} & (V_{DS} < V'_{D0}) \\ I'_{D0} & (V_{DS} \geqq V'_{D0}) \end{cases} \tag{3}$$

with

$$I'_{D0} = I_{D0} \left(\frac{V_{GS} - V_{TH}}{V_{DD} - V_{TH}} \right)^n \tag{4}$$

where V_{GS}, V_{DD}, and V'_{D0} are the gate, supply, and saturation voltage respectively and I_{D0} is the drain current at $V_{GS} = V_{DS} = V_{DD}$. The parameter n is the velocity saturation index that ranges between 2 (long-channel devices) and 1 (short-channel) [11]. The saturation voltage V'_{D0} is given by:

$$V'_{D0} = V_{D0} \left(\frac{V_{GS} - V_{TH}}{V_{DD} - V_{TH}} \right)^m \tag{5}$$

The parameter V_{D0} is the saturation voltage at $V_{GS} = V_{DD}$, while m and V_{TH} are empirical parameters [11]. These equations are mathematically simpler than physically-based MOSFET models such as BSIM3v3 or MM9 with the disadvantage that, in the original model developed by Sakurai and Newton, the relationship between the empirical and the process parameters supplied by manufacturers is not provided. Therefore the variation of nth-power law model predictions with key parameters like the supply voltage are not taken into account in the original formulation performed by Sakurai and Newton, where each parameter must be recomputed if the supply voltage or some device dimension are changed. In this work we use the physical formulation proposed in [13] and used in [14]. This physical formulation provides an analytical relationship between the nth-power law parameters and the more accurate MM9 model parameters (that take into account the parameter variations with the supply voltage, MOSFET dimensions and temperature).

3   Delay Model

We compute the propagation delay when the input voltage switches. The short-circuit and overshooting currents are first neglected and incorporated later. Assuming a linear variation of the input voltage with rise time t_{in} then $V_{in}(t)$ is:

$$V_{in}(t) = V_{TN} + (V_{DD} - V_{TN}) \frac{t - t_n}{t_{in} - t_n} \tag{6}$$

where V_{TN} is the nMOS threshold voltage, t_n is the time when the nMOS chain starts to conduct ($t_n = V_{TN}t_{in}/V_{DD}$) and t_{in} is the input rise time. At the beginning of the transition the nMOS is off and $V_{out} = V_{DD}$. At $t = t_n$, the nMOS starts to conduct and the output voltage is obtained solving:

$$C_L \frac{dV_{out}}{dt} = -I_n \tag{7}$$

where C_L is the total output capacitance of the gate, and I_n is the current through the nMOS transistor. An analytical solution to (7) is possible if the nMOS transistor is assumed to be in the saturation region (valid while $V_{out} > V'_{DO_n}$):

$$V_{out} = V_{DD} - \frac{I_{DO_n}}{C_L} \left(\frac{t - t_n}{t_{in} - t_n}\right)^{n+1} \frac{t_{in} - t_n}{n+1} \tag{8}$$

Equation (8) is used to obtain the propagation delay from the input at $0.5V_{DD}$ to the output at $0.5V_{DD}$.

$$t_{pHL_1} = t_n + \left[\frac{Q_f(n+1)}{I_{DO_n}}\right]^{\frac{1}{1+n}} (t_{in} - t_n)^{\frac{n}{n+1}} - \frac{t_{in}}{2} \tag{9}$$

where $Q_f = C_L V_{DD}/2$ is the charge transferred by the nMOS transistor when the output reaches $V_{DD}/2$ while parameters I_{DO_n} and n are the maximum saturation current and the velocity saturation index of nMOS respectively. Equation (9) is valid for slow inputs (defined when $t_{pHL} \leq t_{in}/2$). If Q_{f_0} is defined as the total charge transferred through the nMOS transistor when $t_{pHL} = t_{in}/2$ then eq. (9) is valid in the interval $Q_f < Q_{f_0}$. The parameter Q_{f_0} is obtained equating $t_{pHL_1} = t_{in}/2$ and solving Q_{f_0}

$$Q_{f_0} = \frac{I_{DO_n}}{(n+1)} (t_{in} - t_n) \tag{10}$$

If the input reaches the supply voltage before the output is at $V_{DD}/2$ then $Q_f > Q_{f_0}$ and eq. (9) is no longer valid. The propagation delay when $t_{pHL} \geq t_{in}/2$ (fast input range) can be obtained by solving (7) for $I_n = I_{DO_n}$ leading to:

$$t_{pHL_2} = \frac{t_{in}}{2} + \frac{Q_f - Q_{f_0}}{I_{DO_n}} \tag{11}$$

Equation (11) is valid when $Q_f > Q_{f_0}$ (fast input range). For simplicity, the proposed model for the propagation delay (eqs. (9) and (11)) does not take into account the fact that the nMOS transistor is in the linear region when $V_{out} > V'_{DO_n}$. For the evaluation of the output fall time (t_f) we first compute the output voltage slope at $V_{DD}/2$ from (8)

$$\left.\frac{dV_{out}}{dt}\right|_{V_{DD}/2} = \begin{cases} -\frac{I_{DO_n}V_{DD}}{2Q_f} \left(\frac{Q_f(n+1)}{(t_{in}-t_n)I_{DO_n}}\right)^{\frac{n}{n+1}} & (Q_f < Q_{f_0}) \\ -\frac{I_{DO_n}V_{DD}}{2Q_f} & (Q_f > Q_{f_0}) \end{cases} \tag{12}$$

For the evaluation of the output fall time we use:

$$t_f = \frac{V_{DD}}{\left.\frac{dV_{out}}{dt}\right|_{V_{DD}/2}} \tag{13}$$

The value of the output fall/rise time is important since this parameter is used as the input fall/rise time of the gates driven by the buffer.

3.1   Including Short-Circuit Currents

The model proposed cannot be used for static CMOS gates since short-circuit currents are not considered. In this work we include the short-circuit current contribution to the delay as an additional charge that must be transferred through the pull-down (pull-up) network during an output falling (rising) transition. This additional charge is computed from the short-circuit power model presented in [14].

Therefore, for an output falling transition, the charge transferred through the nMOS transistor is computed as $Q_f = C_L V_{DD}/2 + q_{sc}^f$, where q_{sc}^f is defined as the short-circuit charge transferred during the falling output transition until $V_{out} = V_{DD}/2$. The analytical derivation of q_{sc}^f is complex given that this parameter depends on the relative switching speed between the input and the output. For an input transition faster than the output, q_{sc}^f can be modeled as the total short-circuit charge transferred (defined as Q_{sc}^f) given that when $V_{out} = V_{DD}/2$ the input is high and the short-circuit current ceased. For an input transition slower than the output then $q_{sc}^f < Q_{sc}^f$. For both cases we assume that $q_{sc}^f = \kappa Q_{sc}^f$, where κ is an empirical parameter that must be optimized for each technology. For the $0.35\mu m$ and the $0.18\mu m$ technologies considered we obtained $\kappa = 0.45$ and $\kappa = 0.73$ respectively.

3.2   Including Overshooting Effects

Similarly to the short-circuit currents case, overshooting currents effects are included into the delay model as an additional charge to be transferred through the nMOS transistor. For fast inputs, the charge injected through the coupling capacitor (C_M) when $V_{out} = V_{DD}/2$ is $Q_{ov} = C_M V_{DD}$. For simplicity we assume the same value for Q_{ov} in the slow-input case. Therefore, the total charge that must be transferred through the nMOS transistor during a falling output transition is:

$$Q_f = 0.5\left[\left(C_L + 2C_M\right)V_{DD}\right] + \kappa Q_{sc}^f \tag{14}$$

Equation (14) must be used in eqs. (9), (11) and (13).

4   Results

We plotted model results vs. HSPICE level 50 simulations for a $0.35\mu m$ and a $0.18\mu m$ technologies. Results show the propagation delay for different values of the input transition time t_{in}, the configuration ratio W_p/W_n and the supply voltage V_{DD}.

In Fig. 2 we plot the propagation delay t_{pHL} vs. the input time t_{in} for different values of the W_n/W_p ratio for a $0.35\mu m$ technology. HSPICE simulations (dots) are compared to the model proposed and to a previous model [3]. Short-circuit currents are not taken into account in [3] leading to an underestimation of the propagation delay. The model in [3] provides a piece-wise solution of the propagation delay: depending on the input transition (fast or slow input transitions) it uses an approximated or an exact expression for the propagation delay.

The approximated propagation delay is used when the nMOS transistor changes from the saturation to the linear region and the input is rising (eq.(11) in [3]), otherwise an exact expression for the output response is used. This pice-wise solution leads to a discontinuity in the propagation delay when changing from one region to the other (see Fig. 2).

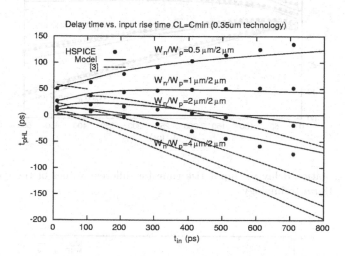

Fig. 2. Propagation delay vs. input rise time for different values of the configuration ratio. Short-circuit currents are not taken into account in [3]

Fig. 3 plots the propagation delay vs. the input rise time for a $0.18\mu m$ technology. When the W_p/W_n ratio is small the propagation delay decreases when increasing the input rise time. The model proposed in this work (solid lines) and the previously-published in [6,7] provide a good approximation to HSPICE simulations (dots).

Fig. 4 is a plot of HSPICE simulations (dots) and model predictions of the propagation delay vs. the supply voltage. The model proposed in this work provides a better fitting than the models in [6,7]. The model in [7] uses Taylor series expansions for the output response. For large supply voltage values the input transition is slow with respect the output response and the output voltage crosses $V_{DD}/2$ when both nMOS and pMOS transistors are in saturation, (called region 3 in [7]) and the output response used to compute the propagation delay is described through a Taylor series expansion. For the voltage range $(0.6V < V_{DD} < 1V)$ the propagation delay is obtained computing the output voltage when the nMOS is saturated, the pMOS off and $V_{in} < V_{DD}$ (region 4 in [7]) and a quadratic Taylor series expansion of the output voltage around time $t_{1-p} = t_{in}\left(1 - \frac{Vtn}{V_{DD}} - \frac{|Vtp|}{V_{DD}}\right)$ is used to compute the propagation delay. As V_{DD} is further reduced, the accuracy of the approximated output voltage decreases because the time point t_{1-p} used in the Taylor expansion is reduced. For low

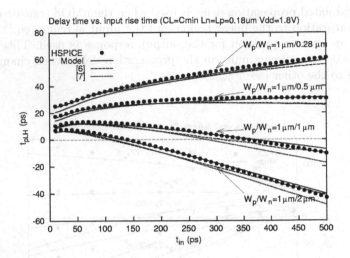

Fig. 3. Propagation delay vs. input rise time for different values of the configuration ratio for a $0.18\mu m$ technology

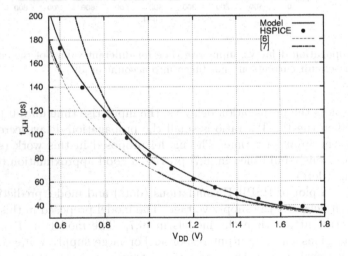

Fig. 4. Propagation delay vs. supply voltage for a $0.18\mu m$ technology.

values of the supply voltage, $V_{DD} < 0.6V$, the propagation delay is computed for the nMOS saturated and $V_{in} = V_{DD}$ (region 5A) and an exact solution for the output response is used. Therefore, there are two discontinuities: between regions 4-5A ($V_{DD} = 0.6V$) and between regions 3 and 4 ($V_{DD} = 1V$).

In Fig. 5 we plot HSPICE simulation of the output transition time t_f for different values of the input rise time t_{in} and the configuration ratio W_p/W_n.

A maximum relative error of 15% is obtained between model predictions and HSPICE simulations. In general a good agreement is obtained with the proposed model.

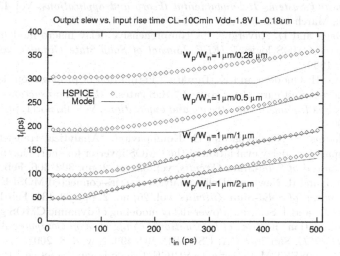

Fig. 5. Propagation delay vs. supply voltage for a $0.18\mu m$ technology.

5   Conclusions

An accurate analytical expression to compute the propagation delay and the output transition time of CMOS buffers has been presented. The main effects present in current submicron CMOS technologies like the input-output coupling capacitance, carriers velocity saturation effects and short-circuit currents are taken into account in the analysis. The model is compared to HSPICE simulations (level 50) and other previously published works for a $0.18\mu m$ and a $0.35\mu m$ process technology reporting a high degree of accuracy. The model represents an improvement with respect to previously published works.

References

1. H.Veendrick, "Short-circuit dissipation of static CMOS circuitry and its impact on the design of buffer circuits," *IEEE J. Solid-State Circuits*, vol SC-19, pp.468–473, 1984.
2. K.O.Jeppson "Modeling the influence of the transistor gain ratio and the input-to-output coupling capacitance of the CMOS inverter delay," *IEEE Journal of Solid-State Circuits*, vol. 29, no. 6, pp. 646–654, June 1994.

3. P. Cocchini, G. Piccinini and M. Zamboni, "A comprehensive submicrometer MOST delay model and its application to CMOS buffers," *IEEE Journal of Solid-State Circuits*, Vol. 32, no. 8, pp. 1254–1262, August 1997.
4. L. Bisdounis, S. Nikolaidis and O. Koufopavlou, "Propagation delay and short-circuit power dissipation modeling of the CMOS inverter," *IEEE Transactions on Circuits and Systems I: Fundamental theory and applications*, Vol 45, no.3, pp. 259–270, March 1998.
5. J. M. Daga and D. Auvergne, "A comprehensive delay macro modeling for submicrometer CMOS logics," *IEEE Journal of Solid-State Circuits*, vol. 34, no.1, January 1999.
6. A. Hirata, H. Onodera and K. Tamaru, "Estimation of propagation delay considering short-circuit current for static CMOS gates," *IEEE Transactions on Circuits and Systems I: Fundamental theory and applications*, Vol. 45, no.11, pp. 1194–1198, Nov. 1998.
7. L. Bisdounis, S. Nikolaidis and O. Koufopavlou, "Analytical transient response and propagation delay evaluation of the CMOS inverter for short-channel devices," *IEEE Journal of Solid-State Circuits*, vol. 33, no. 2, pp. 302–306, Feb. 1998.
8. T. Sakurai and R. Newton, "Delay analysis of series-connected MOSFET circuits," *IEEE Journal of Solid-State Circuits*, vol. 26, no. 2, pp.122–131, Feb 1991.
9. J.L.Rosselló and J. Segura, "Power-delay modeling of dynamic CMOS gates for circuit optimization" in *Proc. of International Conference on Computer-Aided Design (ICCAD 2001)*, San José CA, USA, PP. 494–499, Nov. 4–8, 2001
10. D. Foty, "MOSFET Modeling with SPICE. Principles and practice," Prentice Hall, 1997.
11. T. Sakurai and R. Newton, "A simple MOSFET Model for Circuit Analysis," *IEEE Transactions on Electron Devices*, vol. 38, pp. 887–894, Apr. 1991.
12. T.Sakurai and R. Newton, "Alpha-power law MOSFET model and its implications to CMOS inverter delay and other formulas," *IEEE Journal of Solid-State circuits*, vol 25, no.2, pp. 584–594, April 1990.
13. J.L.Rosselló and J. Segura, "A physical modeling of the alpha-power law MOSFET model" in *Proc. of 15th Design of Circuits and Integrated Systems Conference*, Montpellier, France, Nov. 21-24, 2000 pp. 65–70.
14. J.L.Rosselló and J. Segura, "Charge-based analytical model for the evaluation of power consumption in sub-micron CMOS buffers," *IEEE Transactions on Computer-Aided design*, vol 21, no. 4, pp. 433–448, April 2002.
15. W.Shockley, "A unipolar field effect transistor," *Proc IRE*, vol 40, pp. 1365–1376, Nov 1952.

Output Waveform Evaluation of Basic Pass Transistor Structure*

S. Nikolaidis[1], H. Pournara[1], and A. Chatzigeorgiou[2]

[1] Department of Physics, Aristotle University of Thessaloniki
[2] Department of Applied Informatics, University of Macedonia
54006 Thessaloniki, Greece
hpour@skiathos.physics.auth.gr

Abstract. Pass transistor logic is a promising alternative to conventional CMOS logic for low-power high-performance applications due to the decreased node capacitance and reduced transistor count it offers. However, the lack of supporting design automation tools has hindered the widespread application of pass transistors. In this paper, a simple and robust modeling technique for the timing analysis of the basic pass transistor structure is presented. The proposed methodology is based on the actual phenomena that govern the operation of the pass transistor and enables fast timing simulation of circuits that employ pass transistors as controlled switches without significant loss of accuracy, compared to SPICE simulation.

1 Introduction

Pass transistor logic is being increasingly used in digital circuits due to the advantages that it offers compared to other logic families for a class of logic functions. The use of pass transistors as transfer gates is a promising approach in reducing the physical capacitance being switched in a circuit and in this way offers significant power savings and speed improvement over conventional CMOS implementation. Pass transistor logic styles are very efficient in terms of transistor count for designs that employ the XOR and MUX operation [1] and as a result very compact and fast full adder implementations have been proposed [2], [3], [4].

Although pass transistor logic is attractive for low-power high-performance circuit design it is rarely the logic style of choice for actual designs. The main reason behind this limited application of pass transistor logic, are not the inherent problems of pass transistors such as the threshold drop or the need for level restoring devices, as it is widely believed. Rather, it is the lack of appropriate design automation tools that can support pass transistor implementation during all phases of the system design hierarchy. One aspect of this scarcity in tools can be identified in the limited number of fast timing analysis techniques for pass transistors.

Over the last decade modeling techniques for static CMOS gates, with emphasis on the inverter, have matured to offer significant speed improvement over SPICE-

* This work was supported by AMDREL project, IST-2001-34379, funded by the European Union"

B. Hochet et al. (Eds.): PATMOS 2002, LNCS 2451, pp. 229–238, 2002.
© Springer-Verlag Berlin Heidelberg 2002

based simulators and a level of accuracy, which is acceptable for most applications [5]-[10]. However, the difficulty in solving the circuit differential equations for pass-transistor structures has resulted in a limited number of modeling techniques for such circuits. Among them is a delay-macromodeling technique for transmission gates [11], a simplified analysis of a single pass transistor driven by a step input [12] and a semi-analytical approach in modeling CPL gates by partitioning into smaller sub-circuits [13].

In this paper, the analysis of the basic pass transistor structure, namely an nMOS pass-transistor with one terminal driven by the output of a previous logic stage and the other connected to a single capacitance, will be presented. The operation of this circuit will be analyzed for a constant driving signal and a rising ramp applied at the gate of the pass transistor. This scheme, resembles an often use of pass-transistors in actual designs as a controlled switch which transfers or not the input signal, depending on a control signal, which usually arrives later.

The rest of this paper is organized as follows: Section 2 describes the mode of operation of the pass-transistor and the formulation of the circuit differential equation to be solved. In section 3 the evaluation of the output waveform based on the proposed current model is presented, while simulation results for the proposed method are compared with SPICE in section 4. Finally, we conclude in section 5.

2   Analysis of Operation

The operation of the pass transistor will be studied using the structure shown in Fig. 1. The output capacitance C_L models the gate capacitance of the next level logic gates. Node A is set at logic "0" or "1" and an input ramp is applied to the gate of the pass transistor, node B. Consequently, the pass transistor will either discharge or charge the output capacitance towards logic "0" or "1", respectively.

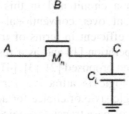

Fig. 1. Basic pass transistor structure

Let us consider first the case when the output capacitance is initially charged while the node A is set to logic low and a ramp input is applied to node B. In this case the output capacitance is discharged with a current flowing from node C to node A so that node A is the source node of the transistor. The operating condition of the transistor corresponds to that of the nMOS transistor of an inverter when a rising ramp is applied to its gate and the contribution of the short-circuiting transistor is ignored. As the input voltage rises, the transistor starts operating in saturation and after some time it moves to the linear region. The differential equations describing the circuit operation at both regions can be solved analytically and the output waveform can be calculated as a function of time. Since such an analysis is well known and uncomplicated it isn't further discussed in this work.

Different operating conditions arise in case the output capacitance is being charged through the pass transistor. The output capacitance is considered initially discharged and node A is set at logic "high" (V_{DD}). The charging current flows through the transistor from node A to node C and thus node C is the source of the transistor. According to the conventional Shockley model [14], since $V_{DS} \geq V_{GS} \ \forall \ t$ the transistor operates always in saturation.

To describe the transistor current, the alpha-power law model, proposed in [5], which takes into account the carrier velocity saturation effect of short-channel devices, is used. According to this model, the transistor current expression in saturation is given by:

$$I_{sat} = k_s \left(V_{GS} - V_{TN} \right)^a \qquad (1)$$

where k_s is the transconductance of the transistor in saturation and α is the velocity saturation index, which both are determined by measurements on the I-V characteristics [5]. V_{TN} is the threshold voltage of the nMOS transistor, which is expressed by its first order Taylor series approximation around $V_{out} = \left(V_{DD} - V_{TO} \right)/2$ as:

$$V_{TN} = d_1 \cdot V_{out} + d_2 \qquad (2)$$

(V_{TO} is the zero-bias threshold voltage [14]).

The fact that the same node (C) serves as output node of the circuit and as source node of the pass transistor makes the analysis of the transistor operation cumbersome, since the differential equation that describes the charging of the output capacitance has the form:

$$C_L \frac{dV_{out}}{dt} = k_s \left(V_{in} - V_{out} - V_{TN} \right)^a \qquad (3)$$

which cannot be solved analytically, since a has a value different than one, even for deep submicron technologies. However, making some reasonable approximations for the transistor current waveform, the output voltage can be modeled with sufficient accuracy.

In order to solve the circuit differential equation two cases for the input ramp are distinguished, namely fast and slow input ramps. In Figs. 2, 3 the model for the current waveform for each case is shown (with respect to the output voltage waveform). For slow input ramps (Fig. 2) the current waveform presents a plateau region where the value of the current remains constant. When a plateau region is not present on the current waveform (Fig. 3) the input ramp should be considered fast. The appearance of the plateau region depends on the slope of the input ramp and the circuit inertia, i.e. the transistor width and the load capacitance. The transistor starts conducting with its current increasing exponentially, according to equation (1). Following the current, the output voltage increases exponentially until time t_p when the rate of $V_{out} - V_{TN}$ increase, equals the rate of input increase. Then the output voltage increases linearly with a constant rate since the transistor current has a constant plateau value. This region of operation continues until the end of the input transition at time point τ where the input voltage reaches its final value and the current starts to decrease since the output voltage increases (V_{GS} decreases). Although the starting point of the plateau, t_p, is not always distinct, because of a smooth transition of the current in this region, the plateau region is easily identified. However, if the slope of the input ramp or the circuit inertia is sufficiently high, as in the case of a large output capacitance, the rate of increase of $V_{out} - V_{TN}$ may not reach the input slope until the end of the input ramp and the current will start decreasing without the appearance of the plateau region (fast input case).

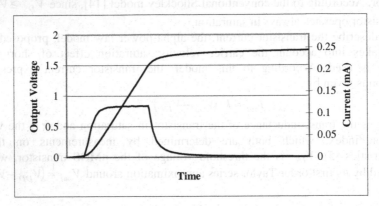

Fig. 2. Output voltage and current waveforms for a slow input case (τ=0.5ns, W=1.8μm, C_L=30fF), V_{DD}=2.5V

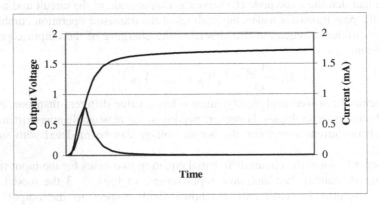

Fig. 3. Output voltage and current waveforms for a fast input case (τ=0.05ns, W=1.8μm, C_L=30fF), V_{DD}=2.5V

3   Output Waveform Evaluation

3.1   Slow Input Ramps

***Region* 1** ($1.3t_n \leq t < t_p$)

According to simulation results it can be safely assumed that the current in this region varies linearly with respect to time and consequently it can be approximated as:

$$I = \gamma \cdot (t - 1.3t_n) \tag{4}$$

where t_n is the time point where the input ramp reaches the threshold voltage of the nMOS transistor ($t_n = V_{TO} \cdot \tau / V_{DD}$). In order to calculate the output waveform with accuracy, two approximations on the starting point of the current and output waveform are being made: a) that the current and output voltage waveform remains equal to zero up to $t=1.3t_n$ as it can be seen from the region boundary and b) that the output voltage can be considered equal to zero up to time point $\tau/4$ ($V_{out}(\tau/4) = V_0 = 0$), fact that it can be safely assumed according to simulation results. By equating the current expression in saturation with the approximated current form in (4) at time point $t=\tau/4$ the coefficient γ can be obtained:

$$k_s \left[V_{in}\left(\frac{\tau}{4}\right) - V_o - V_{TN}\left(\frac{\tau}{4}\right) \right]^a = \gamma \cdot \left(\frac{\tau}{4} - 1.3t_n\right) \tag{5}$$

To increase the accuracy in modeling the current in this region, the value of the transconductance parameter k_s and that of the velocity saturation index α are calculated on the I-V characteristics of the nMOS transistor for very low V_{GS} values and high V_{DS} values, in order to capture the actual operating conditions of the pass transistor in this region. It should be noted, that according to extensive simulation results, these parameters for the alpha-power law model rely heavily on the region of the I-V characteristics on which they are calculated rather than being constant as implied by [5]. In Table I the values of α and k_s as they are extracted for various combinations of V_{GS} and V_{DS} and for the used technology, are given.

Table 1. Values of α and k_s for various V_{GS} and V_{DS} ($W=1,8\mu m$)

V_{GS} (V)	V_{DS} (V)	α	k_s (mA/V)	V_{GS} (V)	V_{DS} (V)	α	k_s (mA/V)
2.4	2.4	1.12	0.74	0.7	2.45	2.02	1.14
2	2	1.12	0.72	0.6	2	2.84	2.41
1.5	2	1.23	0.71	0.7	2	2.17	1.17
1.5	1.5	1.26	0.67	0.6	2.45	2.63	2.20
1	2	1.47	0.72	0.5	2.45	2.93	3.36

The shaded values were used in eq. (5) for the analysis in this region, since, as it was observed by simulation, they are close to the real values of V_{GS} and V_{DS}. The output waveform expression in this region is calculated by solving the circuit differential equation:

$$C_L \frac{dV_{out}}{dt} = \gamma \cdot (t - 1.3t_n) \tag{6}$$

with initial condition $V_{out}[1.3t_n] = 0$.

Region 2 ($t_p \le t < \tau$)

For slow input ramps there will be a time point (t_p) where the output voltage will increase at a rate that will keep the current at a constant value. From this point and until the input reaches its final value this constant current results in a "plateau" for the current waveform (Fig. 2). The time point when this plateau state begins is calculated by equating the slope of the input voltage to the slope of the output, considering the effect of the varying threshold voltage:

$$\frac{d\left(V_{out} + V_{TN}\right)}{dt}\bigg|_{t=t_p} = \frac{dV_{in}}{dt}\bigg|_{t=t_p} \tag{7}$$

The current value during the plateau is simply given by:

$$I_{plateau} = \gamma\left(t_p - 1.3t_n\right) \tag{8}$$

The output voltage is a first order polynomial expression derived from the following differential equation:

$$C_L \frac{dV_{out}}{dt} = I_{plateau} \Rightarrow V_{out}(t) = I_{plateau} \cdot \left(t - t_p\right)/C_L + V\left[t_p\right] \tag{9}$$

Region 3 ($t \geq \tau$)
In this region the input voltage is equal to V_{DD} while the current decreases exponentially with respect to time. We assume that the current is described by an expression of the form:

$$I(t) = I_{plateau} \cdot e^{-\beta(t-\tau)} \tag{10}$$

The output voltage is calculated by solving the circuit differential equation with initial condition the value of the output voltage at time point τ, which is known from the previous region.

In order to calculate the value of β, which is unknown in eq. (10), the charge that is being stored at the output capacitance during region 3 is calculated:

$$Q = \int_{\tau}^{\infty} I_{plateau} \cdot e^{-\beta(t-\tau)} dt = I_{plateau} / \beta \tag{11}$$

This amount of charge can be expressed as the difference between the final charge stored in the output load and that which is stored at time τ.

$$Q = C_L \cdot \left(V_{out}[\infty] - V_{out}[\tau]\right) \tag{12}$$

where the output voltage after infinite time is given by $V_{out}[\infty] = \dfrac{V_{DD} - d_2}{1 + d_1}$ due to

threshold voltage drop across the pass transistor. By equating eq. (11) and (12) the value of β is obtained.

3.2   Fast Input Ramps

In this case only two regions of operation exist since the input voltage reaches V_{DD} before the current enters the plateau state. As a result, the boundary between the two regions is time point τ and the output voltage at each region is calculated exactly as for regions 1 and 3 for slow input ramps.

To determine whether an input corresponds to the fast or slow case, after the calculation of the output voltage expression in the first region, equation (7) is being solved. In case the resulting t_p time point is smaller than the transition time τ, the input is considered slow, otherwise the solution for the fast input case is being followed.

4   Results

The proposed methodology has been validated by comparisons with HSPICE simulation results for a TSMC 0.18 μm technology. To prove the efficiency in modeling the pass transistor current, output voltage and current waveforms generated by the proposed method are compared with SPICE simulation results in Figs 4, 5, for a slow and a fast input case, respectively. The presence of the plateau for the slow input case and the validity of the proposed current model are obvious from these figures.

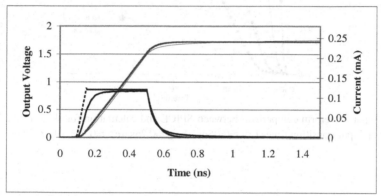

Fig. 4. Output voltage and current waveform comparison between SPICE (solid lines) and calculated results (dashed lines), for a slow input case (τ=0.5ns, W=1.8μm, C_L=30fF).

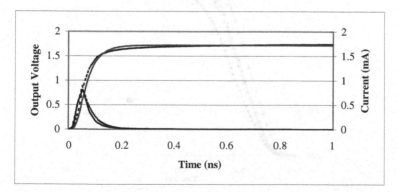

Fig. 5. Output voltage and current waveform comparison between SPICE (solid lines) and calculated results (dashed lines), for a fast input case (τ=0.05ns, W=1.8μm, C_L=30fF).

To illustrate the applicability of the proposed method, comparisons with SPICE results have been performed for a number of input transition times and circuit configurations (pass transistor width and output capacitance). Figs. 6, 7 and 8 show output waveform results for varying input transition times, transistor widths and output capacitances, respectively.

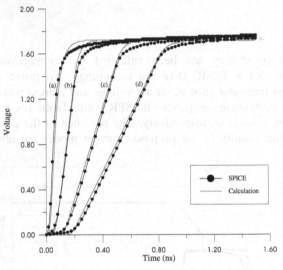

Fig. 6. Output waveform comparison between SPICE and calculation for W=1.8 μm, C_L=30fF and varying input transition times (a) τ=0.05ns, (b)) τ=0.2ns, (c) τ=0.5ns, (d) τ=0.8ns

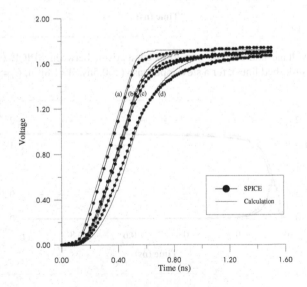

Fig. 7. Output waveform comparison between SPICE and calculation for τ=0.5 ns, C_L=50fF and varying transistor widths (a) W=3.6 μm, (b) W=0.9 μm, (c) W=0.72 μm, (d) W=0.36 μm

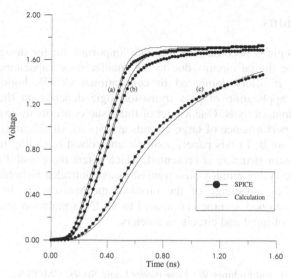

Fig. 8. Output waveform comparison between SPICE and calculation for τ=0.5 ns, W=1.8 µm and varying output capacitance (a) C_L= 50 fF (b) C_L= 100 fF, (c) C_L= 500 fF

Defining propagation delay as the time from the half-V_{DD} point of the input waveform to the half-V_{DD} point of the output, propagation delay results have been obtained and compared with SPICE simulations (Table II). The error between SPICE and the proposed method, in spite of the approximations used in the current model, remains for most of the cases below 7%.

Table 2. Propagation delay comparison between SPICE and the proposed method

	Prop. Delay (SPICE)	Prop. Delay (method)	Error
C_L = 30 fF, W=1.8µm	(ps)	(ps)	(%)
τ(ns)			
0.8	267	275	3.0
0.5	192	182.5	4.9
0.2	95	90	5.3
0.05	51	60	17.6
C_L = 50 fF, τ=0.5 ns			
W (µm)			
3.6	190	177.5	6.6
0.9	250	247.5	1.0
0.72	269	260	3.3
0.36	362	365	0.8
τ= 0.5 ns, W=1.8 µm			
C_L (fF)			
500	770	810	5.2
100	258	247.5	4.1
50	213	200	6.1
30	192	182.5	4.9

5   Conclusions

Pass transistor logic is becoming increasingly important for the design of low-power high-performance digital circuits due to the smaller node capacitances and reduced transistor count it offers compared to conventional CMOS logic. However, the acceptance and application of pass transistor logic depends on the availability of supporting automation tools. One aspect of this issue concerns timing simulators that can analyze the performance of large circuits at a speed, significantly faster than that of SPICE based tools. In this paper, a simple and robust modeling technique for the basic pass transistor structure is presented, which offers the possibility of fast timing analysis for circuits that employ pass transistors as controlled switches. The proposed methodology takes advantage of the physical mechanisms in the pass transistor operation. The obtained accuracy compared to SPICE simulation results is sufficient for a wide range of input and circuit parameters.

References

1. Zimmermann R. and Fichtner W.: Low-Power Logic Styles: CMOS Versus Pass-Transistor Logic, IEEE J. Solid-State Circuits, Vol. 32, (1997), 1079-1090
2. Suzuki M., Ohkubo N., Shinbo T., Yamanaka T., Shimizu A., Sasaki K. and Nakagome Y.: A 1.5-ns 32-b CMOS ALU in Double Pass-Transistor Logic, IEEE J. Solid-State Circuits, vol. 28, (1993), 1145-1150
3. Abu-Khater I.S., Bellaouar A., Elmasry M. I.: Circuit Techniques for CMOS Low-Power High-Performance Multipliers, IEEE J. Solid-State Circuits, vol. 31, (1996), 1535-1546
4. Yano K., Sasaki Y., Rikino K. and Seki K.: Top-Down Pass-Transistor Logic Design, IEEE J. Solid-State Circuits, vol. 31, (1996) 792-803
5. Sakurai T., Newton A.R:Alpha-Power Law MOSFET Model and its Applications to CMOS Inverter Delay and Other Formulas, IEEE J. Solid-State Circuits, Vol. 25, (1990), 584-594
6. Juan-Chico J., Bellido M. J., Acosta A. J., Barriga A., Valencia M.: Delay degradation effect in submicronic CMOS inverters, Proc. of 7th Int. Workshop on Power and Timing Modeling, Optimization and Simulation (PATMOS), (1997), 215-224
7. L. Bisdounis, S. Nikolaidis, O. Koufopavlou, «Analytical Transient Response and Propagation» Delay Evaluation of the CMOS Inverter for Short-channel Devices», IEEE Journal of Solid-State Circuits, Vol. 33, No 2, pp. 302-306, Feb. 1998.
8. Daga J. M. and Auvergne D.: A Comprehensive Delay Macro Modeling for Submicrometer CMOS Logics, IEEE J. Solid-State Circuits, Vol. (34), (1999), 42-55
9. Chatzigeorgiou A., Nikolaidis S. and Tsoukalas I.: A Modeling Technique for CMOS Gates, IEEE Transactions on Computer-Aided Design of Integrated Circuits and Systems, Vol. 18, (1999), 557-575
10. Rossello J. L. and Segura J.: Charge-Based Analytical Model for the Evaluation of Power Consumption in Submicron CMOS Buffers, IEEE Transactions on Computer-Aided Design of Integrated Circuits and Systems, Vol. 21, (2002), 433-448
11. Vemuru S. R.: Delay-Macromodelling of CMOS Transmission-Gate-Based-Circuits, International Journal of Modelling and Simulation, vol. 15, (1995), 90-97
12. Kang S. M. and Leblebici Y.: CMOS Digital Integrated Circuits, Analysis and Design, McGraw Hill, New York (1996)
13. Chatzigeorgiou A., Nikolaidis S. and I. Tsoukalas I.: Timing Analysis of Pass Transistor and CPL Gates, Proc. of 9th Int. Workshop on Power and Timing Modeling, Optimization and Simulation (PATMOS), (1999), 367-376
14. Weste N. H. E. and Eshraghian K., Principles of CMOS VLSI Design, Addison Wesley, Reading (1994)

An Approach to Energy Consumption Modeling in RC Ladder Circuits

M. Alioto[1], G. Palumbo[2], and M. Poli[2]

[1] DII – Dipartimento di Ingegneria dell'Informazione, Università di Siena,
v. Roma n. 56, I-53100 - Siena (Italy)
malioto@dii.unisi.it

[2] DEES – Dipartimento Elettrico Elettronico e Sistemistico, Universita' di Catania,
viale Andrea Doria 6, I-95125 CATANIA - ITALY
Phone ++39.095.7382313; Fax ++39.095.330793
gpalumbo@dees.unict.it, mpoli@dees.unict.it

Abstract. In this communication, an approach to analytically estimate the energy consumption in RC ladder circuits is proposed. Effect of the input rise time on energy dissipated is modeled by assuming a ramp input. The approach is based on an exact analysis of energy dissipated by the network for asymptotic values of the input rise time. Successively, starting from the RC ladder circuits properties, a generalization is provided for arbitrary values of the input rise time. The approach followed leads to a closed-form expression of the energy dissipation. Moreover, this expression is formally equal to that of a first-order RC circuit, and is thus simple enough to be used for pencil-and-paper evaluation. The accuracy of the model has been tested by SPICE simulations. Results show that the energy predicted is in good agreement with simulated values.

1 Introduction

In the design of integrated circuits, energy consumption has become one of the main design goals. For this reason, in the last decade there has been an emphasis on techniques to evaluate energy consumption of logic blocks and interconnections [1]-[3]. Moreover, since circuit simulations are time-consuming and even unfeasible for complex circuit, the formulation of simple yet accurate analytical expressions of energy consumption of fundamental blocks is of concern.

Among the blocks used in CMOS digital ICs, various techniques to model energy consumption of static and dynamic gates have been proposed, and are generally developed from the simple model of the charge of a capacitance [2]-[3]. However, this model may be inadequate in some cases, such as logic gates based on transmission-gates or pass-transistors [2], RC interconnects [2] and adiabatic logic gates [4]. Indeed, these subcircuits are usually modeled by RC ladder circuits with grounded capacitances, as shown in Fig. 1. Hence, energy estimation for such blocks can be carried out by evaluating the energy consumption of their equivalent RC ladder circuits.

B. Hochet et al. (Eds.): PATMOS 2002, LNCS 2451, pp. 239–246, 2002.

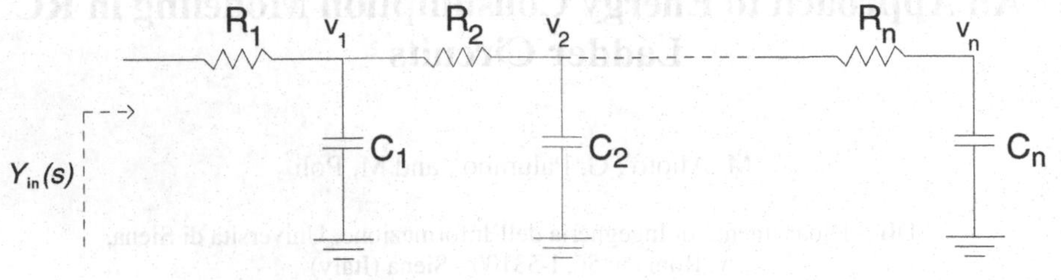

Fig. 1. Topology of an n-th order RC ladder network..

This communication deals with an approach to model energy consumption of RC ladder networks. More specifically, the dependence of energy dissipated on the input signal rise time is analyzed by assuming a ramp input waveform. The circuit analysis developed shows that the energy wasted during an input transition cannot be simply expressed in a closed-form for networks with an order greater than two, since the analytical evaluation of the network poles and zeroes is required. Thus, a preliminary analysis is carried out for asymptotical values of the input rise time, that leads to simple energy expressions that do not involve pole-zero evaluation. Then, the results obtained are extended to intermediate values of the input rise time by exploiting the properties of RC ladder networks. To simplify analysis, an equivalent first-order circuit is introduced and used to derive an energy expression that does not depend on poles and zeroes, but only depends on resistances and capacitances of the original network.

The energy model accuracy has been tested by means of extensive comparison with SPICE simulations. Analysis shows that the model is sufficiently accurate for modeling purposes.

2 Exact Analysis of Energy Consumption

Consider the n-th order RC ladder circuit in Fig. 1. It consists of resistances R_i and grounded capacitances C_i with $i=1\ldots n$, and is driven by an input voltage source, $v_{in}(t)$, whose waveform goes from zero to its amplitude, V_{DD}, with a rise time T. During the input transition, the total energy E_{TOT} provided by the voltage source to the network is equal to the sum of the energy stored in capacitances $C_1\ldots C_n$ at the steady state, E_C,

$$E_C = \frac{1}{2} V_{DD}^2 \sum_{i=1}^{n} C_i = \frac{1}{2} V_{DD}^2 C_T \tag{1}$$

and the energy E_R wasted by resistors $R_1\ldots R_n$, that is the object of our analysis (in (1), C_T is the total network capacitance). By observing that E_{TOT} is the integral of the product of input voltage, v_{in}, and input current, i_{in}, the energy wasted by resistors results as

$$E_R = E_{TOT} - E_C = \int_0^{+\infty} v_{in}(t) i_{in}(t) dt - \frac{1}{2} V_{DD}^2 C_T \tag{2}$$

The input current waveform, i_{in}, can be evaluated through the input admittance of the network, $Y_{in}(s)$, whose poles and zeroes are real, negative and alternatively placed on the frequency axis, with the first zero being in the origin. Analytically, $Y_{in}(s)$ can be written as a function of the time constants associated with the poles, τ_{pi}, and the zeroes, τ_{zi}, of the circuit admittance, or by using partial fraction expansion

$$Y_{in}(s) = sC_T \frac{\prod_{i=1}^{n-1} (s\tau_{z_i} + 1)}{\prod_{i=1}^{n} (s\tau_{p_i} + 1)} = sC_T \sum_{i=1}^{n} \frac{A_i}{s\tau_{p_i} + 1} \tag{3}$$

where, equating the numerator of the two relationships at the right and considering their constant term, the coefficients A_i satisfy the following property

$$\sum_{i=1}^{n} A_i = 1 \tag{4}$$

The input current waveform, i_{in}, can be evaluated by using (3) and applying the inverse Laplace transform to $Y_{in}(s) \cdot V_{in}(s)$, that results to

$$i_{in}(t) = \begin{cases} \dfrac{V_{DD}}{T} C_T \left[\displaystyle\sum_{i=1}^{n} A_i \left(1 - e^{-\frac{t}{\tau_{p_i}}}\right) \right] & \text{if} \quad t \leq T \\[4ex] \dfrac{V_{DD}}{T} C_T \left[\displaystyle\sum_{i=1}^{n} A_i\, e^{-\frac{t}{\tau_{p_i}}} \left(e^{\frac{T}{\tau_{p_i}}} - 1\right) \right] & \text{if} \quad t > T \end{cases} \tag{5}$$

that substituted into relationship (2) leads to energy wasted by resistors in the network

$$E_R = V_{DD}^2 C_T \sum_{i=1}^{n} \frac{A_i \tau_{p_i}}{T} \left[\frac{\tau_{p_i}}{T} \left(e^{-\frac{T}{\tau_{p_i}}} - 1\right) + 1 \right] = C_T V_{DD}^2 \sum_{i=1}^{n} A_i f\left(\frac{T}{\tau_{p_i}}\right) \tag{6}$$

where function f is defined as

$$f(x) = \frac{1}{x} \left[\frac{1}{x}(e^{-x} - 1) + 1 \right] \tag{7}$$

and is plotted versus x in Fig. 2.

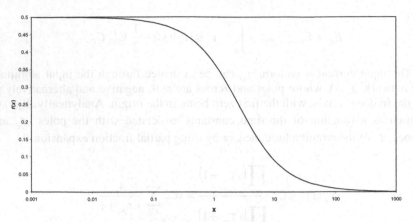

Fig. 2. Plot of function $f(x)$ versus x.

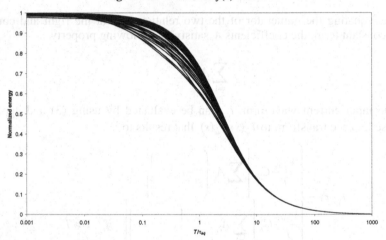

Fig. 3. Simulation results: plot of energy normalized to that with step input versus T/τ_{eq}.

3   Closed-Form Expression of Energy Consumption

General expression of energy wasted (6) depends on the poles and zeroes (or, equivalently, the poles and coefficients A_i), whose analytical evaluation can be accomplished only for very simple networks, such as first- and second-order circuits. Hence, in general a closed-form expression of energy consumption cannot be found, and approximate analysis is needed. For this reason, we first consider energy consumption for asymptotic input rise time values (i.e., $T{\rightarrow}0$ and $T{\rightarrow}\infty$), whose expression does not explicitly depend on poles and zeroes, but only on $R_1...R_n$ and $C_1...C_n$. Then, the expressions obtained are extended to arbitrary values of rise time.

3.1 Energy Consumption for $T \to 0$ and $T \to \infty$

By inspection of Fig. 2, function $f(x)$ tends to 1/2 when $x \to 0$, thus for $T \to 0$ relationship (6) becomes

$$E_R\big|_{T \to 0} = \frac{1}{2}C_T V_{DD}^2 \tag{8}$$

where eq. (4) was used. By inspection of (8), it is apparent that the energy wasted by a generic RC ladder network for a step input is equal to that of a first-order RC circuit with capacitance C_T.

For high values of x, energy consumption can be evaluated by means of the well-known approximation of Elmore delay. Indeed, as demonstrated in [5], approximating the voltage waveforms of an RC circuit by using the Elmore delay leads to an error that asymptotically tends to zero when the input rise time increases. This means that for $T \to \infty$ the transfer function of the voltage across the generic i-th capacitance, v_i, can be approximated by a single-pole function with the Elmore-delay time constant, $T_{D,i}$,

$$\frac{v_i(s)}{v_{in}(s)} \approx \frac{1}{1 + sT_{D,j}} \tag{9}$$

As well known [6-10], in RC circuits the Elmore-delay time constant $T_{D,i}$ (with $i=1...n$) can be evaluated as

$$T_{D,i} = \sum_{k=1}^{n} r_{ik} C_k \tag{10}$$

where r_{ik} is the sum of resistances along the route obtained by intersecting the path from capacitance C_i to the input node with the path from capacitance C_k to the input node. For RC ladder networks, r_{ik} is simply equal to the sum of resistances laying between the input node and the nearest capacitance among C_i and C_k

$$T_{D,i} = \sum_{k=1}^{i}\left(C_k \sum_{j=1}^{k} R_j \right) + \sum_{k=i+1}^{n} C_k \sum_{j=1}^{i} R_j \tag{11}$$

As shown in [11] under the Elmore delay assumption (9), the energy E_{Ri} dissipated by the resistance R_i across capacitances C_{i-1} and C_i in the RC network for $T \to \infty$ is

$$E_{R_i}\big|_{T \to \infty} \approx \frac{V_{DD}^2}{R_i}\frac{(T_{D,i} - T_{D,i-1})^2}{T} = \frac{V_{DD}^2}{T}R_i\left(\sum_{k=i}^{n} C_k\right)^2 \tag{12}$$

where the term $(T_{D,i}-T_{D,i-1})$ was simplified by using (11). It is useful to observe that, in relationship (12), each resistance multiplies the squared sum of capacitances that are at its right.

The overall energy dissipated by the RC ladder network is equal to the sum of contributions given by (12) associated with resistances $R_1...R_n$

$$E_R\big|_{T\to\infty} \approx \frac{V_{DD}^2}{T}\sum_{i=1}^{n} R_i\left(\sum_{k=i}^{n} C_k\right)^2 \tag{13}$$

3.2 Generalization at Intermediate Values of Input Rise Time

To avoid pole-zero evaluation in general expression (6) of energy consumption, it is necessary to introduce some approximation, from understanding of RC ladder networks properties. To this purpose, it is useful to observe that energy consumption (8) and (13) for $T\to 0$ and $T\to\infty$, respectively, can be modeled by a first-order equivalent RC circuit with resistance R_{eq} and capacitance C_{eq}. Indeed, it is apparent that the first-order equivalent circuit dissipates the same energy (8) of the original network if

$$C_{eq} = C_T \tag{14}$$

Analogously, the equivalent first-order circuit energy dissipation for $T\to\infty$, that results $V_{DD}^2 R_{eq} C_{eq}^2 / T$ [4], matches relationship (13) if its equivalent time constant τ_{eq} is

$$\tau_{eq} = R_{eq} C_{eq} = \frac{\sum_{i=1}^{n} R_i\left(\sum_{k=i}^{n} C_k\right)^2}{C_T} \tag{15}$$

Even though the first-order equivalent RC circuit is constructed to match energy dissipation of the original network for $T\to 0$ and $T\to\infty$, it can be approximately extended to arbitrary rise time values. This is because the single-pole approximation can be used even for intermediate values of T, as intuition suggests. This can be shown by considering that the pole-zero map of the n-th order RC network admittance, $Y_{in}(s)$, consists of n poles and n zeroes, that are alternatively placed in the frequency axis. When the poles are far from each other, the dominant-pole approximation holds, while in the cases when two successive poles are close, also the zero included between them tends to be close to them, canceling the effect of one of the two poles considered. This intuitively suggests that, even when poles are not far from each other, their overall effect can be again represented by a single pole. As a result, we approximate energy consumption (6) by that of the equivalent first-order RC circuit above discussed (its energy is simply obtained from (6) with $n=1$)

$$E_R = C_T V_{DD}^2\, f\left(\frac{T}{\tau_{eq}}\right) \tag{16}$$

It is worth noting that relationship (16) does not explicitly depend on poles and zeroes but only on $R_1..R_n$ and $C_1..C_n$.

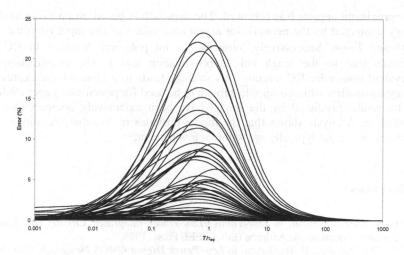

Fig. 4. Model error with respect to simulated curves in Fig. 3.

4 Validation and Simulation Results

The expression of energy dissipated (16) was tested by performing SPICE simulations. To this purpose, 1,000 RC ladder circuits were randomly generated by varying the order n from 2 to 10, while resistances and capacitance were varied by three orders of magnitude higher and lower with respect to a reference value. A ramp input with rise time varying by three orders of magnitude higher and lower with respect to the equivalent time constant was considered.

About 250 curves of simulated energy normalized to that for $T \rightarrow 0$ versus the input rise time normalized to the equivalent time constant, T/τ_{eq}, are reported in Fig. 3 (the curves not shown have a very similar behavior). By inspection of this figure, the exact energy curves show a dependence on the input rise time very similar to that of a single-pole circuit reported in Fig. 2. This confirms that the validity of the single-pole approximation, as intuitively justified in the previous analysis. The error of the model (16) with respect to curves in Fig. 3 is reported versus T/τ_{eq} in Fig. 4, showing a maximum value lower than 25%. Moreover, the typical error is much lower, since it is in the order of a few percent, by inspection of Fig. 4. As a result, the model proposed is accurate enough for modeling purposes, and it provides results very close to exact ones in typical cases.

5 Conclusions

In this paper, an approach to analytically estimate the energy consumption in RC ladder circuits is discussed. The effect of the input rise time on energy dissipated is analyzed by assuming an input ramp waveform. Since exact analysis strictly depends on poles and zeroes evaluation, that can be carried out only for very simple networs, an approximate approach is followed. The approach is based on an exact analysis of energy dissipated by the network for asymptotic values of the input rise time, i.e. for $T \to 0$ and $T \to \infty$. Successively, observations on pole-zero location in RC ladder networks lead to the single-pole approximation and to the construction of an equivalent first-order RC circuit. This strategy leads to a closed-form expression of energy dissipation which is simple enough to be used for pencil-and-paper evaluation.

The results predicted by the model have been extensively compared to SPICE simulations. Analysis shows that the model provides results that are within 25% of simulated ones, and typically are within a few percent.

References

[1] E. Sànchez-Sincencio, in *Low-Voltage/Low-Power Integrated Circuits and Systems*, E. Sànchez-Sincencio, A. Andreou (Eds.), IEEE Press, 1999.

[2] A. Chandrakasan, R. Brodersen, in *Low Power Digital CMOS Design*, A. Chandrakasan, R. Brodersen (Eds.), Kluwer Academic Publisher, 1995.

[3] K. Roy, S. Prasad, *Low-Power CMOS VLSI Circuit Design*, Wiley Interscience, 2000.

[4] W. Athas, "Energy-Recovery CMOS," in *Low Power Digital Design Methodologies*, J. Rabey, M. Pedram, Eds., Kluwer Academic Publisher, 1995.

[5] R. Gupta, B. Tutuianu, L. Pileggi, "The Elmore Delay as a Bound for RC Trees with Generalized Input Signals," *IEEE Trans. on CAD*, vol. 16, no. 1, Jan. 1997.

[6] W. C. Elmore, "The transient response of damped linear networks with particular regard to wideband amplifiers," *J. Appl. Phys.*, vol.19, pp. 55–63, Jan. 1948.

[7] J. Rubinstein, P. Penfield, Jr., and M. A. Horowitz, "Signal delay in RC tree networks," *IEEE Trans. on Computer-Aided Design*, Vol. CAD-2, pp. 202–211, July 1983.

[8] J. L. Wyatt, Jr., "Signal propagation delay in RC models for interconnect," *Circuit Analysis, Simulation and Design*, Part II: *VLSI Circuit Analysis and Simulation*, A. Ruehli, ed., Vol. 3 in the series *Advances in CAD for VLSI*, North-Holland, 1987.

[9] E. G. Friedman, J. H. Mulligan, Jr., "Ramp input response of RC tree networks," *Analog Integrated Circuits and Signal Processing*, vol. 14, no. 1/2, pp.53–58, Sept. 1997

[10] M. Celik, L. Pileggi, "Metrics and Bounds for Phase Delay and Signal Attenuation in RC(L) Clock Trees," *IEEE Trans. on Computer Aided-Design of Integrated Circuits and Systems*, Vol. 18, No. 3, pp. 293–300, March 1999.

[11] M. Alioto, G. Palumbo, "Power Estimation in Adiabatic Circuits: A Simple and Accurate Model," *IEEE Trans. on VLSI Systems*, vol. 9, no. 5, pp. 608–615, October 2001.

Structure Independent Representation of Output Transition Time for CMOS Library

P. Maurine, N. Azemard, and D. Auvergne

LIRMM UMR 5506 Univ. de Montpellier II 161 Rue Ada 34392 Montpellier France

Abstract. Non zero signal rise and fall times significantly contribute to the gate propagation delay. Designers must accurately consider them when defining timing library format. Based on a design oriented macro-model of the timing performance of CMOS structures, we present in this paper a general representation of transition times allowing fast and accurate cell performance evaluation. This general representation is then exploited to define a robust characterization protocol of the output transition time of standard cells. Both the representation and the protocol are finally validated comparing calculated gate input-output transition time values with standard look-up representation obtained from Hspice simulations (Bsim3v.3, level 69, 0.25µm process).

1 Introduction

In deep submicron technologies the propagation delay of any CMOS cell (i) is strongly dependent on the input ramp duration $\tau_{IN}(i)$ applied to its gate, which is the output transition time $\tau_{OUT}(i-1)$ of the preceding gate. As a consequence the accurate characterization of the output transition time of the different cells in their design environment is of prime importance in speed performance verification or optimization steps.

In the standard industrial approach the tabular method is used. The performance of a predefined set of cells is obtained from electrical simulations performed for a limited number of design conditions, such as load and input transition time values [1]. The resulting data are then listed in tables containing typically 25, 49, or 81 operating conditions (number of loading conditions 5,7 or 9 × number of input ramp conditions 5,7 or 9). Intermediate conditions are then directly obtained from a linear interpolation between these predefined points. Due to the non-linear variation, in submicron process, of the propagation delay and transition time with the loading and controlling conditions, this method may induce significant errors when interpolating in the non-linear part of the variation. As an example, in Fig.1 we illustrate the evolution of the output transition time of an inverter designed in a 0.25µm process. Here, this inverter is controlled by a rising linear input ramp of duration τ_{IN} and loaded by 5 times its input gate capacitance. As shown the values of the output transition time, τ_{OUT} interpolated from the look up table, may be underestimated by nearly 15%, compared to the values obtained from Hspice simulations.

B. Hochet et al. (Eds.): PATMOS 2002, LNCS 2451, pp. 247–257, 2002.
© Springer-Verlag Berlin Heidelberg 2002

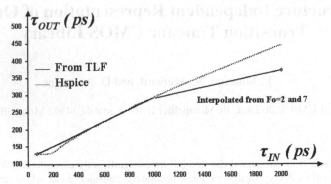

Fig. 1. Comparison of the output transition time values interpolated from the TLF and the simulated ones (Hspice).

It is clear that, for a given cell, the relative accuracy obtained (using Hspice simulation as a reference) with a tabular method strongly depends on the size or the granularity level of the table. Typically the size of the table is limited to 5 to 10 controlling and loading conditions by element, in order to reasonably limit the characterization time to few months.

As a result, the definition of a robust protocol of characterization, based on a uniform representation of the performance of a library is of great interest for cell designers. Indeed it must allow to increase the number of cells in a typical library, and/or to manage the trade off between the accuracy of the performance estimation and the time necessary to characterize a complete library.

Using a design oriented modeling of the CMOS cell timing performance, we propose in section 2, a unified representation of the CMOS cell output transition time, allowing a complete design space representation. Then, in section 3, we deduced from this unified representation a robust characterization protocol of the output transition time of typical CMOS structures. Conclusion is given in section 4.

2   Output Transition Time Modeling

2.1 General Expression

A lot of work has been devoted to the modeling of the output transition time [4-14]. It has been clearly shown that, for CMOS structure, τ_{OUT} can be obtained from the modeling of the charging (discharging) current that flows during the structure switching process.

Moreover it has been demonstrated [2,12] that the evaluation of both the maximum current I_{MAX} that can provide a structure, and the amount of charge ($C \cdot V_{DD}$) to be removed from the output node is sufficient to model the output transition time. More precisely, considering a linear variation of the output voltage, it has been shown that the driving element can be model as a current generator supplying a constant current to the output loading capacitance. Consequently, a simple first order expression of the output transition time τ_{OUT} can then be obtained from

$$\tau_{OUT} = \frac{C \cdot V_{DD}}{I_{MAX}},$$

(1)

where τ_{out} represents the time spent by the output voltage to swing over the full supply voltage value V_{DD}, C is the output loading capacitance, and I_{MAX} the maximum value of the discharging (charging) current.

2.2 Inverter Switching Current and Definition of the Load

As shown in [2] the evaluation of the maximum current available in the structure imposes to consider two controlling conditions: the *Fast* and the *Slow* input ramp domains. The *Fast* input control range is obtained when the input signal reaches its maximum (minimum) value before the output begins to vary. In this case, the switching current exhibits an almost constant value. In the *Slow* input control range the cell output voltage varies in the same time interval than the input one. In this situation, for which a short circuit occurs between the N and P transistors, the switching current available in the cell presents a maximum value smaller than in the *Fast* input case. Moreover this maximum value depends on the value of the input transition time τ_{IN}. Using, for deep submicron process, the Sakurai's representation of the drain-source current with $\alpha = 1$ [3] the evaluation of the current in the *Fast* input range is straightforward for an inverter. Considering the maximum value of the input control voltage we obtain

$$I_{MAX}^{Fast} = K_{N,P} \cdot W_{N,P} \cdot (V_{DD} - V_{TN,P})$$

(2)

for an output falling or rising edge, respectively.

The evaluation of the current value in the *Slow* input range is quite more difficult. However taking advantages of the symmetry properties of the current wave shape [2], the maximum current value can be evaluated from

$$I_{MAX}^{Slow} = \sqrt{\frac{K_{N,P} \cdot W_{N,P} \cdot V_{DD}^2 \cdot C}{\tau_{IN}}}$$

(3)

where τ_{IN} is the transition time of the cell controlling signal, and C is the output load seen by the inverter defined as:

$$C = C_L + C_{PAR} + C_M$$

(4)

where C_L is the sum of the input capacitance of the output loading gates and of the interconnect capacitance, C_{PAR} is the contribution of the cell parasitic capacitance and finally C_M is the contribution of the coupling capacitance [15]. Equation (4) leads to the following definition of the usual fan out factor

$$F_O = \frac{C}{C_{IN}} = \frac{C_L}{C_{IN}} + \frac{C_{PAR}}{C_{IN}} + \frac{C_M}{C_{IN}} \tag{5}$$

$$= F_O^L + F_O^{PAR} + F_O^M$$

Thus Fo is the sum of three contributions.

- The first one is, as defined earlier, mainly due to the logic following the inverter and to the eventual routing capacitance. Thus it is entirely independent of the considered inverter. We call it the logic contribution.

- The second contribution is mainly due to the diffusion capacitance and the gate internal interconnect; its evaluation gives a good indicator of the quality of the cell design.

- The last contribution is due to the accumulation of charge in the channel of the P (N) transistor that must be removed during the switching process. These charges are usually modeled by an equivalent capacitance as proposed by Meyer [15]. Note, that neglecting this contribution may induce an underestimation of 20% of the load for small value of the fan out factor (F_O^L=1).

2.3 Inverter Output Transition Time Model

Finally, with such a definition of the load, the inverter output transition time is directly obtained by replacing I_{MAX} in (1) by its appropriate expression (2 or 3). This gives respectively for an input rising and falling edge

$$\tau_{OUT-HL}(Inv.) = MAX \left\{ \begin{array}{l} \tau_{OUT-HL}^{Fast} \\[2mm] \sqrt{\dfrac{(V_{DD}-V_{TN})}{V_{DD}}} \cdot \sqrt{\tau_{OUT-HL}^{Fast} \cdot \tau_{IN-HL}} \end{array} \right\} \tag{6}$$

$$\tau_{OUT-LH}(Inv.) = MAX \left\{ \begin{array}{l} \tau_{OUT-LH}^{Fast} \\[2mm] \sqrt{\dfrac{(V_{DD}-V_{TP})}{V_{DD}}} \cdot \sqrt{\tau_{OUT-LH}^{Fast} \cdot \tau_{IN-LH}} \end{array} \right\} . \tag{7}$$

Where τ_{OUT}^{Fast} is called the step response of the inverter. Its value can be directly obtained from (1) and (2) as

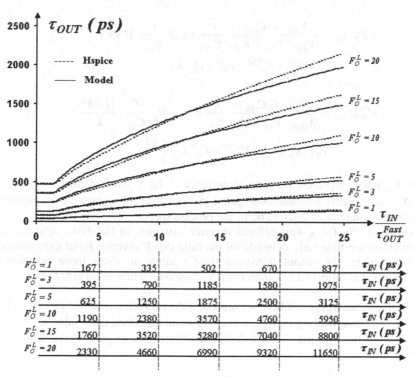

$F_O^L = 1$	167	335	502	670	837	$\tau_{IN}\,(ps)$
$F_O^L = 3$	395	790	1185	1580	1975	$\tau_{IN}\,(ps)$
$F_O^L = 5$	625	1250	1875	2500	3125	$\tau_{IN}\,(ps)$
$F_O^L = 10$	1190	2380	3570	4760	5950	$\tau_{IN}\,(ps)$
$F_O^L = 15$	1760	3520	5280	7040	8800	$\tau_{IN}\,(ps)$
$F_O^L = 20$	2330	4660	6990	9320	11650	$\tau_{IN}\,(ps)$

Fig. 2. Comparison between simulated and calculated (— eq.5) values of the output transition time of an inverter (W_N–0.72µm, L=0.23µm, k=1) for various loading and controlling conditions.

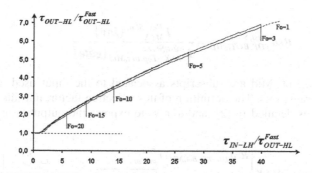

Fig. 3. Illustration of the general representation of the inverter output transition time

$$\tau_{OUT-HL}^{Fast} = \frac{V_{DD} \cdot L \cdot C_{OX}}{(V_{DD}-V_{TN}) \cdot K_N} \cdot \frac{C}{C_N} = \tau_{ST} \cdot \frac{C}{C_{IN}} \cdot (1+k) \qquad (8)$$

$$= \tau_{ST} \cdot \left(F_O^L + F_O^{PAR} + F_O^M\right) \cdot (1+k)$$

$$\tau_{OUT-LH}^{Fast} = \frac{V_{DD} \cdot L \cdot C_{OX}}{(V_{DD}-V_{TP}) \cdot K_P} \cdot \frac{C}{C_P} = \tau_{ST} \cdot R_\mu \cdot \frac{C}{C_{IN}} \cdot \frac{(1+k)}{k}$$

$$= \tau_{ST} \cdot R_\mu \cdot \left(F_O^L + F_O^{PAR} + F_O^M\right) \cdot \frac{(1+k)}{k}$$

where C_N and C_p represent the gate capacitance of the N and P transistors, $C_{IN}=C_N+C_p$, $k=C_p/C_N$ is the cell internal configuration ratio and τ_{ST} appears as a technology dependent parameter characteristic of the process speed.

As shown in (6), for a well-defined inverter structure, in the *Fast* input range, the output transition time only depends on the ratio (load/ inverter input capacitance). In the *Slow* range, the output transition time exhibits an extra input duration time dependency that reflects the complete inter-stage interaction in an array of inverters or gates.

This is illustrated in Fig.2 where we represent the transition time for an output falling edge, for different loading factors, versus the input transition time value τ_{IN}. As shown the values of the transition time calculated with (6) are in good agreement with the simulated one.

2.4   Gate Output Transition Time Model

Similar expression can be obtained for gates, considering in (1) the ratio of maximum available current between an inverter and a gate with identically sized transistors such as

$$Red_{TOP,BOT,MID}^{Fast,Slow} = \frac{I_{MAX}^{Fast,Slow}(Inv)}{I_{MAX-Top,Bot,Mid}^{Fast,Slow}(gate)} \qquad (9)$$

Where the Top, Bot, Mid are subscripts associated to the input used to control the serial array of transistors. The definition of this reduction factor, introduced as logical weight in [16], is detailed in [2], and allows to express the output transition time of the gates as

$$\tau_{OUT-HL}(Gate) = MAX \left\{ \begin{array}{l} Red_{HL-Top,Bot,Mid}^{Fast} \cdot \tau_{out-HL}^{Fast} \\[2mm] Red_{HL-Top,Bot,Mid}^{Slow} \cdot \sqrt{\frac{(V_{DD}-V_{TN})}{V_{DD}}} \cdot \sqrt{\tau_{out-HL}^{Fast} \cdot \tau_{IN-HL}} \end{array} \right\} \qquad (10)$$

for a rising edge applied on one of the input (Top, Bot, Mid). For an input falling edge this becomes

$$\tau_{OUT-LH}(Gate) = MAX \left\{ \begin{array}{l} Red_{LH-Top,Bot,Mid}^{Fast} \cdot \tau_{OUT-LH}^{Fast} \\ Red_{LH-Top,Bot,Mid}^{Slow} \cdot \sqrt{\dfrac{(V_{DD}-V_{TP})}{V_{DD}}} \cdot \sqrt{\tau_{OUT-LH}^{Fast} \cdot \tau_{IN-LH}} \end{array} \right\} \qquad (11)$$

3 Output Transition Time Representation

3.1 Inverters

For simplicity, let us now only consider the case of rising edges applied to the input of gates. As it can be deduced from (6), in the *Fast* input range, τ_{OUT}^{FAST} is characteristic of the inverter structure (gate) and of its load. Considering the sensitivity to the input slope, τ_{OUT}^{FAST} can be used as an internal reference of the output transition time of the considered structure. In this case, (6) becomes

$$\frac{\tau_{OUT-HL}}{\tau_{OUT-HL}^{Fast}}(Inv.) = MAX \left\{ \begin{array}{l} 1 \\ \sqrt{\dfrac{(V_{DD}-V_{TN})}{V_{DD}}} \cdot \sqrt{\dfrac{\tau_{IN-HL}}{\tau_{OUT-HL}^{Fast}}} \end{array} \right\} \qquad (12)$$

We clearly observe in this equation that, using τ_{OUT}^{FAST} as a reference, the normalized inverter output transition time only depends on the input transition time. It has the same value for inverters with different configuration ratio value or loading conditions. This is illustrated in Fig.3 where we represent, using τ_{OUT}^{FAST} as a reference, the output transition time variations displayed in Fig.1. As expected all the curves pile up on the same one, representing the output transition time sensitivity to the input transition time. This is obtained for the complete family of inverters with different values of the configuration ratio and the load. The final value for specific cells is then directly obtained from the evaluation of τ_{OUT}^{FAST}, in (8), that contains the structure and load dependency.

3.2 Gates

As formerly mentioned, the extension to gates is straightforward, multiplying the right part of (6) by a reduction factor representing the ratio of current available in an inverter and a gate implemented with identically sized transistors [2,15].

$$\frac{\tau_{OUT-HL}}{\tau_{OUT-HL}^{Fast}}(Gates) = MAX \left\{ \begin{array}{l} Red_{HL-Top,Bot,Mid}^{Fast} \\ Red_{HL-Top,Bot,Mid}^{Slow} \cdot \sqrt{\dfrac{(V_{DD}-V_{TN})}{V_{DD}}} \cdot \sqrt{\dfrac{\tau_{IN-HL}}{\tau_{OUT-HL}^{Fast}}} \end{array} \right\} \qquad (13)$$

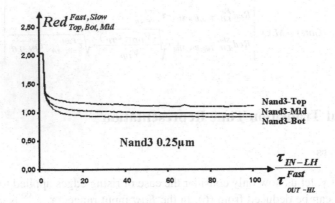

Fig. 4. Input transition time sensitivity of the reduction factor value associated to Nand3 gates designed in the 0.25μm process.

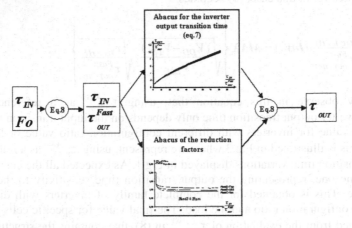

Fig. 5. Synoptic of the evaluation method of the output transition time.

3.3   Output Transition Time Characterization Protocol

From (12) and (13), it appears that it is possible to characterize the output transition time of all the gates of the library, with a reduced set of electrical simulations organized as follows.

- The extraction of τ_{ST} and $R\mu$ values can be done trough the simulation of the step response of an heavily loaded inverter using (8) (the step response being extrapolated from the time spent by the output voltage to switch between 60% and 40% of V_{DD} or inversely).
- The simulation of the output transition time of any inverter can then provide the graph associated to (12).

-   The simulation of the maximum discharging current available in typical Nand and Nor gates and their equivalent inverters, supply the unique representation (Fig.4) of the reduction factor value sensitivity to the input transition time.

Then, the evaluation of the output transition time of any gate of a given library can be process as illustrated by Fig.5. For a specified gate, fan out factor, and duration time value of the controlling input ramp, we determine from (8) the step response of this gate, and the value of the ratio $\tau_{IN}/\tau_{OUT}^{FAST}$. Then we deduce the value of the corresponding reduction factor and ratio $\tau_{OUT}/\tau_{OUT}^{FAST}$ to finally get the corresponding value of the output transition time.

3.4 Validation

In order to validate our approach, we use this protocol to fill the TLF associated to a 0.25μm technology. We then compare the results obtained to the TLF given by the foundry. The relative discrepancies obtained were below 6% for inverters, 10% for Nand2 and Nor2 gates, and 13% for Nand3 and Nor3 gates, validating the proposed unified representation.

However, this validation does not give evidence of the efficiency of the protocol. Indeed, as it needs a specific and reduced set of simulations, we can increase without a great time penalty the number of operating conditions reported in the calculated TLF, obtaining an improved resolution of the output transition time estimation. This means that the time necessary to apply this protocol exhibits a weak sensibility to the size of the table. As an illustration of this fact, we compare, in Fig.6 the value of the output transition time obtained using this method, the usual TLF, and Hspice simulations. This has been performed on an inverter designed in 0.25μm process. As shown, the improvement in accuracy is significant.

Moreover, another great advantage of this protocol is that the time spent to calibrate a library is almost independent of the number of gates it contains. This could be of great benefit in order to define a quasi-continuous sizing of cells or with the advent of on-the-fly-synthesized gate library.

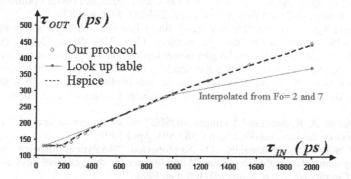

Fig. 6. Comparison of the output transition time value obtained using our protocol, deduced from the usual TLF, and simulated (Hspice).

4 Conclusion

Using an analytical model to evaluate the maximum switching current value we have obtained a simple but accurate design oriented representation of the output transition time of CMOS structures. We have shown its sensitivity to the design parameters, loading factor and input duration time value. Defining *Fast* and *Slow* input ramp controlling domain we have defined a reference τ_{OUT}^{FAST} for the input-output duration time that characterizes the switching cell. It can be used to obtain a unique representation of the timing performances for each category of library cell, independently of their configuration ratio or load. A protocol of characterization of the output transition time of CMOS structure, using only a reduced set of electrical simulations has been developed and validated on a 0.25µm process. This representation appears to be of great help in defining timing library format (TLF) since only one set of simulation by family of cell is necessary to characterize all the gates of different configuration ratio, size and loading conditions. Moreover the full representation obtained in Fig.2 gives a clear idea of the design range to be explored and mostly of the non linear part of the variation where the use of standard look up tables may induce large uncertainties in design performance estimation. The extension of this work to the definition of a characterization protocol for the propagation delay is under progress.

References

[1] Cadence open book "Timing Library Format References" User guide, v. IC445, 2000.

[2] P. Maurine, M. Rezzoug, D. Auvergne "Output transition time modeling of CMOS structures" pp. V-363-V-366, ISCAS 01, Sydney, Australia

[3] T. Sakurai and A.R. Newton, "Alpha-power model, and its application to CMOS inverter delay and other formulas", J. of Solid State Circuits vol.25, pp.584-594, April 1990.

[4] J.R. Burns, "Switching response of complementary symmetry MOS transistor logic circuits" RCA Review, vol. 25, pp627-661, 1964.

[5] N. Hedenstierna and K.O. Jepson, " CMOS circuit speed and buffer optimization", IEEE Trans. Computer-Aided Design, vol. 6, pp. 270-281, March 1987.

[6] I. Kayssi Ayman, A. Sakallah Karem, M. Burks Timothy, " Analytical transient response of CMOS inverters" IEEE Trans. on circuits and Syst. Vol. 39, pp. 42-45, 1992

[7] T. Sakurai and A.R. Newton, "Alpha-power model, and its application to CMOS inverter delay and other formulas", J. of Solid State Circuits vol. 25, pp. 584-594, April 1990.

[8] Santanu Dutta, Shivaling S. Mahant Shetti, and Stephen L. Lusky, "A Comprehensive Delay Model for CMOS Inverters" J. of Solid State Circuits, vol. 30, no. 8, pp. 864-871, 1995.

[9] T. Sakurai, A. R. Newton "A simple MOSFET model for circuit analysis" IEEE Trans. On electron devices, vol.38, n°4, pp. 887-894, April 1991.

[10] L. Bisdounis, S. Nikolaidis, O. Koufopavlou "Analytical transient response of propagation delay evaluation of the CMOS inverter for short channel devices" J. of Solid State Circuits vol. 33, n°2, pp. 302-306, Feb.1998.

[11] A. Hirata, H. Onodera, K. Tamaru "Proposal of a Timing Model for CMOS Logic Gates Driving a CRC π load" in proc. of the Int. Conf. On CAD 1998 (San Jose), pp 537-544.

[12] D. Auvergne, J. M. Daga, M. Rezzoug, "Signal transition time effect on CMOS delay evaluation "IEEE Trans. on Circuit and Systems-1, vol.47, n°9, pp.1362-1369, sept.2000

[13] J. M. Daga, D. Auvergne "A comprehensive delay macromodeling for submicron CMOS logics" IEEE J. of Solid State Circuits Vol.34, n°1, pp.42-55, 1999.

[14] A. Nabavi-Lishi "Inverter Models of CMOS Gates for Supply Current and Delay Evaluation" IEEE Transactions on Computer Aided Design of Integrated Circuits and Systems, Vol. 13, N° 10, October 1994.

[15] J. Meyer "Semiconductor Device Modeling for CAD" Ch. 5, Herskowitz and Schilling ed. mc Graw Hill, 1972.

[16] I. Sutherland, B. Sproull, D. Harris "Logical effort : designing fast CMOS circuits" Morgan Kaufmann Publishers.

A Low Energy Clustered Instruction Memory Hierarchy for Long Instruction Word Processors*

Murali Jayapala[1], Francisco Barat[1], Pieter Op de Beeck[1], Francky Catthoor[2],
Geert Deconinck[1], and Henk Corporaal[2,3]

[1] ESAT/ACCA, Kasteelpark Arenberg 10, K.U.Leuven, Heverlee, Belgium-3001
{mjayapal, fbaratqu, pieter, gdec}@esat.kuleuven.ac.be
[2] IMEC vzw, Kapeldreef 75, Heverlee, Belgium-3001
{catthoor, heco}@imec.be
[3] Department of Electrical Engineering, Delft University of Technology, Mekelweg 4,
2628 CD Delft, The Netherlands

Abstract. In the current embedded processors for media applications, up to 30% of the total processor power is consumed in the instruction memory hierarchy. In this context, we present an inherently low energy clustered instruction memory hierarchy template. Small instruction memories are distributed over groups of functional units and the interconnects are localized in order to minimize energy consumption. Furthermore, we present a simple profile based algorithm to optimally synthesize the L0 clusters, for a given application. Using a few representative multimedia benchmarks we show that up to 45% of the L0 buffer energy can be reduced using our clustering approach.

1 Introduction

Many of the current embedded systems for multimedia applications, like mobile and hand-held devices, are typically battery operated. Therefore, low energy is one of the key design goals of such systems. Typically the core of such systems are programmable processors, and in some cases application specific instruction set processors (ASIPs). VLIW ASIPs in particular are known to be very effective in achieving high performance for our domain of interest [4]. However, power analysis of such processors indicates that a significant amount of power is consumed in the on-chip memories. For example in the TMS320C6000, a VLIW processor from Texas Instruments, up to 30% of the total processor power is consumed in instruction caches alone [6]. Hence, reducing power consumption in the instruction memory hierarchy is important in reducing the overall power consumption of the system.

To this end we present a low energy clustered instruction memory hierarchy as shown in Figure 1. Small instruction memories are distributed over groups of functional units and the interconnects are localized in order to minimize energy consumption. Furthermore, we present a simple profile based algorithm to optimally synthesize the L0 clusters, for a given application.

---

* This work is supported in part by MESA under the MEDEA+ program

B. Hochet et al. (Eds.): PATMOS 2002, LNCS 2451, pp. 258–267, 2002.

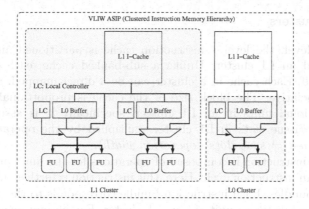

Fig. 1. Clustered instruction memory hierarchy

The rest of the paper is organized as follows. Section 2 describes the operation of the clustered architecture and how energy consumption can be reduced by clustering. Section 3 describes the profile based algorithm to optimally synthesize the L0 clusters. Section 4 positions our work with respect to some of the related work available in the literature, and finally in section 5 we present some of the experimental results and analysis of the clustering approach.

2   The Architecture Template

The fully clustered instruction memory hierarchy is as shown in Figure 1. At the level 1, the conventional instruction cache is partitioned to form the L1 clusters. At the level 0, a special instruction buffer or a cache is partitioned to form L0 clusters. The level 0 buffers are typically small and used during loop execution. Different loop buffer schemes like the decoded instruction buffer scheme [1] or the loop cache scheme [3] or the special filter cache scheme [2], can be adopted in the L0 clusters. Of these loop buffer schemes, for our simulations we specifically consider the decoded instruction buffer scheme [1,8]. In essence the loop buffer operation is as follows. During the first iteration of the loop the instructions are distributed over the loop buffers. For rest of the loop execution the instructions are derived from the L0 buffers instead of the instruction cache. During the execution of non-loop parts of the code instructions are derived from level 1 instruction caches.

Typically, the levels are distinguished based on the access latency to the memory at the corresponding level. However, we distinguish them based on the physical proximity of the buffers. In the sense that, level 0 buffers are placed closer to the functional units than the level 1 caches, while their access latency could still be the same as the access latency of the level 1 instruction cache.

2.1 L1 Clusters

At the top level, the level 1 instruction cache is partitioned and each partition is called an L1 cluster. Unlike a sub-banked cache, each partition is a cache in itself. Each cache in a cluster can be a direct-mapped, set-associative or fully associative. The block size of the cache is proportional to the number of functional units in an L1 cluster, specifically it is assumed to be a multiple of #issue slots in L1 cluster multiplied by the operation width i,e *block size = n * (#issue slots * operation width)*.

A VLIW instruction is composed of operations for the functional units to be executed in an instruction cycle. Here, this instruction is further sub-divided into instruction bundles. Each instruction bundle corresponds to the group of operations for the functional units in an L1 cluster. Furthermore, each instruction bundle is sub-divided into operation bundles. Each operation bundle corresponds to the group of operations for the functional units in an L0 cluster. This categorization is shown in Figure 2. Furthermore, the length of instruction bundles are assumed to be of variable (NOP compressed).

Each L1 cluster has a separate fetch, decode and issue mechanisms. Since, the instruction bundles are variable in length, we assume a scheme similar to the fetch, decode and issue mechanisms in Texas Instruments TMS320C6000 [7]. However, this scheme is applicable to each and every L1 cluster. A fetch from the L1 cache receives an instruction packet, which is the size of the #issue slots in an L1 cluster multiplied by the width of an operation[1] of the L1 cluster.

The fetch mechanisms across the L1 clusters operate asynchronously, while the issue mechanisms are synchronized every instruction cycle. A fetch from the L1 cache of one L1 cluster and a fetch from the L1 cache of another L1 cluster might contain operations to be executed in different instruction cycles. However, we assume that the encoding scheme provides enough information to determine this difference and to issue operations in the correct instruction cycle.

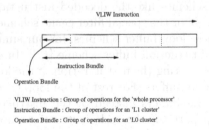

VLIW Instruction : Group of operations for the 'whole processor'
Instruction Bundle : Group of operations for an 'L1 cluster'
Operation Bundle : Group of operations for an 'L0 cluster'

Fig. 2. Instruction format describing the categorization

---

[1] Here, issue width of an L1 cluster is assumed to be equal to number of functional units in an L1 cluster

2.2 L0 Clusters

As shown in the Figure 1, each L0 cluster has an L0 buffer. These buffers are used only during the execution of the loops. Each L0 cluster has a local controller to generate addresses and to regulate the accesses to corresponding loop buffers. Also, at the end of every iteration of a loop the local controllers are synchronized through a synchronization logic. In our earlier work we have presented the details of the local controller and the synchronization logic. For further details we refer the reader to [8].

Instruction Clusters and Datapath Clusters. An L0 cluster is basically an instruction cluster, and in principle an instruction cluster and a datapath cluster can be different. In a datapath cluster, as seen in some of the current VLIW processors like the TMS320C6000 [7], the functional units derive 'data' from a single register file. On the other hand, in an instruction cluster the functional units derive 'instructions' from a single L0 buffer. Even though in both cases the main aim of partitioning is to reduce energy (power) consumption, the principle of partitioning is different [5,9], and the decisions can be taken independently.

From the energy consumption perspective, an instruction cluster could include one or more datapath clusters. Usually, in most datapath organizations, one access to an instruction buffer is followed by three or more accesses to the register files. Hence, the access rate to a register file is at least three times more than the access rate to an instruction buffer. Also, the energy per access of an register file is higher than energy per access of and instruction buffer of the same size, because the register files are multi-ported. Hence, in order to minimize the energy consumption, the datapath clusters should be smaller than the instruction clusters. However, it is still possible that an instruction cluster and a datapath cluster are equivalent (in terms of functional unit grouping).

2.3 Energy Reduction by Clustering

The architectural template presented in the previous sections is inherently low energy in two aspects. Firstly, the energy consumption in the storage can be reduced by employing smaller and distributed memories, which are low power consuming. Secondly, the energy consumption in the interconnect (communication) can be reduced by localizing the possible data (instruction) transfers. In a conventional approach, long and power consuming interconnects are needed to deliver the instructions from a centralized storage to the functional units. However, in a distributed organization like in Figure 1, such long interconnects can be avoided.

Energy Reduction in Storage by Clustering. Clustering the storage at an architectural level aids in reducing the energy consumption in two ways. Firstly, smaller and distributed memory can be employed. Secondly, at the architectural level the access patterns to these memories can be analyzed and the information gathered can be utilized in restricting the accesses to certain clusters. This principle holds for all the levels of memory in the instruction memory hierarchy.

Analytically, the energy consumption of a centralized (non-clustered) organization can be written as

$$E_{centralized} = Naccess * E_{per-acc}$$

where, $E_{centralized}$ represents the energy consumption of a centralized buffer, and

$$E_{clustered} = \sum_{i=1}^{N_{CLUSTERS}} Naccess_i * E_{per-acc_i}$$

where $E_{clustered}$ represents the energy consumption of a clustered organization. By partitioning a centralized memory, $E_{per-acc_i} < E_{per-acc}$, i.e. each partition will have smaller energy per access than a centralized memory. On the otherhand if the accesses to each partition can be restricted such that $Naccess_i < Naccess$, then it is clear that $E_{clustered} < E_{centralized}$. However, in certain cases even if $\sum Naccess_i > Naccess$, accessing a smaller memory ($E_{per-acc_i} < E_{per-acc}$) sometimes pays off in reducing energy.

Typically, the variation of $E_{clustered}$ with number of clusters is as shown if Figure 3. For a certain combination of $access_i$, $E_{per-acc_i}$ and $N_{CLUSTERS}$, the energy consumption is maximally reduced over a centralized organization. The following section describes a clustering scheme where in, given an instruction profile of an application it gives a clustering scheme, which achieves a certain maximal reduction in energy.

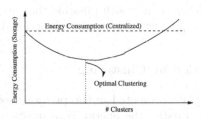

Fig. 3. Typical variation of energy consumption in storage with number of clusters

3   Profile Based Clustering of L0 Buffers

The architecture template has many parameters to be explored both in L0 and L1 clusters. For instance for the L1 clusters, the type of the L1 caches (direct/set-associative), the number of sets and the block size of each L1 cache, etc. Similarly for the L0 buffers the parameters are the size of each L0 buffer, etc. Since, the L0 buffers are accessed exclusively from the L1 caches (except during the initiation phase), the synthesis of the L0 and L1 clusters can be decoupled. The generic approach that we follow to form the L0 clusters is shown in Figure 4. Given an instruction profile during the execution of the loops, the functional units and the L0 buffers are grouped to form the L0 clusters.

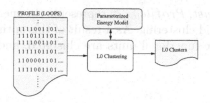

Fig. 4. Profile based L0 clustering approach

3.1   Instruction Profile

The basic observation which aids in clustering is that typically in an instruction cycle not all the functional units are active. Furthermore, the functional units active in different instruction cycles may or may not be the same. Now, the active functional units need to be delivered with instructions. Hence, there is a direct correlation between the access pattern to the instruction memory to functional unit activation. For a given application an instruction profile can be generated which contains the information of functional unit activation every instruction cycle. Because of the direct correlation between the functional unit activation and the access patterns, this profile can be used to form clusters. To remove some effects of data dependency, an average profile can be generated over multiple runs of the application with different input data. Since the L0 buffers are accessed only during the execution of loops, the profile information corresponding to loop execution is used for generating L0 clusters.

3.2   L0 Clustering

Here, for a given an instruction profile (# of functional units and their activation trace) and the centralized L0 buffer's size (# words and width) the L0 buffer is partitioned and the functional units are grouped into clusters so as to minimize energy consumption. This problem is formulated as a 0-1 assignment optimization problem. The formulation is as follows.

$$\min\{\ L0Cost(L0clust,\ Profile_{loops})\ \}$$

$$Subject\ To \quad \sum_{i=1}^{N_{max\ CLUST}} L0clust_{ij} = 1;\ \forall j$$

Where,

$$LOClust_{ij} = \begin{cases} 1 \ if \ j^{th} \ functional \ unit \ is \ assigned \ to \ cluster \ i \\ 0 \ otherwise \end{cases}$$

N_{FU} is the total number of functional units

$N_{maxCLUST}$ is the maximum number of feasible clusters $= N_{FU}$

(At most each functional unit can be an $LOcluster$)

The $LOCost(LOclust, Profile_{loops})$ represents the energy consumption in the L0 buffers for any valid clustering. As a result of this optimization, the L0 buffer is partitioned and the functional units are clustered, and is represented by the matrix $LOClust_{ij}$.

4   Related Work

In the literature the problem of reducing power in instruction memory hierarchy has been addressed at various abstraction levels of the processor, an overview of the different optimizations can be found in [14]. However, we believe that power can be reduced to a large extent by optimizing at higher levels of abstraction, and further optimizing at lower levels. The notion of partitioning or sub-banking a memory block at the logic level is quite well known [13]. However, in our scheme the partitioning is done at the architectural level, and the sub-banking can still be used within each of the clusters.

Other architectural level clustering like the ones in [10,11] are essentially different from our scheme. Firstly, these schemes concentrate specifically on the level 1 caches, while we consider clustering the level 0 buffers, and the interconnect in addition to the level 1 caches. Secondly, they do not consider grouping the functional units into clusters as we do. Furthermore, our approach to clustering is at a higher level than these schemes and in principle they can still be applied to the level 1 caches in the L1 clusters of our architecture (Figure 1).

5   Experimental Results

We performed some simulations using some of the benchmarks from Mediabench. Some of the relevant characteristics are shown in Table 1. These benchmarks were compiled on to a non-clustered VLIW processor with ten functional units, using the compiler in the Trimaran tool suite [12]. Commonly used transformations like predication and software pipelining were applied to increase the instruction level parallelism. The profile as described in the section 3.1 was generated using the instruction set simulator of the Trimaran tool suite.

Currently, of all the possibilities in the instruction memory hierarchy template we assumed a single L1 cluster, since our main goal is to form L0 clusters. Furthermore, the L0 buffers were assumed to be SRAM based buffers, and the number of words in the L0 buffers are assumed to be given. We have explored only the possible grouping of functional units and the partitioning of L0 buffers into L0 clusters.

Table 1. Benchmark characteristics

Benchmark	avg ILP	%exec time loops	ILP loops	%exec time non-loops	ILP non-loops
Adpcm	5.6	95	5.6	5	2.7
Jpeg Decode	3.0	30	2.9	70	3.0
IDCT	3.9	50	5.6	50	2.4
Mpeg2 Decode	3.3	30	5.6	70	2.3

The profile corresponding to the loop execution, for each of the benchmark, was passed through the L0 clustering stage. Here, we assumed a centralized L0 buffer with 64 words[2] depth and 32*10 bits wide (operation width * # issue slots). The energy consumption in the L0 clusters was calculated using the formula described in section 2.3. The energy per access for these buffers was obtained from the parameterized energy models of Wattch [15] (by modeling them as simple array elements), and the number of accesses for each cluster was calculated using the instruction profile.

Table 2. The L0 clustering results

Benchmark	$N_{LOClust}$	$\dfrac{(FUs\ grouping)}{L0\ buffer\ size\ (bits)}$	E_{redn}
Adpcm	5	$\dfrac{(1,2,3,4,5)\quad(6,7)\quad(8)\quad(9)\quad 10}{64*(5*32)\ \ 64*(2*32)\ \ 64*32\ \ 64*32\ \ 64*32}$	30%
Jpeg Decode	7	$\dfrac{(1,2)\quad(3,4)\quad(5)}{64*(2*32)\ \ 64*(2*32)\ \ 64*32}$ $\dfrac{(6,7)\qquad 8\qquad 0\qquad 10}{64*(2*32)\ \ 64*32\ \ 64*32}$	44%
IDCT	4	$\dfrac{(1,2,3,4,5)\quad(6,7)\quad(8,9)\quad 10}{64*(5*32)\ \ 64*(2*32)\ \ 64*(2*32)\ \ 64*32}$	25%
Mpeg2 Decode	4	$\dfrac{(1,2,3,4,5,8)\quad(6,7)\qquad 9\qquad 10}{64*(6*32)\ \ \ 64*(2*32)\ \ 64*32\ \ 64*32}$	22%
Centralized	20Kbit	$\dfrac{(1,2,3,4,5,6,7,8,9,10)}{64*(10*32)}$	

The variation of the L0 buffer energy with the number clusters is shown in Figure 5. With an increase in the number of clusters the energy consumption in the L0 buffers drops to a certain optimal energy consumption and increases with further increase in number of clusters. The resulting L0 clusters for the optimal energy consumption is shown in Table 2. Here, the resulting clusters are specific to each benchmark. Furthermore, the achieved energy reduction in the L0 buffer energy over a single centralized L0 buffer organization, for each benchmark, is also shown in Table 2.

Clearly, the results indicate that by clustering the L0 buffers, instead of arbitrarily partitioning, up to 45% of L0 buffer can be reduced, and the profile based clustering can be used to automatically synthesize the clusters. Furthermore, we have not considered the interconnect energy in our energy costs. We believe that

---

[2] it was observed that number of instructions within the loop body was less than 64 instructions

if that were to be considered in the energy equations, the reduction would be more significant. However, we leave such an effort to our future work.

5.1 Discussion: L1 Clustering

As an interesting exercise, we formulated the L1 clustering similar to the L0 clustering as shown in Figure 4. Here, the problem was to group L0 clusters and partition the level 1 cache into L1 clusters, instead of grouping the functional units and partitioning the L0 buffers into L0 clusters. Since the accesses to the level 1 caches occurs during the execution of the non-loop parts of the code, the corresponding instruction profile was used. Furthermore, the centralized cache was assumed to be a 20KB, direct mapped, with 256 words and a block size of 80bytes (due to the restriction in the architecture section 2). The energy cost for any valid clustering was obtained using the formula described in section 2.3. The energy per access of the caches were obtained from Cacti [16], and the number of accesses to each cache (assuming variable length encoding) was calculated using the instruction profile.

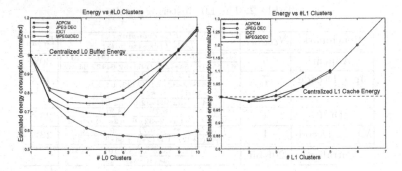

Fig. 5. Variation of energy consumption with number of L0 clusters

The variation of L1 cache energy with the number of clusters is shown in Figure 5. Similar to the L0 clustering stage, the energy consumption drops to a certain optimal energy and then increases with the number clusters. However, unlike the L0 clustering the maximum reduction in energy with L1 clustering was marginal, about 2%. The reduction in marginal mainly because of the overhead of tags in each clusters. In our experiments the centralized organization was assumed to be a direct mapped cache. By partitioning this direct mapped cache, we were essentially replacing it with smaller direct mapped caches, which naturally increases the number of tag bits. Even though the effective storage size is the same the number of tag bits would have increased. If this tag-overhead can be avoided we believe the energy reduction could be more significant. Some authors have proposed ways to design caches with less tag overhead [17]. We intend to evaluate this in our future work.

6 Summary

In summary, we have presented a low energy clustered instruction memory hierarchy for long instruction word processors. Where, small instruction memories are distributed over groups of functional units and the interconnects are localized in order to minimize energy consumption. Furthermore, we also presented a simple profile based algorithm to optimally synthesize the L0 clusters, for a given application. As shown in our experimental results section, by taking into account the access patterns to the instruction memory, visible only at the architectural level, we can achieve significant reduction in energy.

References

1. R. S. Bajwa, et al., "Instruction Buffering to Reduce Power in Processors for Signal Processing", IEEE Transactions on VLSI systems, vol 5, no 4, Dec 1997.
2. N. Bellas, et al., "Architectural and Compiler Support for Energy Reduction in Memory Hierarchy of High Performance Microprocessors", ISLPED 1998.
3. L. H. Lee, et al., "Instruction Fetch Energy Reduction Using Loop Caches For Applications with Small Tight Loops", ISLPED 1999.
4. M .Jacome, et al., "Design Challenges for New Application Specific Processors", Special Issue System Design of Embedded Systems, IEEE Design & Test of Computers, April-June 2000.
5. M. Jacome, et al., "Exploring Performance Tradeoffs for Clustered VLIW ASIPs", ICCAD, November 2000.
6. Texas Instruments Inc, Technical Report, "TMS3206000 Power Consumption Summary", http://www.ti.com
7. Texas Instruments Inc, "TMS320C6000 CPU and Instruction Set Reference Guide", http://www.ti.com
8. M. Jayapala, et al., "Loop Cache (Buffer) Organization: Energy Analysis and Partitioning", Technical Report K.U.Leuven/ESAT, 22 Jan 2002.
9. A. Wolfe, et al., "Datapath Design for a VLIW Video Signal Processor", IEEE Symposium on High-Performance Computer Architecture (HPCA '97).
10. S. Kim, et al., "Power-aware Partitioned Cache Architectures", ISLPED 2001.
11. M. Huang, et al., "L1 Data Cache Decomposition for Energy Efficiency", ISLPED 2001.
12. Trimaran, An Infrastructure for Research in Instruction-Level Parallelism, 1999. http://www.trimaran.org
13. Ching-Long Su, et al., "Cache Design Trade-offs for Power and Performance Optimization: A Case Study", ISLPED 1995.
14. L. Nachtergaele, V. Tiwari and N. Dutt, "System and Architectural-Level Power Reduction of Microprocessor-based Communication and Multimedia Applications", ICCAD 2000.
15. D. Brooks, et al., "Wattch: A Framework for Architectural-Level Power Analysis and Optimizations", ISCA 2000.
16. S.J.E. Wilton and N.P. Jouppi, " CACTI: an enhanced cache access and cycle time model", IEEE Journal of Solid-State Circuits, 31(5):677–688, 1996.
17. P. Petrov and A. Orailoglu, "Power Efficient Embedded Processor IPs through Application-Specific Tag Compression in Data Caches", Design and Test in Europe Conf (DATE), April 2002.

Design and Realization of a Low Power Register File Using Energy Model

Xue-mei Zhao and Yi-zheng Ye

Microelectronics Center, Harbin Institute of Technology No. 92 Xi Da Zhi Jie, P.O.B 313,
Harbin, 150001, P.R.China
zhaoxuemei1@hotmail.com

Abstract. This paper uses a analytical-based characterization model to discuss energy consumption of register files with different circuit techniques that using multi-port SRAM technology. Energy distribution chart of register file with different architectural parameters is acquired according to the calculation results of energy model. How the decoder structure and the dominant component have effect on the power of the register file is demonstrated with emphasise. With this low power method, a 64×32 bits three-port register file operate at a 500MHz frequency with 54mW power dissipation.

1 Introduction

Portable and wearable computing applications require high speed and low energy consumption because for such products longer battery life translates to extended use and better marketability. Register files take a substantial portion in the energy consumption in modern microprocessors [1,2]. Therefore, the research on the power efficient design of register files is significant.

This paper is organized as follows: in section2 we describe the energy model of each component of register file. Then we analyse and calculate the proportion of every component in total energy consumption according to the number of ports and registers. In section3 energy consumption of register file is analysed. An experiment are given in section4. Conclusions are made in section5.

2 Power Model of the Register File

In the overall register file, read and/or write energy is the sum of energy dissipated in the decode logic, memory array, sense amplifiers, energy dissipated in control circuitry that driving the signals to operate the sense amplifiers, precharge circuitry and write drivers. Every part includes three kind of energy dissipation: dynamic, through-current (short-circuit), and leakage energies. The average dynamic energy consumption of each block per CPU cycle is as follows:

$$E = \sum_r (f_r \cdot C_{switch\_r} \cdot V_{dd}^2) \qquad (1)$$

B. Hochet et al. (Eds.): PATMOS 2002, LNCS 2451, pp. 268–277, 2002.

C_{switch_r} is the switch capacitance of the device or gate r; V_{dd} is the work voltage; f_r is the work frequency. In our research, we modify the circuit structure to reduce register file dynamic power by lowering the switching frequency and the switch capacitance in each component.

Schemes to reduce short-circuit power should try to minimize the time of short-circuit of these components. Leakage power has quite small proportion in total energy consumption. It is dissipated irrespective of the switching activity of the circuit, and only relevant to temperature and threshold voltage V_{th}. Consequently, leakage power will be omitted in the following parts.

Our method used in estimating CMOS energy consumption is based on these energy estimations on this CV^2 model for gate-level or macro-level, and then for the entire circuit. This kind of analytical characterization model can be used to obtain power estimates before the practical transistor-level design is implemented.

2.1 Power Model of the Bit Line

The energy dissipated in the bit lines is due to precharge, as well as discharge which is because one of the two bitlines is reduced to a logic low value in the read and write port. It is assumed that the precharging voltage is $1/2*V_{dd}$ and V_{ss} is the voltage swing on the bit lines during sensing phase.

The energy dissipated for precharging is

$$E = \frac{1}{4}V_{dd}^2 N_{reg} N_{read} N_{bit\_w}(C_{metal}H_{cell} + C_{drain}W_{pass,r})$$ (2)

The energy dissipated for driving bit lines in a read access is

$$E - \frac{1}{2}V_{dd}V_{ss}N_{bit\_w}N_{read}f_{zero}N_{reg}(C_{metal}H_{cell} + C_{drain}W_{pass,r})$$ (3)

The bit line capacitance includes the metal line capacitance and the diffusion capacitance of pass transistor. N_{reg} is the number of registers, N_{read} is read port number, N_{bit_w} is the number of bit width, C_{drain} is the diffusion capacitance of pass transistor r. H_{cell} is the cell height. In the double-end sensing scheme, f_{zero} is 1. But in the single-end sensing f_{zero} will be the frequency of storing zero in the cell.

If writing to a cell are usually done by a full-swing signal and consequently bit line is equalized[5], the write energy is

$$E = V_{dd}[(\frac{1}{2}V_{dd} - V_{ss})f_{bit.w} + (\frac{1}{2}V_{dd} + V_{ss})f'_{bit,w}]$$
$$N_{reg}N_{write}(C_{metal}H_{cell} + C_{drain}W_{pass,w})$$ (4)

N_{write} is write port number, $f_{bit,w}$ is the 0–0 and 1–1 transition times on the bit lines and $f'_{bit,w}$ is the 0–1 and 1–0 transition times on the bit lines during writing to the cell arrays. That is, $N_{bit\_w}=f_{bit,w}+f'_{bit,w}$.

2.2 Power Model of the Word Line

In every read or write cycle, a word line will be charged and another one will be discharged. The energy dissipation in the read word line is given by:

$$E = V_{dd}^2 f_{read} N_{bit\_w} (N_{r\_bit} C_{gate} W_{pass,r} + W_{cell} C_{metal}) \tag{5}$$

The word line capacitance is the sum of the gate capacitance of pass transistor and the line capacitance. In equation 5, f_{read} is the frequency of read operation; N_{r_bit} depends on the way of read operation. N_{r_bit} will be 1, if single-end sensing is used. However, in double-end sensing scheme N_{r_bit} will be 2. C_{gate} means the gate capacitance per unit width. $W_{pass,r}$ is the width of the read pass transistor. C_{metal} is the metal line capacitance per unit length. W_{cell} is the storage cell width, which is a function of the read and write port number: N_{read} and N_{write}.

The energy formula of the write word line is similar to that of the read word line. $W_{pass,r}$ is replaced by the width of the write pass transistor, $W_{pass,w}$. f_{write} is the frequency of write operation; N_{w_bit} usually is 2.

$$E = V_{dd}^2 f_{write} N_{bit\_w} (N_{w\_bit} C_{gate} W_{pass,w} + W_{cell} C_{metal}) \tag{6}$$

2.3 Power Model of the Sense Circuitry

The energy dissipated by sensing circuitry closely depends on the type of the sense amplifier used. The inverter usually is used as the sense amplifier in single-end scheme. This type of sense amplifier has short-circuit current while sense amplifier is activated. The short-circuit energy dissipation in inverter is

$$E = N_{bit\_w} N_{read} f_{read} V_{dd} I_{SA} \Delta t \tag{7}$$

In equation 7, I_{SA} is the short-circuit current of sense amplifier, and $\bullet t$ is the sensing time of sense amplifier.

In double-bitline scheme, current-mirror amplifier is a common used component. Energy dissipation of a double-stage current-mirror sense amplifier shown in [6] is due to the DC current.

$$E = N_{read} N_{bit\_w} [2(V_{dd} I_{sc1} \Delta t + 0.5 V_{swing,s1}^2 C_{load,s1}) +$$
$$V_{dd} I_{sc2} \Delta t + 0.5 V_{swing,s2}^2 C_{load,s2})] \tag{8}$$

where Δt is the response time; I_{sc1} and I_{sc2} are the short currents flowing in the current mirror of stage 1 and stage 2 in the sense amplifier respectively; $V_{swing,S1}$ and $V_{swing,S2}$ are the voltage swings of the outputs S1 and S2 respectively; $C_{load,S1}$ and $C_{load,S2}$ are their corresponding capacitive loads.

Latch-type sense amplifier has no short circuit and can sense small bit line swings. So it acquires lower power and higher speed.

$$E = V_{dd}^2 N_{bit\_w} N_{read} f_{read} (C_{gate,n} W_{SA,inn} + C_{gate,p} W_{SA,inp}) \tag{9}$$

2.4 Power Model of the Decoder

Figure 1 shows the basic decoder architecture. Typically, the number of register file is not very large, so two-stage decoder is enough. In the first stage, each predecoder usually consists of some multi-input NANDs and some drivers. In the second stage, each word decoder consists of one NOR and one inverter. In the critical path of the decoder, the dynamic power dissipation during one switching cycle is the total power of each stage and metal line. The energy consumption of stage r includes switching gate load of the next stage $r+1$ and self- load of stage r. All the decoder power is shown:

$$E = \sum_r (V_{dd}^2 N_r f_r (C_{g,r} W_{r+1} + C_{j,r} W_r + L_{m,r} C_{metal})) \quad (10)$$

In equation 10, N_r is the number of gates in stage r. f_r means the frequency of decoder operation. $C_{g,r}$ is the gate capacitance of a device or gate r; $C_{j,r}$ is the junction capacitance of a gate r. W_{r+1} means the width of gate $r+1$. Similarly, W_r means the width of gate r. $L_{m,r}$ is the length of the metal line in stage r; C_{metal} is the capacitance of the metal per unit length. It is reported that the dynamic power, short circuit power, and leakage power all are proportional to the size of the stage [4], and then the total power dissipation in the decoder can be approximately expressed as the sum of the stage sizes. Consequently, power can be reduced when the total width of transistors in the chain is minimized. The minimum power solution has exactly one stage, which is the input driver directly driving the load. But it is not cost-efficient due to the large delays involved. We should adopt a trade-off sizing strategy to reduce power with constraint delay.

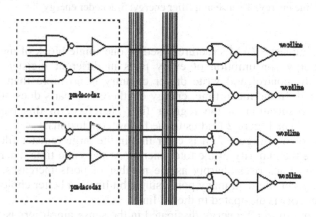

Fig. 1. Decoder structure

3 Analysis of Energy Consumption of the Register File

3.1 Energy Distribution in the Single Block Structure

We are interested in relative power estimates that would allow us to compare power complexity of different architectures. The technology dependent parameters in above 10 equations, such as the gate, diffusion, and interconnect capacitance, are determined through a simulation measurement technique.

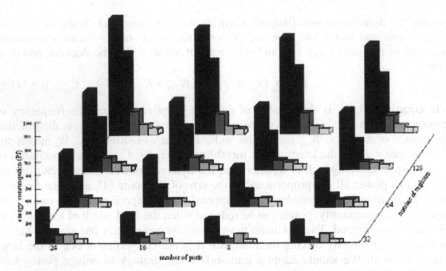

Fig. 2. Energy consumption distribution charts. Bars represent: 1-Total energy; 2-bit line energy; 3- word line energy; 4-sense amplifier energy; 5-decoder energy

We use the simplest SRAM structure based on one single block of memory cells to analyse the energy consumption of every parts in different register files. Energy dissipation in the control and write driver circuitry is no more than 10% of total energy consumed. In the mean time, energy of these two parts does not depend too much on the organization of the register file. Therefore, in this paper we don't concern this part. The Figure 2 lead us to the follow conclusions:

- As is shown in Figure 2, it is clear that in the large register file the bit lines consume the majority of the total energy. At the same time, energy dissipated in the bit lines rises sharply as the number of ports increases. Thus, in the large register file with many ports, single bit line is a better choice because too much energy is dissipated in the bit line.

- According to Fig.2, energy dissipated in the sense amplifiers is the main part of the total energy of the small register file. Also, energy of the sense amplifiers increases slightly when the number of ports and registers adds. Therefore, sensing scheme is significant for the overall power of the small register file. Differential sensing scheme (especially, cross-coupled latch-type sense amplifier) with lower power could be used in the small register file. Single-ended sensing requires higher swing voltage on bit lines than differential sensing. Consequently, single-ended sensing is slower, however it has smaller area because of fewer bit lines. Therefore, as the number of ports and registers grows, single-ended sensing is usually a preferred approach.

- From Fig.2, it is obviously that minority energy dissipated in the word lines. There is a slow upward trend in word line energy when the number of the registers and ports increases.

- As can be seen from Fig.2, the decoders dissipate only a small percentage of the total energy. However, the percentage of the every other partition's energy changes as the structure of the decoder alters. For example, DWL structure makes the length of the bit line half than in the single block structure. Consequently, the parasitical capacitance of the bit line is decreased., which would uses less energy.

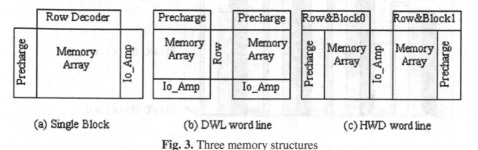

(a) Single Block   (b) DWL word line   (c) HWD word line

Fig. 3. Three memory structures

3.2 Effect of Different Structure on Total Energy

From the above analyses, different component dominates the energy in different scale of register file. But circuit technique of each component could merely change its energy dissipation. Only the overall structure changed will lower total energy essentially. For a specific register file desired, firstly we decide its structure and confirm the energy dominating component, and then determine which form the other component should be. At last, we should select the lower power circuit technique for every part. Thus, total energy of register file will be reduced.

Typically, the number of register file is 32, 64, 128 or 256. It assumed that all of these four register files have three ports and 32 bits per word. Through calculation of energy model, we estimate and compare the energy dissipation in these four kinds of register files with three different decoder structures (shown in Figure 3). The first structure is a single block, the second is dual block with DWL (Double Word Line) technique, and the third is HWD (Hierarchical Word Line structure). The structure of decoder will decide the structure of whole memory. The energy consumption of register files with three structures and four different numbers of registers is shown in Figure 4. In the experimental results, we get:

- In DWL structure, because of the redundant precharging and controlling circuits, more energy will be dissipated compared with single block structure. On the other hand, the length of the bit line is half of that in the single block structure. So the parasitical capacitance of the bit line is decreased, which would uses less energy. Whether DWL structure is energy-efficient depends on the saving energy of short bit line offset by the dissipation of redundant circuits. Compared with DWL structure, the single block has simple decoder and energy-efficient when the number of registers is less than 32. When the number of registers adds to 64,128 or even 256, the advantage of dual block scheme will be obvious bit by bit.
- HWD approach extends the DWL idea by breaking down the word line into additional levels of hierarchy. The block decoder and local word decoder are used to enable the proper sub-array to access the desired data. The advantage of HWD structure is apparent when the number of registers goes very large.

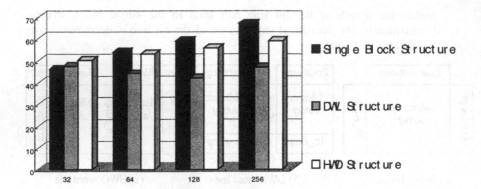

Fig. 4. Energy comparison in three kind of structures

4 Realization of a Register File

We choose the reasonable structure configuration for our register file according to the above thoughts. Our register file, N_{reg}=64, N_{bit_w}=32, N_{read}=2, N_{write}=1, is a typical configuration for RISC CPU. In order to acquire substantial energy saving without sacrificing speed, we adopt a new register file structure configuration with differential sense, the modified cell structure, a better decoder structure, and separate R0 technique.

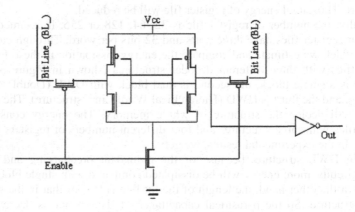

Fig. 5. Latch sense amplifier

4.1 Latch Sense Amplifier

Because amplifier dissipates the most energy in small register file, a low power amplifier should be chosen carefully. Our processor has two read ports and one write port. Even using the double bit lines will not waste the area too much. So we prefer using a differential sense amplifier (such as synchronous latch-type sense amplifier shown in Fig.5) that does not consume short-circuit current. The negative edge of the strobe disconnects the sense amplifier from the ground and turns on the input pass transistors. The sense amplifier is switched on at the rising edge of the strobe. The input pass transistors are switched off, disconnecting the sense amplifier from the bit lines, which provides fast switching of the sense amplifier. During the next positive edge of the strobe, the sense amplifier holds its state. thus, no latch is needed at the sense amplifier output.

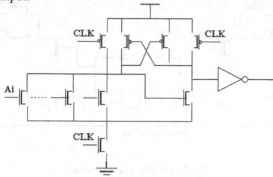

Fig. 6. Source Coupled Logic (SCL) decoder

4.2 Decoder Structure

In our register file we prefer DWL structure, because it is more energy efficient. In order to get fast and low power address pre-decoder, we select Source Coupled Logic (shown in Figure 6) that is one of fastest address decoder technique ever reported [7] instead of traditional NAND. Since a wide fanin NOR can be implemented with very small logical effort in the domino circuit style, a large fanin NAND could be replaced by a NOR. The second decoder uses two level structure.

4.3 Modified Cell Structure

To work together with differential sense amplifier, the modified cell structure is adopted in Figure 7. Two inverter buffers are added to drive the bitline read, So the content of storage cell will not be affected when both of the read ports are operated. But the signals in bit line is inverse with the content of storage cell. Since the storage cell need not drive the bit line, the size of the storage cell can be design to minimum. The modified circuitry has no latency overhead but the modified cell might contribute slight area overhead.

4.4 Separate R0

In register file, R0 is the most frequently referenced register throughout all the benchmarks. It is shown that 38% of register file write operation and 23% of register file read operation are R0. "0" values of R0 are directly sent to output latch by the muxers. If R0 is removed from the register file energy will save 10%.

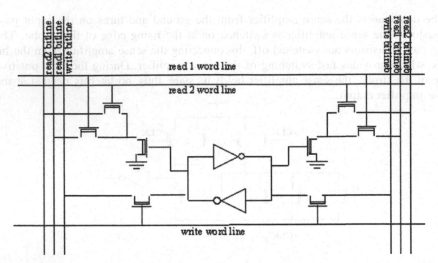

Fig. 7. Modified cell structure

4.5 Experimental Results

Our register file can get fast speed and low power using the above techniques. Simulation and comparison results in TSMC 2.5v 0.25μm CMOS technology are shown in table 1. Its power is enormously reduced, and it achieves better overall performance [10][11].

Table 1. Performance Comparison

	Register file Organization	Area (mm2)	Speed (ps)	Power (mW)
Our register file	64×32 bits 3 ports	0.302× 0.440	680	54@ 500MHz
R.Franch's register file[9]	16×64 bits 3 ports	1.81×0.855	640	670@ 625MHz
T.Suzuki's Cache[8]	576×32 bits 6T SRAM	2.0×0.7		690@ 500MHz

5 Conclusion

In the low power register file design, decoder structure is the key factor of energy consumption. A appropriate decoder structure should be chosen on the base of the number of the registers and the ports. In this preferred structure, the dominating component in the energy dissipation should be found according to the energy distribution chart. Choose a structure to make the dominating component energy efficient, and design the other parts accordingly.

Acknowledgments. The authors thank Chen Chun-xu for his work on layout, Lai Feng-chang and Shi Rui for their advisable suggestions, and many other staff member of microelectronics center for their encouragement.

References

1. Ikeda, Makoto; Asada, Kunihiro, Data bypassing register file for low power microprocessor. IEICE Transactions on Electronics E78-C 10 Oct 1995 Inst of Electronics, Inf & Commun Engineers of Japan p 1470–1472
2. Joshi, R.V.; Hwang, W.; Wilson, S.; Shahidi, G.; Chuang, C.T, Low power 900 MHz register file (8 ports, 32 words × 64 bits) IN 1.8 v, 0.25μm SOI technology. Proceedings of the IEEE International Conference on VLSI Design Jan 3-Jan 7 2000 2000 p 44–49
3. R. Y. Chen, R. M. Owens, M. J. Irwin, and R. S. Bajwa, Validation of an architectural level power analysis technique. In DAC '98, San Francisco, CA, June 1998.
4. Veendrick, H.J.M. et al, Short-circuit dissipation of static CMOS circuitry and its impact on the design of buffer circuits, IEEE JSSC, vol. SC-19, no. 4, pp. 468–473, August 1984.
5. Zyuban, V.; Kogge, P. The energy complexity of register files, Low Power Electronics and Design, 1998. Proceedings. 1998 International Symposium on , 1998 Page(s): 305–310
6. Wilton, S. E., and Jouppi, N., An enhanced access and cycle time model for on–chip caches. DEC WRL Research Report 93/5, July 1994
7. H. Nambu, A 1.8ns Access, 550MHz 4.5Mb CMOS SRAM. in proc. 1998 ISSCC, pp. 360–361.
8. T. Suzuki et. al., Synonym hit ram: A 500MHz 1.5ns CMOS SRAM Macro with 576b parallel comparison and parity check functions. In: 1998IEEE international solid-state circuits conference, pp. 348 349, January 1995.
9. R. Franch et. al., 640-ps, 0.25m CMOS 16×64-b Three-port Register File. IEEE JSSC, vol.32, No.8, Auguest 1997.
10. S.Chao et. al., A 1.3ns 32×32 three port BiCMOS register file. in proc. 1994 BCTM, 1994, pp.91–94.
11. M. Nomura et. al., A 500-MHz, 0.4•mCMOS 32×32 3-port Register File. in proc. 1995 CICC, pp. 151–154.

Register File Energy Reduction by Operand Data Reuse

Hiroshi Takamura, Koji Inoue, and Vasily G. Moshnyaga

Department of Electronics Engineering and Computer Science, Fukuoka University
8-19-1 Nanakuma, Jonan-ku, Fukuoka 814-0180, Japan
{takamura, inoue, vasily}@v.tl.fukuoka-u.ac.jp

Abstract. This paper presents an experimental study of register file utilization in conventional RISC-type data path architecture to determine benefits that we can expect to achieve by eliminating unnecessary register file reads and writes. Our analysis shows that operand bypassing, enhanced for operand-reuse can discard the register file accesses up to 65% as a peak and by 39% on average for tested benchmark programs.

1 Introduction

With the growing popularity of portable applications such as notebook computers, hand-held multimedia wireless devices, portable video phones, etc., energy reduction of microprocessor becomes an important factor for extending the battery lifetime. In today's microprocessors, register files contribute to a substantial portion of energy dissipation [1]. In Motorola's M.CORE architecture, for example, the register file consumes 16% of the total processor power and 42% of the data path power [2].

To the first order, the energy dissipation of a register file may be expressed as the sum over all memory accesses (N) of the energy per access (E_i), i.e. Energy=$\Sigma_{i=1}^{N} E_i$. In register files, most of energy per access (E_i) is burned when driving the high capacitance of the bit-lines, which are heavily loaded with multiple storage cells, and so require a large energy for charging. The number of bit-lines simultaneously activated per access traditionally equals the access bit-width, n_i. Thus, given the average physical capacitance of a bit-line (C) and the supply voltage (V), the register file energy dissipation can be expressed by, Energy= $\Sigma_{i=1}^{N} n_i \times C \times V^2$. Although reducing the energy amounts to all factors (N, n_i, C, V), the savings obtained by lowering the number of accesses (N) and the bit-width per access (n_i), is fairly independent of integration technology and less expensive.

The goal of this paper is to study techniques capable of reusing operand data to eliminate unnecessary register file accesses.

1.1 Related Research

Previous works [2-6] have shown that many register values are used only once [3], indicating that they may be unnecessarily written back to the register file. To reduce

B. Hochet et al. (Eds.): PATMOS 2002, LNCS 2451, pp. 278–288, 2002.

the register file accesses, Hu and Martonosi [4] proposed buffering the results between the functional unit and the register file in a value-aged buffer, in order to read the short-lived values from the buffer not the file. Since the buffer is smaller than a typical register file, it has better energy characteristics. Zyuban and Kogge [5] advocated the register file partitioning, suggesting a register-file split architecture based on the opcode steering. Tseng and Asanovic [6] showed that many operands could be provided by the bypass circuitry [7], meaning that the corresponding register file reads are redundant. The work proposed several techniques, that provide caching of the register file reads, precise read control, latch clock gating, storage cell optimization, bit-line splitting, etc. Although this paper had demonstrated the benefits of the bypass circuitry, it did not exploit the relation between the bypassing and the data reuse, as well as effects of operand bypassing on the register file writes.

Recent works [9,10] have showed the importance of instruction and data reuse for low power microprocessor design. The basic observation here is that many instructions produce the same results repeatedly. By buffering the input and output values of such instruction, the output values of an instruction can be obtained via a table lookup, rather than by performing all the steps required to process the instruction. Simpler forms of reuse that do not require tracking of actual values are shown in [9]. Not performing all the steps of instruction processing can also benefit energy consumption [10]. Instruction reuse can also salvage some speculative work that is otherwise discarded on a branch mis-speculation.

1.2 Contribution

In this paper we experimentally study the architectural enhancement of data-path bypass network in order to reuse operand values and reduce the number of accesses saving register utilization in conventional RISC-type data path architecture to determine benefits that we can expect to achieve by eliminating unnecessary register file accesses during both the read and the write operations. Our analysis shows that techniques such as operand bypassing and selective register-file read allow us to discard 62% of fetched operands while eliminating 39% of the total register file updates on average.

This paper is organized as follows. Section 2 discusses the operand bypassing. Section 4 presents data reusing techniques and their implementation schemes. Section 3 shows the experimental results. Section 5 summarizes our findings and outlines future work.

2 Background

The register file *bypassing* or *forwarding* is widely used in modern pipelined architectures for minimizing the performance impact of the read-after-write (RAW) data hazards [7]. The main idea is to add multiplexors to the input of the ALU, and supply the operands from the pipeline registers instead of the register file. Figure 1 illustrates the bypass network on example of five-stage pipeline, where x1, x2 denote bypass multiplexors, and the bars ID/EX, EX/MEM, MEM/WB denote pipeline registers.

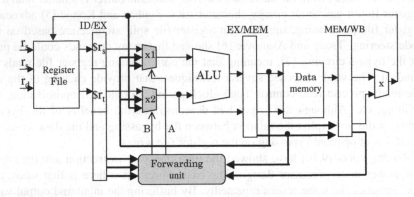

Fig. 1. An example of bypassing hardware

Here ID, EX, MEM, WB stand for instruction decode, execution, memory access, and write back, respectively. The bypassing works as follows. The ALU result from the EX/MEM pipeline register is always fed back to the ALU latches. If the Forwarding Unit detects that the previous ALU operation has written the register corresponding to a source for the current ALU operation, it selects the forwarded result as the ALU input rather than the value read from the register file.

Consider the following instruction sequence, in which each instruction is defined by opcode, destination register (r_d), and two source registers (r_s, r_t), respectively.

$$sub \quad \$2, \ \$1, \ \$3 \qquad \# \ r1 - r3 \rightarrow r2$$
$$and \quad \$4, \ \$2, \ \$5 \qquad \# \ r2 \ \& \ r5 \rightarrow r4$$
$$or \quad \$8, \ \$2, \ \$6 \qquad \# \ r6 \ | \ r2 \rightarrow r8$$
$$add \quad \$9, \ \$4, \ \$2 \qquad \# \ r2 + r4 \rightarrow r9$$

The Forwarding Unit sees that the *sub* instruction writes register $2 while the next *and* instruction reads the register $2, and therefore selects the EX/MEM pipeline register instead of ID/EX pipeline register to get the proper value for register $2. The following *or* instruction also reads $2, so the Forwarding Unit selects the MEM/WB pipeline register for the upper input of the ALU in cycle 5, as shown in Figure 2. The following *add* instruction reads both register $4, the result of *and* instruction, and the register $2, still the target of the *sub* instruction. Therefore in clock cycle 6 the Forwarding unit selects the MEM/WB pipeline register for the upper ALU input and the new WB/IF pipeline register for the lower ALU input. Because the pipeline registers contain both the data to be forwarded as well as the source and the destination register fields, to implement the bypass control we need a comparison of the destination registers of the IR contained in the EX/MEM and MEM/WB stages against the source registers of the IR contained in the ID/EX and EX/MEM registers. Moreover, in conventional RISC architecture, the bypassing control is performed in EXecute

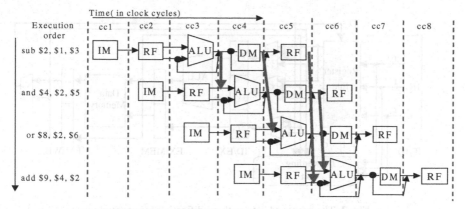

Fig. 2. Pipeline dependencies in the instruction sequence

stage, because the ALU forwarding multiplexors are found in that stage. Thus, we must pass the register numbers from the ID stage via ID/EX pipeline register, to determine whether to forward values.

The conventional bypassing allows us to reuse the data temporally allocated on pipeline registers. However, it does not affect the register-file, which either reads or writes data with each new clock-cycle. If the fetched values are discarded because of bypassing, fetching stale values from the register file creates unnecessary switching activity. Moreover, simulations reveal that almost half of values written to the register-file are short-lived, i.e. used only once usually by the instruction executed immediately after one producing the value. Storing these values also generates large activity. To save power, we have to avoid this redundant activity as much as possible.

3   The Proposed Approach

3.1   An Overview

The key idea of our approach is to combine conventional bypass network with the register file deactivation in order to save energy whenever the operand values are bypassed. Figure 3 exemplifies a five-stage pipelined data-path modified for the data reuse. In comparison to the original design, we propose four architectural enhancements to the data-path. The first one is to move the bypass control prior to the register file enabling; that is to the ID stage. The second enhancement puts the Forwarding Unit in control of the register file by generating signals (s,t) to disable the read word-lines of the file; and signal (d) to disable the write word-lines of the file. The third enhancement allows the Forwarding Unit to swap the inputs of the ALU in order to avoid redundant register file reads. On the register file, this enhancement enforces extra gating at the output of address decoder (Figure 4) and eventually contributes an AND-gate overhead per word-line.

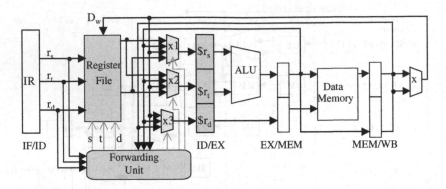

Fig. 3. The proposed data-path modification for operand reuse

Finally, our forth enhancement allocates a special register (s_d) in the data-path to hold data being stored to data memory without writing to the register file. We assume that this register can be controlled either by software, based on the register lifetime information provided by compiler, or by Forwarding Unit hardware, that compares the destination operand of the current memory instruction with the source operands of the next instruction. In the latter case, the Forwarding Unit turns off the MEM and WB stages of the machine pipeline, when the bypass is activated, and thereby reduces energy consumed by writing and reading the same value to the register file. We assume that this register is treated as a temporary state register during interrupt. Ex-

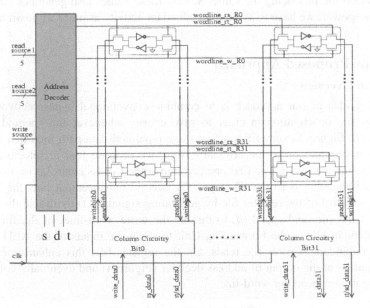

Fig. 4. The modified register file

tending the Forwarding Unit functionality might increase its delay. If it takes longer than the first half of the cycle to finish the data-reuse detection, a latency penalty can be produced. If this latency is too long, we can even consider adding an extra pipe-stage.

3.2 Operation

By giving the Forwarding Unit explicit control of bypass registers and register file, it is possible to reduce register file traffic considerably. As an example consider the following sequence:

```
add   $5, $3, $1        # r3 + r1 → r5
sub   $2, $1, $3        # r1 - r3 → r2
and   $2, $2, $5        # r2 & r5 → r2
or    $6, $2, $6        # r2 | r6 → r6
add   $2, $2, $7        # r2 + r7 → r2
sw    ($3), $2          # r2 → M(r3)
sub   $2, $1, $2        # r1 - r2 → r2
```

The Forwarding Unit sees on decoding the sub instruction that both operands of the instruction match the operand r_t of the previous add instruction, and therefore sets signal s and t to disable the register file read operation in the next clock cycle, while swapping the inputs on the ALU in the EX stage. When the third instruction enters the pipeline, the unit detects that it not only uses the register $r2$ similarly to its predecessor as destination, but also as the second source operand. Because the result of sub spans only two instructions, there is no need to access the register file for writing. In this case, the Forwarding Unit generates signals d and t to eliminate both the register file write and the register file read (requested by instruction sub) by disabling the write word-lines and the read word-lines, respectively. The source operand r_t is by-passed from the EX/MEM. The successive or and add instructions use the result of and instruction as source operand. So here also, the Forwarding Unit activates by-passing in order to reuse data whilst omitting its writing/reading of the register file. When the store (sw) instruction is decoded, the Forwarding Unit moves the value tagged by $2 from the MEM/WB pipeline register to the special register r_d, again avoiding two register file accesses. Thus we access the register file only when the computed data is no more necessary in the successive clock cycles, while reusing the data otherwise. As a result, just on this small example the register file is accessed only 10 times instead of 20!

We should notice that utilizing local information only for data reuse is efficient for the short lifetime values only. To ensure correct operation, we assume that software supervises the data reuse by pointing to Forwarding Unit control on values, which have to be reused. Generally, the maximum lifetime of the reused data in the five-stage pipeline is three clock cycles only for arithmetic–logic instructions and 2 clock cycles for the *load* instruction. Also, operand reuse in load and store instructions may be limited. Consider the following sequence:

```
add   $3,  $1,  $2        # r1 - r2 → r3
lw    $2,  $20($4)        # M(20+r4) → r2
sub   $4,  $1,  $2        # r1 - r2 → r4
```

Because the `sub` instruction uses the result of the `load` instruction as a source operand, a stall has to be inserted in the pipeline. In this case, bypassing the second operand of `add` to be reused in the `sub` instruction becomes impossible. Usually the `lw` instruction decreases data reusability of successive instructions. Experiments show that programs with fewer loads have better reusability

4 Experimental Results

We used Flexible Architecture Simulation Tool (FAST) to evaluate the number of accesses to register file. The tool provided cycle-accurate instruction simulation on a single-issue five-stage pipelined RISC-type microprocessor (similar to MIPS [2]). The simulator traces user-level instructions and records register file access information as well as instruction operands' reuse frequency. We assumed that register-file performs one write and two reads per cycle regardless of the instruction opcode and pipeline state. We experimented with nine typical SPEC95 and MediaBench benchmark programs tested on various data sets. Refer to Table 1 for our benchmark workload and workload. Each benchmark was run to completion The results have been determined in terms of the ratio of reused source operands to the total number of source operands; the reuse frequency for the first and the second source operands, respectively, and the reuse frequency for the register-file writes.

Table 1. Benchmarks and descriptions

Benchmark Name{Data set}	Description	Symbol	Instruction count
adpcm_c	Adaptive PCM voice coding	ade	6,602,451
adpcm_d	Adaptive PCM voice decoding	add	8,024,540
compress{train}	An in-memory version of a UNIX file compression utility	com_n	63,719,628
compress{test}		com_t	4,275,434
compress {big}		com_b	83,180,240,140
go{test}	A go-playing program	go	24,522,085,063
mpeg{mei16v2}	A Mpeg2 video decoding program	mpd_m	62,345,741
mpeg{tennis}		mpd_t	667,957,333
mpeg{verify}		mpd_v	10,711,481
mpeg{trace}		mpd_d	62,343,421
mpeg{clinton}	A Mpeg2 video encoding program	mpe	1,463,074,731
pegwit{my.pub}	A public key generation program	pegc	16,444,080
pegwit{trace}	A public key encryption program	pege	38,408,699
pegwit{pegwit}	A public key decryption program	pegd	21,454,539

Table 2 and Figure 5 show the simulation results for the register-file read. In this figures, *normal* denotes the simplest data-reuse mode provided by traditional bypass network due to reading the pipeline registers instead of register file; *swap* represents the normal mode enhanced by swapping of input operand for data reuse; *stay* utilizes the data-reuse register but without operand swapping; and *stay &swap* is the combination of all these modes. The Rs1 and Rs2 define the reuse ratio of the first and the second operands, respectively, to the total amount of source operands used in the code.

We observe that traditional bypassing (the *normal* mode) provides reuse of almost 1/3 of the total number of source operands, while the impact of Swap and Stay affects only a few percents. The best results (up to 62%) are achieved by the Stay &Swap that combines all the reuse options. Furthermore, because 80-85% of instructions have only one source operand (e.g. load, store), the reuse ratio for the first source operand is much higher (up to 35%) than for the second source operand (up to 10%). This is especially evident for the *pegwit* program, which involves many loads and stores. As Figure 5 shows, the proposed data reuse approach allows us to reduce the total number of register file read accesses from 29.5% (*pegc*) to 62.7% (*com_t*).

Figure 6 shows the reduction rate for the register file writes which we observe in comparison with the conventional bypassing. The results are given regarding to the number of instructions considered for reuse: one (1inst) and two (2 inst). As we see, preserving short lifetime variables from writing to the register file allows us to save up to 62% of total number of the register file writes when two previous instructions are scanned for reuse and 55%, when only one prior instruction is considered.

Table 2. Percentage of operand reuse

Bench-mark	Normal		Swap		Stay		Stay &swap	
	Rs1	Rs2	Rs1	Rs2	Rs1	Rs2	Rs1	Rs2
ade	17.4	8E-05	17.4	2.44	17.4	1.50	29.4	3.94
add	15.9	8E-05	17.6	5.81	15.9	5.86	25.1	7.48
com_n	24.1	0.95	27.2	6.51	26.1	2.09	35.7	8.05
com_t	53.0	0.023	54.3	2.81	53.8	1.57	58.6	4.11
com_b	25.3	0.03	28.1	7.13	27.0	1.63	35.3	8.26
go	26.0	0.015	27.6	7.35	26.5	4.88	32.1	10.1
mpd_m	26.0	0.05	27.5	7.31	26.5	4.86	32.0	10.0
mpd_t	27.1	0.87	30.3	5.60	27.7	3.26	36.5	7.35
mpd_v	26.1	1.05	29.0	5.35	26.5	3.64	36.5	7.23
mpd_d	27.1	0.88	30.3	5.60	27.7	3.26	36.5	7.35
mpe	11.0	2.84	17.9	4.02	11.2	15.8	20.1	16.8
pegc	22.7	0.26	25.2	0.76	22.9	1.33	27.8	1.81
pege	21.5	0.42	24.0	2.52	21.6	2.87	27.3	4.74
Pegd	22.4	0.2	24.3	2.79	22.5	0.81	28.5	4.54

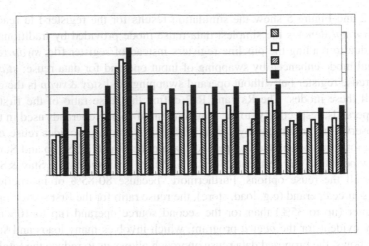

Fig. 5. Reduction rate (%) for the register file read

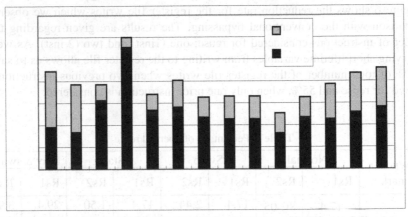

Fig. 6. Reduction rate (%) for the register file write

Figure 7 shows the reduction rate for the register file read and write computed as ratio of the total number of reused operands to the total number of the operands in the code. We see that the proposed data reuse approach allows us to save up 61% of the total number of register file accesses for the *comp_t* program, and 39% on average for all the tested programs.

5 Summary

Our preliminary work indicated that we could effectively leverage data reuse for register file access reduction to save power consumption in RISC-type processors. In order to reuse data values, we propose to gate the register file on bypass, introduce control logic in the instruction fetch stage, and disable writing the short-lived values to

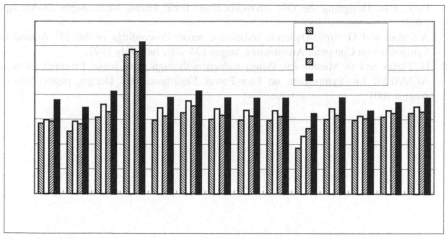

Fig. 7. Reduction rate (%) for the register file read and write

the register file. According to experiments, these techniques can decrease the total number of register-file accesses by 39% on average and by 61% on peak. We currently are working on simulation of a large set of SPEC95 benchmarks and Media-bench programs, as well as extending the experiments to power modeling. Future work will be dedicated to detailed circuit design of a prototype data-path capable of reusing data.

References

1. D. R. Gonzales, Micro-RISC architecture for the wireless market. IEEE Micro, 19(4), pages 30–37, July/August 1999.
2. J. Scott. Designing the low-power M_CORE architecture. Proceedings of the Power Driven Microarchitecture Workshop, hold at 25th Annual Int. Symposium on Computer Architecture, Barcelona, Spain, June 1998
3. M. Franklin and G. S. Sohi. Register traffic analysis for streamlining inter-operation communication in fine-grain parallel processors. Proc. 25th Annual Int. Symposium on Microarchitecture, pages 236–245, December 1992
4. Z. Hu and M. Martonosi. Reducing register file power consumption by exploiting value lifetime characteristics. Proceedings of the Workshop on Complexity-Effective Design, hold at 27th Annual Int. Symposium on Computer Architecture, Vancouver, Canada, June 2000.
5. V. Zyuban and P. Kogge, Split Register File Architectures for Inherently Low Power Microprocessors, Proceedings of the Power Driven Micro-architecture Workshop, hold at 27th Annual Int. Symposium on Computer Architecture, Barcelona, Spain, June 1998.
6. J. Tseng and K. Asanovic. Energy-efficient register access. In Proceedings of the 13th Symposium on Integrated Circuits and System Design, pages 377-382, Manaus, Amazonas, Brazil, September 2000.
7. J.L Hennessy and D.A. Patterson, Computer Architecture: A Quantitative Approach, 2nd Edition, Morgan Kaufmann, 1996.

8. P. Y. Hsu. Designing the TFP microprocessor. IEEE Micro, 14(2), pages 23–33, April 1994.
9. A.Sodani and G. Sohi, Dynamic instruction reuse, Proceedings of the 24th Annual Int. Symposium on Computer Architecture, pages 194–205, June-July 1997.
10. E. Taples and D. Marculescu, Power reduction through work reuse, Proceedings of the ACM/IEEE Int. Symposium on Low-Power Electronics and Design, pages 340–345, August 2001.

Energy-Efficient Design of the Reorder Buffer [1]

Dmitry Ponomarev, Gurhan Kucuk, and Kanad Ghose

Department of Computer Science,
State University of New York, Binghamton, NY 13902-6000
{dima, gurhan, ghose}@cs.binghamton.edu
http://www.cs.binghamton.edu/~lowpower

Abstract. Some of today's superscalar processors, such as the Intel Pentium III, implement physical registers using the Reorder Buffer (ROB) slots. As much as 27% of the total CPU power is expended within the ROB in such designs, making the ROB a dominant source of power dissipation within the processor. This paper proposes three relatively independent techniques for the ROB power reduction with no or minimal impact on the performance. These techniques are: 1) dynamic ROB resizing; 2) the use of low-power comparators that dissipate energy mainly on a full match of the comparands and, 3) the use of zero-byte encoding. We validate our results by executing the complete suite of SPEC 95 benchmarks on a true cycle-by-cycle hardware-level simulator and using SPICE measurements for actual layouts of the ROB in 0.5 micron CMOS process. The total power savings achieved within the ROB using our approaches are in excess of 76% with the average performance penalty of less than 3%.

1 Introduction

Contemporary superscalar microprocessors use extensive execution reordering to maximize performance by extracting ILP. One of the main dynamic instruction scheduling artifacts used in such datapath designs is the Reorder Buffer (ROB), which is used to recover to a precise state when interrupts or branch mispredictions occur. The ROB is essentially implemented as a circular FIFO queue with head and tail pointers. Entries are made at the tail of the ROB in program order for each of the co-dispatched instructions. Instructions are committed from the head of the ROB to the architectural register file, thus preserving the correct (program) order of updates to the program state.

In some designs (such as the Pentium III), the physical registers are integrated into the ROB to support register renaming. An associative addressing mechanism is used in some ROBs (as in AMD K5) to determine if any entries exist in the ROB for an architectural register. This avoids the need to maintain and use an explicit rename table.

[1] Supported in part by DARPA through contract number FC 306020020525 under the PAC-C program, the NSF through award no. MIP 9504767 & EIA 9911099 and IEEC at SUNY Binghamton.

The associative lookup of ROB entries, the writes to the ROB in the course of setting up a new entry or in writing results/exception codes and in reading out data from the ROB during operand reads or commits cause a significant power dissipation. For example, a recent study by Folegnani and Gonzalez estimated that more than 27% of the total power expended within a Pentium-like microprocessor is dissipated in the ROB [4].

We study three techniques for reducing the power dissipation within the ROB. First, we consider dynamic ROB resizing. Here, the ROB is implemented as a set of independent partitions and these partitions are dynamically activated/deactivated based on the demands of applications. Second, we propose the use of fast low-power comparators, as introduced in our earlier work [5], to perform the associative search of an ROB slot containing the source register. Third, we noticed that high percentage of bytes within the data items travelling on the result, dispatch and commit buses contain all zeroes. We exploit this by adding the Zero Indicator (ZI) bit to each byte of data that is read from or written to the ROB to avoid the reading and writing of these bytes, thus saving energy.

The rest of the paper is organized as follows. Section 2 discusses the related work. The superscalar datapath configuration assumed for this study and sources of power dissipation within the ROB are discussed in Section 3. In Section 4, we describe our simulation methodology. Section 5 presents the dynamic ROB resizing strategy. In Section 6, we review the energy-efficient comparator. Section 7 describes the zero-byte encoding mechanism. Simulation results are presented in Section 8 followed by our concluding remarks in Section 9.

2   Related Work

Dynamic ROB resizing driven by the ROB occupancy changes was studied in our earlier work [7] in conjunction with similar resizing of the issue queue and the load/store queue. In this paper, we explore the ROB resizing in isolation and estimate the resulting power savings and effects on performance.

Alternative approaches to dynamic resizing of datapath resources were proposed by several researchers [2, 4] to reduce power dissipation of the issue queues. Unfortunately, the metrics used in these proposals to drive the resizing decisions, such as the ready bits [2] or the number of instructions committed from the "youngest" region of the issue queue [4] make it impossible to extend these techniques to the reorder buffers. Due to the space restrictions, we avoid further comparison of our work with the techniques of [2] and [4] and instead refer the reader to [7], where other limitations of the solutions proposed in [2] and [4] are documented.

In [5], we introduced a design of energy-efficient dissipate-on-match comparator and evaluated its use within the issue queue. In this paper we show how the same comparator can be used to save power in associatively-addressed ROBs. Zero-byte encoding was proposed for use in caches in [9] and issue queues in [5].

Another way of reducing the ROB complexity and energy dissipation was studied in [12], where we proposed a scheme for complete elimination of the ROB read ports needed for reading the source operands.

3 Superscalar Datapath and Sources of Energy Dissipation

We consider a superscalar datapath where the ROB slots serve as physical registers. The ROB entry set up for an instruction at the time of dispatch contains a field to hold the result produced by the instruction - this serves as the analog of a physical register. We assume that each ROB entry may hold only a 32-bit long result, thus requiring the allocation of two ROB entries for an instruction producing a double-precision value. Every ROB entry contains a field for the address of the destination architectural register of the instruction allocated to this entry. A dispatched instruction attempts to read operand values either from the Architectural Register File (ARF) directly if the operand value was committed, or associatively from the ROB (from the most recently established entry for an architectural register), in case the operand value was generated but not committed. Source registers that contain valid data are read out into the issue queue entry for the instruction. If a source operand is not available at the time of dispatch in the ARF or the ROB, the address of the physical register (i.e., the ROB slot) is saved in the tag field associated with the source register in the issue queue entry for the instruction. When a function unit completes, it puts out the result produced along with the address of the destination register for this result on a forwarding bus which runs across the length of the issue queue [6]. An associative tag matching process is then used to steer the result to matching entries within the issue queue. The result is also written into the corresponding ROB slot via a result bus. Result values are committed to the ARF from the head of the ROB in program order.

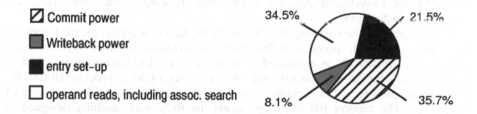

Fig. 1. Energy dissipation components of the traditional ROB (% of total)

The ROB, as used in the datapath described above, is essentially implemented as a large register file. The most recent value generated for an architectural register is located using either an associative addressing capability within the ROB or through the use of an explicit rename table. In a W-way superscalar processor, W sets of internal buses are used to establish, use and commit ROB entries. Where the ROB supports associative lookup without a rename table, 2W address buses are employed for performing an associative search for up to 2 source physical registers for each of the W dispatched instructions.

Energy dissipating events within the multi-ported register file that implements the ROB are as follows:

- Establishment of the entry for dispatched instructions. The resulting energy dissipation has the following components: the energy dissipated in writing the address of the instruction (PC value), flags indicating the instruction type (branch, store, register-to-register), the destination register id, and predicted branch direction (for a branch instruction).

- Instruction dispatch. The energy is dissipated when the address of the corresponding architectural register is driven on the address bus and comparators associated with each ROB entry compare this address against the locally stored architectural register identifiers.
- Reading the valid sources from the ROB slots that implement physical registers.
- Writing the results to the ROB from the function units at the writeback time.
- Committing the ROB entries.
- Clearing the ROB on mispredictions. Compared to all of the other ROB dissipation components, this is quite small.

Figure 1 presents the relative values of some of these energy dissipation components. Results are averaged across all SPEC 95 benchmarks, obtained from the simulated execution of the SPEC 95 benchmarks for a 4-way machine, using the configuration described in Section 4.

4 Simulation Methodology

We used the AccuPower toolset [8] to estimate the energy savings achievable within the ROB. The widely-used Simplescalar simulator [1] was significantly modified (the code for dispatch, issue, writeback and commit steps was written from scratch) to implement *true hardware level, cycle-by-cycle* simulation models for the issue queue, the ROB, the load/store queue, the register files, forwarding interconnections and dedicated transfer links.

For estimating the energy/power for the ROB, the event counts gleaned from the simulator were used, along with the energy dissipations for each type of event described in section 3, as measured from the actual VLSI layouts using SPICE. CMOS layouts for the ROB in a 0.5 micron 4 metal layer CMOS process (HPCMOS-14TB) were used to get an accurate idea of the energy dissipations for each type of transition. The register file that implements the ROB was carefully designed to optimize the dimensions and allow the use of a 300 MHz clock. A Vdd of 3.3 volts is assumed for all the measurements.

The configuration of the 4-way superscalar processor used in this study is shown in Table 1. Benchmarks were compiled using the Simplescalar gcc compiler that generates code in the portable ISA (PISA) format. Reference inputs were used for all the simulated benchmarks. For all of the SPEC 95 benchmarks, the results from the simulation of the first 200 million instructions were discarded and the results from the execution of the following 200 million instructions were used.

5 Dynamic ROB Resizing

It is well-documented [10] that the overall performance, as measured by the IPC, varies widely across applications. One approach to minimizing the power requirements of the datapath is to use a dynamic resource allocation strategy to

Table 1. Architectural configuration of a simulated 4-way superscalar processor

Parameter	Configuration
Machine width	4-wide fetch, 4-wide issue, 4-wide commit
Window size	32 entry issue queue, 128 entry ROB, 32 entry load/store queue
L1 I-cache	32 KB, 2-way set-associative, 32 byte line, 2 cycles hit time
L1 D-cache	32 KB, 4-way set-associative, 32 byte line, 2 cycles hit time
L2 Cache combined	512 KB, 4-way set-associative, 64 byte line, 4 cycles hit time.
BTB	1024 entry, 2-way set-associative
Memory	128 bit wide, 12 cycles first chunk, 2 cycles interchunk
TLB	64 entry (I), 128 entry (D), 4-way set-associative, 30 cycles miss latency
Function Units and Latency (total/issue)	4 Int Add (1/1), 1 Int Mult (3/1) / Div (20/19), 2 Load/Store (2/1), 4 FP Add (2), 1FP Mult (4/1) / Div (12/12) / Sqrt (24/24)

closely track the resource demands of programs that are being executed. Because of its large size, the ROB is especially suitable candidate for such dynamic allocation. If a program exhibits modest resource requirements, the degree of the ROB overcommitment may be quite large. The ROB resizing strategy used in this paper was first proposed in [7], where the ROB was resized in combination with other resources. In this paper, we study the ROB resizing on its own, keeping the sizes of other resources at their maximum values at all times. In the rest of this section, we summarize our resizing strategy and we refer the readers to [7] for a more detailed discussion.

To support dynamic resizing, the ROB is implemented as a number of independent partitions, each of them being a self-standing and independently usable unit, complete with its own prechargers, sense amps and input/output drivers. We assumed that the number of entries in a ROB partition is 16. A number of partitions can be strung up together to implement a larger structure. The lines running across the entries within a partition (such as bitlines, lines to drive the architectural register addresses for comparison etc.) can be connected to a common through line via the bypass switches to add (i.e., allocate) the partition to the current ROB and extend its effective size. Similarly, the partition can be deallocated by turning off the corresponding bypass switches. The circular FIFO nature of the ROB requires some additional considerations in the resizing process: in some situations partitions can be activated and deactivated only when the queue extremities coincide with the partition boundaries [7].

Some efforts (such as the one in [2]) focussing on the dynamic resizing of issue queues, use the (commit) IPC values as measured on a cycle-by-cycle basis to trigger the downsizing or upsizing actions. We found, however, that dispatch rates and IPCs are not always positively correlated with the ROB occupancy, measured as the number of valid entries, and hence, with the resource demands of the applications. To illustrate this, Figure 2 shows the behavior of two floating point benchmarks from SPEC 95 suite – *apsi* and *hydro2d*. Samples of dispatch rate and the average occupancy of the ROB were taken every 1 million cycles. These samples reflected the activity within the most recent window of 1 million cycles. Results are shown for the execution of 200M instructions.

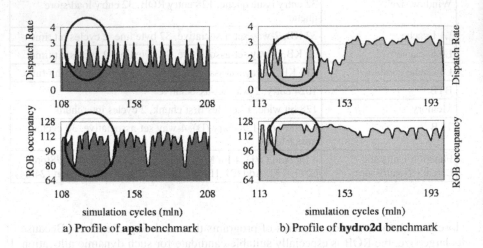

a) Profile of **apsi** benchmark   b) Profile of **hydro2d** benchmark

Fig. 2. Dynamic behavior of two SPEC95 floating point benchmarks.

As these representative results show, the average number of active entries within the ROB changes significantly throughout the course of execution. These graphs also suggest that it is hardly possible to efficiently control the allocation of the ROB partitions based solely on the dispatch rate (or the IPC). Our resizing strategy does not rely on the IPC or dispatch rate as a measure of resource requirements, instead we use the actual ROB occupancy as sampled at periodic intervals as an indication of the application's needs.

The decision of downsizing the ROB is made periodically – once every *update_period* cycles. During this period, the ROB occupancy, as measured by the number of allocated entries within the active region of the ROB, is sampled several times with the frequency of once every *sample_period* cycles. The average of these samples is taken as the *active_size* in the current update period. At the end of every update period, the difference (*diff=current_size-active_size*) is computed, where *current_size* is the size of the ROB at the end of the update period. If *diff* is less than the size of an ROB partition, no resizing takes place. Otherwise, one ROB partition is turned-off. More aggressive designs – not studied here – can be used where more than one partition can be turned-off at the same time, as determined by *diff*.

By sampling the resource usage periodically instead of on a continuous basis, as done in [2,4] for the issue queue, we conserve dissipations within the monitoring logic for the downsizing. Our results also reveal that the error arising from the computations of the average occupancy by taking the average occupancy measured at discrete points (at every sample interval) is tolerable since significant energy savings are achieved using our approach. Although the current paper shows the savings on dynamic/switching power, dynamic deactivation of partitions also saves leakage power that would be otherwise dissipated within the ROB.

The second phase of the resizing strategy is implemented to scale the ROB size back up once the demands of the application begin to require more resources. Here, we use the ROB overflow counter (*rob_overflow*), which counts the number of cycles when dispatch is blocked because of the absence of a free entry in the ROB to hold the new instruction. This counter is initialized to zero every *update_period* cycles and is incremented by one whenever instruction dispatch is blocked because of the unavailability of a free ROB entry. Once this overflow counter exceeds a preset value (*overflow_threshold*), the ROB size is incremented by turning one currently inactive partition on. After that, the counter is re-initialized. Notice, that for performance reasons, the process of increasing the size of the ROB is more aggressive than the process of shrinking the ROB. To avoid the occurrence of several such incremental updates in one update period, we reset the update period counter when the increase of the ROB size is triggered. This is to avoid the instability in the system associated with the fact that ROB is sampled for different maximum sizes in the same update period.

6 Use of Energy-Efficient Comparators

As shown in Figure 1, the power dissipated within the ROB during instruction dispatch, particularly during the associative look-up for the ROB slot containing the source register, is a significant portion of the total ROB power. We assume that the architectural register file consists of 32 integer registers and 32 floating point registers, thus requiring 6-bit comparators to perform the associative look-up. The typical comparator circuitry used for the associative architectural register address matching in the ROB is a dynamic pulldown comparator or a set of 8-transistor associative bitcell pulling down a common precharged line on a mismatch in any bit position. All of these comparators dissipate energy on a *mismatch* in any bit position. The energy is thus wasted in comparisons that do not locate matching entries, while little energy (in the form of precharging) is spent in locating matching entries.

The data collected for the simulated execution of SPEC 95 benchmarks on our system indicates that during the associative lookup of source values from the ROB, only up to 6 of the comparators in valid entries within the ROB produce a match per instruction on the average. This is clearly an energy-inefficient situation, as more energy is dissipated due to mismatches (compared to the number of matches) with the use of traditional comparators that dissipate energy on a mismatch. This situation can be remedied by using Domino-style compatators, that predominantly dissipate energy only on a full match. Such a comparator design was proposed in [5]. For the ROB, we use a variation of the circuit shown in [5], that compares two 6-bit comparands. The basic circuit is a two-stage domino logic, where the first stage detects a match of the lower 4 bits of the comparands. The following stage does not dissipate any energy when the lower order 4 bits do not match.

Table 2 shows how comparand values are distributed and the extent of partial matches in the values of two adjacent bits in the comparands, averaged over the simulated execution of SPEC 95. The lower order 2 bits of both comparands are equal in roughly 25% of the case, while the lower order 4 bits match 16% of the time. A full match occurs 11% of the time on the average. Equivalently, a mismatch occurs in the lower order 2 bits of the comparator 75% of the time (=100 − 25). The behavior

Table 2. Comparator Statistics.

Number of bits matching—>	% of total cases		
	2 LSBs	4 LSBs	Match
Avg, SPECint 95	23.0	14.0	12.0
Avg, SPECfp 95	26.0	16.0	11.0
Avg, all SPEC 95	25.0	16.0	11.0

LSB = least significant bits

depicted in Table 2 is a consequence of the localized usage of the architectural registers in a typical source code. As a result, the likelihood of a match of the higher order bits of the register addresses (i.e., the comparands) is higher. Our comparator design directly exploits this fact by limiting dynamic energy dissipation due to partial matches to less than 16% of all cases when the lower order 4 bits match; no energy dissipation occurs in the more frequent cases of the higher order bits matching. The overall energy dissipation due to partial or complete mismatches is thus greatly reduced.

7 Use of Zero-Byte Encoding

A study of data streams within superscalar datapaths revealed that significant number of bytes are all zeros within operands on most of the flow paths (ROB to the issue queue, issue queue to function units, functions units to issue queue and ROB, ROB to ARF etc.). On the average, about half of the byte fields within operands are all zeros for SPEC 95 benchmarks. This is really a consequence of using small literal values, either as operands or as address offsets, byte-level operations, operations that use masks to isolate bits etc. Considerable energy savings are possible when bytes containing zero are not transferred, stored or operated on explicitly. Other work in the past for caches [8], function units [11] and *scalar* pipelines [3] have made the same observation. We extend these past work to the ROB. By not writing zero bytes into the ROB at the time of dispatch and writeback, energy savings result as fewer bitlines need to be driven. Similarly, further savings are achieved during commit by not reading out implied zero valued bytes. This can be done by storing an additional Zero Indicator (ZI) bit with each byte that indicates if the associated byte contains all zeros or not. The circuit details of the zero encoding logic used within the ROB is similar to that of [8] and [5].

8 Results and Discussions

Power savings achieved within the ROB using techniques proposed in this paper are summarized in Figures 3-5. Figure 3 shows the effects of the energy-efficient comparator on the power dissipated during the process of comparing the architectural register addresses. The total ROB energy reduction is 41% on the average across all SPEC95 benchmarks in this case. However, the total ROB power is reduced by only about 13% because the comparator does not have any effect on the energy dissipated during entry setup as well as during writeback and commitment (Figure 4).

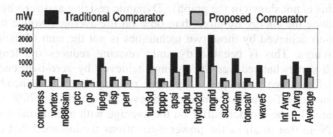

Fig. 3. Power dissipation within the ROB at the time of source operand reads.

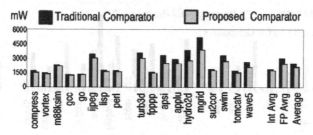

Fig. 4. Total Power dissipation within the ROB with the proposed and the traditional comparators.

Dynamic resizing and zero-byte encoding address the power reduction in exactly these two components (writeback and commitment). For the results of dynamic resizing presented below, both *overflow_threshold* and *update_period* were kept at 2K cycles. We selected the value of 2K cycles for the update_period as a reasonable compromise between too large a value (in which case the resizing algorithm will not react quickly enough to the changes in application's behavior) and too small a value (in which case the resizing overhead increases). For comparison, Table 3 summarizes achieved power savings and performance drop for a range of *overflow_threshold* values. Results shown in this table are the averages across all executed benchmarks.

Table 3. ROB power savings and performance drop for various values of *overflow_threshold*

overflow_threshold	128	256	512	1024	2048
power savings(%)	56.1	57.3	59	60.5	62.2
IPC drop (%)	0.06	0.27	0.78	1.87	3.14

Figure 5 summarizes the overall power savings in the ROB realizable using our techniques. Dynamic resizing results in about 60% energy reduction in the ROB on the average. Average power savings attributed to the use of zero-byte encoding are about 17% (this is not shown in the graph). Dynamic resizing and zero-byte encoding in concert reduce the energy by more than 66% on the average. Notice that the total power reduction achieved by these two techniques is not the sum of their individual respective savings. This is because dynamic resizing reduces the lengths of the bitlines which somewhat reduces the savings achieved by zero-byte encoding. The combination of all three techniques – dynamic resizing, zero-byte encoding and the use of fast, power-efficient comparators achieves a remarkable 76% reduction in power dissipated by the reorder buffer on the average with a minimal impact on the performance.Note that in all of the power estimations we assumed that comparators are enabled only for ROB entries that are valid. If spurious comparisons are allowed, the power savings reported here will go up further.

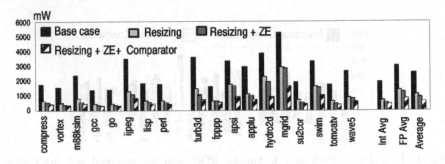

Fig. 5. Total power dissipation within ROB using dynamic resizing, power efficient comparators and zero encoding.

To summarize, the dynamic ROB resizing provides the highest power savings within the ROB. Fundamentally, this is because the ROB resizing immediately exploits its overcommitments by turning off the complete partitions, effectively turning off many comparators and reducing the capacitances of all bitlines and address lines which results in very significant power savings. In the future, as processor designers tend to increase the issue widths and employ larger ROBs, the effect of dynamic resizing will become even more dramatic. The drawbacks of the dynamic resizing are in the form of some performance loss and in its requirement of a fair amount of modifications to the datapath, although we tried to keep the latter to the minimum in our design. The use of energy-efficient comparators and zero-byte encoding can also achieve substantial *additional* energy savings, as seen from our results.

9   Concluding Remarks

We studied three relatively independent techniques to reduce the energy dissipation in the reorder buffer used in modern superscalar processors. First, we used a technique to dynamically resize the ROB based on its actual occupancy. Second, we used fast comparators in the associative search logic that dissipate the energy mainly on a full match of the architectural register addresses. For the technology studied, the associative lookup of the most recent value of an architectural register using these comparators is more energy-efficient than using a rename table lookup. Third, we considered the use of zero-byte encoding to reduce the number of bitlines that have to be driven during instruction dispatch, writeback and commitment. Combined, these three mechanisms reduce the power dissipated by the reorder buffer by more than 76% on the average across all SPEC 95 benchmarks with the average penalty on the IPC below 3%.

The ROB power reductions are achieved without compromising the cycle time and only through a modest growth in the area of the ROB (about 12%, including the new comparators, ZE logic and dynamic resizing logic). Our ongoing studies also show that the use of all of the techniques that reduce the ROB power can also be used to achieve reductions of a similar scale in other datapath artifacts that use associative addressing (such as issue queues and load/store queues). As the power dissipated in instruction dispatching, issuing, forwarding and retirement can often be as much as half of the total chip power dissipation, the use of dynamic resizing, the energy-efficient comparators and zero-byte encoding offers substantial promise in reducing the overall power requirements of contemporary superscalar processors.

References

[1] Burger, D., and Austin, T. M., "The SimpleScalar tool set: Version 2.0", Tech. Report, Dept. of CS, Univ. of Wisconsin-Madison, June 1997 and documentation for all Simplescalar releases (through version 3.0).

[2] Buyuktosunoglu, A., Schuster, S., Brooks, D., Bose, P., Cook, P. and Albonesi, D., "An Adaptive Issue Queue for Reduced Power at High Performance", in PACS Workshop, November 2000.

[3] Canal R., Gonzalez A., and Smith J., "Very Low Power Pipelines using Significance Compression", in Proceedings of International Symposium on Microarchitecture, 2000.

[4] Folegnani, D., Gonzalez, A., "Energy-Effective Issue Logic", in Proc. of ISCA, July 2001.

[5] Kucuk, G., Ghose, K., Ponomarev, D., Kogge, P., "Energy-Efficient Instruction Dispatch Buffer Design for Superscalar Processors", in Proc. of ISLPED, 2001.

[6] Palacharla, S., Jouppi, N. P. and Smith, J.E., "Quantifying the Complexity of Superscalar Processors", Technical report CS-TR-96-1308, Dept. of CS, Univ. of Wisconsin, 1996.

[7] Ponomarev, D., Kucuk, G., Ghose, K., "Reducing Power Requirements of Instruction Scheduling Through Dynamic Allocation of Multiple Datapath Resources", in Proc. of MICRO-34, 2001.

[8] Ponomarev, D., Kucuk, G., Ghose, K., "AccuPower: an Accurate Power Estimation Tool for Superscalar Microprocessors", in Proc. of 5th Design, Automation and Test in Europe Conference, 2002.

[9] Villa, L., Zhang, M. and Asanovic, K., "Dynamic Zero Compression for Cache Energy Reduction", in Proceedings of International Symposium on Microarchitecture, 2000.

[10] Wall, D.W., "Limits on Instruction Level Parallelism". In Proceedings of ASPLOS, November, 1991.

[11] Brooks, D. and Martonosi, M., "Dynamically Exploiting Narrow Width Operands to Improve Processor Power and Performance", Proc. HPCA, 1999.

[12] Kucuk, G., Ponomarev, D., Ghose, K., "Low-Complexity Reorder Buffer Architecture", in Proc. of the 16th International Conference on Supercomputing, 2002.

Trends in Ultralow-Voltage RAM Technology

Kiyoo Itoh

Central Research Laboratory, Hitachi Ltd.,
Kokubunji, Tokyo, 185-8601, Japan.
k-itoh@crl.hitachi.co.jp

Abstract. This paper describes ultralow-voltage RAM technology for stand-alone and embedded memories in terms of signal-to-noise-ratio designs of RAM cells and subthreshold-current reduction. First, structures and areas of current DRAM and SRAM cells are discussed. Next, low-voltage peripheral circuits that have been proposed so far are reviewed with focus on subthreshold-current reduction, speed variation, on-chip voltage conversion, and testing. Finally, based on the above discussion, a perspective is given with emphasis on needs for high-speed simple non-volatile RAMs, new devices/circuits for reducing active-mode leakage currents, and memory-rich SOC architectures.

1 Introduction

Stand-alone memories and embedded memories (e-memories) for a system-on-a-chip (SOC) [1] have rapidly evolved, and thereby both dramatically reduced costs and greatly improved the performance of PCs, workstations, and servers. Consequently, high-density stand-alone memories have achieved a 4-Gb DRAM [2], a 32-Mb SRAM, and a 1-Gb flash memory [3] at R&D level. The power dissipations have also decreased by reducing the standard power-supply voltage (V_{DD}) to as low as 1.8V to 3.3V. High-speed e-memories, which are forecast to occupy over 90% of a low-power SOC [4], have also rapidly developed, as exemplified by a 300-MHz 16-Mb DRAM macro [5], a 1-GHz 24-Mb L3-SRAM cache [6], and a 1.2-V-read 16-Mb flash memory [7]. To enable applicability in minimized-size devices the V_{DD} of e-memories has been reduced in accordance with the rapidly reducing V_{DD} of MPUs [8] to below 1.5V. Recently, however, in accordance with the rapidly-growing power-aware-systems market, lowering the V_{DD} has further accelerated reducing e-memory power. Such is the case even for stand-alone memories that are applied to a low-power system-in-a-package (SIP), which has rapidly become widely used. As the V_{DD} reaches sub-V levels, however, two design issues arise [9]: the high signal-to-noise-ratio (S/N) design for memory cells for stable operations; and reduction of the ever-increasing leakage(gate-tunnel/subthreshold) currents of MOSTs that are seen when reducing the gate-oxide thickness (t_{ox}) and the threshold voltage (V_T). In particular, reducing the leakage current is vital for all future LSIs. Thus, some high-k gate insulators have been intensively developed to reduce the tunnel current. Moreover, new devices such as

B. Hochet et al. (Eds.)· PATMOS 2002, LNCS 2451, pp. 300–313, 2002.

fully-depleted (FD) SOI and the dynamic-V_T (DTMOST) [10] using a partially-depleted SOI have been developed to reduce the subthreshold current. However, the subthreshold swing (i.e., S-factor) of FD SOI is still larger and DTMOST limits the V_{DD} to be around 0.4V at the most due to a rapidly increased pn-junction forward current [11]. Thus, circuit solutions are vital.

This paper mainly discusses low-voltage memory circuits, focusing on RAMs because they cover large parts of low-voltage flash-memory circuits. First, recent developments for DRAM and SRAM cells are discussed. Next, peripheral logic circuits are investigated, including other design issues for low-V_{DD} operations such as speed variations, on-chip supply-voltage converters, and a testing methodology. Finally, a perspective is given with emphasis on needs for non-volatile RAM cells, subthreshold-current reduction circuits for use in active mode, and memory-rich SOC architectures.

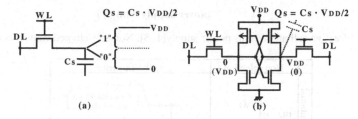

Fig. 1. Memory-cell circuits of DRAM(a) and SRAM(b).

2   Current RAM Cells

The lower limit of the V_{DD} is determined by the S/N of the memory cell, the unscalable V_T of the memory-cell MOST, and the unscalable V_T-mismatch between cross-coupled/paired MOSTs in a large number of DRAM sense amps (SAs) and SRAM cells [9]. The S/N degrades at a lower V_{DD} because the signal charge ($Q_S = C_S V_{DD}/2$, C_S: storage capacitance) for DRAM and SRAM cells (Fig.1) decreases and thus causes a smaller signal voltage on the data line (DL) in a noisy memory array as well as larger soft errors. The unscalable V_T is a result of the specifications set to satisfy a required refresh time (t_{REFmax}) of the DRAM or the data-retention current of the SRAM. Both the V_T and the V_T-mismatch increase with increasing memory capacity and decreasing device size, while Q_S decreases [9].

2.1 DRAM Cells

2.1.1 1-T Cells for Stand-Alone DRAMs

For a given cell-signal voltage ($\approx C_S V_{DD}/2C_D = Q_S/C_D$, C_D: DL-capacitance) of approximately 200mV read out on each DL, minimizing the cell size and the overhead areas for reducing the memory-chip size is the first priority for stand-alone DRAMs.

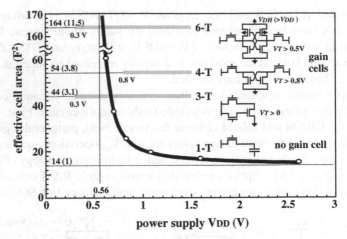

Fig. 2. Effective cell area versus power supply[1, 8]. Non-self-aligned contact is used.

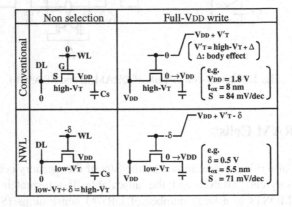

Fig. 3. Conventional and negative word-line schemes[1].

It results in a large C_D because as many memory cells as possible are connected to each DL-pair to reduce the overhead area at each DL-division. A large Q_S (e.g., a large C_S) is thus needed for a necessary signal voltage, which has been realized by using sophisticated vertical(stacked/trench)-capacitors[9,41] and high-k thin-films. Relaxing the sophistication of the process/device requires reducing the C_S, which is realized by having a high V_{DD} and a smaller C_D through the multi-divided DL [9] using multi-level metal wiring. Applying self-aligned contact to memory cells only reduces the cell area despite a speed penalty being inflicted due to the increased contact resistance. A 6-4F² (F: feature size) trench-capacitor vertical-MOST cell[12,13] and a 6F² stacked-capacitor open-DL cell, in which array noise must be reduced by a low-impedance array [14], are leading developments of R&D. Even for such high-Q_S cell, however, the effective cell area, the sum of the genuine cell area and the overhead area at a DL-division, sharply increases with a decreasing V_{DD} [1,8] (Fig.2), because no gain of the

1-T cell requires more DL-divisions at a lower V_{DD} to maintain the necessary signal voltage.

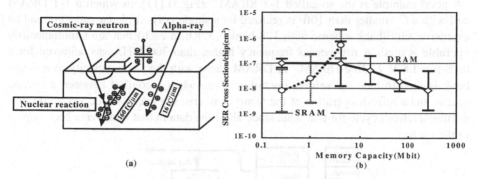

(a)

(b)

Fig. 4. Mechanisms(a) and cross section(b) of soft errors[15].

Adjusting the potential profile of the storage node to suppress the *pn*-leakage current makes *tREFmax* longer, and preserves the refresh busy rate even for larger memory-capacity DRAMs [9], or it lowers the data retention current in standby mode. The subthreshold current caused by the resultant low V_T is cut by the negative word-line (NWL) scheme (Fig.3) [9] with a δ gate-source back-bias during the non-selected period. Fortunately, reducing the word-line voltage by δ in a full-V_{DD} write-operation allows using a thinner-t_{ox} MOST for a given stress voltage, enabling a low-voltage operation with the resultant small *S*-factor. The DRAM cells are resistant against cosmic-ray neutron-hitting [15] (Fig.4), which generates about ten times as many charges as alpha particles. Soft errors gradually decrease with decreasing device size due to a rather large Cs and spatial scaling that reduces collection of charges generated by the hitting.

2.1.2 1-T Cells for E-DRAMs

Using a high-speed logic compatible process and a small sub-array are keys. Such a process is realized by using a non-self-aligned contact and a MOS-planer capacitor, as it was commonly done in the mid1970's. This is attractive as long as the resultant cell is quite a lot smaller than the 6-transistor (6-T) full CMOS SRAM cell [8]. A small sub-array allows even the resultant small C_S to develop enough signal voltage with an extremely low C_D that is realized by increasing the number of DL-divisions for a relatively high V_{DD}. The small sub-array also reduces array-relevant line delays that are major delays on the access/cycle path, achieving a high speed. A multi-bank inter-leaving with many small sub-arrays, pipeline operations, and direct sensing [9] solve the problem of the row-cycle of DRAM being slow, while enabling a faster access time than SRAM due to the smaller physical size of the DRAM array for a given memory capacity. The small sub-array also allows the *tREFmax* to be drastically shortened for a given junction temperature. It may even accept the open-DL cell, which is noisy but smaller than the traditional folded-DL cell, because the small sub-array causes little noise. Here, the soft-error issue that is unavoidably arisen due to a small C_S could be solved by using an error checking and correcting (ECC) circuit [16].

A good example is the so-called 1-T SRAM™ (Fig.5) [17], in which a 1-T DRAM cell with a C_S smaller than 10fF is realized by a single poly-Si planar capacitor and an extensive multi-bank scheme with 128 banks (32Kb in each) that are simultaneously operable is used. A row access frequency higher than 300-MHz was achieved for a 0.18-μm 1.8-V 2-Mb e-DRAM. A DRAM cache with the same capacity as a single bank is used to hide the refresh operation: Even when a conflict between a normal access and a refresh operation at the same bank arises, a cache hit occurs while permitting a refresh cycle for that bank since all of the data in that bank have been copied to the cache.

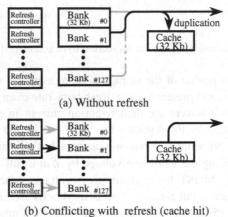

(a) Without refresh

(b) Conflicting with refresh (cache hit)

Fig. 5. Read operations of multi banks of 1-T SRAM™[17].

Low-voltage high-speed sensing is also essential. Although the standard mid-point (a half-V_{DD}) sensing [9] halves the DL power without a dummy cell, it slows down the sensing operation. In overdrive sensing [18,19] this problem is solved by applying a higher voltage solely to the SA-inputs with isolating the data line from the SA or with coupling capacitively. Using additional capacitors may be acceptable in e-DRAMs for which area is not a concern. Even full-V_{DD} (or ground) sensing with a dummy cell for a 1.5-V 16-Mb e-DRAM [5] was recently reported on, which is a revival of the sensing conducted in the NMOS DRAM era in 1970's.

2.1.3 Gain Cells

Gain cells would solve the above-mentioned problems of the 1-T cell, i.e., increasing the effective cell area at a lower V_{DD} and with sophisticated structures. The 3- and 4-transistor (3-T and 4-T) DRAM cells, and the 6-T SRAM cell, as shown in Fig.2, require no special capacitor. In addition, they are all gain cells that can develop a sufficient signal voltage without increasing the number of DL-divisions even at a lower V_{DD}, and thus provide a fixed effective cell area that is independent of the V_{DD}. Actually, however, the V_{DD} has a lower limit for each cell. For the 3-T cell, it would be around 0.3V if the V_T of the storage MOST is chosen to be around 0V. For the 4-T cell, it would be as high as 0.8V because the V_T of cross-coupled MOSTs must be higher than 0.8 V to ensure t_{REFmax} by eliminating the subthreshold current. For the 6-T SRAM cell, it would be around 0.3V if a raised supply voltage (V_{DH}) higher than 0.5V is supplied from an on-chip charge pump, as explained later. Consequently, the 3-T cell would be the smallest at less than 0.6-VV_{DD}. Recently, a small poly-Si vertical-transistor 2-T 5F^{2}gain cell [20] has been proposed despite a small current drivability of the transistor.

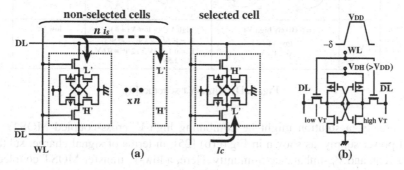

Fig. 6. Data pattern of cells along the data line(a), and low-voltage SRAM cell(b)[9].

2.2 SRAM Cells

Reducing the cell area is the biggest concern regarding SRAMs, as an on-chip L3 cache shows [6]. The loadless CMOS 4-T SRAM [21] is attractive because the cell area is only 56% of that of the 6-T cell. However, it suffers from the data-pattern problem, which we discuss below, and difficulty in controlling the non-selected word-line voltage precisely to maintain the load current. At present, the 6-T cell is the best despite its large area because it enables a simple process and design provided by the cell's wide-voltage margin. Even for this cell, subthreshold currents increase the retention current with lowering the V_T, reaching as much as 10A even for a 1-Mb array for $V_T = 0V$ [9]. This limits the reduction of V_T tightly. The voltage margin of the cell decreases with decreasing V_{DD} and V_T, and requires a decrease in the ratio of transconductance of the transfer MOST to that of the driver MOST. It further decreases if the V_T and V_T-mismatches between paired MOSTs vary, and if the V_T of the transfer MOST is low. A low-V_T transfer MOST may cause a read failure when the total subthreshold current from transfer MOSTs of multiple non-selected cells along DL is greater than the selected cell's current in the worst cell data pattern (Fig.6). A hierarchical DL scheme partly solves this problem by limiting the number of memory cells connected to DL [22]. Using a data equalizer to compensate for the current [23] is also effective, despite an occurring area penalty. Moreover, a gate-source offset-driving memory cell allows using a low-V_T transfer MOST for a high-speed and negligible DL current [24], although eleven transistors are required for each cell.

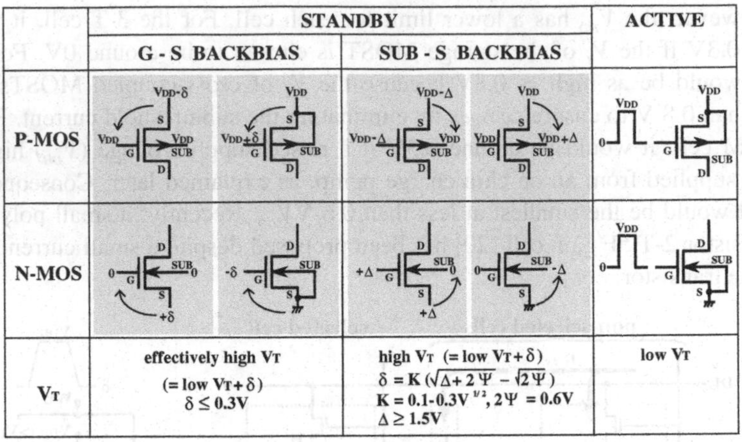

	STANDBY			ACTIVE
	G - S BACKBIAS		**SUB - S BACKBIAS**	
P-MOS				
N-MOS				
V_T	effectively high V_T ($= $ low $V_T + \delta$) $\delta \le 0.3V$		high V_T ($=$ low $V_T + \delta$) $\delta = K(\sqrt{\Delta + 2\Psi} - \sqrt{2\Psi})$ $K = 0.1\text{-}0.3V^{1/2}, 2\Psi = 0.6V$ $\Delta \ge 1.5V$	low V_T

Fig. 7. Dynamic-V_T schemes[9].

Eventually, a solution might be combining high-V_T cross-coupled MOSTs and a raised power supply, as shown in Fig.6 (b) [25], in terms of signal charge, subthreshold current, and V_T-imbalance immunity. Here, a low-V_T transfer MOST coupled with an NWL scheme achieves high speed while reducing the leakage currents from cells along DL. A similar concept incorporated for a 0.4-V-V_{DD} 0.18-μm 256-Kb SRAM [26] showed excellent performances of an active power of only 140 μW at 4.5 MHz and a standby current of 0.9μA at a substrate back-bias of 0.4V at $V_{DH} - V_{DD} = 0.1V$ and $V_T = 0.4$-$0.5V$. Even for the raised power supply, SRAM cells may inevitably require either an additional capacitor at the storage node [27] or the use of an ECC technique to prevent soft errors in the future. This is because the SRAM soft errors rapidly increase with device scaling (Fig.4), due to its small parasitic node capacitance.

3 Peripheral Logic Circuits

3.1 Subthreshold-Current Reduction

Memory peripheral circuits favor subthreshold-current reduction [9]: They consist of multiple iterative circuit blocks, such as row/column decoders and drivers, each of which has quite a large total-channel width involving the subthreshold current, and all circuits in the block, except for a selected one, are inactive even during an active period, enabling simple and effective subthreshold-current control with a small area penalty, which is discussed later. In addition, a slow memory cycle allows each circuit to be active only for a short period within the "long" memory cycle, enabling additional time for the subthreshold-current control. Moreover, the circuits are input-predictable, enabling designers to predict all node voltages in a chip, and to prepare the subthreshold-current reduction scheme in advance.

Two reduction methods have been proposed. One method is the dynamic V_T scheme (Fig.7) [9]. The V_T is low enough to enable high speed in active mode due to no back-bias connection. In standby mode it is raised to reduce the subthreshold current by changing the bias condition. This is further categorized as gate-source (G-S) back-biasing and substrate-source (Sub-S) back-biasing. The other method is the static V_T scheme (Fig.8), which is categorized as a power-switch and multi-V_T scheme. Both methods are effective for standby mode. Even so, they are insufficient for active mode unless a high-speed reduction control necessary at active mode is achieved. In this sense, the Sub-S back-biasing and the power switch are not applicable to active mode because a large voltage swing is involved.

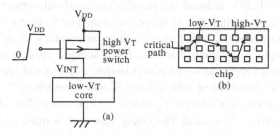

Fig. 8. Power switch(a) and multi-V_T scheme[9].

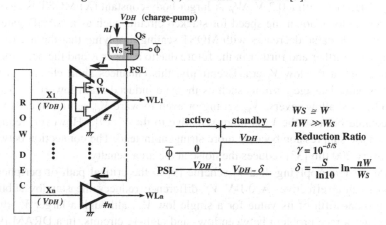

Fig. 9. G-S backbias applied to DRAM word drivers[29, 30].

3.1.1 Standby Mode

G-S Back-Biasing: S-control with a fixed G, and G-control with a fixed S are well known. In this scheme, a small voltage swing (δ) suffices for the reduction and enables a high-speed recovery. Note that an effectively high V_T is established as the sum of a low-actual V_T and δ and the subthreshold current is reduced to 1/10 with a V_T-increment of only 0.1V (i.e. $S = 0.1$V/decade at 100℃). In particular, S-control applied to iterative circuit blocks such as a word driver block is extremely important for memory designs. For example, a low-V_T PMOS switch [29, 30] (Qs in Fig.9) shared with n word drivers enables the common power line (PSL) to drop by δ as a result of the total subthreshold current flow of nI when the switch is off in standby mode. It provides each PMOS driver (Q) with a δ self-back-bias so the subthreshold current (I) eventually decreases. Hence, even if an on-chip charge pump for a raised supply V_{DH} necessary for the DRAM word-bootstrapping suffers from the poor output-current drivability, the V_{DH} is well regulated. In active mode the selected word line is driven after connecting the PSL to a supply voltage (V_{DH}) by turning on Q_S. Here, the Q_S-channel width can be reduced to an extent comparable to that of the Q-channel width without a speed penalty because only one of the n drivers is turned on. For a 256-Mb chip, a δ as little as 0.25V reduced the standby subthreshold-current by 2-3 decades without inflicting penalties in terms of speed and area. The S-control is also applicable to dynamic NAND decoders, which are common in memory, even without a level keeper [31]. Examples of G-control are the NWL scheme, as discussed before, G-S offset drive [9], and an application to a power switch, as explained later.

Sub-S Back-Biasing: Sub-control with a fixed S, and S-control with a fixed Sub are well known methods [9]. For a larger V_T-change (ΔV_T), a large Sub-swing or S-swing (i.e., $\Delta(V_{SUB} - V_S) = \Delta V_{BB}$) is needed. The swing, however, is quite large. For example, existing MOSTs with a 0.2-$V^{1/2}$ body constant require a 2.5-V ΔV_{BB} to reduce the current by 2 decades with a 0.2-V ΔV_T. A larger body-constant (K) MOST is also needed. However, it slows down the speed for stacked circuits such as a NAND gate. On the contrary, the K value decreases with MOST scaling, implying that the necessary ΔV_{BB} will increase further and further in the future due to a lower K and the need for a larger ΔV_T escalated at the low-V_T era. Eventually, this enhances short-channel effects and increases other leakage currents such as the gate-induced drain-lowering (GIDL) current [32]. A shallow reverse V_{BB} setting or even forward V_{BB} setting in active mode is also required because the V_T is more sensitive to the V_{BB} [9]. However, requirements for V_{BB}-noise suppression become more stringent instead. The connection between Sub and S every 200 μm [33] reduces the noise at the area penalty.

Multi-Static V_T: Applying dual-V_T scheme [9] to the critical path of peripheral logic circuits is quite effective. A 0.1-V V_T-difference reduces the standby subthreshold current to one fifth of its value for a single low V_T, although a larger V_T-difference might cause a race problem between low- and high-V_T circuits. In a DRAM [9] multi-V_T can easily be produced by applying internal supply voltages (lowered and raised from V_{DD}) generated by on-chip voltage converters. A combination of dual V_T and dual V_{DD} has also been proposed for making a 1-V e-SRAM [34].

Power Switch: A high-V_T PMOST power switch [9], applied to a low-V_T internal core, completely cuts the subthreshold current from the core by turning off the switch in standby mode. The drawbacks are a large V_{DD} swing at the internal power line, causing a large power/spike current and slow recovery at mode transitions, and need for a large PMOST. Using a low-V_T instead reduces the size for a given transconductance, while the subthreshold current at standby mode is shut down if the G-control is applied.

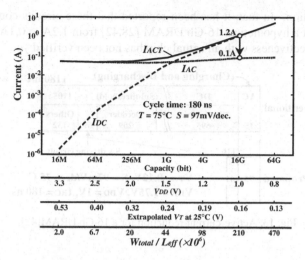

3.3	2.5	2.0	1.5	1.2	1.0	0.8
			V_{DD} (V)			

0.53	0.40	0.32	0.24	0.19	0.16	0.13
			Extrapolated V_T at 25°C (V)			

2.0	6.7	20	44	98	210	470
			W_{total} / L_{eff} ($\times 10^6$)			

Fig. 10. Trends in DRAM active current[28, 42].

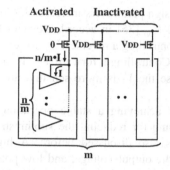

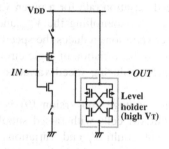

Fig. 11. Partial activation of multi-divided block[42].

Fig. 12. Power switch with a level holder[42].

3.1.2 Active Mode

In the future, with further reducing V_T, the subthreshold current (I_{DC}) will exceed the capacitive current (I_{AC}) and eventually dominate the total active current (I_{ACT}) of the chip (Fig.10), as pointed out as early as 1993 [28, 42]. G-S back-biasing applied to an iterative circuit block, which is divided into m sub-blocks, each consisting of n/m circuits(Fig.11), confines the currents to that of a single selected sub-block [42]. This is because all *non*-selected sub-blocks have no substantial subthreshold current due to G-S back-biasing as in Fig.9 when the switch of the selected sub-block including the selected word line is turned on while the others remain off. The above-mentioned multi-static V_T also reduces the currents. The subthreshold currents of low-V_T circuits on the critical path are reduced by combining power switches and high-V_T level holders [42](Fig.12). The power switch is off just after evaluating the input of the low-V_T circuit and holding the evaluated output at the holder. This prevents the output from discharging, allowing the switch to quickly turn on at a necessary timing for preparing the next evaluation. In fact, it has been reported that these circuits could reduce the active current of a hypothetical 16-Gb DRAM [28,42] from 1.2A to 0.1A (Figs.10,13), although the effectiveness with an actual chip has not been verified yet.

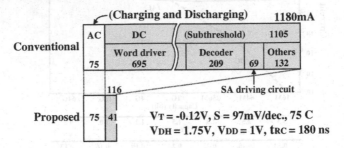

Fig. 13. Active-current reduction for a 16-Gb DRAM[42].

3.2 Other Low-Voltage Relevant Issues

The speed-variation rate for a given variation in design parameters increases by lowering the V_{DD}. Controlling the V_{SUB} and the internal V_{DD} [35] in accordance with the parameter variations reduces the speed variation. Controlling a reverse V_{SUB}, for example, reduced the variation of logic circuits by 20%. Controlling a forward V_{SUB} is more effective for reducing speed variations [36] because the V_T is more sensitive to the V_{SUB}.

On-chip voltage conversion [9] is essential for achieving a stable operation of DRAM/SRAM cells with raised supply voltages, and for reducing the subthreshold current with multi V_T and variations in speed, as also discussed before. A high-conversion efficiency, high degree of accuracy in the output voltage, and low power during the standby period are the key issues. The series regulator, which is widely used in modern DRAMs, offers a highly accurate output voltage despite a small output current and a poor conversion efficiency of around 50%.

Testing low-voltage RAMs is problematic. A large subthreshold current makes discriminating defective and non-defective I_{DDQ} currents difficult. Testing the I_{DDQ} by applying a reverse V_{SUB} [37] is effective when low-temperature measurements and multi-V_T designs are combined. The temperature dependence of speed reduced at a lower V_{DD} [38] is another concern.

4 Perspectives

High S/N designs for RAM cells and arrays, including on-chip ECC circuits to cope with the ever-increasing soft-error rate of RAM cells, and reductions of the subthreshold currents in the active mode are needed to realize low-voltage memories. In the long run, high-speed high-density non-volatile RAMs are attractive for use as low-voltage RAMs. In particular, non-destructive read-out and non-charge-based operations, and the simple planar structures that they could provide, are important to achieving fast cycle and stable operations, even at a lower V_{DD}, at low cost. In this sense, MRAM (magnetic RAM) [39] and OUM (Ovonic Unified Memory) [40] are attractive. For MRAMs, one major challenge is to reduce the magnetic field needed to switch the magnetization of the storage element, while for an OUM, the management of the proximity heating of the cell is an issue. At present, however, the scalabilities and stability reguired to ensure non-volatility still remain unknown, as developments are still in the early stages.

The reduction of subthreshold currents in the active mode may possibly be achieved by using low-power techniques for "old circuits", such as bipolar, BiCMOS, E/D MOS, capacitive boosting, I^2L, and CML circuits. Stand-alone RAMs and e-RAMs, however, could reduce the current by improving the above-mentioned CMOS circuits with the help of new devices, such as the multi-V_r FD SOI. Despite such low-power e-RAMs, however, an SOC will suffer from an incredibly high power dissipated by the random logic gates in the SOC, since control of the subthreshold currents from the random logic gates at a sufficiently high speed may remain impossible. Hence, the number of the gates must be reduced. This presumption implies that new SOC architectures will be required, such as memory-rich SOCs, which effectively reduce the subthreshold currents.

5 Conclusion

Ultralow-voltage RAM technology were reviewed with an emphasis on the signal charge of memory cells and subthreshold current reduction. Through the discussion, the need for high-speed non-volatile RAMs, new device/circuits for reducing active-mode subthreshold currents, and memory-rich architectures was discussed.

References

[1]   K. Itoh and H. Mizuno, "Low-Voltage Embedded-RAM- Technology: Present and Future," *Proc. of the 11th IFIP Int. Conf. on VLSI*, pp.393–398, Dec. 2001.

[2]   H. Yoon *et al.*, "A 4Gb DDR SDRAM with Gain-Controlled Pre-Sensing and Reference Bitline Calibration Schemes in the Twisted Open Bitline Architecture," *ISSCC 2001*, pp. 378–379.

[3]   T. Cho *et al.*, "A 3.3V 1Gb Multi-Level NAND Flash Memory with Non-Uniform Threshold Voltage Distribution," *ISSCC 2001*, pp.28–29.

[4]   International Technology Roadmap for Semiconductors, 2001 Edition, System Driver (Fig.12).

[5]   J. Barth *et al.*, "A 300MHz Multi-Banked eDRAM Macro Featuring GND Sense, Bit-Line Twisting and Direct Reference Cell Write," *ISSCC 2002*, pp.156–157.

[6]   D. Weiss *et al.*, "The On-chip 3MB Subarray Based 3rd Level Cache on an Itanium Microprocessor," *ISSCC 2002*, pp.112–113.

[7]   T. Ditewig, *et al.*, "An embedded 1.2V-read flash memory module in a 0.18µm logic process," *ISSCC 2001*, pp. 34–35.

[8]   K. Itoh et al., "Reviews and Prospects of High-Density RAM Technology," CAS 2000, Sinaia (Romania), Oct. 2000.

[9]   K. Itoh, VLSI Memory Chip Design, Springer-Verlag, March 2001.

[10]   F. Assaderaghi et al., "A novel silicon-on-insulator (SOI) MOSFET for ultralow voltage operation," 1994 Symp. Low Power Electronics, pp. 58–59.

[11]   M. Miyazaki et al., "A 175 mV Multiply-Accumulate Unit using an Adaptive supply voltage and Body Bias (ASB) Architecture," ISSCC 2002, pp. 58–59.

[12]   C. J. Radens et al., "A 0.135µm² 6F² Trench-Sidewall Vertical Device Cell for 4Gb/16Gb DRAM," 2000 Symp. on VLSI Technology, pp. 80–81.

[13]   F. Hofmann and W. Rosner, "Surrounding gate select transistor for 4F² stacked Gbit DRAM," ESSDERC 2001, pp. 131–134.

[14]   T. Sekiguchi et al., "A Low-Impedance Open-Bitline Array for Multigigabit DRAM," IEEE JSSC, vol. 37, pp. 487–498, April 2002.

[15]   E. Ibe, "Current and Future Trend on Cosmic-Ray-Neutron Induced Single Event Upset at the Ground down to 0.1-Micron-Devices," The Svedberg Laboratory Workshop on Applied physics, Uppsala, May 3, 2001.

[16]   H. L. Kalter et al., "A 50-ns 16-Mb DRAM with a 10-ns Data Rate and On-chip ECC," IEEE JSSC, vol. 25, pp. 1118–1128, Oct. 1990.

[17]   W. Leung et al., "The Ideal SoC Memory: 1T-SRAM," 13th Annual IEEE Int. ASIC/SOC Conference, Sept. 2000.

[18]   T. Kawahara, et al., "A Circuit Technology for Sub-10ns ECL 4Mb BiCMOS DRAMs," 1991 Symp. on VLSI Circuits, pp. 131–132.

[19]   H. Mizuno, et al., "CMOS-Logic-Circuit-Compatible DRAM Circuit Designs for Wide-Voltge and Wide-Temperature-Range Applications," 2000 Symp. on VLSI Circuits, pp. 120–121.

[20]   K. Nakazato, et al., "Phase-State Low-Electron-Number-Drive Random-Access Memory(PLED)," ISSCC 2000, pp. 132–133.

[21]   K. Takada, et al., "A 16Mb 400MHz Loadless CMOS Four-Transistor SRAM Macro," ISSCC 2000, pp. 264–265.

[22]   K. Zhang, et al., "The Scaling of Data Sensing Schemes for High Speed Cache Design in Sub-0.18µm," 2000 Symp. on VLSI Circuits, pp. 226–227.

[23]   K. Agawa, et al., "A Bit-Line Leakage Compensation Scheme for Low-Voltage SRAM's," 2000 Symp. on VLSI Circuits, pp. 70–71.

[24]   R. Krishnamurthy, et al., "A 0.13µm 6GHz 256x32b Leakage-tolerant Register File," 2001 Symp. on VLSI Circuits, pp. 25–26.

[25]   K. Itoh, et al., "A Deep Sub-V, Single Power-Supply SRAM Cell with Multi-Vt, Boosted Storage Node and Dynamic Load," 1996 Symp. on VLSI Circuits, pp. 132–133.

[26]   M. Yamaoka, et al., "0.4-V Logic Library Friendly SRAM Array Using Rectangular-Diffusion Cell and Delta-Boosted-Array-Voltage Scheme," 2002 Symp. on VLSI Circuits.

[27]   T. Wada, et al., "A 500MHz Pipelined Burst SRAM with Improved SER Immunity," ISSCC 1999, pp. 196–197.

[28]   T. Sakata, et al., "Two-Dimensional Power-Line Selection Scheme for Low Subthreshold-Current Multi-Gigabit DRAM's," IEEE JSSC, vol. 29, pp. 887–894, August 1994.

[29]   T. Kawahara, et al., "Subthreshold current reduction for decoded-driver by self-reverse biasing," IEEE JSSC, vol. 28, pp. 1136–1144, Nov. 1993.

[30]   M. Horiguchi, et al., "Switched-Source-Impedance CMOS Circuit For Low Standby Subthreshold Current Giga-Scale LSI's," IEEE JSSC, vol. 28, pp. 1131–1135 Nov. 1993.

[31]   A. Alvandpour, et al., "A Conditional Keeper Technique for Sub-0.13µ Wide Dynamic Gates," 2001 Symp. on VLSI Circuits, pp. 29–30.

[32] A. Keshavarzi, *et al.*, "Effectiveness of Reverse Body Bias for Leakage Control in Scaled Dual Vt CMOS ICs," *ISLPED 2001*, pp. 207–212.

[33] H. Mizuno, *et al.*, "18-μA-Standby-Current 1.8-V 200MHz Microprocessor with Self Substrate-Biased Data Retention Mode," *ISSCC 1999*, pp. 280–281.

[34] I. Fukushi *et al.*, "A Low-Power SRAM Using Improved Charge Transfer Sense Amplifiers and a Dual-Vth CMOS Circuit Scheme," *1998 Symp. on VLSI Circuits*, pp. 142–145.

[35] T. Kuroda, *et al.*, "A 0.9V, 150MHz, 10mW, 4mm^2, 2-D Discrete Cosine Transform Core Processor with Variable-Threshold-Voltage (VT) Scheme," *ISSCC 1996*, pp. 166–167.

[36] M. Miyazaki, *et al.*, "1000-MIPS/W Microprocessor using Speed-Adaptive Threshold-Voltage CMOS with Forward Bias," *ISSCC 2000*, pp. 420–421.

[37] T. Miyake, *et al.*, "Design Methodology of High Performance Microprocessor using Ultra-Low Threshold Voltage CMOS," *CICC 2001*, pp. 275–278.

[38] K. Kanda *et al.*, "Design Impact of Positive Temperature Dependence on Drain Current in Sub-1-V CMOS VLSIs," *IEEE JSSC*, vol. 36, pp. 1559–1564, Oct. 2001.

[39] P. K. Naji *et al.*, "A 256kb 3.0V 1T1MTJ Nonvolatile Magnetoresistive RAM," *ISSCC 2001*, pp. 122–123.

[40] M. Gill *et al.*, "Ovonic Unified Memory - A High-Performance Nonvolatile Memory Technology for Stand-Alone Memory and Embedded Applications," *ISSCC 2002*, pp. 202–203.

[41] K. N. Kim *et al.*, "A 0.13μm DRAM technology for giga bit density stand-alone and embedded DRAMs," *2000 Sym. on VLSI Technology*, pp. 10–11.

[42] T. Sakata *et al.*, "Subthreshold-Current Reduction Circuits for Multi-Gigabit DRAM's," *1993 Symp. on VLSI Circuits*, pp. 45–46.

Offline Data Profiling Techniques to Enhance Memory Compression in Embedded Systems*

Luca Benini[1], Alberto Macii[2], and Enrico Macii[2]

[1] Università di Bologna, DEIS, Bologna, ITALY 40136
[2] Politecnico di Torino, DAUIN, Torino, ITALY 10129

Abstract. This paper describes how profile-driven data compression, a very effective approach to reduce memory and bus traffic in single-task embedded systems, can be extended to the case of systems offering multi-function services.

Application-specific profiling is replaced by static data characterization, which allows to cover a larger spectrum of the system's input space; characterization is performed by either averaging several profiling runs over different application mixes, or by resorting to statistical techniques. Results concerning memory traffic show reductions ranging from 10% to 22%, depending on the adopted data characterization technique.

1 Introduction

Electronic systems are going through a rapid and continuous evolution. Nowadays, they offer a large variety of services that range from communication and computing to commodity and amusement. Electronic products can be separated in two main categories: (i) General-purpose systems (i.e., systems that are able to perform a large variety of different tasks and that contain very powerful and highly-flexible programmable processors); (ii) Embedded systems (i.e., systems that are devoted to very specific functions, whose architectures and implementations are optimized for the target application).

As integration capabilities increase, the boundaries between general-purpose and embedded systems start overlapping. It is now quite common facing devices that are able to offer simultaneously a large variety of functions and services: Image and video processing, sound processing, data and voice communication. System operation is user-driven by means of simple programming operations, and tasks are carried out by a processor-based architecture (possibly containing more than one embedded core processor) implemented within a single chip ("System-on-Chip" – SoC).

From the design view-point, most optimization techniques adopted for the development of traditional (i.e., single-function) embedded systems exploit the intrinsic predictability of function executions. For example, system optimization is usually tailored towards specific execution profiles (instructions, data) collected through *a priori* simulations and analysis. This paradigm, which has proved to

---

* This work was supported in part by HP Italiana S.p.A. under grant n. 398/2000.

B. Hochet et al. (Eds.): PATMOS 2002, LNCS 2451, pp. 314–322, 2002.

be very effective, needs to be revisited when dealing with multi-function embedded systems. Although some degrees of predictability in the data profile may still exist, capturing and exploiting them during system design raises new challenges.

In this paper, we focus our attention to a specific aspect of system design, namely the optimization of the processor-to-memory communication path. In particular, we consider the problem of reducing the amount of information that is exchanged between the processor(s) of an SoC and the available storage resources (i.e., caches and main memories).

As communication tends to become *the* major bottleneck in most systems [1], its optimization must be addressed from different angles and perspectives. One viable solution which has been devised recently consists of reducing the amount of required processor-to-memory communication bandwidth by compressing the information that needs to be transfered and stored.

The peculiarity of methods used to compress information being stored in caches or memories is that compression and decompression are subject to very tight performance and latency constraints, since they must comply with data access rates (i.e., read and write) of fast processors. Compression speed becomes the primary cost measure to be used for assessing the quality of a compression scheme; this automatically rules out any software-based solution, and makes HW-assisted approaches the only viable option in this context, where high compression ratios are traded for faster compression units.

Besides speed, also compressor complexity and power consumption must be controlled, since in modern SoCs such units are implemented near to the processor or between cache and memory, where area is at a premium and power density in critical.

The problem of designing efficient compression units has been studied in the past (see, for example, [2,3] for a survey of existing literature). Recently, successful attempts have also been made to exploit HW compression for limiting memory and bus energy consumption in systems containing embedded processors. In fact, energy minimization in the processor-to-memory communication can be seen as a by-product of a decrease in memory and bus traffic. The focus of previous work has been on code compression techniques, thus targeting energy savings in I-caches [5,6] and program memory [7,8]. Extensions to the more general case of data compression have also been studied [4,9,10].

The algorithm of [9] adopts a fixed-dictionary scheme. Its main characteristics is that of exploiting data profiling information to selectively compress cache lines before they are written back to the main memory, and to quickly decompress them when a cache refill operation is started. This approach is particularly suited to single-function embedded systems, where the collection of data statistics to be used by compressor and decompressor is very predictable. The obtained reductions of memory traffic were around 42%, for a significant number of benchmark programs.

The approach of [10] generalizes that of [9] by introducing in the architecture the capability of adaptively updating the compression dictionary according to the current data statistics. This innovation allows to remove from the fixed-dictionary method of [9] the main limitation constituted by the need of off-line data profiling. In other words, the adaptive architecture of [10] is applicable to general-purpose systems (and thus to multi-function embedded systems), were several programs need to be executed and thus ad-hoc data profiling information cannot be collected before the system is started. The achieved memory traffic reductions are, on average, around 31%.

In this paper, we investigate how the profile-driven approach can be extended in order to be applicable to embedded systems offering multi-function services. More specifically, we propose to replace the application-specific profiling step required by the method of [9] with different techniques of data characterization, ranging from random sampling of empirical execution traces to statistical and spectral analysis of the data set.

The results concerning the reduction of memory and bus traffic of a processor-based system that we have obtained go from an average of 10% to an average of 22%, depending on the adopted data characterization method. We can thus conclude that, when the application mix to be executed by a multi-function embedded system is predictable enough to allow the application of one of the data characterization techniques of this paper, the compression architecture of [9] is preferable to that of [10], as it provides comparable results in terms of memory traffic reductions, but with a more limited HW overhead.

2   Profile-Driven Compression

The compression scheme proposed in [9] is dictionary-based. It moves from the assumption that, given the data set of an embedded program, a few data words occur much more often than all the others. Consider the set $\mathcal{F}_N$ of the N most frequent words in the data set. Then, it is possible to encode each word in $\mathcal{F}_N$ with $log_2 N$ bits.

Based on this idea, compression of a cache line can be performed as follows. For each word in the line, check if it belongs to set $\mathcal{F}_N$. If so, compress and store it, then annotate in a header field of the compressed line the information that this word has been compressed; otherwise, store it in uncompressed format and properly annotate this information in the header. After all the words in the line have been considered, check whether the length of the compressed cache line fits size S. If yes, the compressed line is written to the compressed memory area. Otherwise, it is stored in uncompressed format in regular memory.

The choice of N is critical. For smaller values, the number of bits required by a compressed word decreases, but the number of compressible data words also decreases. On the other hand, larger values of N tend to reduce compression (the ratio $(log_2 N)/W$ increases), but more words are compressible.

Since all words in $\mathcal{F}_N$ need to be stored to enable compression and decompression, in order to keep size and complexity of blocks LC and LD of the CDU under control, it is advisable to choose a relatively small value of N. Extensive experimentation lead to a value of $N = 64$. This implies that each compressed word in a cache line takes only 6 bits.

3 Adaptive Compression

The main shortcoming of the profile-driven method described in Section 2 is that it is application and data dependent. Compression and decompression tables are obtained by profiling runs of the target application on pre-specified data sets.

The adaptive compression scheme of [10], does not require off-line profiling and table construction; thus, it is application and data independent and achieves significant memory traffic reductions, but with a more complex architecture of the compression unit with respect to that needed by the profile-driven method. The rationale behind the approach is intuitive. Instead of using a single compression/decompression table with fixed contents, the algorithm maintains two tables, namely a *master* and a *slave*. At any given data transfer, the master table is used to compress and decompress the data, while the slave table is updated with on-line profiling information (learning). When the slave table is "mature", learning is stopped, and the slave table becomes the master. At this point, the old master table becomes the new slave. However, learning cannot start if the compressed memory region still contains lines that have been compressed with the old master table (an additional bit in the CLAT identifies which table has to be used for decompression). The old master table must remain intact to guarantee correct decompression on refill for all compressed lines that have been compressed using its content. When all these lines have been fetched from compressed memory at least once, the content of the new slave table can be discarded and learning can begin.

4 General-Purpose Profiling

In this section we discuss a compression solution that lies in between the profile-driven method of Section 2 and the adaptive technique of Section 3. Although the new approach still relies on data profiling, it is no longer constrained to *a priori* analysis of the execution trace of a single application. Instead, it considers the data profile of the complete application mix that a multi-function embedded system may be capable of running. In the sequel, we present four techniques for constructing set $\mathcal{F}_N$ of data words to be compressed; they have increasing complexity and information capturing capabilities; as such, they require a higher computational effort and they ensure better results.

4.1 Average Profiling

The most intuitive way of determining the set $\mathcal{F}_N$ consists of executing the applications the system is able to support for a random number of times, generating the corresponding average profile, and then picking for compression the N data words that have occurred more often in the trace. We call this simple approach AvgProf, indicating that set $\mathcal{F}_N$ is generated by looking at the "average" profile of all the applications.

The problem with the AvgProf method is that, should some of the applications be more data-intensive than the others, the average profile would become highly biased in favor of such applications, and this would imply a loss of generality of the chosen set $\mathcal{F}_N$.

4.2   Random Profiling

The biasing problem mentioned above can be addressed in different ways. One possibility is the following. Starting from the average profile data trace, we create the set ω of all data words whose occurrence count is within a certain percentage of the most frequent data word of the whole trace; then, we build set $\mathcal{F}_N$ by randomly selecting a total of N data words from set ω. The advantage of this approach, that we call RndProf, is that the random sampling procedure distributes the data words to be included in set $\mathcal{F}_N$ over a wider space, that is, chances are higher that all applications are somehow represented within set $\mathcal{F}_N$.

4.3   Stratified Profiling

In order to provide equal opportunities to all the applications to be represented in set $\mathcal{F}_N$, we can enhance random sampling with some stratification mechanism [11]. We divide set ω into subsets (called *strata*), one for each application for which profiling was performed, and then we randomly select from each stratum an equal number of data words such that the total number of words chosen for compression equals N. This solution, that we call StrProf, further smooths out the biasing effects by enabling a more uniform selection of the data words to be compressed.

4.4   Spectral Profiling

The last technique we have devised for constructing set $\mathcal{F}_N$ from the average profile takes advantage of some analysis of the execution trace performed in the spectral domain [12]. We consider set ω as an integer-valued function, we compute the discrete Fourier transform Ω of such a function, and we select some spectral coefficients from it. This implies generating the spectrum of a new function, called Ω^A, which well approximates the original function. Then, we go back to the time domain by computing the inverse transform of Ω^A, that is, ω^A; the latter contains exactly N data words, and it is thus used as set $\mathcal{F}_N$. The idea behind this method, that we call Spectral, is that of identifying the "components" of ω that are most suitable for being compressed. Key, in the process of generating set $\mathcal{F}_N$, is the choice of the coefficients in the spectral domain, that is, the creation of function Ω^A from Ω. The strategy we have adopted, detailed below, is constrained by the fact that, after the function in the frequency domain is transformed back to the time domain, only N points must remain in ω^A (that is, $\mathcal{F}_N$). In addition, the approximate function ω^A must resemble as much as possible to the original function ω. The first constraint leads to an automatic calculation of the critical frequency of the spectrum; the second constraint forces the elimination from Ω of all the coefficients that are not multiples of the critical frequency.

We are given the representation $\Omega(k)$ of $\omega(n)$ in the frequency domain. The spectral coefficients of $\omega(n)$ (i.e., the samples of $\Omega(k)$), can be interpreted as the coefficients of a linear combination of sinusoids that approximates function $\omega(n)$. $\Omega(0)$ represents the DC-value, that is, the mean value of $\omega(n)$ over the period P of $\omega(n)$. The sinusoid for a given coefficient $k > 0$ has a period of $\frac{N}{k}$, and it is modulated according to the coefficient $\Omega(k)$. The multiples $m \cdot k$ of a given k have a shorter period $\frac{N}{m \cdot k}$.

For our purposes, it is key to remember that the function in the time domain which is obtained by summing a fundamental sinusoid (i.e., the sinusoid at frequency k) to all its harmonics (i.e., the sinusoids at frequencies $m \cdot k$) is still periodic with period $\frac{1}{k}$. By selecting some value k_c as the fundamental frequency, and by suppressing all the remaining frequencies but the multiples (i.e., the harmonics) of k_c, we obtain a sampled function $\Omega^A(k)$ in the frequency domain (i.e., the partial spectrum of function $\omega(n)$) whose counterpart $\omega^A(n)$ in the time domain is a periodic function with period $P = \frac{N}{k_c}$.

The interpretation that can be given of the result above is that $\omega^A(n)$ constitutes the function with period smaller than P (i.e., a sub-multiple) that best approximates function $\omega^A(n)$ with respect to the quantity of information it carries. Then, it is possible to use *one period* of $\omega^A(n)$ to represent *one period* of $\omega(n)$.

As mentioned above, in our context $\omega(n)$ represents the set of all data words whose occurrence count is within a certain percentage of the most frequent data word of the whole average trace; therefore, function $\omega^A(n)$ we obtain from $\Omega^A(k)$ through the inverse DFT provides us with the integer-valued function that best approximates that of ω. In addition, since the period P of $\omega(n)$ is actually the cardinality of the whole average trace, one period of $\omega^A(n)$ directly corresponds to a smaller set. Consequently, the choice of the fundamental frequency k_c uniquely determines the size of set $\mathcal{F}_N$, provided that the partial spectrum $\Omega^A(k)$ is derived from $\Omega(k)$ by removing all the coefficients that are not multiples of k_c (the DC-coefficient $\Omega(0)$ is obviously kept in $\Omega^A(k)$).

From the discussion above it results clear that the selection of the fundamental frequency k_c allows a flexible trade-off between the size of set $\mathcal{F}_N$ and its representativity. By picking a smaller value of k_c, we include in the partial spectrum a large number of frequencies, and therefore we guarantee a finer approximation of the original integer-valued function, at the price of a smaller set $\mathcal{F}_N$. Conversely, by choosing a higher value of k_c we include in $\Omega^A(k)$ fewer frequencies, thus privileging the generation of a shorter stream, at the cost of an increased estimation error.

In principle, after the desired size of $\mathcal{F}_N$ has been fixed, the choice of the fundamental frequency can be done arbitrarily, without any constraint or limitation. The theory of the Fourier transform, however, provides us with a criterion on how the value of k_c should be picked to maximize the effectiveness of the

320   L. Benini, A. Macii, and E. Macii

Table 1. Experimental Results.

Benchmark	ProfDriv [%]	Adaptive [%]	AvgProf [%]	RndProf [%]	StrProf [%]	Spectral [%]
AdaptFilt	42.44	31.25	6.92	8.21	11.20	16.38
Butterfly	27.61	17.18	19.62	26.16	24.29	26.85
Chaos	48.62	39.51	6.14	10.62	11.63	17.23
DTMFCod	32.70	23.76	10.67	12.04	11.57	22.73
IirDemo	43.48	29.39	10.64	11.84	18.16	21.83
Integr	51.66	37.81	6.08	6.24	22.20	24.67
Interp	46.97	36.19	12.09	17.16	16.75	27.78
Loop	38.91	21.62	11.33	12.56	18.05	19.63
Scramble	52.39	31.97	2.93	3.08	5.08	14.11
Upsample	51.13	33.27	12.68	20.40	21.62	24.08
Average	43.59	30.19	9.91	12.83	16.05	21.68

data selection procedure. In fact, to guarantee the exact periodicity of $\omega^A(n)$, it is required that the largest frequency N is also a multiple of k_c, that is, $\Omega^A(k)$ should contain the coefficient at frequency $k = \frac{N}{k_c}$. If this is not the case, the inverse transform of $\Omega^A(k)$ yields a function $\omega^A_{approx}(n)$ in the time domain which is *quasi-periodic*. The use of the latter instead of $\omega^A(n)$ within the data selection procedure introduces an error that decreases as the similarity between the two functions increases.

5   Experimental Results

All the compression schemes considered in this paper have been implemented within the Simplescalar [13] simulation framework. As benchmark programs, we have used a set of DSP-oriented C routines taken from the Ptolemy package [14].

As mentioned in Section 2, the cardinality of set $\mathcal{F}_N$ was set to $N = 64$, implying that compressed data words occupy 6 bits. For the generation of the average data profile, the 10 applications we used for the experiments were ran for a random number of times ranging from 15 and 25, in a random order and on randomly chosen input data sets. This was done to guarantee the highest possible degree on randomness to the average data profile used as starting point by all the methods of Section 4. For the construction of set ω, we considered all the data words in the average profile with an occurrence count within 20% of the most frequent data word.

Table 1 summarizes the results we have obtained. In particular, for each compression approach, it provides the percentage reduction of memory traffic (both read and write) with respect to a system with no data compression. Column labeled ProfDriv provides data for the profile-driven method of Section 2, where the data profile used to build set $\mathcal{F}_N$ refers only to some runs of the specific benchmark to which compression is applied. Clearly, this method provides the highest traffic reductions (around 43%), but its applicability has no generality, as set $\mathcal{F}_{\mathcal{N}}$ is tailored only to the specific application for which profiling was performed. Column labeled Adaptive reports on the usage of the adaptive solution of Section 3. Average savings drop to approximately 30%. Finally, columns labeled AvgProf, RndProf, StrProf and Spectral show the results obtained with the four methods described in Section 4. As expected, the absolute results are inferior to those achieved with method ProfDriv, and, in the case of Spectral method, comparable to those produced by method Adaptive; however, as already discussed in the paper, these methods have the major advantage of being much more general than the profile-driven approach and much less demanding in terms of the hardware compressor than the adaptive approach.

Among the four proposed methods, Spectral is the one that performs best on average (21% reduction); on the other hand, AvgProf and RndProf are the solutions whose results exhibit the largest variance, due to the biasing effect introduced by the average data profile.

6   Conclusions

In this paper we have presented different methods for performing profile-driven data compression with the objective of reducing memory and bus traffic in multi-function embedded systems, i.e., systems which are able to offer simultaneously a large variety of functions and services.

The advantage of the proposed solutions is that significant data traffic reductions (ranging from 10% to 22%, depending on the chosen data characterization approach) can be achieved with limited HW overhead. Results of comparable quality are, in fact, obtained by the application of adaptive approaches, for which the complexity of the compression unit becomes non-negligible.

References

1. J. Rabaey, "Addressing the System-on-a-Chip Interconnect Woes Through Communication-Based Design," *DAC-38*, pp. 667–672, Las Vegas, NV, June 2001.
2. S. Bunton, G. Borriello, "Practical Dictionary Management for Hardware Data Compression," *Communications of the ACM*, Vol. 35, No. 1, pp. 95–104, 1992.
3. B. Abali, *et al.*, "Performance of Hardware Compressed Main Memory," *HP Journal*, pp. 73–81, 2001.
4. J.-S. Lee, W.-K. Hong, S.-D. Kim, "Design and Evaluation of a Selective Compressed Memory System," *ICCD-99*, pp. 184–191, Austin, TX, October 1999.
5. H. Lekatsas, J. Henkel, W. Wolf, "Code Compression for Low Power Embedded Systems," *DAC-37*, Los Angeles, CA, pp. 294–299, June 2000.
6. L. Benini, A. Macii, A. Nannarelli, "Cached-Code Compression for Energy Minimization in Embedded Processors," *ISLPED-01*, Huntington Beach, CA, pp. 322–327, August 2001.
7. Y. Yoshida, B.-Y. Song, H. Okuhata, T. Onoye, I. Shirakawa, "An Object Code Compression Approach to Embedded Processors," *ISLPED-97*, Monterey, CA, pp. 265–268, August 1997.

8. L. Benini, A. Macii, E. Macii, M. Poncino, "Selective Instruction Compression for Memory Energy Reduction in Embedded Systems," *ISLPED-99*, San Diego, CA, pp. 206–211, August 1999.

9. L. Benini, D. Bruni, A. Macii, E. Macii, "Hardware-Assisted Data Compression for Energy Minimization in Systems with Embedded Processors," *DATE-02*, pp. 449–453, Paris, France, March 2002.

10. L. Benini, D. Bruni, B. Riccò, A. Macii, E. Macii, "An Adaptive Data Compression Scheme for Memory Traffic Minimization in Processor-Based Systems," *ISCAS-02*, pp. IV-866-IV-869, Scottsdale, AZ, May 2002.

11. C-S. Ding, C-T. Hsieh, Q. Wu, M. Pedram, "Stratified Random Sampling for Power Estimation," *ICCAD-96*, pp. 577–582, San Jose, CA, November 1996.

12. A. Macii, E. Macii, M. Poncino, R. Scarsi, "Stream Synthesis for Efficient Power Simulation Based on Spectral Transforms," *IEEE Transactions on VLSI Systems*, Vol. 9, No. 3, pp. 417–426, June 2001.

13. D. C. Burger, T. M. Austin, S. Bennett, *Evaluating Future Microprocessors – The Simplescaler Toolset*, Tech. Rep. 1308, Univ. of Wisconsin, Dept. of CS, 1996.

14. J. Davis II, *et al.*, *Overview of the Ptolemy Project*, Tech. Rep. UCB/ERL No. M99/37, Univ. of California, Dept. of EECS, 1999.

Performance and Power Comparative Study of Discrete Wavelet Transform on Programmable Processors

N.D. Zervas[1], G. Pagkless[1], M. Dasigenis[2], and D. Soudris[2]

[1] Alma Technologies S.A.,
2 Marathonos Av. Pikermi, Attica, GR19009, Greece
zervas@alma-tech.com
http://www.alma-tech.com
[2] Dept. of Electrical and Computer Eng.,
Democritus University of Thrace, 67100 Xanthi, Greece
{mdasyg, dsoudris}@ee.duth.gr

Abstract. The Discrete Wavelet Transformations (DWT) are data intensive algorithms. Energy dissipation and execution time of such algorithms heavily depends on data memory hierarchy performance, when programmable platforms are considered. Existing filtering operations for the 1D-DWT, employ different levels of data accesses locality. However locality of data references, usually comes at the expense of complex control and addressing operations. In this paper, the two main scheduling techniques for the 1D-DWT are compared in terms of energy consumption and performance. Additionally, the effect of an in-place mapping scheme, which minimizes memory requirements and improves locality of data references for the 1D-DWT, is described and evaluated. As execution platform, two commercially available general purpose processors are used.

1 Introduction

The inherent time-scale locality characteristics of the Discrete Wavelet Transformations has established them as powerful tools for numerous applications such as signal analysis, signal compression and numerical analysis. This has lead numerous research groups to develop algorithms and hardware architectures to implement the DWT. In [1], [2], [3] and [4] interesting VLSI architectures for the 1D and 2D DWT have been proposed. Additionally, exhaustive comparisons among the various scheduling algorithms for the DWT, regarding their efficiency when the DWT is mapped in custom VLSI architectures, has been performed in [5] and [6].

Nowadays, the rapid advances in the area of programmable processors have made them an attractive solution even for real-time complex DSP applications, since they offer enough processing power combined with flexibility and small time-to-market. However very little has been done, to determine which scheduling algorithms for the DWT perform better when mapped on a DSP or general-purpose or embedded programmable processor. This paper is the first step of

B. Hochet et al. (Eds.): PATMOS 2002, LNCS 2451, pp. 323–331, 2002.

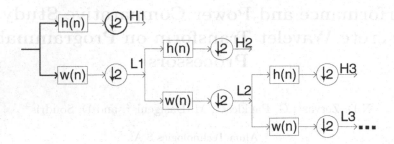

Fig. 1. The DWT decomposition

an attempt to fill this gap. Specifically, the two main scheduling algorithms for the 1D-DWT, namely the pyramid (PA) and the recursive pyramid (RPA) algorithms, are compared in terms of throughput. Additionally, two alternative methods to apply the *in-place* optimization [7] in the case of 1D-DWT are described and their effect is evaluated.

The rest of this paper is organized as follows: Section 2 briefly describes the DWT multiresolution decomposition, together with PA and RPA algorithms. In Section 3, the two alternative ways to apply the in-place optimization on 1D-DWT, are described. In Section 4, experimental results are presented and analyzed, while Section 5 summarizes the conclusions and presents future research directions of our work.

2   Basic Background

The DWT can be viewed as the multiresolution decomposition of a sequence [8]. It takes a length N sequence $IN[n]$, and generates and output sequence of length N. The output is a multiresolution representation of $IN[n]$. The highest resolution level is of length $N/2$, the next resolution level is of length $N/4$, and so on. We denote the number of frequencies or resolutions levels or levels with the symbol J. The DWT filter bank structure, realizing the DWT decomposition, is illustrated in Fig. 1. Typically, filters $h[n]$ and $w[n]$ shown in Fig. 1 are FIR filters. We denote N_H and N_W the number of taps of the high-pass ($h[n]$) and low-pass ($w[n]$) filters respectively, and define $N_F = \max\{N_H, N_W\}$.

To implement the decomposition of Fig. 1 on a single processor architecture, two main algorithms have been proposed until now. The first is the classical pyramid algorithm (PA) [8], while the second is a reformulation of the PA called recursive pyramid algorithm (RPA) [3]. The two algorithms differ on the filtering operations scheduling. This imposes different memory requirements and different input access schedules. In the next subsections the two algorithms are briefly described.

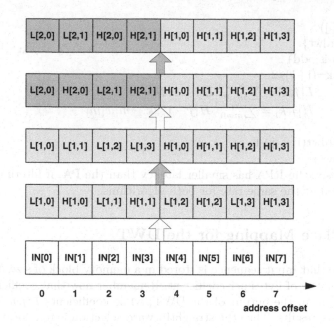

Fig. 2. In-place for the 1D-DWT with reordering

2.1   Pyramid Algorithm

The pyramid algorithm is the direct implementation of the decomposition of Fig. 1. Specifically, filtering along level $j + 1$ is initiated after the completion of filtering along level j. The pseudo-code for the PA follows:

```
begin{Direct Pyramid}
    for( j=1 to J)
        for(n=1 to 2j-1)
```
$$L[j, n] = \sum_{m=0}^{N_F - 1} L[j - 1, 2n - m]w[m]$$
$$H[j, n] = \sum_{m=0}^{N_F - 1} H[j - 1, 2n - m]w[m]$$
```
    end{Direct Pyramid}
```

2.2   Recursive Pyramid Algorithm

The basic idea behind the recursive pyramid algorithm is the following:*Proceed to next level filtering ASAP.* This means that filtering along level j is interleaved, when enough coefficients to perform next filtering along level $j + 1$ are produced. The pseudo-code for the RPA follows:

```
    begin{Recursive Pyramid}
        for(i=1 to N-1)
            rdwt(i, 1);
    end{Recursive Pyramid}
```

rdwt(i, j)
begin{rdwt}
   if(i is odd)
      k=(i+1)/2

$$L[j, k] = \sum_{m=0}^{N_F - 1} L[j - 1, 2k - m]w[m]$$
$$H[j, k] = \sum_{m=0}^{N_F - 1} H[j - 1, 2k - m]w[m]$$

   else
      rdwt(i/2, j+1)
end{rdwt}

Obviously, the RPA has smaller latency than the PA, if filtering operations are executed at the same rate for both algorithms.

3   In-Place Mapping for the DWT

We assume that input sequence is stored in a memory block of size $N \times d$, where d is the number of bytes per coefficient. Remember now that coefficients $H[j, n]$ and $L[j, n]$ form the output of the DWT, while coefficients $L[j, n]$ $(j \neq J)$ are intermediate results. Thus the straightforward selection is to allocate a separate memory block to store the N output coefficients, and $J - 1$ memory blocks to store the $N/2^j d$ intermediate results of each level. Hence, the total storage requirements for the 1D-DWT is:

$$\left(N + N + \sum_{j=1}^{L-1} N/2^j \right) \times d = N \times \left(3 + 1/2^{L-1} \right) \times d \qquad (1)$$

The amount of memory required for the DWT can be significantly reduced by applying an in-place optimization [7]. The key concept of such an optimization is to *store output and intermediate data in-place, of input data that are no longer needed.*

For example consider the 1D-sequence of Fig. 2 and assume a 5/3 DWT. Additionally assume that the N_F coefficients to be filtered are first fetched to local FIFO buffer, called filtering FIFO and denoted as FF. The pair of coefficients $L[1, 0]$, $H[1, 0]$ is produced by filtering the 3 first input coefficients, after performing a symmetrical mirroring. Since, input coefficients $IN[0]$ and $IN[1]$ are currently in the FF and will not fetched again from the input memory, we can store $L[1, 0]$, $H[1, 0]$ in their place (address offsets 0 and 1). In the same way, coefficients $L[1, i]$, $H[1, i]$, can be stored at address offsets $2 \cdot i$ and $2 \cdot i + 1$ respectively. A direct effect of this in-place mapping is that after the completion along the input, the low-frequency coefficients of level 1, that will be consumed to produce the coefficients of level 2, are not stored in consecutive addresses in the memory.

One way to overturn this problem is to to reorder the coefficients so that the first $N/2$ addresses store the coefficients $L[1, i]$, and the addresses from $N/2$ up to N store the coefficients $H[1, i]$. In this way to perform filtering along level 1,

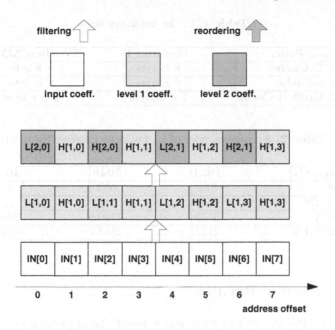

Fig. 3. In-place for the 1D-DWT without reordering

coefficients from consecutive memory addresses will be fetched. Generally, after reordering coefficients $L[j, i]$ and $H[j, i]$, are stored in place of coefficients $IN[i]$ and $IN[N/2^j + i]$, respectively. However, reordering the coefficients of level j after the completion of filtering along level $j - 1$, requires $N/2^j$ extra memory accesses. This result to a total of $2N(1 - 1/2^L)$ overhead in terms of memory accesses, to compute the L-level DWT of an input of size N.

An alternative method to implement the in-place optimization of the DWT, is to slightly modify addressing equations, instead of employing reordering (Fig. 3).Specifically, if reordering is not performed, then the low-frequency coefficients $L[j, i]$, are stored in place of the input coefficients: $IN[2^j \cdot i]$, while the high-frequency coefficients $H[j, i]$, are stored in in place of the input coefficients: $IN[2^j \cdot i + 1]$. In this way the overhead of $2N(1 - 2/2^L)$ accesses is replaced by the overhead of $N(1 - 2/2^L)$ multiplications of index i times a power of two (2^j), which are reduced to binary shifts.

It is evident that the later approach to apply the in-place optimization is the most efficient, since the $2N(1 - 2/2^L)$ memory accesses dissipate more energy and require more clock cycles than $N(1 - 2/2^L)$ binary shifts in any DSP or general purpose programmable processor [9]. Since the result of a comparison among the two alternative ways to apply the in-place mapping is easily predictable, no experimental results will be provided for this approach, in this paper.

Table 1. Cache memories sizes

Algo./Proc.	Pentium I	Pentium MMX
L1 D-Cache	8 KByte	8 Kbyte
L1 I-Cache	8 KByte	8 KByte
L2 Unified Cache	256 KByte	256 KByte

Table 2. Average number of L1 D-Cache misses (Pentium)

Algo./N	1024	2048	10240
PA	332	669	3617
PA_IN	164	414	5263
RPA	3143	47602	241105
RPA_IN	2122	28685	143504

4 Experimental Results

The PA and RPA algorithms with and without the application of the in-place optimization have been implemented in C language. All four source codes have been mapped on two Intel's processors, namely a Pentium, and Pentium MMX. Table 1 illustrates the cache memory configuration for both processors. Compilation of the C sources have been performed using the Intel's C/C++ compiler v4.0. All four codes has been fed with input sequence of length $N = 1024, 2048$ and 10240, and executed considering $J = 3, 4, 5$ and 6 decomposition levels. Intel's Vtune environment [10] has been used to acquire profiling data, in terms of execution cycles and number of cache misses.

Tables 2 and 3 give the average, in terms of decomposition layers, number of level 1 (L1) data-cache misses for Pentium I and Pentium MMX respectively. It must be stressed here that the number of data-cache misses increases with J. Specially for the J=10240 case, this phenomenon is more intense, due to the large size of the higher decomposition layers. It is noted that we have chosen to present experimental results in the compact format of Tables 2, 3 for space economy purposes. Table 4 illustrates the number of L1 instruction cache misses, which is constant with respect to J and N. As far as execution speed is concerned the relative results for the four codes are almost identical for both processors. For this reason in Fig. 4 the average number of cycles for both processors is depicted.

Comparing Table 4 to Tables 2 and 3 it can be deducted that the number of L1 instruction cache misses is very small compared to the number of data cache misses. This is due to the fact that for all cases the code fits in the L1 instruction cache, and thus only compulsory misses occur. In the following experimental results analysis the instruction cache misses will be ignored, due to their minor role.

Table 3. Average number of L1 D-Cache misses (Pentium MMX)

Algo./N	1024	2048	10240
PA	282	539	3702
PA_IN	152	284	5220
RPA	160	317	240343
RPA_IN	158	320	143445

4.1 PA versus RPA

As shown in Fig. 4, the PA algorithm has been proved to be faster than the RPA algorithm in all cases. This is due to its better data reference locality characteristics. Specifically, the PA algorithm linearly traverses along each level while on the other hand the RPA algorithm continuously interchanges filtering along layers. Hence, since PA scans sequential addresses, it is less valuable to conflict misses than the RPA, which has a less canonical data memory access trace. This fact causes the D-cache to perform better for the PA than the RPA especially for large values of N.

Furthermore, the RPA employs a more complex control than the PA algorithm. As a result the RPA employs a greater number of executed instruction than the PA.

4.2 In-Place

As described in section 3, applying the in-place optimization on the one hand reduces memory requirements, and thus can improve data cache performance, but on the other hand increases the complexity of the addressing equations. Thus, it is very difficult to predict its effect on performance in a programmable platform.

In the case of the PA algorithm, the application of the in-place optimization worsens execution speed for all cases. This can be justified by the fact that memory requirements minimization did not succeed in reducing further the already small number of cache misses, while on the other hand increased the number of executed instructions.

In the case of RPA, the application of the in-place optimization has contradictory effect for $J = 3$, 4 and $J = 5$, 6; in the first case it improves execution speed, while in the latter case it worsens it. This can be explained as follows: The majority of data cache misses occurs due to the interchange that happens on the lower levels. This is because the smallest the level is, the greater the level's size becomes. On the other hand the overhead of more complex addressing equations remains the same of decomposition of the DWT. Thus the execution speed benefits from localizing the accesses by in-place mapping offset the addressing complexity penalty, for small number of decomposition levels. On the other hand addressing complexity penalty dominates for great number of decomposition levels of the DWT.

Table 4. Number of L1 I-Cache misses

Algo./Proc.	Pentium I	Pentium MMX
PA	35	27
PA_IN	26	23
RPA	37	33
RPA_IN	36	29

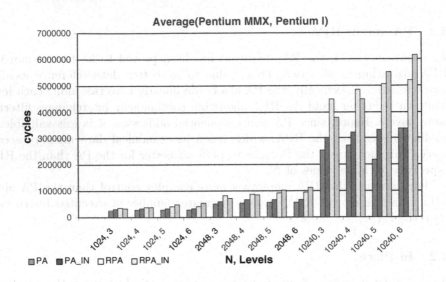

Fig. 4. Average number of cycles

5   Conclusions and Future Work

The two main scheduling algorithms for the computation of the 1D-DWT, have been mapped on two general-purpose programmable processors and compared in terms of execution speed. Experimental results indicate that the classic PA algorithm which employs the lowest control and addressing complexity, has also better data reference locality characteristics than the RPA algorithm. This makes the PA to perform better on a programmable platform than the RPA, which is however favorable for VLSI hardware implementations. Additionally, two alternative methods to apply the in-place optimization have been described. The application of this optimization results in a significant reduction of memory requirements at the cost of slightly more complex addressing. This fact makes the in-place mapping always favorable for VLSI implementations of the DWT. However, experimental results indicated that in a programmable domain the memory size savings comes at the cost of slower execution in most cases.

Hence, at least as far as DWTs are concerned, it can be said that scheduling algorithms and optimizations developed targeting custom hardware platforms, can not migrate directly to the programmable domain. The focus of our future research is on theoretical exploration as well as on further experimentation regarding the 1D and 2D-DWT. The final aim is to identify which algorithms and optimizations, under which conditions perform better when executed on a programmable platform.

References

1. G. Knowles, "*VLSI architecture for the discrete wavelet transform*", Elec. Letters, vol. 26, no. 5, pp. 1184–1185, July 1990.
2. K. K. Parhi, T. Nishitani, "*VLSI Architectures for Discrete Wavelet Transforms*", IEEE Trans. on VLSI Systems, Vol. 1, No. 2, pp. 191–202, June 1993.
3. C. Chakrabarti, R. M. Owens, M. J. Irwin, "*VLSI Architectures for the Discrete Wavelet Transform*", IEEE Trans. on Circuits and Systems, Vol. 42, pp. 305–316, May 1995.
4. J.T. Kim et al, "*Scalable VLSI architectures for lattice structure-based discrete wavelet transform*", IEEE Trans. on Circuits and Systems II, Vol. 46, No. 8, pp. 1031–1043, Aug. 1998.
5. C. Chakrabarti, M. Vishwanath, R. M. Owens, "*Architectures for Wavelet Transform: A Survey*", Journal of VLSI Signal Processing, Kluwer Academic Publishers, Vol. 14, Issue 2, pp. 171–192, 1996.
6. G. Lafruit, L. Nachtergaele, J. Bormans, M. Engels, I. Bolsens, "*Optimal Memory Organizations for Scalable Texture Codecs in MPEG-4*", IEEE Tran. on Circuits and Systems for Video Technology, Vol. 9, No. 2, pp. 218–242, March 1999.
7. F. Catthoor, K. Danckaert, C. Kulkarni, E. Brockmeyer, P.G. Kjeldsberg, T. van Achteren, "Data Access and Storage Management for Embedded programmable Processors," Kluwer Academic Publishers, Boston, 2002.
8. S. Mallat, "*Multifrequency channel decompositions of images and wavelet models*", IEEE Tran. on Acoustics, Speech, Signal Process, Vol. 37, no 12, pp. 2091–2110, Dec. 1989.
9. V. Tiwari, S. Malik and A. Wolfe, "*Instruction level power analysis and optimization of software*", Journal of VLSI Signal Processing Systems, 13: 223–228,1996.
10. http://developer.intel.com/vtune.

Power Consumption Estimation of a C Program for Data-Intensive Applications

Eric Senn, Nathalie Julien, Johann Laurent, and Eric Martin

L.E.S.T.E.R., University of South-Brittany, BP92116
56321 Lorient cedex, France
{eric.senn, nathalie.julien, johann.laurent, eric.martin}@univ-ubs.fr
http://lester.univ-ubs.fr:8080/

Abstract. A method for estimating the power consumption of an algorithm is presented. The estimation can be performed both from the C program and from the assembly code. It relies on a *power model* for the targeted processor. Without compilation, several targets can be compared at the C-level in order to rapidly explore the design space. The estimation can be refined afterwards at the assembly level to allow further code optimizations. The power model of the Texas Instrument TMS320C6201 is presented as a case study. Estimations are performed on real-life digital signal processing applications with average errors of 4.2 % at the C-level, and 1.8 % at the assembly level.

1 Introduction

Power consumption is currently a critical parameter in system design. The fact is that the co-design step can lead to many solutions: there are many ways of partitioning a system, and many ways of writing the software even once the hardware is chosen. It is now well-known that the software has a very strong impact on the final power consumption [1]. To find the best solution is not obvious. Indeed, it is not enough to know whether the application's constraints are met or not; it is also necessary to be able to compare several solutions to seek the best one. So, the designer needs fast and accurate tools to evaluate a design, and to guide him through the design space. Without such a tool, the application power consumption would only be known by physical measurement at the very last stage of the design process. That involves to buy the targeted processor, the associated development tools and evaluation board, together with expansive devices to measure the supply current consumption, and that finally expands the time-to-market.

This work demonstrates that an accurate power consumption estimation can be conducted very early in the design process. We show how to estimate the power consumption of an algorithm directly from the C program without execution. Provided that the power model of the processor is available, there is no need for owning the processor itself, nor for any specific development tool, since it is not even necessary to have the code compiled. Thus, a fast and cheap comparison of different processors, or of different versions of an algorithm, is

B. Hochet et al. (Eds.): PATMOS 2002, LNCS 2451, pp. 332–341, 2002.
© Springer-Verlag Berlin Heidelberg 2002

possible [2]. In the following step of the design flow, a method for refining the estimation at the assembly level is also provided. This way, hot points can be definitely located in the code, and further optimizations can be focused on these critical parts.

Some power estimation methods are based on cycle-level simulations like in Wattch or SimplePower [3,4]. However, such methods rely on a low-level description of the processor architecture, which is often unavailable for off-the-shelf processors. Another classical approach is to evaluate the power consumption with an instruction-level power analysis [5]. This method relies on current measurements for each instruction and couple of successive instructions . Its main limitation is the unrealistic number of measurements for complex architectures [6]. Finally, recent studies have introduced a functional approach [7,9], but few works are considering VLIW processors [8]. All these methods perform power estimation only at the assembly-level with an accuracy from 4% for simple cases to 10% when both parallelism and pipeline stalls are effectively considered. As far as we know, only one unsuccessful attempt of algorithmic estimation has already been made [10].

Our estimation method relies on a *power model* of the targeted processor, elaborated during the *model definition* step. This model definition is based on the *Functional Level Power Analysis* of the processor architecture [12]. During this analysis, functional blocks are identified, and the consumption of each block is characterized by physical measurements. Once the power model is elaborated, the *estimation process* consists in extracting the values of a few parameters from the code; these values are injected in the power model to compute the power consumption. The estimation process is very fast since it relies on a static profiling of the code. Several targets can be evaluated as long as several power models are available in the library.

As a case study, a complete power model for the Texas Instruments TMSC6201 has been developed. It is presented in section 2 with details on the model definition. The estimation process, and the method to extract parameters from both the C and the assembly codes is exhibited in section 3. Estimations for several digital signal processing algorithms are presented in section 4. Estimations are performed both at the assembly and C-level, and compared with physical measurements. The gap between estimation and measure is always lower than 3.5% at the assembly-level, and than 8% at the C-level.

2   Model Definition

The model definition is done once and before any estimation to begin. It is based on a *Functional Level Power Analysis (FLPA)* of the processor, and provides the *power model*. This model is a set of consumption rules that describes how the average supply current of the processor core evolves with some *algorithmic* and *configuration parameters*. Algorithmic parameters indicate the activity level between every functional block in the processor (parallelism rate, cache miss rate

...). Configuration parameters are explicitly defined by the programmer (clock frequency, data mapping ...).

The model definition is illustrated here on the TI C6x. This processor was initially chosen to demonstrate the methodology on a complex architecture. Indeed, it has a VLIW instructions set, a deep pipeline (up to 11 stages), and parallelism capabilities (up to 8 operations in parallel). Its internal program memory can be used like a cache in several modes, and an External Memory Interface (EMIF) is used to load and store data and program from the external memory [11]. The use of an instruction level method for such a complex architecture would conduct to a prohibitive number of measurements.

The FLPA results for the TI C6x are summarized on the Figure 1. Three blocks and five parameters are identified. These parameters are called *algorithmic parameters* for their value actually depends on the algorithm. The parallelism rate α assesses the flow between the FETCH stages and the internal program memory controller inside the IMU (Instruction Management Unit). The processing rate β between the IMU and the PU (Processing Unit) represents the utilization rate of the processing units (ALU, MPY). The activity rate between the IMU and the MMU (Memory Management Unit) is expressed by the program cache miss rate γ. The parameter τ corresponds to the external data memory access rate. The parameter ε stands for the activity rate between the data memory controller and the Direct Memory Access (DMA). The DMA may be used for fast transfer of bulky data blocks from external to internal memory ($\varepsilon = 0$ if the DMA is not used).

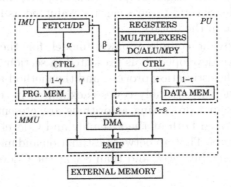

Fig. 1. FLPA for the C6x

To the former algorithmic parameters are added three *configuration parameters*, that also strongly impact on the processor's consumption: the clock frequency F, the memory mode MM, and the data mapping DM.

The influence of F is obvious. The C6x maximum frequency is *200 MHz*, but the designer can tweak this parameter to adjust consumption and performances.

The memory mode *MM* illustrates the way the internal program memory is used. Four modes are available. All the instructions are in the internal memory in the *mapped mode* (MM_M). They are in the external memory in the *bypass mode* (MM_B). In the *cache mode*, the internal memory is used like a direct mapped cache (MM_C), as well as in the *freeze mode* where no writing in the cache is allowed (MM_F). Internal logic components used to fetch instructions (for instance tag comparison in cache mode) actually depends on the memory mode, and so the consumption.

The data mapping impacts on the processor's consumption for two reasons. First, the logic involved to access a data in internal or in external memory is different. Secondly, whenever a data has to be loaded or stored in the external memory, or whenever two data in the same internal memory bank are accessed at the same time, the pipeline is stalled and that really changes the consumption.

Hence the final *power model* for the TI C6x, presented in the Figure 2.

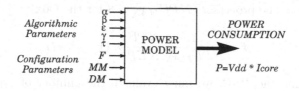

Fig. 2. Power Model

This model comes with a set of *consumption rules* that gives the power consumption of the processor, given the former parameters' values. To determine these rules, the parameters were made to vary, with the help of small assembly programs. Variations of the processor's core supply current were measured, and mathematical functions were obtained by curve fitting. For this processor, no significant difference in power consumption was observed between an addition and a multiplication, or a read and a write in the internal memory. Moreover, the effect of data correlation on the global power consumption appeared lower than 2%. More details on the consumption rules and their determination can be found in [12].

3   Estimation Process

To estimate the power consumption of a program with our power model, we must determine the value of all its input parameters. We will first explain how to precisely compute these parameter values from the assembly code, and then how to predict them directly from the C code without compilation.

3.1   Parameters Extraction from the Assembly Code

In the C6x, eight instructions are fetched at the same time. They form a *fetch packet*. In this fetch packet, operations are gathered in *execution packets* depending on the available resources and the parallelism capabilities. The parallelism rate α can be computed by dividing the number of fetch packet (NFP) with the number of execution packet (NEP) counted in the code. However, the effective parallelism rate is drastically reduced whenever the pipeline stalls. Therefore, the final value for α must take the number of pipeline stalls into account. Hence, a *pipeline stall rate* (PSR) is defined, and α is computed as follows:

$$\alpha = \frac{NFP}{NEP} \times (1 - PSR) \tag{1}$$

Identically, the PSR is considered to compute the processing rate β, with NPU the average number of processing unit used per cycle (counted in the code), and NPU_{MAX} the maximum number of processing units that can be used at the same time in the processor ($NPU_{MAX} = 8$ for the C6x):

$$\beta = \frac{1}{NPU_{MAX}} \frac{NPU}{NEP} \times (1 - PSR) \tag{2}$$

To determine the PSR, we must evaluate the number of cycles where the pipeline is stalled (NPS), and divide it by the total number of cycles for the program to be executed (NTC):

$$PSR = \frac{NPS}{NTC} \tag{3}$$

Pipeline stalls have several causes:

- a delayed data memory access: if the data is fetched in external memory (related to ε) or if two data are accessed in the same internal memory bank (related to the data mapping DM)
- a delayed program memory access: in case of a cache miss for instance (related to the cache miss rate γ), or if the cache is bypassed or freezed (related to the memory mode MM)
- a control hazard, due to branches in the code: we choose to neglect this contribution because only data intensive applications are considered.

As a result, NPS is expressed as the sum of the number of cycles for stalls due to an external data access NPS_τ, for stalls due to an internal data bank conflict NPS_{BC}, and for stalls due to cache misses NPS_γ.

$$NPS = NPS_\gamma + NPS_\tau + NPS_{BC} \tag{4}$$

Whenever a cache miss occurs, the cache controller, via the EMIF, fetch a full instruction frame (containing 8 instructions) from the external memory. The number of cycles needed depends on the memory access time T_{access}. As a result, with $NFRAME$ the number of frames causing a cache miss:

$$NPS_\gamma = NFRAME \times T_{access} \qquad (5)$$

Similarly, the pipeline is stalled during T_{access} for each data access in the external memory. That gives, with $NEXT$ the number of data accesses in external memory:

$$NPS_\tau = NEXT \times T_{access} \qquad (6)$$

A conflict in an internal data bank is resolved in only one clock cycle. So, NPS_{BC} is merely the number of bank conflicts $NCONFLICT$.

$$NPS_{BC} = NCONFLICT \qquad (7)$$

The three numbers $NEXT$, $NCONFLICT$ and $NFRAME$ can be computed from the assembly code. In fact, $NFRAME$ is also needed to compute the cache miss rate γ, with the total number of instruction frames in the code, and the cache size. However, since the assembly code for digital signal processing applications generally fits in the program memory of the C6x, γ is often equal to zero (as well as $NFRAME$).

The numbers $NEXT$ and $NCONFLICT$ are directly related to the data mapping. This mapping is expressed in the power model through the configuration parameter DM.

The number of DMA accesses can be counted in the assembly code. The DMA access rate ε is computed by dividing the number of DMA accesses by the total number of data accesses in the program.

At last, external data accesses are fully taken into account through $NEXT$ which includes the parameter τ; indeed, τ does not appear explicitly in our set of consumption rules.

3.2   Parameters Prediction from the C Code

In the previous section, the parameters needed to actually estimate the power consumption of an application were extracted from the assembly code. In this section, we show how to determine these parameters directly from the C code, without compilation.

As stated before, the pipeline stall rate PSR is needed to compute the values of the parameters α and β. To calculate the PSR, we need the number of external data accesses $NEXT$, the number of internal data bank conflicts $NCONFLICT$, and the number of instruction frames that involve cache misses $NFRAME$ (Equations 3-7).

It is remarkable that the two numbers $NEXT$ and $NCONFLICT$ can be determined directly from the C program. Indeed, they are related to the data mapping which is actually fixed by the programmer (by the mean of explicit compilation directives associated to the C sources) and only taken into account

by the compiler during the linkage. Accesses to the DMA are explicitly pro-
grammed as well. Because the programmer knows exactly the number of DMA
accesses, he can easily calculate the DMA access rate ε without compilation.

Nevertheless, it is not possible to predict $NFRAME$ at the C-level. Indeed, the
assembly code size is needed to be compared with the cache size; a compilation is
necessary. As explained before, the C6x cache is however large enough for most
of the digital signal processing applications, and in these cases $NFRAME$ and γ
equal zero. Whenever $NFRAME$ and γ are not known in the early step of the
design process, it is still possible to provide the designer with *consumption maps*
to guide him in the code writing [2].

To determine α and β at the C-level, the three parameters NFP, NEP
and NPU must be predicted from the algorithm (instead of being counted in
the assembly code). It is clear that this prediction must rely on a model that
anticipates the way the code is executed on the target. According to the processor
architecture and with a little knowledge on the compiler, four *prediction models*
were defined:

The *sequential model* (SEQ) is the simplest one since it assumes that all the
operations are executed sequentially. This model is only realistic for non-parallel
processor.

The *maximum model* (MAX) corresponds to the case where the compiler
fully exploits all the architecture possibilities. In the C6x, 8 operations can be
done in parallel; for example 2 loads, 4 additions and 2 multiplications in one
clock cycle. This model gives a maximum bound of the application consumption.

The *minimum model* (MIN) is more restrictive than the previous model since
it assumes that load and store instructions are never executed at the same time -
indeed, it was noticed on the compiled code that all parallelism capabilities were
not always fully exploited for these instructions. That will give a reasonable
lower bound for the algorithm's power consumption.

At last, the *data model* (DATA) refines the prediction for load and store
instructions. The only difference with the MAX model is to allow parallel loads
and stores only if they involve data from different memory banks. Indeed, there
are two banks in the C6x internal data memory, which can be accessed in one
clock cycle.

Assuming data-intensive applications, the prediction is performed by applying those models for each significant loop of the algorithm. Operations at the
beginning or at the end of the loop body are neglected. As illustration, we present
below a simple code example:

```
For (i=0; i<512; i++) {Y=X[i]*(H[i]+H[i+1]+H[i-1])+Y;}
```

In this loop nest, there are 4 loads (LD), and 4 other operations (OP): 1
multiplication, and 3 additions. In our example, the final store for Y, only done
once at the end of the loop, is not considered. Here, our 8 operations will always
be gathered in one single fetch packet so $NFP = 1$. Because no NOP operation
is involved, $NPU = 8$ and α and β parameters have the same value. In the SEQ
model, instructions are assumed to be executed sequentially. Then $NEP = 8$,
and $\alpha = \beta = 0.125$. Results for the other models are summarized in Table 1.

Table 1. Prediction models for the example

model	EP1	EP2	EP3	EP4	$\alpha = \beta$
MAX	2LD	2LD,4OP	-	-	0.5
MIN	1LD	1LD	1LD	1LD,4OP	0.25
DATA	2LD	1LD	1LD,4OP	-	0.33

Of course, realistic cases are more elaborated: the parameter prediction is done for each part of the program (loops, subroutines ...) for which local values are obtained. The global parameters values, for the complete C source, are computed by a weighted averaging of all the local values. Such an approach permits to spot "hot points" in the program. In the case of data-dependent algorithms, a statistic analysis should be performed to get those values.

4 Application

The estimation at the assembly level was already validated by direct comparison with measurements in [12]. In this section, the same process is applied at the C-level, and the two approaches are finally compared.

The estimation is performed for several digital signal processing algorithms: a FIR filter, a FFT, a LMS filter, a Discrete Wavelet Transform (DWT) with two different image sizes, an Enhanced Full Rate (EFR) Vocoder for GSM, and a MPEG1 Decoder (1600 lines in the C program; 4000 lines in the assembly code). The results are presented for different memory modes (mapped, cache and bypass) and data mappings (*EXT*ernal or *INT*ernal memory).

In Table 2, the value of the power model parameters extracted from the assembly code, and from the C code assuming the DATA prediction model, are presented. For these applications, $\gamma = 0$ since all the code is contained in the internal program memory, and $\varepsilon = 0$ since the DMA is not used. The PSR measured value (PSR^m), obtained with the help of the TI development tool, is used for estimation at the assembly level (but the calculated value could be used as well). The average error between the estimated (PSR) and the measured (PSR^m) pipeline stall rates is 3.2%. It never exceeds 5.5% which indicates the PSR estimation accuracy.

The power consumption of the algorithm is computed from those parameters' values. The relative error between estimation and measurements is given in Table 3. Results are given for the assembly level and for the four prediction models at the C-level.

Of course, the SEQ model gives the worst results since it does not take into account the architecture possibilities (parallelism, several memory banks etc.). In fact, this model has been developed to explore the estimation possibilities without any knowledge about the architecture of the targeted processor. It seems that such an approach cannot provide enough accuracy to be satisfying.

It is remarkable that, for the LMS in bypass mode, every model overestimates the power consumption with close results. This exception can be explained by

Table 2. Parameters estimations $(F = 200MHz)$

	Configuration		Assembly level			C-level		
Application	*MM*	*DM*	α	β	PSR^m	α	β	PSR
FIR	MM_M	INT	0.492	0.454	0	0.5	0.5	0
FFT	MM_M	INT	0.099	0.08	0.64	0.119	0.113	0.604
LMS-1	MM_B	INT	-	0.029	0.93	-	0.0312	0.95
LMS-2	MM_C	INT	0.625	0.483	0.25	0.76	0.475	0.24
DWT-1 (64*64)	MM_M	INT	0.362	0.287	0.0027	0.365	0.324	0.0269
DWT-2 (64*64)	MM_M	EXT	0.0915	0.0723	0.755	0.105	0.0932	0.713
DWT-3 (512*512)	MM_M	EXT	0.088	0.0695	0.765	0.1	0.089	0.726
EFR	MM_M	INT	0.594	0.472	0.225	0.669	0.479	0.219
MPEG	MM_M	INT	0.706	0.715	0.108	0.682	0.568	0.09

the fact that, in this marginal memory mode, every instruction is loaded from the external memory and thus pipeline stalls are dominant. As the SEQ model assumes sequential operations, it is the most accurate in this mode.

For all the other algorithms, the MAX and the MIN models always respectively overestimates and underestimates the application power consumption. Hence, the proposed models need a restricted knowledge on the processor architecture; but they guaranty to bound the power consumption of a C algorithm with reasonable errors.

Table 3. Power estimation vs measurements

	Measurements	Estimation vs Measure (%)				
Application	$P(W)$	*Asm*	*SEQ*	*MAX*	*MIN*	*DATA*
FIR	4.5	2.3	-38	5.5	-24.3	5.5
FFT	2.65	2.5	-10	28.5	-1	2.87
LMS-1	4.97	3.5	1.4	2.8	2	2.8
LMS-2	5.66	-1.8	-50	6.4	-15.2	6.4
DWT-1	3.75	1.9	-27	4.7	-13.2	4.7
DWT-2	2.55	-0.2	-10	3.4	-4.2	3.4
DWT-3	2.55	-1	-10.4	2.4	-4.7	2.4
EFR	5.07	-2.8	-50	11.1	-24	1.5
MPEG	5.83	0.7	-54	10	-33	-8
Average errors:		1.8	27.8	8.3	-13.5	4.2

The DATA model is the more accurate since it provides a maximum error of 8 % against measurements. After compilation, the estimation can be performed at the assembly level where the maximum error is decreased to 3.5%.

5   Conclusion

The main interest in this work is to propose an accurate and fast estimation of a C program without compilation. This estimation relies on a prediction that includes parallelism capabilities and pipeline stalls, which strongly impact on the power consumption. The method is therefore suitable to complex processors. Whenever the compiled code is too large for the target's internal program memory, the cache miss rate γ is hardly predictable. In this case, consumption maps are proposed, to summarize the variations of the power consumption [2].

Current works include the development of an on-line tool and the extension of the power models library to other processors. Future works will address the prediction of the execution time from the algorithm, to also achieve energy estimation at the C-level.

References

1. M. Valluri, L. John, "Is Compiling for Performance = Compiling for Power?," presented at the 5th Annual Workshop on Interaction between Compilers and Computer Architectures INTERACT-5, Monterey, Mexico (2001)
2. N. Julien, E. Senn, J. Laurent, E. Martin, "Power Consumption Estimation of a C Algorithm: A New Perspective for Software Design", in Proc. of the Sixth Workshop on Languages, Compilers, and Run-Time Systems for Scalable Computers, ACM LCR'02 (2002)
3. D. Brooks, V. Tiwari, M. Martonosi, "Wattch: A Framework for Architectural-Level Power Analysis and Optimizations" in Proc ISCA (2000)
4. W. Ye, N. Vijaykrishnan, M. Kandemir, M.J. Irwin "The Design and Use of SimplePower: A Cycle Accurate Energy Estimation Tool" in Proc. Design Automation Conf. (2000)
5. V. Tiwari, S. Malik, A. Wolfe, "Power analysis of embedded software: a first step towards software power minimization" IEEE Trans. VLSI Systems, vol.2 (1994)
6. B. Klass, D.E. Thomas, H. Schmit, D.F. Nagle, "Modeling Inter-Instruction Energy Effects in a Digital Signal Processor," presented at the Power Driven Microarchitecture Workshop in ISCA (1998)
7. S. Steinke, M. Knauer, L. Wehmeyer, P. Marwedel, "An accurate and Fine Grain Instruction-Level Energy Model Supporting Software Optimizations," in Proc. PATMOS (2001)
8. L. Benini, D. Bruni, M. Chinosi, C. Silvano, V. Zaccaria, R. Zafalon, "A Power Modeling and Estimation Framework for VLIW-based Embedded Systems," in Proc. PATMOS (2001)
9. G. Qu, N. Kawabe, K. Usami, M. Potkonjak, "Function-Level Power Estimation Methodology for Microprocessors," in Proc. Design Automation Conf. (2000)
10. C. H. Gebotys, R. J. Gebotys, "An Empirical Comparison of Algorithmic, Instruction, and Architectural Power Prediction Models for High Performance Embedded DSP Processors," in Proc. ACM Int. Symp. on Low Power Electronics Design (1998)
11. TMS320C6x User's Guide, Texas Instruments Inc. (1999)
12. J. Laurent, E. Senn, N. Julien, E. Martin, "High Level Energy Estimation for DSP Systems," in Proc. Int. Workshop on Power And Timing Modeling, Optimization and Simulation PATMOS (2001)

A Low Overhead Auto-Optimizing Bus Encoding Scheme for Low Power Data Transmission[*]

Claudia Kretzschmar, Robert Siegmund, and Dietmar Müller

Dpt. of Systems and Circuit Design
Chemnitz University of Technology
09126 Chemnitz, Germany
{clkre,rsie}@infotech.tu-chemnitz.de

Abstract. In this paper a new adaptive system bus encoding technique is presented which reduces the power dissipation on a bus by minimizing the weight of the encoded data stream. It is based on dynamically auto-optimizing code tables which are used to map data words unambiguously to code words. The proposed encoding scheme is highly efficient due to its ability to adapt to varying statistical parameters of the data streams to be encoded. Unlike other adaptive techniques presented to date, which are infeasible to implement into hardware due to their tremendous overhead in self dissipated power, the implementation of our encoding technique requires less hardware overhead and does not modify the system bus interface. The fundamentals of the encoding scheme and a hardware-efficient implementation are given. Experimental results showed a reduction in bus transition activity of up to 38%.

1 Introduction

In recent years, the complexity of digital systems has reached up to 20 million gates and standards for electronic devices such as mobile communication systems are changing rapidly. There are two aspects in the focus of the design of portable communication systems: low power dissipation in order to maximize stand-alone operation time and flexibility in functionality in order to be able to adapt to new standards. A growing number of systems are realized on reconfigurable hardware in order to adapt to new functional requirements. A trend can be observed to emerge towards a combination of reconfigurable hardware and ASIC cores such as DSPs or micro-controllers on a single chip. The advantages of a reconfigurable hardware over pure software reconfiguration are obvious: A corresponding system implementation provides the flexibility to be adaptable to new functional requirements while meeting constraints for critical system parameters such as data throughput rates and latencies. A typical example is a wireless network receiver module which must be compliant to the latest telecommunication protocols. However, especially in the domain of mobile applications the minimization of on-chip power dissipation becomes extremely important. The power dissipated on a system bus of a CMOS circuit can be approximated by $P_V = \frac{1}{2}V_{dd}^2 f \sum_{i=0}^{n-1} C_{L_i} \alpha_i$, where n is the bus width, f the clock frequency, V_{dd} the operating voltage, C_{L_i} the capacitance and

---

[*] This work is sponsored by the DFG within the joint project VIVA under MU 1024/5-2.

B. Hochet et al. (Eds.): PATMOS 2002, LNCS 2451, pp. 342–352, 2002.
© Springer-Verlag Berlin Heidelberg 2002

α_i the transition activity of bus line i, respectively. Capacitances of bus lines exceed module internal capacitances by some orders of magnitude, therefore up to 80 % of the total power on a chip are dissipated on system buses. Usually, at higher levels of design abstraction the designer has no influence on operating voltage, clock frequency and intrinsic capacitances. The only parameter that can be optimized is the transition activity.

In this paper a new adaptive method for system bus encoding in order to efficiently minimize bus line transition activity as well as total power dissipation is presented. The method is based on dynamic reconfiguration of transition minimizing code tables at *system runtime* with a code that is periodically computed from the current statistical parameters of the data stream. We refer to it as *Adaptive Minimum Weight Codes (AMWC)*. The development of the encoding scheme has been targeted to a low hardware overhead implementing the coder-decoder system. For AMWC, no extra bus lines are required as opposed to redundancy increasing codes, such as Businvert, and data is transmitted over the bus each cycle (e.g. we do not exploit spatial redundancy) with a delay of one clock cycle. In contrast to all published static encoding schemes the adaptability of our encoding method to changing characteristics of the transmitted data stream eliminates the necessity of a priori knowledge of its statistical parameters for selecting an appropriate encoding rule. Therefore our method is especially suited for system buses that transport data streams with unknown or strongly varying distribution of transition activity. For such data streams our method yielded a reduction in transition activity of up to 38 %.

The paper is structured as follows: Section 2 gives an overview of related work and the motivation of this work. The theory of AMWC encoding is detailed in Sect. 3. An implementation of a corresponding coder-decoder system is given in Sect. 3.2. Experimental results are shown in Sect. 4. Section 5 summarizes the paper.

2   Related Work and Motivation

The data stream encoding method presented in this paper is based on a technique named *probability based mapping* (PBM), which is described in [1]. PBM uses the probability distribution function (pdf) of data words in order to determine an unambiguous code with the property that more frequently occuring data words are mapped on code words with a lower number of ones, so that the average number of ones in the coded data stream is minimized. A decorrelator at the coder output translates a one in the code word in a transition while a zero is translated in a no-transition, so that PBM effectively reduces the number of transitions on the bus. For n bit wide buses, PBM requires a code table of size $2^n \times n$.

Another approach is described in [2]. Here mapping rules from data words to code words are based on the joint probability distribution function (jpdf) of two successive data words, therefore taking into account temporal correlation of data words. The encoding scheme yields a higher reduction in switching activity, but the size of the code table for a n bit bus is $2^{2n} \times n$ and becomes prohibitively large for realistic bus widths of e.g. 32 bits. A further technique which reduces the weight of the encoded data stream by introducing redundancy has been published in [3].

The encoding techniques mentioned so far are static encoding schemes in the sense that the code tables are constructed from example data stream pdf's and are statically compiled into the coder-decoder hardware. The pdf is usually determined by system simulation using a characteristic data stream. However, the statistical parameters of real life bus data streams are often non-stationary or even a priori unknown. In this case, system simulations which are restricted to small sections of data streams, yield pdf's which differ from those of the data streams transmitted during operation of the system in real-life. Two pdf's which have been extracted from different data streams are visualized in Figure 1. Our experiments showed that static encoding schemes work

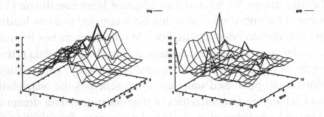

Fig. 1. Probability distribution profiles

efficiently only for data streams they are particularly optimized for, while the encoding efficiency regarding transition reduction is decreased dramatically when data streams with differing statistical parameters are applied, as it may be the case in e.g. an image processing system. We also demonstrated that this misery can be tackled using adaptive encoding techniques such as [4], [2], [5] which do not require the a-priori knowledge of statistical parameters of the data streams to be encoded, but observe them periodically at system run time and modify the encoding rule accordingly.

Adaptive encoding schemes reduce switching activity often very efficiently due to their adaptability to varying statistical parameters. So the method presented in [5] achieves an average reduction of 22 % compared with the uncoded data stream while the majority of adaptive schemes showed a less effective average reduction. However, the efficient reduction is achieved at the cost of tremendous hardware requirements for the coder-decoder system which over-compensates the savings on the bus.

For that reason, we have investigated into the development of a bus activity minimizing encoding technique with a high reduction potential and whose coder-decoder system is optimized for low hardware overhead. The encoding technique presented in this paper is based on the principles of APBM published in [5] which maps data words probability based on code words. But in contrast to that AMWC uses a new method for data stream observation as well as for the computation of a new encoding rule which efficiently minimizes the hardware expense of the encoding scheme. The resulting coder-decoder system can be implemented in a hardware-efficient fashion with a standard cell library as well as in dynamically reconfigurable logic which provide RAM-based look-up tables (LUT) that can be modified during system run time.

3   Adaptive Minimum Weight Codes (AMWC)

3.1   Overview

The Adaptive Minimum Weight Codes (AMWC) scheme is based on mapping data words ϕ unambiguously on code words ψ in order to reduce switching activity in the encoded data stream $\hat{X}$, which is transmitted over the system bus. The block diagram in Figure 2 visualizes the encoding principle. The weight minimizing code assignment Ψ is stored in the code table of the coder. At the decoder the original data stream is reconstructed from the received code words using the inverse code Ψ^{-1}, which is implemented in the code table of the decoder. A decorrelator which is attached to the output of the coder,

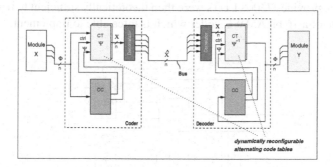

Fig. 2. AMWC Coder-Decoder system using reconfigurable Code Tables

transforms a one in the code word into a transition on the bus and a zero in the code word into an no-transition by XORing the code word with the previous bus state. The decorrelator transfers the problem of transition minimization to the minimization of the average weight of code words in the encoded data stream X. So frequently occuring data words are mapped to code words containing a minimum number of ones while data words with a low probability are mapped to code words with a high number of ones, causing more transitions on the bus. AMWC avoids the introduction of additional bus lines by assigning code words to data words, which have the same length but a different binary representation resulting in a minimum weight of the coded bus X. A transition minimizing code has then the following property:

$$\overline{\alpha}_{\hat{X}} = \sum_{\chi \in X} W(\chi) \cdot p_\phi \rightarrow min \tag{1}$$

On decoder side the transitions and no-transitions are re-transformed into ones and zeros by a correlator. The code words are decoded using the inverse code Ψ^{-1} in the code tables of the decoder.

Because the pdf of the data streams may be a priory unknown or non-stationary, as depicted in Figure 1, AMWC has the ability to adapt the encoding rule in certain intervals to the statistics. For that reason the code computation block (CC) observes the probability of data words by windowing the uncoded data stream. Subsequently the pdf is extracted and a new unambiguous, minimum weight code is computed, which is used to reconfigure the code tables. In order to avoid the necessity of transmitting the new code over the bus, the code computation block is implemented in analogous fashion within coder and decoder.

3.2 Coder Implementation

RAM-based look-up tables (LUT) are suited for the code storage since at most one data word is assigned to the corresponding code word per clock cycle. A LUT for a non-redundant code which is used by AMWC has a size of $2^n \times n$, where n represents the number of bus lines. In order to store the code table of a 32 bit bus, a $17GByte$ memory would be required in both the coder and decoder. A XCV3200E XILINX VIRTEX E FPGA which provides 73008 LUTs were then theoretically sufficient to realize the code table storage of a 16 bit AMWC coder which is infeasible to implement.

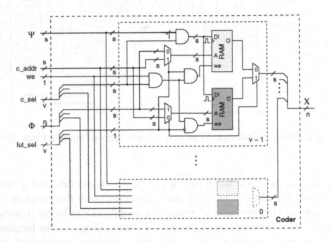

Fig. 3. AMWC Sub Coder implementation

The width of the system bus under consideration does not only influence the complexity of code tables but also the window length and the accuracy of the computed code. In order to extract a pdf that corresponds to the current probability distribution a window should contain at least as many samples as possible data words which results in 4 Giga words for a 32 bit bus. Windows which are shorter take into account temporary variations within the pdf. In contrast to that larger windows will average the pdf over changes in statistics. However, if a code is applied to a data stream segment whose pdf varies from that the code is optimized for, a decrease in encoding efficiency or even an increase regarding the uncoded data stream is achieved.

Therefore, a feasible realization of AMWC requires the splitting of large buses into smaller groups of bus lines and separate encoding of each group accepting the loss of correlation between the sub buses. A corresponding architecture for an AMWC coder is shown in Figure 3. The system bus is splitted into v sub buses of equal width s. Each of the v sub coders implements two alternating code tables in order to ensure error-free decodability during the reconfiguration cycle. One of the $2^s \times s$ wide tables contains the current code Ψ while the code of the second table is optimized for the statistics of the sampling window by the code computation block. CC also controls the selection of the code tables and their update. At the output of the coder all encoded sub buses are joined to the system bus which is transmitted to the decoder after being decorrelated.

The decoder is structured in analog fashion. The original data words are retranslated from the received code words using an inverse code table which is constructed by swapping code Ψ and address "c_addr" in Figure 3.

3.3 Code Computation Block CC

The Code Computation Block (CC) observes statistical parameters of the uncoded data stream and accordingly adapts the encoding rule within certain intervals. Furthermore it provides the control signals for the coder. It is similarly implemented within coder and decoder in order to avoid the necessity of code transmission between coder and decoder. For all sub (de)coders only one CC is implemented, which sequentially updates the CTs. Figure 4 shows the block diagram of CC. It consists of a sampling table of the size

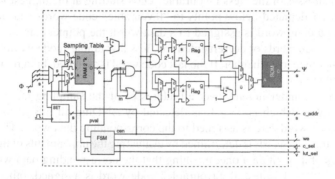

Fig. 4. Code computation block CC

$2^s \times k$ bit, a ROM, which stores the $2^s \times s$ code words in ascending order concerning their weight, two ROM address registers and a finite state machine. The function can be partitioned into two stages: sampling and code computation.

Since CC sequentially updates the code tables of all sub (de)coders, a multiplexer at the input of CC selects the sub bus to be observed during the sampling period. The frequency of each data word is recorded by incrementing the corresponding entry of the sampling table. Usually for data sampling a window of fixed size is used. In contrast to that, our new and innovative sampling approach takes a varying amount of data samples into account. The sampling period continues until one of the counters reaches its maximum counter value of $2^k - 1$, which depends on the depth k of the sampling table. The encoding efficiency especially benefits from that approach if statistical parameters vary over time. The code adaption period is shortened if one or a few data words occur very frequently. If, on the other hand, data words are more or less uniformly distributed, the observation period will enlarge while the previously computed code is still used. The influence on coding efficiency will be negligible as long as the range of code words and data words are equivalent. Since all code words, respectively data words, will then occur nearly uniformly, equation 1 will not achieve a minimum, so that encoding will only slightly influence the switching activity.

During the code computation cycle code words are assigned to data words in order to achieve a minimum weight on the encoded bus X for the current pdf. If the optimum code word should be assigned to each data word the probability values would have to be sorted which consumes time and power. Therefore AMWC uses a fixed mapping rule which is based on dividing code words in two classes according to their weight: "favourable" and "less favourable" ones. The first class contains code words with low weight, leading to a minimum number of transitions on the bus. They should be mapped to frequently occuring data words. The second class covers code words with a high number of ones. All code words are stored within the ROM table in ascending order concerning their weight, so that "favourable" code words are stored in the lower part and "less favourable" ones in the upper part. Since the pdf is varying over time, the two classes cannot be separated at a certain value. For that reason a register for each class is introduced which address the next unmapped code word. The upper register in Figure 4 contains addresses of the "less favourable" class starting at the highest ROM address and the register depicted below points to "favourable" code words starting at address 1. Each time a code word is assigned to a data word, the pointers are incremented for "favourable" code words or decremented for "less favourable" code words, so that both addresses approach to each other. If the last code word is assigned, both registers address the same code word.

An address counter iterates over all addresses of the sampling table. The highest probability value, which is $2^k - 1$, is detected by ANDing all bits of the probability value. The code word zero is assigned to the corresponding data word. The assignment of the rest of the data words is determined by evaluating the upper m bits of the probability value. If these bits contain a one, meaning that the corresponding data word appeared more than $2^{k-m} - 1$ times, a "favourable" code word is assigned, otherwise a "less favourable" one is selected. The new code is available after only one iteration.

At decoder side the original data words are reconstructed under the assumption that the same initial code is used by coder and decoder, the code of the sub (de)coders is updated using the same sequence, the observation periods are synchronized with each other and the transmission is error-free.

3.4   System Bus Segmentation

As already mentioned the system bus has to be splitted into a number of sub buses which are separately encoded within the sub (de)coders. The two reasons are: reduction of the complexity of the code tables and keep the code adaption periods within reasonable limits by using relatively small window sizes.

As all sub (de)coders share a single code computation block, the system bus has to be splitted into v sub buses of equal width s. A sub data word represents a whole address of the corresponding code table of the sub (de)coder. A typical data bus contains 32, 64 or 128 lines, which could be splitted into a number of 16, 8 or 4 bit wide sub buses. A 16 bit wide sub bus would consist of 65536 possible data words. The corresponding code table requires a memory of 131 KByte which is not power-efficiently realizable. In contrast to that for the code assignment of the 256 different data words of a 8 bit wide sub bus a 256 Byte RAM is sufficient. Even less demanding is the splitting into 4 bit wide sub buses. A complete code table can be stored within a 8 Byte RAM. Also the required time for the code table update using the latter 2 variations is moderate: 256 cycles for the 8 bit wide sub bus and only 16 cycles for the 4 bit wide sub bus.

So we investigated in splitting system buses into 8 or 4 bit wide sub buses. Realizing the first alternative for a 32 bit wide bus requires 4×2 alternating code tables with an address space of 8 bit each. The total memory results in 2 KByte RAM. Using the 4 bit wide sub buses the total memory is reduced to 128 Byte RAM composed of 8×2 alternating code tables with a complexity of $2^4 \times 4$ Bit each.

4   Experimental Results

We implemented the proposed AMWC coder-decoder system as synthesizable VHDL model for a bus width of 32 bit and determined the switching activity by simulation runs using the following set of test data streams:

- **art**: A random data stream, generated with Mathematica®, with an over time varying switching activity profile.
- **eps**: A file in Encapsulated Postscript format.
- **gzip**: Example for an executable file.
- **gauss**: White Gaussian noise.

The proposed encoding scheme was realized with 4 (AMWC$_4$) or 8 bit (AMWC$_8$) wide sub (de)coders. The window size of AMWC depends on the depth k of the sampling table and the probability distribution of data words. As soon as one of the counters reaches the maximum value of $2^k - 1$, the observation of the pdf is stopped. For our experiments we varied k from 2 to 7. The variable m, which determines how many bits of the probability values are considered for code assignment, was changed between 1 and $k - 1$. The best results were achieved using k=4, m=2 for AMWC$_4$ and k=4, m=3 for AMWC$_8$. We compared the results with two other adaptive encoding schemes: APBM [5] and APBI [4] implemented for a 32 bit bus and a window size of 32 samples.

Table 1. Relative reduction in switching activity

Seq	α_Φ	$\alpha_{\hat{x}}$			
		AMWC$_{4,k4m2}$	AMWC$_{8,k4m3}$	APBM	APBI
art	160896	38.3 %	32.9 %	40.6 %	8.6 %
eps	221309	19.7 %	23.7 %	21.0 %	10.5 %
gzip	154620	20.0 %	15.6 %	27.4 %	8.8 %
gauss	4086760	-0.8 %	-1.7 %	0.5 %	7.1 %
Avg.		19.3 %	17.8 %	22.4 %	8.8 %

Table 1 shows the relative reduction in transitions using the test data streams. As expected, the most effective reduction in switching activity was achieved for the **art** data stream with a strongly varying activity profile. The results achieved by the two variations of AMWC demonstrate the impact of the code adaption frequency and the considered amount of data word correlation on coding efficiency. Since the width of the sub bus determines the update frequency of the AMWC sub coders, the adaption periods of $AMWC_8$ are prolonged in comparism to $AMWC_4$ while a higher amount of correlation considered for $AMWC_8$. As the results indicate, the rate of code update has a higher influence on coding efficiency than the consideration of higher correlation coefficients by encoding wider sub buses. Only a slight deterioration in coding efficiency regarding APBM is found using $AMWC_4$. The results have been additionally compared with the Adaptive Partial BusInvert (APBI) encoding scheme [4] which is a 1 bit redundant scheme. The schemes proposed in this paper achieve a more efficient reduction in transitions for the most data streams. On average they outperform APBI by factor 2.

The results of the proposed encoding schemes have been compared with PBM as a static scheme. As all static schemes, PBM has to be optimized for the specific data stream in order to achieve an effective reduction in switching activity on the bus. We investigated in a second experiment in the effect, when PBM is optimized for one data stream but used to encode a different one. The results are shown in Table 2. If the statistics is stationary and a priori known, PBM, optimized for the dedicated data stream, reduces transitions very efficiently as confirmed by the figures in bold. But encoding data steams with different statistical parameters, the number of transitions is often increased compared to the uncoded data stream, which leads on average to an increased transition activity. In contrast to that, AMWC is able to adapt to varying parameters which yields to a good overall performance.

Table 2. Variations in transitions: static vs. adaptive schemes

Seq	$\alpha_{\hat{\phi}}$				
	$AMWC_4$	PBM_{art}	PBM_{eps}	PBM_{gzip}	PBM_{gauss}
art	38.3 %	**47.7 %**	-16.5 %	-14.0 %	- 2.3 %
eps	19.7 %	-33.5 %	**21.1 %**	-4.0 %	-24.3 %
gzip	20.0 %	-77.4 %	- 5.1 %	**28.6 %**	-54.3 %
gauss	-0.8 %	-0.1 %	0.5 %	0.3 %	**1.5 %**
Avg.	19.3 %	-15.8 %	0.0 %	2.7 %	-19.9 %

In a third experiment we synthesized the encoding schemes for a XILINX VIRTEX XCV1000. We implemented the 4 bit wide sub buses using the RAM-based look-up tables. For the 8 bit wide sub buses we instantiated the Block RAMs as 8×512 bit dual port RAMs, which are able to accommodate both alternating code tables of a 8 bit sub (de)coder. Table 3 outlines the hardware requirements for coder and decoder. The static encoding scheme consumes the fewest resources while APBM requires nearly ten times as much area. In contrast to that the hardware expense of $AMWC_4$ could be reduced to 44 % of an APBM implementation. The $AMWC_8$ variation using Block RAMs further reduced the logic within CLBs of up to 32 % plus Block RAM compared to APBM. We used the VIRTEX Power estimation worksheet in order to evaluate the power dissipated by coder and decoder. The development of AMWC was primarily focused on a low hardware expense for the implementation of the coder-decoder system. We expected a decreased power dissipation from the reduced hardware expense which was confirmed by the power estimation for $AMWC_4$. In contrast to that the power dissipated by the BlockRAMs which is shown after the plus sign is much higher than expected so that using the VIRTEX technique no effective reduction can be expected. It has to be evaluated if $2^8 \times 8$ RAMs realized in other technologies also dissipate such a high amount of power.

Table 3. Hardware Effort and Power Estimation

Seq	Module	FG	CY	DFF	MEM16	BRAM	P in mW
$AMWC_4$	cod	222	0	126	68	0	7
	dec	225	0	130	68	0	8
$AMWC_8$	cod	147	21	161	0	6	6+41
	dec	150	21	161	0	6	6+41
APBM	cod	663	0	140	96	0	12
	dec	714	31	173	128	0	14
PBM	cod	32	0	32	32	0	4
	dec	32	0	32	32	0	4
APBI	cod	679	9	337	0	0	14
	dec	532	9	307	0	0	12

5   Conclusions

The conducted experiments confirm the high efficiency of AMWC concerning the reduction in switching activity on system buses. Compared to other encoding methods probability based encoding schemes in general yield a higher reduction in switching activity, if a few data words are more probable than others. Due to its adaptability the proposed encoding scheme outperforms the relative reduction in transitions achieved by static encoding schemes if data streams with varying statistics are transmitted. Adaptive encoding schemes usually consume much more hardware and power than static schemes. However, the hardware overhead compared to other adaptive schemes such as APBM or APBI was enormously reduced by the presented implementation of AMWC. Also the power dissipated by $AMWC_4$ could be significantly decreased in comparism to the other adaptive encoding schemes. AMWC is especially suited for data streams with over time varying pdf's, while it works less effective for data streams with uniform pdf's. First measurements of power consumption done for real-life applications mapped into a FPGA confirmed the results of the power estimation for $AMWC_4$.

References

1. Sumant Ramprasad, Naresh R. Shanbhag, and Ibrahim N. Hajj. A Coding Framework For Low-Power Address And Data Busses. *IEEE Transactions on VLSI Systems*, June 1999.

2. L. Benini, A. Macii, E. Macii, M. Poncino, and R. Scarsi. Synthesis of Low-Overhead Interfaces for Power-Efficient Communication over Wide Buses. In *36th Design Automation Conference DAC*, 1999.
3. M. Stan and W. Burleson. Limited-weight Codes for Low-power I/O. In *Int. Work. on Low Power Design*, Apr. 1994.
4. C. Kretzschmar, R. Siegmund, and D. Mueller. Adaptive Bus Encoding Technique for Switching Activity Reduced Data Transfer over Wide System Buses. In *Workshop on Power and Timing Modeling and Optimization PATMOS*, pages 66–75, September 2000.
5. C. Kretzschmar, R. Siegmund, and D. Mueller. Auto-optimizing Bus Encoding for reduced Power Dissipation in Dynamically Reconfigurable Hardware. In *International Conference on Enginieering of Reconfigurable Systems and Algorithms ERSA*, pages 71–77, June 2001.

Measurement of the Switching Activity of CMOS Digital Circuits at the Gate Level

C. Baena, J. Juan-Chico, M.J. Bellido, P. Ruiz de Clavijo, C.J. Jiménez and M. Valencia[1]

Instituto de Microelectrónica de Sevilla-CNM / Universidad de Sevilla
Avda. Reina Mercedes s/n, 41012-Sevilla, SPAIN
Phone: +34-95-505-66-66; Fax: +34-95-505-66-86;baena@imse.cnm.es

Abstract. Accurate estimation of switching activity is very important in digital circuits. In this paper we present a comparison between the evaluation of the switching activity calculated using logic (Verilog) and electrical (HSPICE) simulators. We also study how the variation on the delay model (min, typ, max) and parasitic effects affect the number of transitions in the circuit. Results show a variable and significant overestimation of this measurement using logic simulators even when including postlayout effects. Furthermore, we show the contribution of glitches to the overall switching activity, giving that the treatment of glitches in conventional logic simulators is the main cause of switching activity overestimation.

1 Introduction

Evaluating the switching activity in CMOS digital circuits is a key point to calculate its power consumption [1, 2]. In mixed-signal circuits, switching activity of the digital part creates a switching noise that is transferred to the analog part [3, 4, 5]. Furthermore, as digital circuits become faster and larger, the influence of glitches in the switching activity grows because there are more and more input collisions [6, 7, 8, 9]. Thus, evaluation of switching activity is today a major topic in the design process of both pure digital, and mixed-signal integrated circuits.

Measuring the switching activity in a digital circuit concerns three important questions: The first one is referred to determining the representative input stimuli that must be obtained in order to get an accurate estimation of the switching activity. The second one is concerned to the timing simulator. In timing simulation of digital circuits, standard gate-level logic simulators (like Verilog [10]) are able to handle very large circuits and they are commonly used by circuit designers. Otherwise, accurate evaluation of the switching activity is possible by using electrical simulators (like HSPICE [11]), but these simulators are limited to rather small circuits, they spend lots of computational resources, and they are not used in a typical digital design flow. The third issue focuses on the origin of the logic transitions at the nodes of the circuit. Input changes cause two types of logic transitions: First, proper operation generates

1. This work has been sponsored by MCYT of Spain under Projects TIC2000-1350 and TIC2001/2283

B. Hochet et al. (Eds.): PATMOS 2002, LNCS 2451, pp. 353-362, 2002.
© Springer-Verlag Berlin Heidelberg 2002

functional transitions and second, the generation and propagation of spurious transitory signal pulses (glitches) cause non-functional transitions.

The basic method to estimate the power consumption at logic levels consists in obtaining the final value of it by summing up the power contribution each node has every time it makes a transition. So, it is necessary to calculate the total number of transitions in the circuit besides the use of a power model to estimate the consumption at each node. Tools that use this method obtain an overestimation in the power consumption. In order to correct this result, new power models are proposed in [12,13].

In this communication, we demonstrate that the switching activity can be greatly overestimated when calculated with conventional logic simulators like Verilog. This overestimation is mainly due to an inaccurate propagation and elimination of glitches, which happens regardless the model used among those provided by the foundry, or the inclusion of postlayout information.

This contribution is organized as follows: The method used for switching activity computation applied to ISCAS'85 benchmark circuits is summarized in Section 2. Simulation results are presented and analysed in the section 3. Finally, in section 4, we draw some conclusions on switching activity evaluation.

2   Switching Activity Measurement Procedure

In this section, we describe a method to obtain the switching activity in a circuit. To illustrate the method the ISCAS´85 benchmark circuits are considered [14].

As said above, we compare different measures of switching activity of a circuit using two kinds of simulators, logic and electrical. The procedures are very similar in all of the cases and a scheme of them are presented in Fig 1. We start with the description of the circuit, provided by the ISCAS´85 benchmarks document. This description must be translated to another format suitable for the design environment Design FrameWork II (DFWII) in our case [15].

To do this translation, a software parser has been written using the PERL language [16]. The parser takes the original description of the circuit as supplied with the set of benchmarks, and produces the corresponding Verilog netlist. The parser also needs a simple mapping library which assigns the right cell for the current technology to each logic operator. In our case, circuits are implemented in a CMOS 0.35 μm technology. Once the circuit description is loaded in DFWII, we can generate HSPICE netlists in order to do electrical simulation, or run a Verilog logic simulation.

At this point, we follow three different paths, but before that, we need to study which and how many vectors of test must be applied to get an accurate and realistic evaluation of the switching activity.

Switching activity inside a circuit is highly input-pattern dependent [17], thus, simulation results are directly related to the specific input patterns used. The two main objectives when selecting a set of input patterns are to generate an "average" switching activity and to use a number of patterns that is small enough in order to limit the cost in computational resources. The method described in the following points accomplish both objectives:

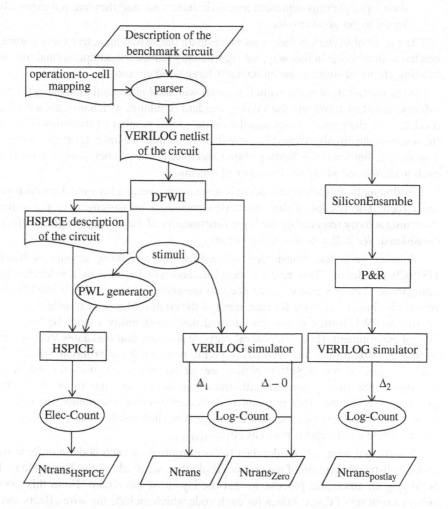

Figure 1 . A scheme of the method for switching activity computation

- First we run a Verilog simulation on 1000 random input patterns and get the number of transitions in the whole circuit. We have checked that for such a number of random patterns, similar switching activity is obtained (within 2%) for any set of patterns, thus, the result is an average measure of the switching activity.
- Several 1000 random vectors simulations are run, and the number of transitions per input vector calculated. The mean value of all measurements is taken as a standard value of the circuit's switching activity.
- Then, we simulate the circuit several times using only 50 random test vectors in order to find a set of input patterns that generated a number of transitions per stimuli within the 3% of the mean value previously determined. Thus, these 50 ran-

dom input patterns represent a generic input case and they are not expressly selected to get good results.

These set of vectors is then used to compare the switching activity using logic and electrical simulators. In this way, we significantly reduce the computational time when running electrical simulations on medium-large sized circuits.

After the selection of the stimuli we start with the logic simulation using the timing information of each cell and the Verilog standard simulator, which uses an inertial delay model. From the results of this simulation the global number of transitions (i.e., all of the nodes in the circuit, Ntrans) is computed. To do that, we have developed a program, Log-Count, that scans the Verilog output and returns the number of logic transitions in each node, as well as the total number of transitions.

Additionally, this same procedure is done using a zero delay model for each gate in the circuit. The results of this simulation provide a measurement of the minimum switching activity required by the logic functionality of the circuit. No glitch effects are considered. We will note this result $Ntrans_{Zero}$.

Another procedure considered to measure the switching activity is based on HSPICE simulation. This result is accurate and will be used as a reference in our comparison. For this purpose, we need to translate the same stimuli used before to piece-wise-linear functions for each input in the circuit. Manually translating the input vectors to PWL format is not feasible and this functionality cannot be found in the design environment. Hence, a general-purpose program that translates Verilog vectors to SPICE PWL format has been developed to generate the appropriate stimuli. These stimuli joined to the HSPICE netlist are all the necessary data for the HSPICE simulator. The files generated with these simulations are the input to a software program, Elec-Count. This program is dedicated to count the number of times each node in the circuit crosses the Vdd/2 voltage. The final result is the switching activity for the whole circuit and is noted as $Ntrans_{HSPICE}$.

Finally, we want to consider the effects of routing on each node in order to have a more realistic description of the circuit. To do that, we used Silicon Ensemble tool and following an automatic process we get the layout of the circuit. From this view we obtain a new set of delay values for each node which include the wire effects and new capacities values. After that, we run another logic simulation using this new information and, following the same procedure we used before for the logic simulation, we obtain the number of logic transitions in each node, as well as the total number of transitions and called it $Ntrans_{postlay}$.

3   Results

In this section, the whole method will be applied to nine of ISCAS'85 benchmark circuits, in order to compare the switching activity obtained with logic simulation including pre and postlayout against the "intrinsic" switching activity (zero delay) and the "accurate" value obtained with HSPICE.

Table 1: ISCAS85 benchmark circuits

	no. of inputs	no. of outputs	total gates
c432	36	7	160
c499	41	32	202
c880	60	26	383
c1355	41	32	546
c1908	33	25	880
c2670	233	140	1193
c3540	50	22	1669
c5315	178	123	2307
c6288	32	32	2416
c7552	207	108	3512

In Table 1, we list the circuits selected and their complexity. For each one, the number of transitions for the simulation of 50 test vectors using Verilog considering inertial delay model (minimum, typical and maximum values) for each gate ($Ntrans_{min}$, $Ntrans_{typ}$, $Ntrans_{max}$), and Zero delay model ($Ntrans_{Zero}$) is shown in Table 2. The result of the simulation under the same conditions using HSPICE ($Ntrans_{HSPICE}$) is also included. As can be seen in the table, if we compare minimum and maximum delays to the typical, the differences in the total number of transitions are less than 1%. Furthermore, in some cases, the number of transitions using minimum value for the delay is bigger than the result using maximum delays but in other cases, it happens the opposite.

In Table 3, we represent the relative error between $Ntrans_{typ}$ and $Ntrans_{HSPICE}$ ($\%err_{\Delta1-Hsp}$) and the relative error between $Ntrans_{Zero}$ and $Ntrans_{HSPICE}$ ($\%err_{Zero-Hsp}$). As can be seen, in each example, the relative error between $Ntrans_{typ}$ and $Ntrans_{HSPICE}$ is very different. It varies between 3% for c880 and 115% for c6288. Another important conclusion can be drawn from Table 3: when we compare $Ntrans_{Zero}$ to $Ntrans_{HSPICE}$, we observe that a high contribution to the switching activity is due to the glitches generated and propagated inside the circuit. In all of the cases, the contribution of the glitches is between 20% and 50%, except for the case of c6288 for which this contribution is even greater than 77%. This is mainly due to the size of this circuit and specially, the high number of levels (123, [13]) the circuit has.

Table 2: Number of transitions using 50 input vectors

Circuit	$\text{Ntrans}_{\text{HSPICE}}$	$\text{Ntrans}_{\text{Zero}}$	$\text{Ntrans}_{\text{min}}$	$\text{Ntrans}_{\text{typ}}$	$\text{Ntrans}_{\text{max}}$
c432	4517	3637	4719	4735	4753
c499	6196	4868	6423	6417	6421
c880	11033	7707	11353	11337	11331
c1355	13960	10420	16318	16190	15990
c1908	25873	18465	32393	32411	32441
c2670	38655	26279	45095	44979	45029
c3540	52303	28799	61044	60920	60376
c5315	79803	48899	100589	100295	100327
c6288	194784	44378	421909	418815	416729
c7552	144535	76315	174682	174292	174062

Table 3: Relative errors with respect to $\text{Ntrans}_{\text{HSPICE}}$

Circuit	$\%\text{err}_{\Delta1\text{-Hsp}}$	$\%\text{err}_{\text{Zero-Hsp}}$
c432	4.8	19.5
c499	3.5	21.4
c880	2.7	30
c1355	16	25.3
c1908	25.2	28.6
c2670	16.3	32
c3540	16.5	45
c5315	25.7	38.7
c6288	115	77.2
c7552	20.6	47.2

Table 4: Number of transitions pre-postlayout

Circuit	Ntrans$_{min}$			Ntrans$_{typ}$			Ntrans$_{max}$		
	pre	post	%	pre	post	%	pre	post	%
c432	4719	4499	4.7	4735	4505	4.8	4753	4511	5.1
c499	6423	5937	7.6	6417	6260	2.4	6421	5977	6.9
c880	11353	11495	-1.2	11337	11493	-0.1	11331	11485	-1.3
c1355	16318	15447	5.3	16190	15383	2.2	15990	15297	4.3
c1908	32393	32976	-1.8	32411	32880	-1.4	32441	32782	-1
c2670	45095	50122	-11.1	44979	50106	-11.4	45029	50072	-11.2
c3540	61044	61465	-0.7	60920	60979	-1	60376	61005	-1
c5315	100589	108587	-7.9	100295	108441	-8.1	100327	108015	-7.7
c6288	421909	365702	13.3	418815	362870	13.3	416729	361019	12.1
c7552	174682	166908	4.4	174292	166806	4.3	174062	166709	4.2

In order to be more realistic doing logic simulation, in the Table 4 we show the number of transitions obtained after considering the parasitic effects in each node using minimum, typical and maximum delay for the gate and we compare them to the results obtained previously, before layout. For the three values of the delay we can say that the relative deviation between pre and postlayout are not very significant, less than 10% except in the case of c2670 (11%) and the case of c6288 (13%). As we said for table 2, in some of the circuits the error is positive and in other examples is negative. After analysing the results presented in tables 2 and 4, we can conclude that the post-layout information does not improves in general the computation of the switching activity when using logic simulators.

In table 5, we present the relative error in the number of transitions after a logic simulation having used postlayout typical delays versus HSPICE simulation[1]. Generic conclusions are similar we did when we compared the results with the prelayout logic simulation (Table 3). The more simple circuits (c432, c499 and c880) have a relative error is really close to the HSPICE value, but in others the difference can reach the 86% as the case of c6288. It is important to notice that there are cases in which the postlayout results are further to the reality. In effect, although c6288 decreases its relative error from 115% (prelayout) to 86% (postlayout), in the case of c2670 the change is from 16.3% (pre) to 29.6% (post) making postlayout worse than prelayout result. Then, postlayout values do not guarantee an accurate measurement of the switching activity.

---

1. Unfortunately, for this technology we haven't had available the necessary data to run postlayout electrical simulation.

Table 5: Relative errors number of transitions postlayout vs HSPICE

Circuit	$\%\text{err}_{\text{postlay-Hsp}}$
c432	-0.2
c499	1
c880	4.2
c1355	10.2
c1908	27.1
c2670	29.6
c3540	16.6
c5315	35.9
c6288	86.3
c7552	15.4

The results obtained in the different tables make us conclude that Verilog simulation is not an appropriate way to measure the switching activity in a circuit. Specially for two main reasons, the first one because the relative error can be very high in some cases, in the case of the circuit c6288, the result is not valid at all; and the second idea to emphasise is the great variation in the percentage among the different examples that makes the results for the switching activity not reliable in comparison to HSPICE.

From these results, it can be concluded that the deviation in the power consumption estimation of a circuit obtained from logic simulators is derived from the overestimation in the switching activity. So, the way to improve this result in this kind of tools is getting a more accurate switching activity estimation through the use of new delay models with a better treatment of glitch generation and propagation [18].

Finally, in Table 6 we show the approximate of CPU time spent in each simulation. From these results, we can point out, the well known conclusion, that electrical simulators are limited to rather small circuits because their cost is high in computational resources and CPU time. These kind of tools are restricted to critical parts of a digital circuit.

Table 6: Simulation CPU time

	CPU time (s) VERILOG simulation	CPU time (s) HSPICE simulation
c432	6.4	2714
c499	7.2	8087
c880	8.3	15240
c1355	8.3	28411
c1908	9.9	83989
c2670	16.7	260106
c3540	14.8	722935
c5315	24.1	1518849
c6288	34.2	836644
c7552	31.3	4577727

4   Conclusions

Some results of switching activity estimation in digital CMOS circuits when they are measured using standard simulators has been presented. In order to be impartial, benchmark circuits has been selected as circuits under test, and random medium length stimuli have been applied.

Generally, activity due to glitches (i.e., $Ntrans_{Zero}$) has a remarkable contribution (form 19% for c432 to 77% for c6288) to the overall switching activity. Thus, it can be emphasized the great importance of adequately handling the glitch generation and propagation effects by timing simulators.

When the results of standard logic simulation (Verilog) are compared to accurate data (HSPICE) (i.e. Ntrans vs. $Ntrans_{HSPICE}$), it is observed that the overestimation of the Ntrans varies appreciably, i.e. from 3% for c880 to 115% for c6288. That overestimation persists even when minimum and maximum values are used and postlayout effects are taking into account. The greatest variation between min/max is 1% and between pre and postlayout is 13%. Both deviations are much smaller than the average value of the overestimation. Hence, logic simulators are neither precise nor reliable at measuring switching activity. It is due to the fact that they are not accurate at simulating glitch propagation.

References

1. A. Ghosh, S. Devadas, K. Keutzer, and J. White: "Estimation of Average Switching Activity in Combinational and Sequential Circuits". Proc. 29th Design Automation Conference, pp. 253–259. June 1992.

2. J. Monteiro, S. Devadas, and B. Lin: "A Methodology for Efficient Estimation of Switching Activity in Sequential Logic Circuits". Proc. 31th Design Automation Conference, pp. 12–17. Jun. 1994.

3. X. Aragonès, J. L. González and A. Rubio, "Analysis and Solutions for Switching Noise Coupling in Mixed-Signal ICs". Kluwer Academic Publishers, 1999.

4. D.J. Allstot, S-H. Chee and M. Shrivastawa, "Folded Source-Coupled Logic vs. CMOS Static Logic for Low-Noise Mixed-Signal ICs", IEEE Trans. Circuits and Systems I, Vol. 40, pp. 553–563, Sept. 1993.

5. Y. Tsividis, "Mixed Analog-Digital VLSI Design and Technology". McGraw-Hill, 1995.

6. E. Melcher, W. Röthig, M. Dana. "Multiple Input Transitions in CMOS Gates". Microprocessing and Microprogramming 35 (1992) pp. 683–690. North Holland.

7. C. Metra, M. Favalli, B. Riccò. "Glitch power dissipation model". In Proc. PATMOS'95. pp. 175–189

8. M. Eisele, J. Berthold. "Dynamic Gate Delay Modeling for Accurate Estimation of Glitch Power at Logic Level". In Proc. PATMOS'95. pp. 190–201. 1995.

9. M.J. Bellido, J. Juan-Chico, A.J. Acosta, M. Valencia and J.L. Huertas: "Logical modeling of delay degradation effect in static CMOS gates". IEE Proc. Circuits Devices Sist., Vol. 147, Nº2, pp. 107–117. April 2000.

10. "Verilog-XL Reference Manual". Cadence Design Framework II V. 4.43, 1999.

11. "Star-Hspice Manual". Avant! Corporation, June 1999.

12. G. Jochens, L. Kruse and W. Nebel: "Application of toggle-based power estimation to module characterization". In Proc. PATMOS'97. pp. 161–170. 1997.

13. D. Rabe, G. Jochens, L. Kruse and W. Nebel. "Power-simulation of cell based ASICs: accuracy-and performance trade-offs". In Proc. DATE'98. pp. 356–361. 1998.

14. D. Bryan and H. Fujiwara: "A neutral netlist of 10 Combinational Benchmark Circuits and a target Translator in Fortran". IEEE International Symposium on Circuits and Systems, pp. 695–698. June, 1985.

15. Design Framework II. Version 4.4.3. Cadence, 1999.

16. L. Wall, T. Christiansen and R. L. Schwartz, "Programming PERL". O'Reilly, 1996.

17. F. N. Najm: "A Survey of Power Estimation Techniques in VLSI Circuits". IEEE Transactions on VLSI Systems, Vol. 2, num. 4, pp. 446–455. Dec. 1994.

18. J. Juan-Chico, M.J. Bellido, P. Ruiz-de-Clavijo, A.J. Acosta and M. Valencia: "Gate-Level Modeling of the Delay Degradation Effect". Proc. 15th DCIS, pp. 537–542, Nov. 2000.

Low-Power FSMs in FPGA: Encoding Alternatives

G. Sutter[1], E. Todorovich[1], S. Lopez-Buedo[2], and E. Boemo[2]

[1] INCA, Universidad Nacional del Centro, Tandil, Argentina
http://www.exa.unicen.edu.ar/inca/
{gsutter, etodorov}@exa.unicen.edu.ar

[2] Computer Engineering School, Universidad Autónoma de Madrid, Spain
http://www.ii.uam.es
{sergio-lopez.buedo, eduardo.boemo}@ii.uam.es

Abstract. In this paper, the problem of state encoding of FPGA-based synchronous finite state machines (FSMs) for low-power is addressed. Four codification schemes have been studied: First, the usual binary encoding and the One-Hot approach suggested by the FPGA vendor; then, a code that minimizes the output logic; finally, the so-called Two-Hot code strategy. FSMs of the MCNC and PREP benchmark suites have been analyzed. Main results show that binary state encoding fit well with small machines (up to 8 states), meanwhile One-Hot is better for large FSMs (over 16 states). A power saving of up to the 57 % can be achieved selecting the appropriate encoding. An area-power correlation has been observed in spite of the circuit or encoding scheme. Thus, FSMs that make use of fewer resources are good candidates to consume less power.

Keywords: Low-Power, Finite State Machine, FPGA, One-Hot, State Encoding.

1 Introduction

Low-power design is nowadays a central point in the construction of integrated systems. It allows expensive packaging to be avoided, chip reliability to be increased, cooling to be simplified, and the autonomy of batteries be extended (or their weight to be reduced). The dynamic power dissipated in a CMOS circuits can be expressed by the well-known formula:

$$P = \sum_{all\ nodes} c_n f_n V_{DD}^2 \qquad (1)$$

where, c_n is the load capacitance at the output of the node n, f_n the frequency of switching and V_{DD} supply voltage. The dominant source of power dissipation in CMOS circuits is the dynamic power: the energy required in each cycle to charge and discharge each node capacitance. It is also referred as the capacitive power dissipation.

Main idea in the design of low-power FSMs is minimize Hamming distance of the most probable state transitions. However, this solution usually increases the required

B. Hochet et al. (Eds.): PATMOS 2002, LNCS 2451, pp. 363–370, 2002.

logic to decode the next state. Then, a tradeoff between switching reduction and extra capacitance exists. This paper addresses the state encoding problem in LUT based programmable logic, using Xilinx 4K-series FPGAs as technological framework. In Section II, the basic definitions are summarized, and a review of the traditional approaches is presented. In the next section, the characteristics of the benchmark circuits are highlighted. Finally, the main experimental results are summarized.

2 Preliminaries

A finite state machines is defined by a 6-tuple $M = (\Sigma, \sigma, Q, q_0, \delta, \lambda)$, where Σ is a finite set of input symbols, $\sigma \neq \varnothing$ is a finite set of output symbols, $Q \neq \varnothing$ is a finite set of states, $q_0 \in Q$ is the "reset" state, $\delta(q, a) : Q \times \Sigma \to Q$ is the transition function, and $\lambda (q, a) : Q \times \Sigma \to \sigma$ is the output function.

The 6-tuple M can be described by a state transition graph (STG), where nodes represent the states, and directed edges, labeled with the input and output values, describe the transition relation between states. In hardware materializations, each state corresponds to a binary vector stored in the state register. From the current state and input values, the combinational logic computes the next state and the output function. The binary values of the inputs and outputs of the FSM are usually fixed by the particular application, while the state encoding can be defined by the designer.

2.1 Traditional Approaches for State Encoding

The traditional methods used to generate state machines result in highly-encoded states. This type of machines typically has a minimum number of flip-flops but require implementing wide combinatorial functions.

Early research on FSM state encoding intended to minimize area or delay. For example, the NOVA tool implements an optimal two level state encoding [3], while the MUSTANG state assignment system [4] is targeted to multilevel networks. The JEDI tool [5] is a general symbolic encoding program (i.e., for encoding inputs, outputs, and states) targeted for multi-level implementations. This tool is included in the SIS system [6].

2.2 Approaches for Low Power State Encoding

Main works in low-power FSMs compute first the switching activity and transition probabilities [7]. The key idea is the reduction of the average activity by minimizing the bit changes during state transitions. In [8], a probabilistic description of the state machines is used. Then, the state assignment minimizes the Hamming distance between states with high transition probability. To obtain the probabilistic behavior of a general FSM, the STG is modeled as a Markov Chain, and the state algorithm problem is solved using $\lceil \log_2 n \rceil$ bits, where n is the number of states. A spanning tree based state encoding algorithm is implemented in [9]. The most important characteristic is that the representation is not limited to $\lceil \log_2 n \rceil$. The resulting encoding can be ranging from $\lceil \log_2 n \rceil$ to n bits. Other interesting contribution are in [2], [21], [22].

Table 1. Original and state minimized benchmark circuits.

circuits	Original Machine				Minimized Mach.			
	inp	outp	rul	#st	inp	outp	rul	#st
bbara	4	2	60	10	4	2	42	7
bbsse	7	7	56	16	7	7	208	13
bbtas	2	2	24	6	2	2	24	6
beecount	3	4	28	7	3	4	20	4
cse	7	7	91	16	7	7	91	16
dk14	3	5	56	7	3	5	56	7
dk15	3	5	32	4	3	5	32	4
dk16	2	3	108	27	2	3	108	27
dk17	2	3	32	8	2	3	32	8
dk27	1	2	14	7	1	2	14	7
dk512	1	3	30	15	1	3	30	15
donfile	2	1	96	24	2	1	4	1
ex1	9	19	138	20	9	19	233	18
ex2	2	2	72	19	2	2	56	14
ex3	2	2	36	10	2	2	20	5
ex4	6	9	21	14	6	9	21	14
ex5	2	2	32	9	2	2	16	4
ex6	5	8	34	8	5	8	34	8
ex7	2	2	36	10	2	2	16	4
keyb	7	2	170	19	7	2	170	19
kirkman	12	6	370	16	12	6	370	16
lion9	2	1	25	9	2	1	16	4
mark1	5	16	22	15	5	16	180	12
opus	5	6	22	10	5	6	29	9
planet	7	19	115	48	7	19	115	48
prep3	8	8	29	8	8	8	29	8
prep4	8	8	78	16	8	8	78	16

2.3 FPGA State Encoding

The research line described above was targeted to gate arrays or cell-based integrated circuits. FPGA manufacturers and synthesis tools use One-Hot as default state encoding [10], [11]. This assignment allows the designer to create state machine implementations that are more efficient for FPGA architectures in terms of area and logic depth (speed). FPGAs are plenty of registers but the LUTs are limited to few bits wide. One-Hot increases the flip flop usage (one per state) and decreases the width of combinatorial logic. In addition, the Hamming distance of One-Hot encoding is always two in spite of the machine size. It make easy to decode the next state, resulting attractive in large FSMs. However, a better implementation of small machines can be obtained using binary encoding.

4 Experiments

In this paper, each circuit was encoded in four ways: binary, One-Hot, Two-Hot, and a style proposed by JEDI [5], named "out-oriented" in this paper. This last algorithm uses a binary state encoding that minimizes the output logic. Two-Hot reduces flip-flop usage maintaining at the same time easy-decoding characteristic of One-Hot. Binary and "out-oriented" are highly encoded techniques, whereas One-Hot and Two-Hot can be considered sparse encodings.

All the experiments use the MCNC91 benchmark set [12] together with two FSMs extracted from the former PREP consortium [13]. The original MCNC FSMs are defined using the KISS2 format [6]. So, the first step has been to write a KISS format translator into VHDL. It takes the KISS file, infers a Mealy or Moore machine, and finally writes the corresponding code. The program also generates a file containing an entity with the machine, and another with a top-level VHDL code with tri-states buffers in the pads to measure the off-chip current separately.

The benchmark FSMs were first minimized with STAMINA [14]. The number of inputs, outputs, next state rules and states (for both, the original circuit and the minimized one) are presented in Table 1. Then, each description was translated into VHDL. The resulting code was compiled using FPGA Express [15] and Xilinx Foundation tools [16] into a XC4010EPC84-1 FPGA sample. All circuits have been implemented and tested under identical conditions. That is, all the electrical measurements are related to the same FPGA sample, output pins, tool settings, printed circuit board, input vectors, clock frequency, and logic analyzer probes. Random vectors were utilized to stimulate the circuit. At the output, each pad supported the load of the logic analyzer, lower than 3pf [17].

The circuits were measured at 100 Hz, 2MHz, and 4 MHz to extrapolate the static power. All prototypes include a tri-state buffer at the output pads to measure the off-chip power [18]. Other alternatives to measure power are reviewed in [19][20].

5 Experimental Results

Table 2 shows the area, delay and power obtained for each benchmark circuit. Area is expressed in CLBs, but the number FF utilized is also indicated. The delay, expressed in ns, corresponds to the critical path. Finally, the dynamic power is shown in mW/MHz.

Power Saving: Fig. 1 points out the power saving comparison: (a) OH (One-Hot) vs. binary encoding and (b) OH vs. "out-oriented". Positive values indicate power reduction obtained using OH encoding. The x axis represents the number of states for the FSM. The figure can be separated in three zones. For machines with up to eight states, binary encoding must be utilized to reduce power. For machines with more than 16 states always OH is the best choice. Finally, between 8 and 16 states, there is not clear the relation, but "out-oriented" is better than pure binary. On the other hand, TH (Two-Hot) encoding consume more than OH in almost all cases, but it is better than "out-oriented" and Binary for big FSMs.

Circuits	FSM characterist.				Area Bin		Area OH		Area Out-O		Area T-H		Delay (ns)				Power mW/MHz			
	inputs	outputs	rules	states	CLBs	FF	CLBs	FF	CLBs	FF	CLBs	FF	Bin	OH	Out-O	T-H	Bin	OH	Out-O	T-H
bbara	4	2	42	7	11	3	8	7	10	3	15	4	30.0	25.6	29.4	31.2	1.39	1.38	1.87	1.46
bbsse	7	7	208	13	36	4	26	13	27	4	36	5	43.1	36.2	34.6	40.1	4.02	3.37	3.14	3.43
bbtas	2	2	24	6	4	3	4	6	3	3	4	3	16.8	12.7	16.7	15.5	1.08	0.95	0.77	0.97
beecoun	3	4	20	4	7	2	10	4	7	2	12	4	21.1	18.6	16.4	28.9	1.33	1.62	1.33	2.36
cse	7	7	91	16	52	4	42	16	48	4	53	5	54.9	39.1	47.4	47.9	3.73	3.50	2.99	3.83
dk14	3	5	56	7	27	3	26	7	25	3	27	4	34.1	32.5	31.7	37.8	4.15	3.88	4.08	3.92
dk15	3	5	32	4	18	2	20	4	20	2	20	4	29.2	28.2	25.6	32.8	3.32	3.02	3.28	3.85
dk16	2	3	108	27	59	5	31	27	50	5	57	7	52.1	35.0	43.3	44.0	8.09	3.73	6.67	6.64
dk17	2	3	32	8	12	3	10	8	13	3	14	4	24.2	27.8	27.3	24.5	2.30	1.94	2.27	2.28
dk27	1	2	14	7	3	3	4	7	3	3	4	4	12.6	20.2	18.6	18.8	0.88	1.08	0.95	1.36
dk512	1	3	30	15	14	4	10	14	9	4	16	5	20.8	20.4	26.0	23.9	2.46	1.54	1.85	2.48
ex2	2	2	56	14	21	4	17	11	12	4	22	5	31.0	21.3	24.4	27.4	3.60	2.03	1.88	3.23
ex3	2	2	20	5	6	3	8	5	7	3	7	3	19.2	18.1	16.7	13.7	1.38	1.52	1.51	1.44
ex4	6	9	21	14	22	4	15	14	19	4	18	5	31.2	29.4	27.0	27.2	2.51	1.66	2.10	2.11
ex5	2	2	16	4	1	2	5	4	4	2	7	4	8.8	20.1	17.7	25.8	0.55	1.26	0.98	1.39
ex6	5	8	34	8	34	3	28	8	29	3	35	4	40.0	31.4	33.6	47.6	4.25	3.59	3.71	4.86
ex7	2	2	16	4	2	2	5	4	2	2	7	4	10.2	14.5	9.5	18.3	0.62	1.16	0.64	1.49
keyb	7	2	170	19	57	5	42	19	50	5	53	6	58.1	41.7	54.9	62.3	6.55	5.05	4.43	6.02
kirkman	12	6	370	16	45	4	43	16	45	4	57	5	38.3	36.2	38.9	36.6	4.14	4.00	3.73	5.21
lion9	2	1	16	4	2	2	2	4	2	2	5	4	8.8	15.1	8.8	25.5	0.44	0.54	0.43	1.04
mark1	5	16	180	12	19	4	15	12	17	4	17	5	30.2	24.6	24.1	30.5	2.50	1.79	2.11	2.41
opus	5	6	29	9	23	4	15	9	20	4	18	4	31.1	33.0	27.8	28.1	2.95	1.74	2.16	2.45
planet	7	19	115	48	113	6	65	48	106	6	99	10	60.6	41.3	54.3	61.1	14.4	6.23	13.2	11.7
prep3	8	8	29	8	13	3	14	8	12	3	18	4	33.3	26.9	26.5	30.9	1.66	2.04	1.42	1.99
prep4	8	8	78	16	39	4	37	16	35	4	41	5	45.9	31.4	41.5	37.7	5.47	5.29	4.37	4.92

Table 2. Area, Time and Power for the benchmark set.

States-Power relationship: For any state encoding, the power is linearly correlated with the number of states. The coefficient R^2 for the different regression analysis is over 0.85 (Fig. 2). Power is even more correlated ($R^2 \cong 0.87$) respect to $n+i$ (number of states plus number of inputs).

States-Area relationship: In this case, the correlation is similar to the previous analysis, with a $R^2 \cong 0.80$.

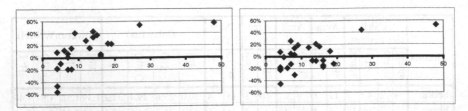

Fig. 1. Power Saving on account of state encoding. a) One-Hot versus Binary. b) One-Hot versus Out-oriented

Time-Power: The relationship is shown graphically in Fig. 4. The linear correlation is $R^2 \cong 0.7$. The experiments do not follow the FPGA rule-of-thumb that indicates that faster circuits consume less power .

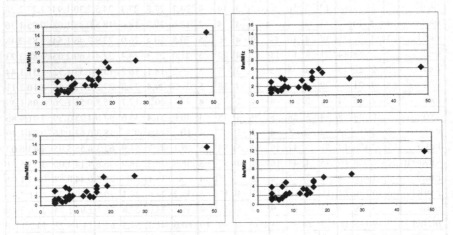

Fig. 2. Power per FSM states: a) Binary; b) One-Hot; c) Out-oriented; and d) Two-Hot.

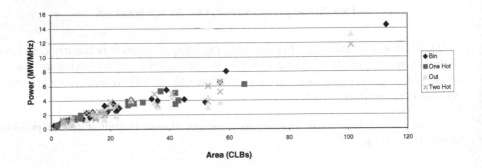

Fig. 3. Area-power relationship

Area-Power: The correlation is important ($R^2 \cong 0.91$) and it can be used as a primary approach to decide for a state assignment. The Fig. 3 represents this distribution. A comparison between area and power shows that the 77% of the benchmark circuits, the smaller circuit consume lower power.

Other correlation like States-Delay are not visible (R^2 lower than 0.6). Area, time and power correlation with the others FSM parameters (inputs, outputs, rules) and combinations of this parameters, neither produce significant results.

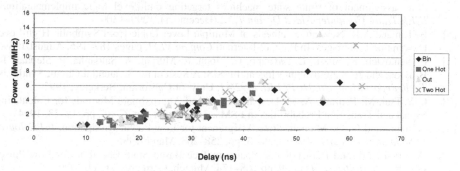

Fig. 4. Delay-Power relationships.

6 Conclusion

This paper has presented an analysis of the state encoding alternatives for FSMs. The main conclusions are that in small state machines (up to 8 states), area, speed and power is minimized using binary state encoding. On the contrary, One-Hot state encoding is better for large machines (over 16 states). A comparison between 26 test circuits shows important differences in power consumption. Depending on the state encoding, reaching up to 57% of power saving can be obtained. The Two-Hot approach do not offer advantages over One-Hot, nevertheless it is better than binary for big FSMs. The Out-oriented is a binary encoding that's minimize the decode logic and its in average better than pure binary. Finally, a clear area-power relationship exists. It can be used to estimate power during the design cycle using the information provided for the synthesis tool.

Acknowledgments. Ministry of Science of Spain, under Contract TIC2001-2688-C03-03, has supported this work. Additional funds have been obtained from Projects 658001 and 658004 of the *Fundación General de la Universidad Autónoma de Madrid.*

References

[1]   S. Lopez-Buedo, J. Garrido and E. Boemo, *Thermal Testing on Reconfigurable Computers, IEEE Design & Test of Computers*, pp.84–90, January-March 2000.

[2]   X. Wu, M. Pedram, and L. Wang, Multi-code state assignment for low power design, *IEEE Proceedings-Circuits, Devices and Systems*, Vol.147, No.5, pp.271–275, Oct. 2000.

[3]   T.Villa, A.Sangiovanni-Vincentelli, "NOVA: State assignment for finite state machines for optimal two-level logic implementation", *IEEE TCAD, Vol.9-9*, p.905, Sept. 1990.

[4]   Devadas, S., Ma, H., Newton, A., and Sangiovanni-Vincentelli, A. 1988. MUSTANG: State assignment of finite state machines targeting multilevel logic implementations. *IEEE Trans. Computer-Aided Design* 7, 12 (December), 1290–1300.

[5]   B. Lin and A.R. Newton. Synthesis of Multiple Level Logic from Symbolic High-Level Description Languages. In Proc. of Internat. *Conf.on VLSI*, pages 187–196, August 1989.

[6]   E. Sentovich, K. Singh, L. Lavagno, C. Moon, R. Murgai, A. Saldanha, P. Stephan, R. Brayton, and A. Sangiovanni-Vincentelli. SIS: A System for Sequential Circuit Synthesis. *Tech. Report Mem. No. UCB/ ERL M92/41*, Univ. of California, Berkeley, 1992.

[7]   C-Y Tsui, M. Pedram, A. M. Despain, Exact and Approximate Methods for Calculating Signal and Transition Probabilities in FSMs, *31st Design Aut. Conf.*, pp. 18–23, 1994.

[8]   L.Benini and G. De Micheli. State Assignment for Low Power Dissipation. *IEEE Journ. of Solid State Circuits*, Vol. 30, No. 3, pp. 258–268, March 1995.

[9]   Winfried Nöth and Reiner Kolla. Spanning Tree Based State Encoding for Low Power Dissipation. *In Proc of Date99*, pp 168–174, Munich, Germany, March 1999.

[10] Xilinx software manual, Synthesis and Simulation Design Guide: Encoding State. Xilinx inc, 2000

[11] FPGA Compiler II / FPGA Express VHDL Reference Manual, Version 1999.05, Synopsys, Inc.,May 1999

[12] Bob Lisanke. "Logic synthesis and optimization benchmarks". *Technical report, MCNC, Research Triangle Park*, North Carolina, December 1988.

[13] PREP Benchmarks (Programmable Electronics Performance Company), see: *http://www.prep.org.*

[14] G.D. Hachtel, J.-K. Rho, F. Somenzi, and R. Jacoby. Exact and Heuristic Algorithms for the Minimization of Incompletely Specified State Machines. *In Proc. of the European Conference on Design Automation*, pages 184–191, Amsterdam, Holland, Feb 1991.

[15] FPGA Express home page. Synopsis, inc.; *http://www.synopsys.com/products/ fpga/fpga_express.htm*

[16] Xilinx Foundation Tools F3.1i, information available at *www.xilinx.com/support/ library.htm*

[17] Tektronix inc., "TLA 700 Series Logic Analyzer User Manual, available at *http://www.tektronix.com.*

[18] E. Todorovich, G. Sutter, N. Acosta, E. Boemo and S. López-Buedo, End-user low-power alternatives at topological and physical levels. Some examples on FPGAs, *Proc. DCIS'2000*, Montpellier, France, November 2000.

[19] J. Alcalde, J. Rius and J. Figueras, Experimental techniques to measure current, power and energy in CMOS integrated circuits, Proc. DCIS'00, Montpellier, France, Nov. 2000.

[20] L. Mengíbar, M. García, D. Martín, and L. Entrena, Experiments in FPGA Characterization for Low-power Design, *Proc. DCIS'99*, Palma de Mallorca, 1999.

[21] Chi-Ying Tsui, Massoud Pedram, Chih-Ang Chen, and Alvin Despain, Low Power State Assignment Targeting Two- and Multi-level Logic Implementations, *Proceedings of ACM/IEEE International Conf. of Computer-Aided Design*, pp. 82–87, November 1994

[22] M. Martínez, M. J. Avedillo, J. M. Quintana, M. Koegst, ST. Rulke, and H. Susse: Low Power State Assignment Algorithm, *Proc. Design of Circuits and Integrated Systems Conf. (DCIS'00)*, pp. 181–187, 2000.

Synthetic Generation of Events for Address-Event-Representation Communications

Alejandro Linares-Barranco[1], Gabriel Jiménez[1], Antón Civit[1], and
Bernabé Linares-Barranco[2].

[1]Arquitectura y Tecnología de Computadores. Universidad de Sevilla.
Av. Reina Mercedes s/n, 41012-Sevilla, SPAIN
{alinares, gaji, civit}@atc.us.es
http://icaro.eii.us.es
[2]Instituto de Microelectrónica de Sevilla. CSIC.
Av. Reina Mercedes s/n. Edificio CICA. 41012-Sevilla. SPAIN
bernabe@imse.cnm.es
http://www.imse.cnm.es

Abstract. Address-Event-Representation (AER) is a communications protocol
for transferring images between chips, originally developed for bio-inspired
image processing systems. Such systems may consist of a complicated
hierarchical structure with many chips that transmit images among them in real
time, while performing some processing (for example, convolutions). In
developing AER based systems it is very convenient to have available some
kind of means of generating AER streams from on-computer stored images. In
this paper we present a method for generating AER streams in real time from
images stored in a computer's memory. The method exploits the concept of
linear feedback shift register random number generators. This method has been
tested by software and compared to other possible algorithms for generating
AER streams. It has been found that the proposed method yields a minimum
error with respect to the ideal situation. A hardware platform that exploits this
technique is currently under development.

1 Introduction

Address-Event-Representation (AER) was proposed in 1991 by Sivilotti [1] for
transferring the state of an array of analog time dependent values from one chip to
another. It uses mixed analog and digital principles and exploits pulse density
modulation for coding information. Fig. 1 explains the principle behind the AER
basics. The Emitter chip contains an array of cells (like, for example, a camera or
artificial retina chip) where each pixel shows a continuously varying time dependent
state that changes with a slow time constant (in the order of *ms*). Each cell or pixel
includes a local oscillator (VCO) that generates digital pulses of minimum width (a
few nano-secconds). The density of pulses is proportional to the state of the pixel (or
pixel intensity). Each time a pixel generates a pulse (which is called "Event"), it
communicates to the array periphery and a digital word representing a code or address
for that pixel is placed on the external inter-chip digital bus (the AER bus). Additional
handshaking lines (Acknowledge and Request) are also used for completing the
asynchronous communication. The inter-chip AER bus operates at the maximum
possible speed. In the receiver chip the pulses are directed to the pixels whose code or
address was on the bus. This way, pixels with the same code or address in the emitter

B. Hochet et al. (Eds.): PATMOS 2002, LNCS 2451, pp. 371–379, 2002.
© Springer-Verlag Berlin Heidelberg 2002

and receiver chips will "see" the same pulse stream. The receiver pixel integrates the pulses and reconstructs the original low frequency continuous-time waveform. Pixels that are more active access the bus more frequently than those less active.

Fig. 1. Illustration of AER inter-chip communication scheme

Transmitting the pixel addresses allows performing extra operations on the images while they travel from one chip to another. For example, inserting properly coded EEPROMs allows shifting and rotation of images. Also, the image transmitted by one chip can be received by many receiver chips in parallel, by properly handling the asynchronous communication protocol. The peculiar nature of the AER protocol also allows for very efficient convolution operations within a receiver chip [2].

There is a growing community of AER protocol users for bio-inspired applications in vision and audition systems, as demonstrated by the success in the last years of the AER group at the Neuromorphic Engineering Workshop series [3]. The goal of this community is to build large multi-chip and multi-layer hierarchically structured systems capable of performing complicated array data processing in real time. The success of such systems will strongly depend on the availability of robust and efficient development and debugging AER-tools. One such tool is a computer interface that allows not only reading an AER stream into a computer and displaying it on its screen in real-time, but also the opposite: from images available in the computer's memory, generate a synthetic AER stream in a similar manner as would do a dedicated VLSI AER emitter chip [4-6].

2 Synthetic AER Stream Generation

One can think of many software algorithms that would transform a bitmap image into an AER stream of pixel addresses. At the end, the frequency of appearance of the address of a given pixel must be proportional to the intensity of that pixel. Note that the precise location of the address pulses is not critical. The pulses can be slightly shifted from their nominal positions because the AER receivers will integrate them to recover the original pixel waveform.

Whatever algorithm is used, it will generate a vector of addresses that will be sent to an AER bus that connects to the input of an AER receiver chip. Let us call this vector of addresses the "period". If we have an image of NxM pixels and each pixel can have up to k grey levels, one possibility would be to place each pixel address in the "period" as many times as the value of its intensity (from 0 to k). In the worst case (all pixels with value k), the "period" would be filled with $NxMxk$ addresses. Each algorithm would implement a particular way of distributing these addresses within the "period". Let us consider 3 different algorithms. The "uniform" method, the "scan" method, and the "random-distribution" method which we propose in this paper.

The "uniform" method: in this method the image is scanned pixel by pixel one time. For each pixel, its intensity $x_{ij} \in [0, k]$ is read and x_{ij} pulses are distributed over the "period" at equal distances. As the "period" is getting filled the algorithm may want to place addresses in slots that are already occupied. In this case, it will put the pulse in the nearest empty slot of the "period". This method will make more mistakes at the end of the process and the execution time grows considerably at the end.

The "scan" method: in this method the image is scanned many times. For each scan, every time a nonzero pixel is reached its address is put on the "period" in the first available slot, and the pixel value is decremented by one. This method is very fast, because it does not need to look for empty slots, although the image needs to be scanned many times (k times in the worst case). However, with this method, the pixels with highest values will appear very frequently at the end of the "period".

The "random distribution" method: this method is similar to the uniform method, but instead of having the algorithm trying to place the addresses in equally distant slots, it will take the slot number from a random number generator. The generator is based on Linear Feedback Shift Registers. Consequently, each slot number is generated only once. If a pixel in the image has a value p, then the method will take p values from the random number generator and place the pixel address in the corresponding p slots of the "period". They will not be perfectly equidistant, but in average they will be reasonably well spaced. This method is fast, because the image is swept only once, and because the algorithm does not need to perform searches for empty slots. Next Sections explains in more details the implementation issues for this method.

3   Random Distribution Method

This method is an implementation of Linear Feedback Shift Register (LFSR) based random number generators [7]. Linear feedback shift register random number generators are based on a linear recurrence of the form:

$$x_n = (a_1 x_{n-1} + \ldots + a_k x_{n-k}) \bmod 2$$

where $k>1$ is the order of the recurrence, $a_k=1$, and $a_j \in \{0,1\}$ for each j. This recurrence is always purely periodic and the period length of its longest cycle is 2^k-1 if and only if its characteristic polynomial

$$P(z) = -\sum_{i=0}^{k} a_i z^{k-i}$$

is a primitive polynomial over the Galois field with 2 elements.

With these premises and limiting the length of the "period" to the maximum number of address events necessary to transmit an image, we know the number of bits needed for the LFSR and the primitive polynomial.

In the software program shown bellow, a maximum of 22 bits are used in the LFSR, although not all random numbers are always used. Only those numbers that are below a limit are used to generate the AER addresses. This limit is different for each image and corresponds to the following formulation:

Let us suppose that **Im** is a matrix that represents the image, that **N** is the number of rows and **M** the number of columns, and that **l** is the length of the "period". Then

$$l = \sum_{i=0}^{N-1} \sum_{j=0}^{M-1} \text{Im}(i, j)$$

The characteristics polynomial **P(z)** used for 22 bits is:

$$P(z) = z^{20} + z^{19} + 1$$

which corresponds to the LFSR of Fig. 2, where bits 21, 16, 11 and 6 are set to 1 as a seed value. The two most significant bits are obtained with a counter. This way, the generated random numbers are distributed along quarters of the "period". Consequently, strictly speaking, it is a pseudo-random method.

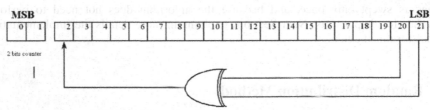

Fig. 2. Left Shift Register for random distribution with a counter for ensurement of the separation

4   Software Simulation Results

Using pseudo-randomization to distribute the address events in time, is at the end, a model of the behaviour present in an AER emitter chip with an array of VCOs. Let us check the error introduced by this model. In an ideal AER stream all events for one pixel and image would be equidistant in time. We will now evaluate how much a synthetically generated AER stream deviates from the ideal stream. The three methods in Section II will be compared.

Lets suppose d_{ij} the ideal distance between events of a pixel.

$$d_{i,j} = \frac{n^\circ\, eventos}{P(i,j)} = \frac{\sum\limits_{a}^{128}\sum\limits_{b}^{128} P(a,b)}{P(i,j)}$$

Then we can measure the error transmission for a pixel as the difference between the ideal distance and the average of real distances of events of a pixel.

$$e_{i,j} = d_{i,j} - \frac{\sum\limits_{k=1}^{V_{i,j}-1}\left(p_{k+1}(i,j) - p_k(i,j)\right)}{V_{i,j}}$$

Finally, we can define a matrix with same dimensions of the image, where for each pixel position it represents the error normalized respect to the ideal distance:

$$NE = \begin{pmatrix} e_{1,1}\big/d_{1,1} & e_{1,2}\big/d_{1,2} & \cdots & e_{1,128}\big/d_{1,128} \\ e_{2,1}\big/d_{2,1} & e_{2,2}\big/d_{2,2} & \cdots & e_{2,128}\big/d_{2,128} \\ \vdots & \vdots & \ddots & \vdots \\ e_{128,1}\big/d_{128,1} & e_{128,2}\big/d_{128,2} & \cdots & e_{128,128}\big/d_{128,128} \end{pmatrix} => ne_{i,j} = e_{i,j}\big/d_{i,j}$$

Consider the example image in Figure 3. First, let us compute for each pixel what would be the ideal spacing of its address in the "period". The result is shown in Figure 4(a), normalized to '1'. For each of the synthetic AER generation methods ("uniform", "scan", and "random"), we generate by software the corresponding "period" and compute for each pixel address the average spacing of events. Then we obtain the difference with respect to the ideal case of Figure 4(a). The resulting relative error is shown in figures 4(b)-(d) for the methods "scan", "uniform", and "random", respectively. As can be seen, the "random" method proposed in this paper yields a significant improvement over the other methods. To our knowledge, no other synthetic methods have been reported so far.

Fig. 3. Reference image to convert into AER format via software.

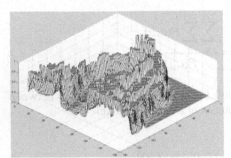

Fig. 4a. Pixels Event Distances for ideal AER stream (normalized to 1).

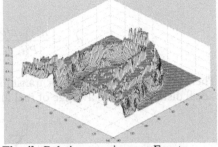

Fig. 4b. Relative error in mean Event Distances using the "scan" method, respect to ideal distribution. Max error of 0,8146.

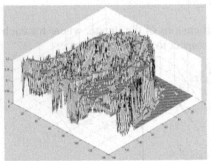

Fig. 4c. Relative error in the mean Event Distances using the "uniform" method, respect to ideal distribution. Max error of 0,1874

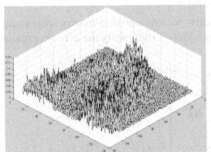

Fig. 4d. Relative error in the mean Event Distances using the "random" method, respect to ideal distribution. Max error of 0,1282

5   Computer-AER Interface

A complete computer-AER interface that exploits the above principle is currently under development. It has two components. One is the software that runs on a standard PC system and manages the conversion of the input image to AER format by generating the corresponding "period". This "period" is then fed to a hardware interface that connects physically to AER chips. The input image for the PC can be a file, a TV signal or a WebCam signal. Let us explain in more detail the hardware and the software interfaces.

5.1   Hardware Interface

The hardware printed circuit board whose block diagram is shown in Figure 5 is currently under development. It can read AER streams into the computer as well as generate AER streams from the computer to feed AER receiver chips. It interfaces the computer through its PCI bus. The PCI Switch Core is the interface between the board and the PC, via the PCI bus. The SR Register is a status register, which can be read under a read operation of its PCI address, and the SR Read Control Process will manage the processes.

The interface supports write operations into the AER bus, and read operations from it. For a write operation the OFIFO memory saves an amount of Address Events from the PCI bus and takes control of the AER bus protocol signals. For read operations a process is always waiting for events in the AER input bus and puts them into the IFIFO. When the IFIFO is filled, an interrupt is generated in the PCI bus and the PCI bus reads the IFIFO. A read operation from an AER bus has to have priority over the write operation because data may be lost. This is assured by using interruptions for reading the IFIFO.

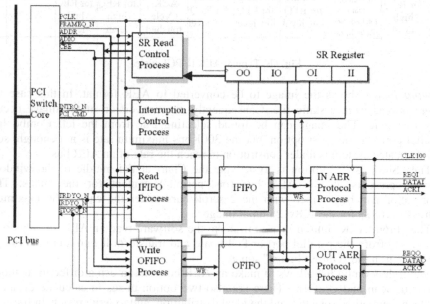

Fig. 5. Block diagram of the PCI-AER hardware bridge.

When AER data has to be written into the AER output bus, the REQ signal has to be activated at the same time the address is placed on the AER bus, and it has to be active until the ACK signal is activated by the receiver chip. Then both the REQ and Address will be cleared, and the bus will be ready for a new address event cycle.

When AER data is waiting in the AER input bus the process will be waiting until REQ is activated. Then the address is given to the IFIFO and the process activates the ACK signal and waits for the REQ low edge to deactivate the ACK signal. Figure 6 shows a typical AER cycle with these 8 clock steps.

5.2 Software Interface

The software interface is shown in Figure 7.

1	2	3	4	5	6	7	8	
Init Read RAM operation	Wait low level ACK signal	Write in AER bus RAM data. Set high REQ	Wait ACK from Receiver	Wait ACK from Receiver	Wait ACK from Receiver	Arrives high level ACK	Set low REQ. Cycle finished	Init Read RAM operation

Fig. 6a. Typical PCI to AER cycle

1	2	3	4	5	6	7	8	
Waiting for REQ high level	REQ detected. Data latched. Set ACK high	Waiting for REQ low level	Waiting for REQ low level	Waiting for REQ low level	Set low ACK. Cycle finished	Waiting for REQ high level	Waiting for REQ high level	Waiting for REQ high level

Fig. 6b. Typical AER to PCI cycle.

Input Image shows the image to be converted to AER format. In this case the image is read from a static file. Consequently, the conversion of the image occurs only one time. The image can be found selecting File from the left Combo list. Parallel port was our first option, but the 300Kbps maximun rate is not enough, so a PCI dedicated system is under constrution to reach the rate of an AER bus.

The "Show AER" option is to activate the Chart at the middle of the window, which represents all the AER bus (or "period") content for the image. This information can be zoomed with the 2 scroll bars under the chart. The maximum address is 16383 for the 128x128 input image.

The "From AER" button is used to make the software read an AER stream stored in memory from the parallel port or PCI board and transform it into a bitmap image, and place it in the "Output Image" square.

The right Combo list allows to make an AER conversion using different methods: Random, Scan and Uniform. There are also two option boxes to calculate distances between the method selected and the ideal distribution, and to write results in files.

When the "To AER" button is clicked the "Input Image" is transformed into an AER stream in the temporal "period", using the method selected. The "period" content is then sent through the hardware board to the AER bus, writing all the events from position 0 to end of the "period". When a period position has no information, a pause is reached for that absent address.

6 Conclusions

A windows based application has been developed to convert bitmaps to AER format using pseudo-random number generators to distribute equal events over time, and to compare this distribution with the scan and uniform methods. A dedicated hardware is proposed to write and read to/from an AER bus. An FPGA and RAM memories based implementation is currently under development.

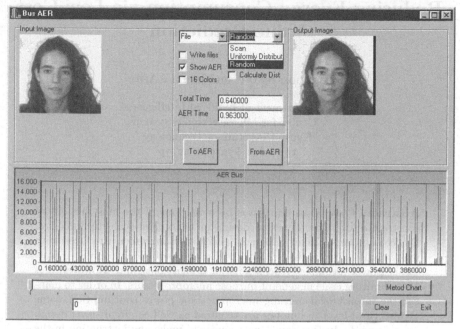

Fig. 7. Software interface.

References

[1]   M. Sivilotti, *Wiring Considerations in analog VLSI Systems with Application to Field-Programmable Networks*, Ph.D. Thesis, California Institute of Technology, Pasadena CA, 1991.

[2]   Teresa Serrano-Gotarredona, Andreas G. Andreou, Bernabé Linares-Barranco. "AER Image Filtering Architecture for Vision-Processing Systems". IEEE Transactions on Circuits and Systems. Fundamental Theory and Applications, Vol. 46, N0. 9, September 1999.

[3]   A. Cohen, R. Douglas, C. Koch, T. Sejnowski, S. Shamma, T. Horiuchi, and G. Indiveri, *Report to the National Science Foundation: Workshop on Neuromorphic Engineering*, Telluride, Colorado, USA, June-July 2001. [www.ini.unizh.ch/telluride]

[4]   Kwabena A. Boahen. "Communicating Neuronal Ensembles between Neuromorphic Chips". Neuromorphic Systems. Kluwer Academic Publishers, Boston 1998.

[5]   Charles M. Higgins and Christof Koch. "Multi-Chip Neuromorphic Motion Processing". January 1999.

[6]   VLSI Analogs of Neuronal Visual Processing: A Synthesis of Form and Function. Thesis by Misha Mahowald. California Institute of Technology Pasadena, California 1992.

[7]   Pierre L'Ecuyer, François Panneton. "A New Class of Linear Feedback Shift Register Generators". Proceedings of the 2000 Winter Simulation Conference.

Reducing Energy Consumption via Low-Cost Value Prediction

Toshinori Sato[1,2] and Itsujiro Arita[1]

[1] Department of Artificial Intelligence
[2] Center for Microelectronic Systems
Kyushu Institute of Technology
tsato@ai.kyutech.ac.jp

Abstract. Power consumption is becoming one of the most important constraints for microprocessor design in nanometer-scale technologies. Device engineers, circuit designers, and system architects are faced with many challenges. In the area of mobile and embedded computer platforms, power has already been a major design constraint. However, it is also a limiting issue in general-purpose microprocessors. In order to manage the impact of increasing microprocessor power consumption, some architectural-level techniques are required as well as circuit-level design improvements. In this paper, we propose to make any instruction in the program execution flow non-critical by using a low-cost value predictor in order to improve energy efficiency. Based on simulations, we find that up to 11.4% of energy reduction in functional units can be attained by utilizing value prediction.

1 Introduction

The increasing popularity of portable and mobile computer platforms such as laptop PCs and smart cell phones is a driving force in investigation of high-performance and power-efficient microprocessors. For example, Java-2 MicroEdition (J2ME) works on cell phones [9], allowing users to download several applications such as 3D-animated games and play them on their cell phones. Guides for travellers and flight ticket reservations are also available, and mobile banking and trading are provided. Therefore, it is required that embedded processors have high performance, and their designers have begun to include features that are traditionally found in general purpose processors. For example, modern embedded microprocessors support out-of-order execution [10]. As the computing power of microprocessors for mobile devices increases, however, their power consumption also increases. In addition, while power is already a major design constraint in the area of mobile and embedded computer platforms, it has also become a limiting issue in general-purpose microprocessors.

The active power P_{active} and gate delay t_{pd} of a CMOS circuit are given by

$$P_{active} = fC_{load}V_{dd}^2 \qquad (1)$$

B. Hochet et al. (Eds.): PATMOS 2002, LNCS 2451, pp. 380–389, 2002.

$$t_{pd} \propto \frac{V_{dd}}{(V_{dd} - V_{th})^\alpha} \tag{2}$$

where f is the clock frequency, C_{load} the load capacitance, V_{dd} the supply voltage, and V_{th} the threshold voltage of the device. α is a factor dependent upon the carrier velocity saturation and is approximately 1.3–1.5 in advanced MOS-FETs [6]. Based on Eq. (1), it can easily be found that a power-supply reduction is the most effective way to lower power consumption. However, Eq. (2) tells us that reductions in the supply voltage increase gate delay, resulting in a slower clock frequency, and thus diminishing the computing performance of the microprocessor.

In order to mitigate the performance loss, we can exploit information regarding circuit criticality. There are a lot of device-level techniques for the design tradeoffs such as transistor size optimizations [5], multiple supply voltages [20], and multiple [22] and variable [6] threshold approaches. Power reduction without performance loss is achieved by selecting non-critical paths as candidates for low-power design. In other words, performance-oriented design is used only in speed critical paths. The same philosophy can be applied to architecture-level design. Only instructions on critical paths are executed on fast and power-hungry functional units, and non-critical instructions can be executed on slow and power-efficient units [13,15,17]. In this paper, we propose to translate any instructions into non-critical ones by using a low-cost value predictor [16] and to execute the translated non-critical instructions slowly. It is expected that processor performance is not be diminished if the value prediction accuracy is considerably high.

The organization of the rest of this paper is as follows: Section 2 surveys related work. Section 3 explains how to improve energy efficiency by generating non-critical instructions. Section 4 describes our evaluation environment. Section 5 shows simulation results. Finally, Section 6 concludes the paper.

2   Related Work

Value prediction [11] is a speculative technique which executes instructions using predicted data values. It breaks data dependence chains speculatively, resulting in critical path reduction. Many studies have proposed value prediction mechanisms, some of which achieve predictability rates as high as 80% [21]. However, these predictors such as 2-level and hybrid predictors require considerable hardware cost for realizing their high predictabilities. Some studies were performed to reduce hardware cost [2,12,16]. Morancho et al. [12] proposed to reduce hardware cost by classifying instructions based on their value predictability. Easily predictable instructions use simpler predictors, whose hardware cost is low. High-cost predictors are used only for hard-to-predict instructions. Calder et al. [2] proposed filtering instructions based on their level of criticality. Only instructions regarding critical paths are held in the value predictor, resulting in capacity saving. We proposed 0/1-value predictor [16], which generates only values 0 and

1. Since 22.6% of dynamic instructions on average generate the value 0 or 1, processors benefit from the 0/1-value predictor.

The critical path is a chain of dependent instructions, which determines the number of cycles executing the program. And thus, the performance of the processor is limited by the speed at which it executes the instructions along the critical path. If we can identify which instructions are critical, we can accelerate their execution by any means. Critical path prediction [4,19] is such a technique for identifying critical instructions dynamically. Combination of the critical path predictor and the value predictor enables to break critical paths, resulting in boosting processor performance. Fields et al. [4] found that mispredictions can be reduced by up to 500% by using the information regarding critical instructions.

Exploiting information regarding instruction criticality is effective not only for improving processor performance but also for reducing energy consumption [3,13,15,17]. Casmira et al. [3] studied the potential benefit of exploiting instruction criticality, which they call slack, for power reduction. They found there is significant availability of non-critical instructions, but did not mention any practical mechanism that utilize the results for power reduction. Pyreddy et al. [13] use profiled-based heuristics proposed by Tune et al. [19] for identifying critical instructions. From a profile run, each instruction is marked as critical or non-critical. When the program is executed, the critical instructions are executed on fast and power-hungry functional units and the non-critical ones are executed on slow and power-efficient units. Unfortunately, they could attain little power savings and caused significant performance loss. In contrast, Sato et al. [15] and Seng et al. [17] utilized a dynamic mechanism. They proposed to use the critical path predictor to identify non-critical instructions. Seng et al. [17] focused on eliminating hot spots of power density in general-purpose high-performance microprocessors. They reported significant gains in the ratio of performance and power density, but did not mention how energy consumption can be reduced. We evaluated the potential of energy reduction via this criticality-based instruction scheduling [15] because we are interested in embedded processors as well as general-purpose processors, and found that approximately 40% of energy consumed in functional units can be reduced.

3   Energy Reduction

This section explains how to reduce energy consumption by value prediction.

Because every value prediction requires verification whether it is correct or not, the number of instructions that are executed and committed is not reduced. On the contrary, it might increase when mispredictions are considered. Therefore, only utilizing value prediction can not reduce energy consumption. The key idea of energy reduction via value prediction is the verification should not be fast if it is expected that its prediction is highly accurate. In other words, predicted instructions are not on critical paths and thus they can be executed on low frequency functional units. Such units can be operate by low supply volt-

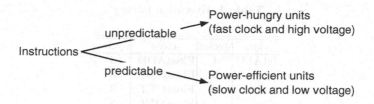

Fig. 1. Instruction execution policy

age, resulting in significant power reduction according to Eq (1). If a prediction is correct, there are no penalty on execution cycles and thus energy consumption is reduced. Otherwise, there are considerable penalty due to long latency of each mispredicted instruction and re-execution of each misspeculated instruction, resulting in degrading energy efficiency. In summary, only unpredictable instructions are executed on fast and power-hungry functional units, and predictable instructions can be executed on the slow and power-efficient units, as shown in Figure 1.

The potential of energy reduction is estimated as follows. We decide that clock frequency and supply voltage for the slow units are half of those for the fast units. We assume that half of instructions are predictable and will be executed on the slow units. Energy consumption is calculated as follows. The number of total instructions are unchanged, and thus the percentage of instructions executed on the slow units determines the energy reduction. When the original energy consumption is defined as E, the fast units consumes $\frac{1}{2}E$, because its power consumption is unchanged and the number of instructions executed there becomes half. On the other hand, the slow units consumes $2 * \frac{1}{2} * \frac{1}{2} * (\frac{1}{2})^2 E$, because the number of instructions executed there, its clock frequency, and its supply voltage are all half while execution time is 2 times increased. Thus, total energy consumption is $\frac{5}{8}E$. In other words, 37.5% of energy reduction is attained.

It should be noted that the value predictor consumes power. If its power consumption is considerably large, the proposal is not effective on energy reduction. Therefore, we should use any low-cost value predictor whose power consumption is significantly small. In this study, we use the 0/1-value predictor [16] as a very low-cost predictor.

4   Methodology

In this section, we describe our evaluation environment by explaining a processor model and benchmark programs.

4.1   Processor Model

We use two types of simulators for this study. One is a functional simulator for evaluating predictability and the other is a timing simulator for evaluating processor performance. The timing simulator models wrong path execu-

Table 1. Execution latency

class	cycles	class	cycles
IntALU	1	FloatADD	3
IntMUL	3	FloatCMP	3
IntDIV	12	FloatCVT	3
		FloatMUL	3
		FloatDIV	12
		FloatSQRT	12

tion caused by misspeculations. We implemented the simulators using the SimpleScalar/PISA tool set (ver.3.0a)[1]. SimpleScalar/PISA instruction set architecture (ISA) is based on MIPS ISA.

The timing simulator models a 2-way out-of-order execution embedded processor, which is primary based on NEC VR5500 [10]. Due to unavailability of the detailed information and to limitation of the tool set, several configurations of our processor model does not completely match those of VR5500. Our model has three integer ALUs, a floating-point unit, and a load-store unit. Branch instructions are executed on the ALUs, while VR5500 has a dedicated branch unit. The latency for execution is summarized in Table 1. We assume that the ALUs dynamically change its execution mode between fast and slow mode. This would be impractical assumption. However, we use this assumption as the first step toward exploiting dual-voltage pipeline. In the fast mode, they can execute most integer operations in one cycle as shown in Table 1, and in the slow mode they execute operations in two cycles. We assume that the ALUs are not pipelined in the slow mode to maintain their throughput in the fast mode. Thus, they diminishes their throughput by half in the slow mode.

A single-port, non-blocking, 32KB, 32B block, 2-way set-associative L1 data cache is used for data supply. It has a load latency of 1 cycle after the data address is calculated and a miss latency of 24 cycles. No L2 cache is used. A memory operation that follows a store whose data address is unknown cannot be executed. A 32KB, 32B block, 2-way set-associative L1 instruction cache is used for instruction supply. For control prediction, a simple branch predictor that has 4K 2-bit saturation counters, and an 8-entry return address stack are used. The branch predictor is updated at the instruction commit stage. A mispredicted branch generates a 6 cycle penalty. Out-of-order execution is based on a register update unit which has 32 entries, while VR5500 uses 20-entry reservation station.

For value prediction, we use the 0/1-value predictor [16]. Figure 2 presents the block diagram of the 0/1-value predictor. Its main structure is the Value History Table (VHT). The VHT is a direct-mapped table and has 1-bit **Data Value** and 2-bit **Conf** fields. As determined by a previous study [16], the 0/1-value predictor does not have tag field because it does not always contribute to predictor performance. The **Data Value** field stores the last result of the associated instruction. When an instruction produces the value 0 or 1, the **Data Value** field simply holds the value. Otherwise, it keeps the value 0. In other

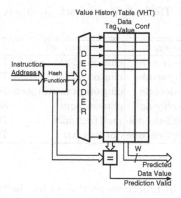

Fig. 2. 0/1-value predictor

words, the 0/1-value predictor has a priority on the value 0. This decision is proper since the value 0 is the most frequently generated value [16]. When the same instruction is encountered, the `Data Value` is used for its predicted value. The `Conf` field is a saturating up-down counter that determines whether or not the instruction should be predicted. The counter is incremented or decremented whenever a prediction is correct or incorrect, respectively. In our study, 2-bit saturating counters are used. When a misspeculation occurs, it is necessary to revert the processor state to a safe point where the speculation is initiated. We use an instruction reissue mechanism which selectively flushes and reissues misspeculated instructions. We have already proposed a practical instruction reissue mechanism [14].

Table 2. Supply voltage and frequency scaling

	clock frequency / supply voltage
TM5800	400MHz/0.93V — 800MHz/1.3V
XScale	500MHz/1.15V — 1.0GHz/1.8V

We will evaluate two scalings for supply voltage and clock frequency; one is based on Transmeta TM5800 [18], and the other is based on Intel XScale [7]. These scalings are summarized in Table 2.

4.2   Benchmark Programs

The MediaBench[8] is used for this study. We use original input files provided by UCLA. Table 3 lists the benchmarks and the input sets. All programs are compiled by the GNU GCC (version 2.6.3) with the optimization options specified by UCLA. Each program is executed to completion. Table 3 also shows the

Table 3. Benchmark programs

program	input set	%cand
g721-decode	clinton.g721	71.7%
g721-encode	clinton.g721	72.1%
gsm-decode	clinton.pcm.run.gsm	74.5%
gsm-encode	clinton.pcm	83.3%
mpeg2-decode	mei16v2.m2v	70.2%
mpeg2-encode	options.par	78.6%

percentage of candidate instructions predicted by the 0/1-value predictor, when the programs are executed on the timing simulator. The candidate instructions for value prediction are register-writing ones, and do not include branch and store instructions. We count only committed instructions.

5   Results

In this section, we present simulation results. First, we evaluate predictability. We define predictability as the number of instructions that are (correctly and incorrectly) predicted by a value predictor over that of all register-writing instructions. We also define prediction coverage as the number of instructions correctly predicted over that of all register-writing instructions. On the other hand, prediction accuracy, which is the percentage of instructions correctly predicted over all predicted instructions, can be easily obtained using the following equation.

$$(Prediction\ accuracy) = \frac{(Prediction\ coverage)}{(Predictability)}$$

After that, processor performance and energy efficiency will be evaluated.

There is a tradeoff between predictability and prediction accuracy. In order to execute instructions on the slow units as much as possible, predictability should be high. On the other hand, in order not to diminish energy efficiency, prediction accuracy also should be high. In general, increasing predictability reduces prediction accuracy, because hard-to-predict instructions must be included. Therefore, we should investigate the tradeoff point carefully. In this study, we examine the threshold value for initiating prediction and the prediction table capacity. The threshold value is varied between 1 and 2, and the capacity is varied between 1K and 32K entries. Figure 3 shows the results. Note that the functional simulator is used in these evaluations. Each bar in Figure 3 is divided into two parts. The lower part (black) indicates the percentage of the instructions whose data value is correctly predicted. The upper part (gray) indicates the percentage that is mispredicted. That is, the lower part is the prediction coverage, while the sum of the two parts is the predictability. For each group of three bars, the first one indicates the result for the case where threshold value is 1. The rest bars are for the cases where threshold values are 2 and 3, respectively. From the results,

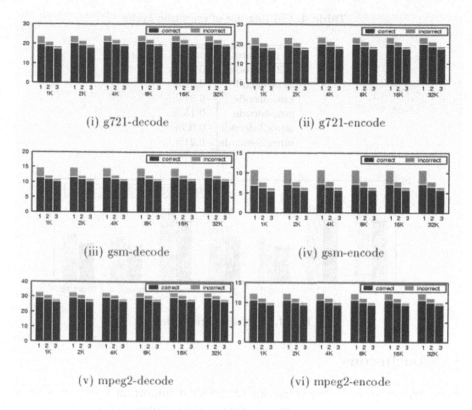

(i) g721-decode   (ii) g721-encode

(iii) gsm-decode   (iv) gsm-encode

(v) mpeg2-decode   (vi) mpeg2-encode

Fig. 3. (%) Value predictability

it can be observed that the threshold value of 2 and the table capacity of 1K entries are in the good tradeoff point. In this configuration, hardware cost of the prediction table is only 384 byte and is less than 0.6% of that of caches. Hence, its power consumption is less considerable and will be ignored in the rest of this paper.

Next, we evaluate processor performance and energy efficiency. Now, we use the timing simulator for the evaluations. Table 4 presents the percent increase of the execution cycles. Because penalty of each misprediction becomes large, value prediction can not contribute to processor performance. However, it is fortunately that the large penalty does not have serious impact on performance. On the whole, processor performance remains the same before and after using value prediction. Figure 4 shows energy reduction of the functional units according to the assumptions described in Table 2. It is observed that energy consumption of the functional units is reduced by up to 9.4% and 11.4% in the cases of TM5800 and XScale models, respectively. These results confirm that energy efficiency can be improved by using value prediction.

Table 4. (%) Increase of execution cycles

program	increase
g721-decode	0.38%
g721-encode	0.32%
gsm-decode	−0.37%
gsm-encode	0.15%
mpeg2-decode	−0.03%
mpeg2-encode	−0.21%

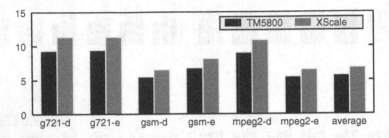

Fig. 4. (%) Energy reduction

6   Conclusions

Power consumption is becoming one of the most important constraints for microprocessor design. In this paper, we proposed to reduce energy consumption by making any instruction in the program execution flow non-critical via the 0/1-value predictor. Predictable instructions are executed on the slow and power-efficient units, resulting in reducing energy consumed in the functional units. From the detailed simulation results, we found that up to 11.4% of energy reduction can be attained by utilizing value prediction.

Acknowledgments. This work is supported in part by a Grant-in-Aid for Scientific Research (B) (#12780273) from the Japan Society for the Promotion of Science.

References

1. Burger D., Austin T.M.: The SimpleScalar tool set, version 2.0. ACM SIGARCH Computer Architecture News, 25(3) (1997)
2. Calder B., Reinman G., Tullsen D.M.: Selective value prediction. 26th Int. Symp. on Computer Architecture (1999)
3. Casmira J., Grunwald D.: Dynamic instruction scheduling slack. Kool Chips Workshop (2000)
4. Fields B.A., Rubin S., Bodik R.: Focusing processor policies via critical-path prediction. 28th Int. Symp. on Computer Architecture (2001)

5. Hashimoto M., Onodera H.: Post-layout transistor sizing for power reduction in cell-based design. IEICE Trans. Fundamentals, E84-A(11) (2001)
6. Hiramoto T., Takamiya M.: Low power and low voltage MOSFETs with variable threshold voltage controlled by back-bias. IEICE Trans. Electronics, E83-C(2) (2000)
7. Intel Corporation: Intel$^{(R)}$ XScale^{TM} technology. http://developer.intel.com/design/intelxscale/ (2002)
8. Lee C., Potkonjak M., Mangione-Smith W.H.: MediaBench: a tool for evaluating and synthesizing multimedia and communications systems. 30th Int. Symp. on Microarchitecture (1997)
9. Levy M: Java to go: part 1: Microprocessor Report, 15(2) (2001)
10. Levy M. NEC processor goes out of order. Microprocessor Report, 15(9) (2001)
11. Lipasti M.H., Wilkerson C.B., Shen J.P.: Value locality and load value prediction. 7th Int. Conf. on Architectural Support for Programming Languages and Operation Systems (1996)
12. Morancho E., Llaberia J.M., Olive A.: Split last-address predictor. 8th Int. Conf. on Parallel Architectures and Compilation Techniques (1998)
13. Pyreddy R., Tyson G.: Evaluating design tradeoffs in dual pipelines. Workshop on Complexity-Effective Design (2001)
14. Sato T.: Evaluating the impact of reissued instructions on data speculative processor performance. Microprocessors and Microsystems, 25(9–10) (2002)
15. Sato T., Koushiro T., Chiyonobu A., Arita I.: Power and performance fitting in nanometer design. 5th Int. Workshop on Innovative Architecture for Future Generation High-Performance Processors and Systems (2002)
16. Sato T., Arita I.: Low-cost value predictors using frequent value locality. 4th Int. Symp. on High Performance Computing, LNCS2327 (2002)
17. Seng J.S., Tune E.S., Tullsen D.M.: Reducing power with dynamic critical path information. 34th Int. Symp. on Microarchitecture (2001)
18. Transmeta corporation: Crusoe^{TM} processor model TM5800. Product Brief (2001)
19. Tune E., Liang D., Tullsen D.M., Calder B.: Dynamic prediction of critical path instructions. 7th Int. Symp. on High Performance Computer Architecture (2001)
20. Usami K., Horowiz M.: Clustered voltage scaling technique for low-power design. Int. Symp. on Low Power Design (1995)
21. Wang K., Franklin M.: Highly accurate data value prediction using hybrid predictors. 30th Int. Symp. on Microarchitecture (1997)
22. Wet L., Chen Z., Johnson M., Roy K.: Design and optimization of low voltage and high performance dual threshold CMOS circuits. Int. Design Automation Conf. (1998)

Dynamic Voltage Scheduling for Real Time Asynchronous Systems

Mohammed Es Salhiene, Laurent Fesquet, and Marc Renaudin

TIMA laboratory – CIS group, 46 Avenue Félix Viallet,
38031 Grenoble France
{Mohammed.Essalhiene, Laurent.Fesquet, Marc.Renaudin}
@imag.fr
http://tima.imag.fr/cis

Abstract. Power consumption is becoming a major issue for embedded systems design. High power consumption reduces battery life and affects system cost and performances. This paper introduces a new power reduction technique that combines an asynchronous processor and a low power operating system (OS). The asynchronous processor is ideal for embedded applications: it is low power and functional within a wide supply voltage range. According to the tasks requirements, the OS regulates the processor operating voltage and so the computational power at run-time. This ensures minimum energy consumption. Simulation results show that low power OS – asynchronous processor combination reduce drastically power consumption in a real-time embedded system.

1 Introduction

Clock speed and integration density have been used as the performance metric of electronic devices for a long time. Designers have mainly focused their efforts on increasing both transistor density and clock frequency. Power consumption was of secondary importance. However, with the proliferation of mobile computing and portable devices (such as cellular phones), reducing power consumption becomes one of the major challenges in electronic systems design [1], [2]. In these applications, there is a trend towards high performance computation and service integration which increase power demand. The future 3G mobile phone will support both voice and MPEG4 video transmission as well as other data intensive applications and will be connected all the time [3].

Reducing power consumption is required to keep autonomy and weight reasonable in battery powered devices, to ensure reliability, and to reduce heat dissipation and system cost [1].

Several hardware and software techniques have been developed over the last years to manage the electrical consumption. Nevertheless, as devices become much more powerful and sophisticated, power requirements increase continuously [4]. Therefore new power management techniques will have to be investigated.

B. Hochet et al. (Eds.): PATMOS 2002, LNCS 2451, pp. 390–399, 2002.

In this paper, we are considering a new method that combines asynchronous processors and a low power operating system. While the literature on power management exposes a lot of research on what to do at hardware or at software level, our approach investigates at both levels. The ASPRO216 is a clockless processor developed at TIMA laboratory. It is low power, low noise and supports wide voltage ranges down to 0.65 VDC. In cooperation with a power management policies that we developed, the OS adjusts the performance level of the processor to the tasks requirements at runtime by controlling the processor operating voltage. This scheme exploits the ability of the asynchronous processor to self-regulate its processing speed with respect to the supply voltage [5].

The rest of the paper is organized as follows: Section 2 gives an overview of the power reduction at operating system level; Section 3 presents the ASPRO architecture; Section 4 proposes a foundation for low power RTOS for asynchronous processors; Dynamic voltage scheduling algorithms and the simulation results of adopted power management strategies are described; and Section 5 concludes the paper.

2 Power Reduction in Operating Systems

For current CMOS integrated circuits, power dissipation is dominated by the switching power that arises from the charging and discharging of the loading capacitance and can be expressed as:

$$P = C V^2 f. \tag{1}$$

where C is the load capacitance, V is the supply voltage and f is the clock frequency [1], [6], [7]. This equation suggests that minimizing the load capacitance, reducing supply voltage or slowing the clock can reduce power consumption.

While the load capacitance can only be affected during chip design (for example by minimizing on chip routing capacitance and reducing external components access [8]), voltage scalable processor and power controllable peripheral devices make possible to reduce power at operating system level. In cooperation with scheduling policies, operating systems can vary processor frequency and voltage (dynamic voltage scaling) and put devices in low power sleep states (dynamic power management) [9].

Indeed a typical embedded system consists of processor, memory, communication links and other peripheral devices which are not always used. Thus an OS can control the power states of peripheral devices in the system according to workload. When a device is idle, it can be put in a low power sleeping state after a long enough idle period to compensate time and energy overhead of shutting down and waking up [2], [10]. A predictive wake-up eliminates wait for the wake-up delay.

To determine whether a device can sleep, time-out, predictive and stochastic policies have been developed [11], [12]. In the time-out based policy, after a device is idle for a time-out value, it remains idle for at least a certain time. Predictive policy uses the past idle periods or both the current and the past idle period to predict the length of future idle periods eliminating the time-out periods wasted energy. While in the stochastic policy, requests and devices are modeled by stochastic processes, such as Markov processes. Power minimization is achieved by resolving a stochastic optimization problem. Choosing a policy for a given application depends on prediction accuracy, power savings and resources requirements such as memory and computation [2], [11], [12].

Shutdown mechanism can be applied to put the processor into idle mode when not in use [2]. However, a more fruitful way to save power is to run slower at reduced voltage according to computational load demands [13]. A technique called dynamic voltage scaling (DVS) allows processors to dynamically alter their voltage and speed at run-time under the control of voltage scheduling algorithms [14]. These algorithms predict future workload and set the processor voltage and speed based on this prediction.

Interval based voltage scheduler called PAST [15] assumes that the processor utilization of the next interval will be like the previous one and updates the processor speed accordingly. If the last interval was busier than idle, speed is increased. Similarly, if it was mostly idle, speed is decreased. A comparative simulation of PAST and other proposed algorithms points out that simple smoothing algorithms can be more effective than sophisticated prediction techniques [16].

Recent studies take the real-time constraints [17], [18] into account. The processor speed is estimated considering workload prediction and tasks deadlines. The goal is to complete tasks before or on deadlines. However, since the processor is often running at reduced speed, all these studies assume missed deadlines as a trade-off between power saved and deadlines missed.

3   ASPRO Overview

We have designed a CMOS standard-cell Quasi-Delay-Insensitive (QDI) 16-bit asynchronous microprocessor using a 0.25 μm technology from STMicroelectronics [19], [20]. ASPRO-216 has been developed for embedded applications. It can be customized both at the hardware and software levels to fit specific application requirements. It is a scalar processor which issues instructions in-order and completes their execution out-of-order. Its architecture extensively uses an overlapping pipelined execution scheme involving de-synchronized units.

The performance of ASPRO-216 is 140 peak MIPS, 0.5 Watt, at 2.5 Volts. It is functional between 0.65V and 2.5V. As an illustration of the wide supply voltage range that can be applied to the processor, Figure 1 gives the performance in MIPS with respect to the supply voltage when the processor is executing a Finite Impulse Response filter routine.

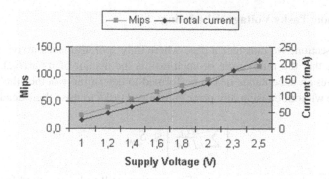

Fig. 1. Mips vs. voltage

4 Dynamic Voltage Scheduling Algorithms

This section presents a foundation for low power RTOS for asynchronous processor. Considered tasks model and dynamic voltage scheduling algorithms are described.

4.1 Tasks Model

A real time system has often to manage periodic and sporadic tasks. While periodic tasks are commonly used to process sensor data and update the current state of the system, sporadic tasks are required to process asynchronous events. However, most of the voltage scheduling schemes presented in the literature consider systems with periodic tasks only. No attention has been dedicated to a system with sporadic tasks except recently in [17]. We consequently consider in this paper both periodic and sporadic tasks. Each task can be characterized by a 3-tuple $<NIi, Di, Ti>$ where NIi is the number of instructions of the task, Di is its deadline and Ti is its period or its minimum inter-arrival time. We assume that:

– tasks are independent and their parameters become known when arriving,

– periodic tasks have deadlines equal to their periods and tasks periods are different,

– at the maximum supply voltage (at highest processor speed), all considered periodic and sporadic tasks can be processed,

– the overhead due to the context switching is negligible.

Since the computation can continue during the voltage switching, we assume that the overhead associated with the voltage scheduling is negligible.

4.2 Sporadic Tasks Voltage Scheduling

In this subsection, we consider a case where only sporadic tasks arrive to the system. We assume that the ready time of each task is the instant of its arrival. When a new task τ_i arrives an acceptance test is performed to determine whether the task can meet its deadline without causing any prior guaranteed tasks to miss their deadlines:

$$\left(\sum_{j \leq i} \frac{\overline{NI}_j}{td_j - t} \right) \leq S_{MAX} \tag{2}$$

where $\overline{NI}_j$ denotes the number of instructions still to be executed for the task τ_j, td_j denotes its deadline, t denotes the current time and S_{MAX} denotes the highest processor speed in Mips at the maximum supply voltage. If τ_i is accepted, it is inserted into a priority task queue according to the Earliest-Deadline-First (EDF) order and the voltage scheduling algorithm updates the processor speed to complete all tasks in the tasks queue before or on their deadlines. This speed is given by:

$$S = \underset{l \in Q}{MAX} \left[\frac{\overline{NI}_l + \sum_{j \neq l / Pl \leq Pj} \overline{NI}_j}{td_l - t} \right] \tag{3}$$

where Q denotes the stream of the sporadic tasks existing in the task queue and P_k denotes its priority according to EDF policy. For each task τ_i in the task queue, including the new task, the voltage scheduler computes the required speed S_i to finish the task τ_i considering all priority tasks. Then, it set the processor speed to the maximum value of S_i. The processor speed is updated whenever a task is added or removed from the task queue. Compared to the voltage scheduling proposed in [17] our algorithm avoids an overestimation of the processing requirements and leads to a higher power saving. This is because our approach takes only runnable and ready tasks into account to compute the operating voltage.

To illustrate the effectiveness of the proposed voltage scheduler, we consider three tasks as shown in figure 2-a with arrival time and deadlines assigned to each task. Speed and power are normalized to their values at the maximum supply voltage: S_{MAX} = 1 Mips and P_{MAX} = 1 mW. So when no power reduction technique is applied, the processor runs at 1 Mips and consumes 1 mW. All the speed and power figures reported are based on real measurements performed with an ASPRO mother board supplied with different voltages.

Using shutdown technique, the processor can be stopped on completion of task $\tau 2$, waked up at task $\tau 3$ arrival time and stopped when this task completes. The consumed power is then 42% P_{MAX}. Since asynchronous processor can instantly be stopped and waked up without any time overhead all tasks meet their deadlines. In synchronous system this technique is ineffective. Because sporadic tasks have random arrival times,

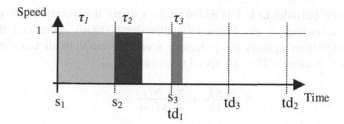

a. Example of sporadic tasks.

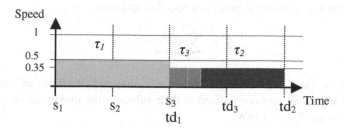

b. Voltage scheduling illustration.

Fig. 2. How sporadic tasks voltage scheduling reduces power consumption.

it is difficult to predict the future idle times. Thus, tasks can miss their deadlines. Furthermore shutting down and waking up synchronous processors cause a time and energy overhead.

Voltage scheduling policy is more effective to reduce power consumption in comparison with shutdown technique. When it is applied, the processor runs at variable voltage and speed as shown in figure 2-b. In this example, the processor runs at 50% S_{MAX} until task $\tau1$ completes. Then the processor speed is reevaluated and set at 35% S_{MAX} for tasks $\tau2$ and $\tau3$. The consumed power is then 14% P_{MAX}.

When no task is running, the processor enters a sleeping state in which it consumes no power.

4.3 Periodic Tasks Voltage Scheduling

In this subsection, we assume that all tasks are periodic (i.e no sporadic tasks arrive to the system). We also assume that at t = 0, a set of n periodic tasks are ready and sorted into a priority task queue according to the Earliest-Deadline-First (EDF) order . The processor speed is set to:

$$S = \sum_{j=1}^{n} \frac{NI_j}{D_j} \tag{4}$$

When a new periodic task τ_i is added to the system, it is inserted into the priority task queue according to the Earliest-Deadline-First (EDF) order and the voltage scheduling algorithm updates the processor speed to complete all tasks in the tasks queue on their deadlines. The new speed is given by:

$$S = \frac{NI_i}{D_i} + \sum_{j=1}^{n} \frac{\overline{NI}_j}{td_j - t} \tag{5}$$

Where $\overline{NI}_j$ denotes the number of instructions still to be executed for the task τ_j, td_j denotes its deadline and t denotes the current time. Similarly, if a periodic task is removed from the system, the processor speed is updated:

$$S = \sum_{j=1}^{n=n-1} \frac{\overline{NI}_j}{td_j - t} \tag{6}$$

Consider the tasks set in table 1 with ready time and deadlines assigned to each task. Speed and power are normalized to their values at the maximum supply voltage: $S_{MAX} = 1$ Mips and $P_{MAX} = 1$ mW.

Table 1. Example Periodic Tasks Set

Tasks	NI_i	D_i	T_i	ready time
t_1	$0{,}25.10^6$	2	2	0
t_2	1.10^6	5	5	0
t_3	$0{,}5.10^6$	3	3	4

In figure 3a, when the tasks are scheduled running at the highest processor speed, 1 Mips, some waiting states exist between the end of a task and the arrival time of the next task. The system then wastes power waiting for the next task. Therefore the processor speed can be lowered by reducing the supply voltage such that the tasks make full use of the CPU time.

In figure 3b, the processor speed is reduced to achieve power reduction. In the beginning, it is set to 33% S_{MAX} according to equation 4. At t = 4, task τ3 becomes ready and the processor speed is updated, according to equation 5, to 49% SMAX. The consumed mean power is then 14% P_{MAX}.

4.4 Periodic and Sporadic Tasks Voltage Scheduling

In this subsection, we are considering a system that deals with both periodic and sporadic tasks. The voltage scheduling algorithm is based on the two previous algorithms. In the beginning, there are n periodic tasks in the task queue and no sporadic tasks. In this context, the processor speed is set to $\sum_{i=1}^{n} \frac{NI_i}{D_i}$.

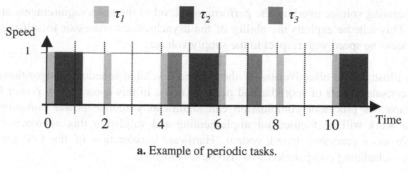

a. Example of periodic tasks.

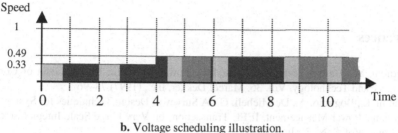

b. Voltage scheduling illustration.

Fig. 3. How periodic tasks voltage scheduling reduces power consumption.

Whenever a sporadic task arrives or completes, the processor speed is updated according to:

$$S = \underset{l \in Q}{MAX} \left[\frac{\overline{NI}_l + \sum\limits_{j \neq l/Pl \leq Pj} \overline{NI}_j}{td_l - t} \right] + \sum_{i=1}^{n} \frac{\overline{NI}_i}{td_i - t} \qquad (7)$$

Where Q denotes the stream of the sporadic tasks existing in the task queue and $\sum\limits_{i=1}^{n} \dfrac{\overline{NI}_i}{td_i - t}$ the fraction of the processor speed required to process the uncompleted periodic tasks.

5 Conclusions and Future Work

Reducing power consumption is becoming a major challenge in electronic systems design. Over the last years, several techniques have been developed to manage the power consumption. Most of them investigate at hardware or software level. In contrast with these researches, we introduce in this paper a new power reduction technique that combines an asynchronous processor and a low power operating system. We have proposed 3 voltage scheduling algorithms that dynamically alters the processor operating voltage to adjust the performance level to the tasks requirements at run-time. This scheme exploits the ability of the asynchronous processor to self-regulate its processing speed with respect to the supply voltage.

To illustrate the effectiveness of the proposed voltage scheduling algorithms, we have considered sets of sporadic and periodic tasks. In this context, low power OS – asynchronous processor combination can significantly reduce power consumption. Future work will be focused on implementing and validating this approach on an asynchronous processor based system. Hardware acceleration of the OS dynamic voltage scheduling computation will also be investigated.

References

1. Pedram, M.: Design Technologies for Low Power VLSI. In: Encyclopedia of Computer Science and Technology, Vol. 36. Marcel Dekker, Inc., (1997) 73–96
2. Benini, L., Bogliolo, A., De Micheli, G.: A Survey of Design Techniques for System-Level Dynamic Power Management. IEEE Transactions on Very Large Scale Integration (VLSI) Systems, Vol. 8, N° 3, (June 2000) 299–316
3. Mudge, T.: Power: A First-Class Architectural Design Constraint. IEEE Computer, (2001) 52–58
4. Gowan, M.K. , Biro, L.L., Jackson, D.B.: Power Considerations in the Design of the Alpha 21264 Microprocessor. 35th Design Automation Conference, San Francisco, CA USA (1998) 726–731
5. Nielsen, L.S., Niessen, C., Sparso, J., Van Berkel, J.: Low Power Operation Using Self-Timed Circuits and Adaptative Scaling of the Supply Voltage. In: IEEE Transaction on Very large Scale Integration (VLSI) Systems, Vol. 2, N°4, (December 1994)
6. Chandrakasan, A.P., Sheng, S., Brodersen, R.W.: Low Power CMOS Digital Design. In: IEEE Journal of Solid-State Circuits, Vol. 27, N° 4, (April 1992) 473–484
7. Burd, T., Brodersen, R.W.: Energy Efficient CMOS Microprocessor Design. Proc. 28th Hawaï Int'l Cof. on System Sciences, Vol. 1, (Jan 1995) 288–297
8. Smit, G.J.M., Havinga, P.J.M.: A survey of energy saving techniques for mobile comput-ers. Moby Dick technical report (1997)
    http://www.cs.utwente.nl/~havinga/papers/energy.ps
9. Simunic, T., Benini, L., Acquaviva, A., De Micheli, G.: Dynamic Voltage Scaling for Portable Systems. 38th Design Automation Conference, Las Vegas (18-22 June 2001) 524–529
10. Lu, Y.H., Benini, L., De Micheli, G.: Operating-System Directed Power Reduction. ISLPED'00, Rapallo, Italy (2000) 37–42
11. Lu, Y.H., De Micheli, G.: Comparing System-Level Power Management Policies. IEEE Design & test of Computers, (2001) 10–19
12. Srivasta, M.B., Chandrakasan, A.P., Brodersen, R.W.: Predictive System Shutdown and Other Architectural Techniques for Energy Effcicient Programmable Computation. IEEE Transactions on very Large Scale Integration (VLSI) Systems, Vol. 4, N°1, (March 1996) 42–55
13. Flautner, K.: Automatic Monitoring for Interactive Performance and Power Reduction. Dissertation, Michigan University (2001)

14. Pering, T., Burd, T., Broderesen, R.: Dynamic Voltage Scaling and the Design of a Low-Power Microprocessor System. In: Power-Driven Microarchitecture Workshop, in conjunction with Intl. Symposium on Computer Architecture, Barcelona, Spain, (June 1998)
15. Weiser, M., Welch, B., Demers, A., Shenker, S.: Scheduling for reduced CPU energy. In: USENIX Symposium on Operating Systems Design and Implementation, (1994) 13–25
16. Govil, K., Chan, E., Wassermann, H.: Comparing algorithms for dynamic speed-setting of a low-power CPU. In: ACM International Conference on Mobile Computing and Networking, (1995) 13–25
17. Pering, T., Burd, T., Brodersen, R.: Voltage Scheduling in the lpARM Microprocessor System. ISLPED '00, Rapallo, Italy (2000) 96–101
18. Kumar, P., Srivastava, M.: Predictive Strategies for Low-Power RTOS Scheduling. IEEE (2000) 343–348
19. Renaudin, M., Vivet, P., Robin, F.: ASPRO : an Asynchronous 16-Bit RISC Microprocessor with DSP Capabilities. ESSCIRC 99, Duisburg, Germany, (21-23 Sept.1999) 28–31
20. Renaudin, M., Vivet, P., Robin, F.: ASPRO-216 : a standard-cell Q.D.I. 16-bit RISC asynchronous microprocessor. Proc. of the Fourth International Symposium on Advanced Research in Asynchronous Circuits and Systems, IEEE (1998) 22–23

Efficient and Fast Current Curve Estimation of CMOS Digital Circuits at the Logic Level*

Paulino Ruiz-de-Clavijo, Jorge Juan, Manuel J. Bellido, Alejandro Millán, and David Guerrero

Instituto de Microelectronica de Sevilla – Centro Nacional de Microelectronica
Av. Reina Mercedes, s/n (Edificio CICA) - 41012 Sevilla (Spain)
Tel.: +34 955056666 – Fax: +34 955056686
http://www.imse.cnm.es
Departamento de Tecnologia Electronica - Universidad de Sevilla
Av. Reina Mercedes, s/n (E. T. S. Ingenieria Informatica) – 41012 Sevilla (Spain)
Tel.: +34 954550974 - Fax: +34 954552764
http://www.dte.us.es
{paulino, jjchico, bellido, amillan, guerre}@imse.cnm.es

Abstract. This contribution presents a method to obtain current estimations at the logic level. This method uses a simple current model and a current curve generation algorithm that is implemented as an attached module to a logic simulator under development called HALOTIS. The implementation is aimed at efficiency and overall estimations, making it suitable to switching noise evaluation and current peaks localisation. Simulation results and comparison to HSPICE confirm the usefulness and efficiency of the approach.

1 Introduction

Switching noise is becoming a major problem in current mixed signal circuits. The noise induced through substrate and power lines coupling by the digital part reduces the performance of the analog part which is built in the same substrate, as is the case of the resolution of A/D converters [1].

From the analog part, design and layout techniques have been employed to reduce the impact of this kind of noise, like guard rings. From the digital part, some work has been devoted to design low-switching-noise digital CMOS circuits [2,3,4] and to develop techniques to evaluate how noisy digital circuits are [5,6]. At the circuit level, supply and substrate currents are taken as a measure of the noise generation [5,6] while at the logic level, it is the switching activity density which has been taken as an adequate noise generation profile [5].

We believe that just the switching activity information is of limited use, since it does not distinguishes between light or heavy loaded gates, making it difficult to spot the most important parts of the noise estimation, like the current peaks.

---

* This work has been partially supported by the MCYT MODEL project TIC 2000-1350 and MCYT VERDI project TIC 2002-2283 of the Spanish Government.

B. Hochet et al. (Eds.): PATMOS 2002, LNCS 2451, pp. 400–408, 2002.

We think however that it is possible to obtain good switching noise profiles at the logic level, provided we use accurate simulation techniques and adequate models.

In this paper we propose a fast algorithm to evaluate the supply current spent by a digital circuit. At this time the motivation is its use as a method to estimate the switching noise generated by the circuit, so we use a very simple current model targeted at efficiency and peak current evaluation. Nevertheless, it is also the basis for other applications like accurate instant and average power estimations at the logic level. The algorithm is integrated as part of the HALOTIS logic timing simulator [7], so it inherits benefits of the event-driven technique (fast simulation) and of the accurate delay models it implements. Regarding the delay model, HALOTIS uses the *Degradation Delay Model* (DDM) which provides with a very accurate and efficient way to handle the generation and propagation of *glitches* [8,9,10,11]. This property is of special interest in this work, since these glitches contributes an important part of the switching activity and hence the average current [12].

In the next section, the current model is presented. Sect. 3 describes the implementation of the algorithm in HALOTIS logic simulator. Sect. 4 is devoted to simulations results, and we will finish by summarising the main conclusions.

2   Transition Based Simple Current Model

In this section we present a simple model to evaluate the current spent in a signal transition. The model is suitable to be implemented in the already mentioned HALOTIS logic timing simulator, or in any other logic timing tool that uses variable slope linear ramps to represent digital signal transitions.

Given a linear signal transition in node N, we evaluate the average current I_0 provided by the source as

$$I_0 = \frac{V_2 - V_1}{t_2 - t_1} C_L \tag{1}$$

where V_1 and V_2 are the initial and final transition voltages, t_1 and t_2 the initial and final transition times and C_L is the total capacitance of the node.

The simplest model to evaluate the current produced by a transition is just considering a constant current I_0 that lasts during the transition evolution, as shown in Fig. 1. It is important to note that V_1 and V_2 may be any voltage value between the supply rails, since the simulation engine is able to handle non-fully switched transitions.

It will be shown later in this paper that, despite its simplicity, the presented model, when combined with the appropriate delay model, is useful to to obtain good current profiles, specially when focusing on the determination of current peaks and the overall current waveform.

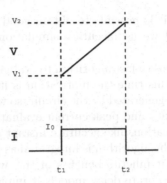

Fig. 1. Simple transition based current model

3   Model Implementation

The current model is implemented in HALOTIS as a separate module that processes node signal transitions on the fly as they are eliminated from the main simulation queue. The process, depicted in Fig. 2, takes place as follows: when processed transitions get out the simulation engine, they enter the *current pulse calculator*, which discards falling transitions, since only current driven by the power supply is of interest, while rising transitions are processed using the model described in the previous section, and a current pulse is obtained. Current pulses are then passed to the *cumulative graph* module which adds each new pulse to the current curve, which is this way dynamically generated. Current curve points can then be directed to the display or printer using conventional plotting software, both on the fly at simulations time or after the simulation process ends.

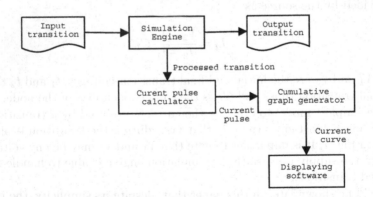

Fig. 2. Current generation algorithm implementation.

4   Results

To validate the model and algorithm, we have simulated two circuits using HSPICE [13] as a reference and HALOTIS. As we stated above, the use of glitch aware delay models is of great importance when evaluating the current. In order to show this point, HALOTIS simulations are made using the Delay Degradation Model (DDM), which accurately handles the propagation of glitches, and a *Conventional Delay Model* (CDM) which treats the propagation and elimination of glitches by using an *inertial delay* [14], like most standard logic simulators do.

The first example consists of a chain of regular inverters that propagates a fast train of pulses (Fig. 3). It is used as a simple test to check how the current calculation algorithm and the DDM works. Fig. 4 shows the voltages at even nodes of the chain using the three kinds of simulation. We can see that the use of the DDM is able to take account of the signal *degradation*, leading to very similar results when compared to HSPICE, but at the logic level. On the other hand, the CDM is not able to render this effect, producing an inaccurate result both in the timing and switching activity aspects.

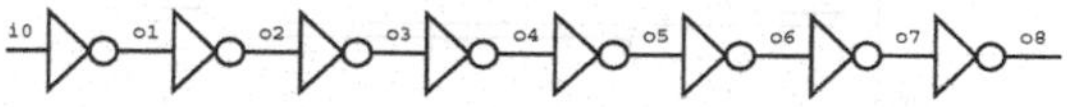

Fig. 3. Inverter chain.

The current figures obtained in the same three simulation cases are depicted in Fig. 5. Since only a few signal transitions are simulated during a short period of time, signal quantisation is quite evident in logic simulation results, in part due to the simplicity of the current model we are using. Nevertheless, DDM results quite well reflects the current evolution by reproducing the main parts of HSPICE's current curve: initial current rising, high current region followed by a middle current region, including a spurious current sink, and a final low current part. On the contrary, CDM results shows an average behaviour during the whole simulation, apart from the initial current rising and final current decreasing. This behaviour is a direct consequence of the overestimated switching activity obtained with conventional models, as was shown in Fig. 4.c.

A much more interesting result is obtained from the simulation of a rather complex circuits, like a 4x4 bit multiplier. In this case, many signal transitions take place at almost the same time showing up clearly its impact in the production of current peaks.

Fig. 6 and Fig. 7 show the results of two sequences simulated with HSPICE, HALOTIS-DDM and HALOTIS-CDM. From these curves we can see that the current generation algorithm implemented in HALOTIS is able to render the current profile of the circuit, to a point that it is possible to distinguish where current peaks are located, its approximate amplitude and how long the switching activity of the circuit lasts after an input pattern is applied. In both cases, not considering the degradation effect (CDM results) yields to an overestimation of

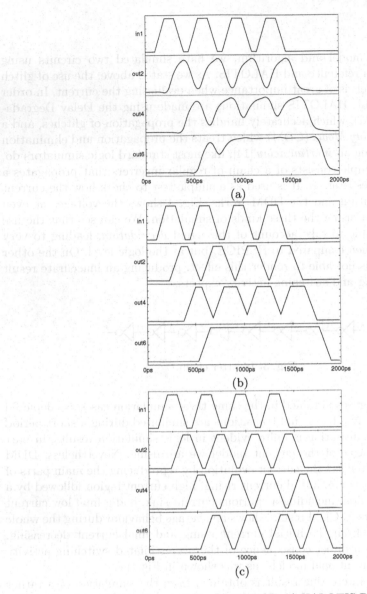

Fig. 4. Inverter chain voltage results: a) HSPICE, b) HALOTIS-DDM, c) HALOTIS-CDM.

the average current as a direct consequence of the overestimation of the switching activity already pointed out. This effect is specially obvious in the last pattern of the first sequence (Fig. 6) and in the fist and third patterns of the second sequence (Fig. 7).

As we noted above, current curve evaluations using the proposed method are done entirely at the logic level, thus important CPU time improvements

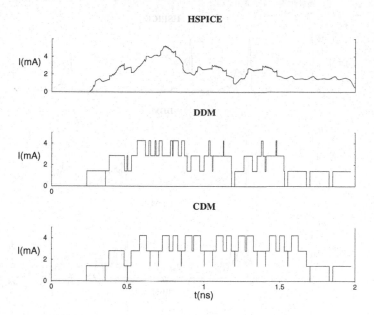

Fig. 5. Inverter chain current results using HSPICE, HALOTIS-DDM and HALOTIS-CDM.

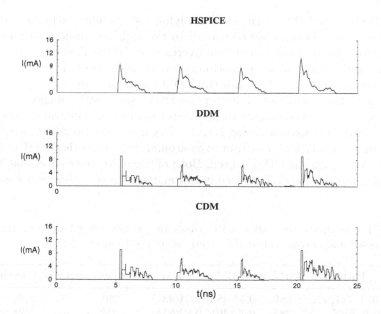

Fig. 6. Simulation results of sequence 0x0, 7x7, 5xA, Ex6 and FxF.

over HSPICE simulations are obtained. In Table 1 we show the CPU times spent in each type of simulation, as well as the speed up of logic simulations

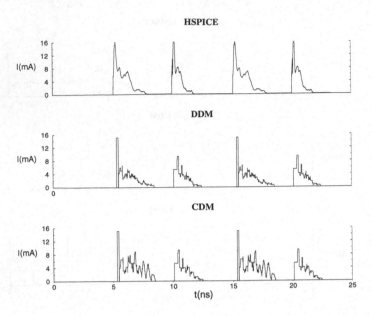

Fig. 7. Simulation results of sequence 0x0, FxF, 0x0, FxF, 0x0

over HSPICE and the overhead of applying the current calculation algorithm in the logic simulator, when compared to the logic simulation without current calculation. The current calculation overhead in HSPICE is negligible. As it is clear from the table, logic simulation speed up over HSPICE is in the order of 400 times, as expected. Although the DDM use slightly more complex formulas than the CDM, simulation times for the DDM are usually smaller than for the CDM, since the last one makes the simulator to process more transitions, derived from an overestimated switching activity. Regarding the overhead introduced by the current calculation algorithm, it is around 20% when the DDM is used and around 30% when the CDM is used. Both of them are quite affordable from the logic simulation perspective, since it is a matter of seconds in most cases.

Table 1. Multiplier simulation CPU times in seconds (speed up over HSPICE in parenthesis) and current calculation overhead in the logic simulation.

Pattern	HSPICE	DDM	CDM	DDM overhead	CDM overhead
0x0, 7x7, ...	280	0.58(483)	0.64(437)	23%	33%
0x0, FxF, ...	281	0.70(401)	0.80(351)	21%	33%

5 Conclusions

It has been presented a simple and fast algorithm to estimate the supply current at the logic level, with application to measure digital switching noise generation. Simulation results show that the algorithm is able to accurately locate current peaks and activity regions with improved speed over HSPICE within 2-3 orders of magnitude. It has been shown that the use of the degradation delay model plays an important role, since this model provides with the method to take into account only "real" transitions, and filter non-existent glitches propagated by conventional logic simulation tools. The current curve calculation module adds little (around 25%) overhead to the logic simulation process.

We believe that the proposed method is the basis of valuable tool for the digital designer willing to get fast current/noise estimations at the logic level in the early stages of the design.

References

1. Brandt, B.P., Wooley, B.A.: A 50-MHz Multibit Sigma-Delta Modulator for 12-b 2-MHz A/D Conversion IEEE Journal of Solid-State Circuits **26(12)** (December 1991) 1746–1756
2. Allstot, D.J., Chee, S., Kiaie, S., SHrivastawa: Folded Source-coupled Logic vs. CMOS Static Logic for Low-Noise Mixed-Signal ICs' IEEE Tr. on Circuits and Systems-I **40(9)** (September 1993) 553–563
3. Gonzalez, J.L., Rubio, A.: Low Delta-I noise CMOS Circuits Based on Differential Logic and Current Limiters IEEE Transactions on Circuits and Systems I **46(7)** (July 1999) 872–876
4. Acosta, A.J., Parra, P., Valencia, P.: Reduction of Switching Noise in Digital CMOS Circuits by pin swapping of Library Cells. Power and Timing Modeling, Optimization PATMOS'2001 **25(6)** (December 1990) 1588–1590
5. Aragones, X., Gonzalez, J.L., Rubio, A.: Analysis and Solutions for Switching Noise Coupling in Mixed-Signal ICs. Kluwer Academic Publishers, (1999)
6. Heijningen, M., Badaroglu, M., Donnay, S., Engels, M., Bolsens, I.: High-Level Simulation of Substrate Noise Generation Including Power Supply Noise Coupling. 37th Design Automation Conference (DAC), Los Angeles (USA), (June 2000)
7. Ruiz-de-Clavijo, P., Juan, J., Bellido, M. J., Acosta, A. J., Valencia, M.: HALO-TIS: High Accuracy Logic Timing Simulator with Inertial and Degradation Delay Model. Design, Automation and Test in Europe (DATE) Conference and Exhibition, Munich (Germany), (March 2001)
8. Bellido-Diaz, M. J., Juan-Chico, J., Acosta, A. J., Valencia, M., Huertas, J. L.: Logical modelling of delay degradation effect in static CMOS gates. IEE Proc. Circuits Devices and Systems **147(2)** (April 2000) 107–117
9. Juan-Chico, J., Ruiz-de-Clavijo, P., Bellido, M. J., Acosta, A. J., Valencia, M.: Inertial and degradation delay model for CMOS logic gates. In Proc. IEEE International Symposium on Circuits and Systems (ISCAS), Geneva, (May 2000) I–459–462
10. Juan-Chico, J., Ruiz-de-Clavijo, P., Bellido, M. J., Acosta, A. J., Valencia, M.: Degradation delay model extension to CMOS gates. In Proc. Power and Timing Modelling, Optimization and Simulation (PATMOS) (September 2000) 149–158

11. Juan-Chico, J., Bellido, M. J., Ruiz-de-Clavijo, P., Baena, C., Valencia, M.: AUTODDM: AUTOmatic characterization tool for the Delay Degradation Model. In Proc. 8th IEEE International Conference on Electronics, Circuits and Systems (ICECS), Malta, (September 2001) 1631–1634
12. Baena, C., Juan-Chico,J., Bellido M.J., Ruiz-de-Clavijo P., Jimenez, C.J.,Valencia, M.: Simulation-driven switching activity evaluation of CMOS digital circuits. In Proc. XVI Conference on Design of Circuits and Integrated Systems (DCIS), Porto, (November 2001) 608–612
13. HSPICE User's Manual. Meta-Software (1999)
14. Unger, S.H.: The Essence of Logic Circuits. Prentice-Hall International, Inc. 1989.

Power Efficient Vector Quantization Design Using Pixel Truncation

Kostas Masselos, Panagiotis Merakos, and Costas E. Goutis

VLSI Design Laboratory, Department of Electrical and Computer Engineering, University of Patras, Rio 26500, Greece
{masselos, merakos, goutis}@ee.upatras.gr

Abstract. Vector quantization image encoding requires a huge amount of computation and thus of power consumption. In this paper a novel method is proposed for the reduction of the power consumption of vector quantization image processing by truncating the least significant bits of the image pixels and the codewords elements during the nearest neighbor computation. Experimental results prove that at least 3 pixels/elements bits can be truncated without affecting the picture quality. This results in an average 65% reduction of bus power consumption and in an average 62% reduction of the power consumed in major data path blocks.

1 Introduction

The rapid advances in the area of multimedia and wireless communications lead to powerful portable systems with multimedia capabilities such as PDAs, wireless video phones and hand held digital video cameras. Real time video and image compression is required in such systems for transmission bandwidth and/or storage reduction. However, it is a very power consuming procedure.

The most significant power savings can be achieved at the algorithmic level [4]. At that level there are different design parameters that need to be trade-off [5]. Some of them are related to algorithmic performance such as output picture quality (in terms of PSNR) and compression ratio (in bits per pixel). Other parameters are related to implementation issues like power consumption and hardware complexity (circuit area).

In this paper an approach for power consumption reduction of vector quantization [3] image compression is proposed. Most of the existing approaches for optimization of vector quantization image compression target performance [1] or hardware complexity [2]. Approaches for power optimization of image/video processing systems at the algorithmic level have been presented in [5–7]. Approaches proposed in [6], [7] are based on algorithm selection. The approach presented in [5] modifies the conventional vector quantization image compression scheme. The approach proposed in this paper is based on the trade-off between output image quality and data words bitwidths.

B. Hochet et al. (Eds.): PATMOS 2002, LNCS 2451, pp. 409–418, 2002.
© Springer-Verlag Berlin Heidelberg 2002

The rest of the paper is organized as follows: In section 2 the vector quantization image compression is described. The pixel truncation method is presented in section 3. In section 4 experimental results are presented. Conclusions are given in section 5.

2 Vector Quantization Image Compression

Vector quantization [3] is an efficient image coding technique, achieving low bit rates i.e. lower than 1 bit per pixel. Vector quantization is described as

$$Q : R^k \to C$$
$$C = \{y_1, y_2, \dots y_N\}, y_i \in R^k \; \forall i = 1,2,\dots, N \tag{1}$$
$$y_i = Q(x) \, if \, d(x, y_i) \le d(x, y_j) \, for \, i, j = 1,2,\dots, N.$$

where x is a k-dimensional input vector belonging to the k-dimensional space R^k, C is the codebook of N k-dimensional words y_i and d is the distortion criterion used.

In vector quantization, a vector, which is a block of pixels, is approximated by a representative vector (codeword) of the codebook, which minimizes the distortion (usually mean square error) among all the codevectors in the codebook. Compression is achieved by transmitting or storing the codeword address (index) instead of the codeword itself. The generic data flow of vector quantization image compression is shown in fig.1. Depending on the performance requirements of the application a number of parallel data paths may be used in combination with multi-port memories or distributed codebook memory organization.

The basic disadvantage of vector quantization image coding is its computational complexity. This complexity causes problems with respect to power consumption and hardware implementation, especially in real-time applications. Assuming an image of NxM pixels of IPB bits divided to blocks of BxB pixels, a codebook of C BxB elements codewords of CWB bits the computational complexity of vector quantization image compression is given by the following equations:

$$Number \; of \; accesses \; to \; image \; memory = N \times M \times C \times f(IPB) \tag{2}$$

$$Number \; of \; accesses \; to \; codebook \; memory = N \times M \times C \times f(CWB) \tag{3}$$

$$Number \; of \; subtractions = N \times M \times C \times f(IPB + CWB) \tag{4}$$

$$Number \; of \; multiply \; accumulations = \tag{5}$$
$$N \times M \times C \times f(2 \times (\max(IPB, CWB) + 1))$$

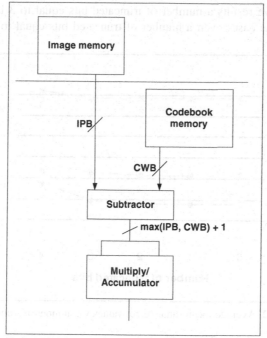

Fig. 1. Generic data flow of vector quantization image compression

3 Pixel Truncation

In this section the power consumption reduction of the vector quantization kernel using fixed bit truncation of the image pixels and codeword elements is explored. The conventional full search vector quantization [3] is used to demonstrate the concept while the proposed approach can be applied to other variants of vector quantization such as tree search vector quantization etc [3]. Simulation results prove that vector quantization output image quality (in terms of PSNR) is only slightly reduced when the LSBs of the image pixel values and codeword elements are truncated.

The optimal number of truncated bits is determined using simulation. Several benchmark images and codebooks produced from them have been simulated. The impact of the number of truncated bits on the distortion criterion calculation has been monitored. It is assumed that the same number of bits is truncated from both image pixels and codeword elements. Results from test images "Lena", "Man", "Baboon", "Bridge", "Peppers" and "Airplane" are presented. Codebooks have been generated using the LBG algorithm [3] and the corresponding image as training sequence. The simulation results for different numbers of truncated bits are presented in figure 2. It is observed that the average mean square error is almost flat and starts increasing when the number of truncated bits is equal to 4. For the case of 3 truncated bits the % increase of the mean square error varies from 1% to 12%. When the number of truncated bits becomes 4 the % increase of the mean square error varies from 5% to 54%.

Based on the above results a number of truncated bits equal to 3 is reasonable to be used while in some cases even a number of truncated bits equal to 4 may be considered.

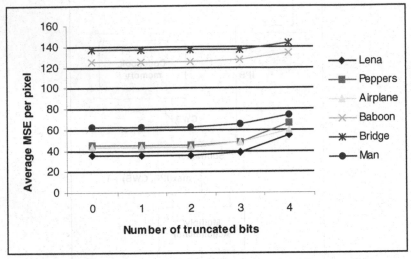

Fig. 2. Average mean square error values vs. number of truncated bits

Truncating the LSBs of the image pixels and the codeword elements leads to less hardware and switching activity thus reducing power consumption but also area. Figure 3 shows the profile of the switching activity in the image pixels input of the processing unit. Figure 4 shows the profile the profile of the switching activity in the codeword elements input of the processing unit. As shown in figure 3 image data follow the Dual Bit Type model [8]. Codebook data produced by image data during codebook generation [3] do not follow the Dual Bit Type model since they are not pure image data.

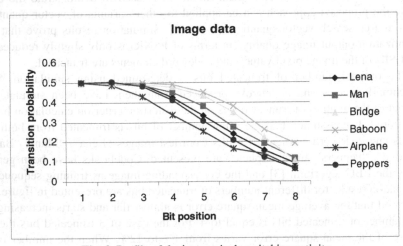

Fig. 3. Profile of the image pixels switching activity

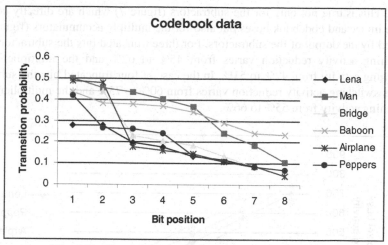

Fig. 4. Profile of the codeword elements switching activity

The important point is that the transition activity is much higher for the least significant bits both for image and codebook data. By truncating the LSBs of the image pixels and the codeword elements the percentage of the switching activity reduction is expected to be higher than that of the bit truncation. Thus further power reduction is achieved in addition to the reduction in hardware complexity.

4 Experimental Results

In this section the effect of the proposed bit truncation approach for the reduction of the power consumption of vector quantization is explored. Simulations using the test images "Lena", "Man", "Baboon", "Bridge", "Peppers" and "Airplane" have been carried out. A constant compression ratio of 0.5 bits is assumed corresponding to 256-word codebook. Codebooks have been generated using the LBG algorithm [3] and the corresponding image as training sequence.

The effect of bit truncation on the switching activity in the bus connecting the memory storing the image under compression to the processing units is presented in figure 5. The bus switching activity which is directly proportional to bus power consumption is linearly decreasing with the number of truncated bits. For three truncated bits the image pixels bus switching activity reduction varies from 45% to 60%. In the case of four truncated bits the image pixels bus switching activity reduction varies from 60% to 75%. Similar conclusions can be derived for the switching activity in the bus connecting the codebook memory to the processing elements. The bus switching activity is slightly less linearly decreasing with the number of truncated bits this time in comparison to the image pixels bus case. For three truncated bits the image pixels bus switching activity reduction varies from 45% to 68%. In the case of four truncated bits the image pixels bus switching activity reduction varies from 58% to 78%. Bit truncation leads to reduction of the switching activity at the input of the functional units. This is true not only for the subtractors (figure 7) which are directly connected to the image and codebook buses but also for the multiply accumulators (figure 8) that are fed by the output of the subtractors. For three truncated bits the subtractor's inputs switching activity reduction varies from 45% to 62% and the multiplier's inputs switching activity from 42% to 51%. In the case of four truncated bits the subtractor's inputs switching activity reduction varies from 60% to 75% and the multiplier's inputs switching activity from 55% to 66%.

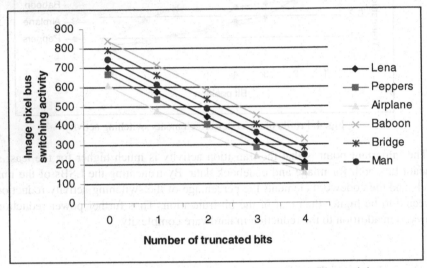

Fig. 5. Effect of bit truncation on image pixel bus switching activity

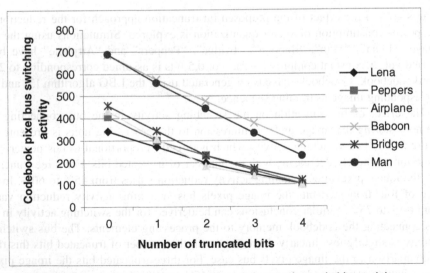

Fig. 6. Effect of bit truncation on codeword elements bus switching activity

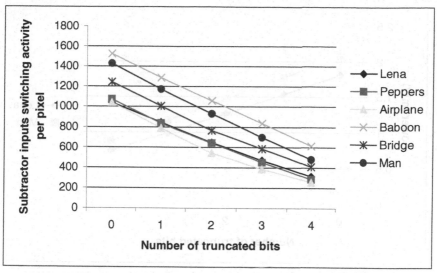

Fig. 7. Effect of bit truncation on the switching activity at the subtractor's inputs

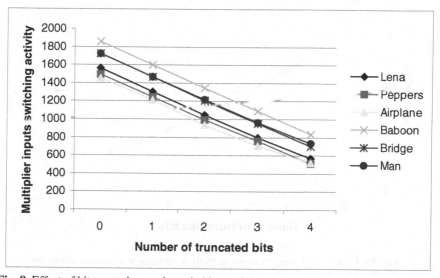

Fig. 8. Effect of bit truncation on the switching activity at the multiply accumulator's inputs

For the estimation of the power consumed in the functional units the Synopsys Design Compiler utility for static probabilistic power estimation has been used. A ripple carry architecture has been realized for the subtractors while an array architecture has been used for the multiplier of the multiply-accumulator unit. The circuits have been realized on a 0.18 μm, six metal layers technology of UMC. The effect of the pixel truncation approach on the power consumed in subtractors and multipliers is presented in figures 9 and 10 respectively.

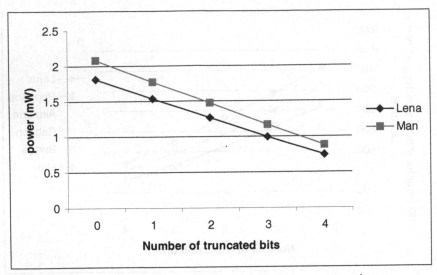

Fig. 9. Effect of bit truncation on subtractor power consumption

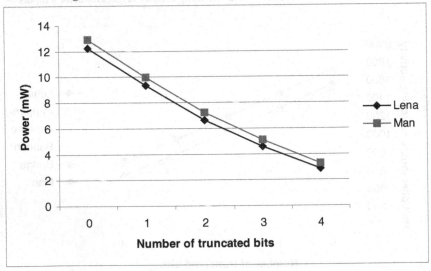

Fig. 10. Effect of bit truncation on multiply accumulator power consumption

The power consumption of the subtractor is linearly reduced with the number of truncated bits. For three truncated bits the subtractor's power is reduced by approximately 45% for two test images. In the case of four truncated bits the subtractor's power is reduced by approximately 58% for two test images. The power consumption of multiply accumulator is slightly less linearly decreasing with the number of truncated bits in comparison to the subtractor. For three truncated bits the multiply accumulator's power is reduced by approximately 62% for two test images. In the case of four truncated bits the multiply accumulator's power is reduced by approximately 75% for two test images.

The effect of bit truncation on the output image quality in terms of PSNR (db) is presented in figure 11. The PSNR remains almost flat up to the point of three truncated bits and start decreases for four truncated bits. This decrease is sometimes significant (>0.5 db) and the result in output image quality is noticeable to human perception and in other cases small and unnoticeable to human perception.

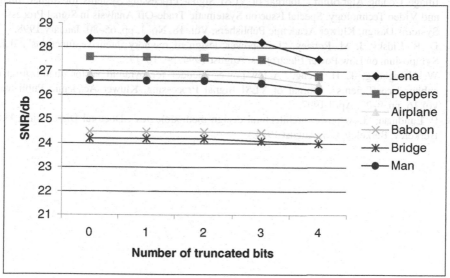

Fig. 11. Effect of bit truncation on output image quality

5   Conclusions

In this paper the trade-off between the prediction accuracy and the power consumption in the vector quantization image compression module has been investigated. A pixel/element truncation method has been proposed to reduce power consumption. Experimental results prove that the truncation of 3 bits from the image pixels and the codeword elements leads to an average 65% reduction of bus power consumption and in an average 62% reduction of the power consumed in major data path blocks with almost no degradation of the PSNR.

References

1. W-C. Fang, C-Y. Chang, B. J. Sheu, O. T-C. Chen, J. C. Curlaander, "VLSI Systolic Binary Tree-Searched Vector Quantizer for Image Compression", IEEE Transactions on Very Large Scale Integration (VLSI) Systems, Vol. 2, No 1, pp. 33–43, March 1994.
2. Y-H. Hu, "Optimal VLSI Architecture for Vector Quantization", proc. 1995 IEEE International Conference in Acoustics Speech and Signal Processing, pp. 2853–2856.

3. A. Gersho, R. M. Gray, "Vector Quantization and Signal Compression", Kluwer Academic Publishers 1992.
4. J. M. Rabaey, M. Pedram, "Low Power Design Methodologies", Kluwer Academic Publishers 1995.
5. K. Masselos, P. Merakos, T. Stouraitis, C. E. Goutis, "Trade-Off Analysis of a Low-Power Image Coding Algorithm", Journal of VLSI Signal Processing Systems for Signal, Image and Video Technology, Special Issue on Systematic Trade-Off Analysis in Signal Processing Systems Design, Kluwer Academic Publishers, Vol. 18, No. 1, pp. 65–80, January 1998.
6. D. B. Lidsky, J. M. Rabaey, "Low-power design of memory intensive functions", IEEE Symposium on Low Power Electronics, August 1994, pp. 16–17.
7. W. Namgoong, T. H. Meng, "A low-power encoder for pyramid vector quantization of subband coefficients", Journal of VLSI Signal Processing, Kluwer Academic Publishers, vol.16, pp. 9–23, April 1997.
8. P. Landman, "Low power architectural design methodologies", Doctoral Dissertation, U.C. Berkeley, Berkeley, CA, August 1994.

Minimizing Spurious Switching Activities in CMOS Circuits

Artur Wróblewski, Florian Auernhammer, and Josef A. Nossek

Insitute for Circuit Theory and Signal Processing
Munich University of Technology
Arcisstr. 21, 80333 Munich, Germany
arwr@nws.e-technik.tu-muenchen.de

Abstract. In combinatorial blocks of static CMOS circuits transistor sizing can be applied for delay balancing in order to guarantee synchronously arriving signal slopes at the input of logic gates. Since the delay of a logic gate depends directly on transistor sizes, the variation of channel-widths and -lengths (W and L) allows to equalize different path delays without influencing the total propagation delay of the circuit. Unfortunately not only the delay, but also the total capacitance and the short-circuit power consumption of a circuit depend on the transistor sizes. In order to take this fact into account, the method is formulated as a multiobjective optimization problem, where the path delay differences and the power consumption are the design objectives. To achieve optimal results, transistor lengths have to be increased, which results in both increased gate capacitances and area. Splitting the long transistors counteracts this negative influence and reduces the power dissipated. Moreover, the so called "Zero-Delay" paths can be avoided by introducing additional delays.

1 Introduction

Optimal sizing of MOS-transistors is a widely investigated method for the design of CMOS-circuits with restricted area, propagation delay or power consumption. A large number of the previously published approaches aim at area and power optimization under given delay constraints [1,2,3]. Optimal area utilization is still important, but since the substantial progress in development of deep submicron techniques, power dissipation has become the main limiting factor. The power consumption models often include only the power consumed for charging transistor gate- and drain/source-capacitances. The power models in [1] include also the dissipation caused by short-circuit currents that occur during transition when both P- and N-transistors of a CMOS-stage are conducting. Delay balancing for reducing the glitching activities in combinational Wallace-trees and array multipliers has been introduced in [5].

Unlike for most approaches that focus on maximizing the speed of a circuit by variation of transistor widths, the method presented here allows also the transisitor lengths to be variable. Reducing speed for delay balancing is only

B. Hochet et al. (Eds.): PATMOS 2002, LNCS 2451, pp. 419–428, 2002.

allowed for parts of the circuit that are not in the critical path. In [6] a method is presented, where all transistor widths outside the critical path are reduced in order to reduce the total capacitance of the circuit. However, delay balancing may not be possible if only the widths are variable because the limit here is the minimum feature size. Further speed reduction can then be achieved by buffer insertion [4] or increasing the transistor length as presented here. In order to keep track of the conflicting design objectives like increasing transistor sizes for delay balancing, and at the same time reducing the total power consumption caused by charging capacitances, the method is formulated as a multiobjective optimization problem.

In the following we consider circuits in which increasing transisitor lengths is necessary. Decreasing of W to make the gate slower, which is the usual approach, results in smaller area and less power consumption. On the contrary increasing L provides slower gates, but influences both, the area and power dissipation, negatively. Thus, increasing L represents the worst case approach to transistor sizing. Therefore, the power savings presented here reflect only the benefits of a delay balanced circuit due to reduced glitch activity. Of course **GliMATS**, a program that has been implemented for automated circuit optimization, is not limited by this artificial constraint.

2   Delay and Power Models

The delay and power models used for the transistor sizing method presented here are defined at gate level. Although transistor level models may offer more degrees of freedom and allow individual sizing of each single transistor, it turns out to be more desirable to have a low dimensional optimization problem in order to be able to optimize larger circuits within acceptable computation time. When modeling a circuit at gate level (*macromodeling*), the relatively large number of local parameters that describe every single transistor is reduced to a set of scale factors for each gate. In the considered case the number of variables is reduced to one specific W and one specific L for each gate. If W and/or L are varied, all transistor widths and/or lengths within the gate are scaled by the same factor simultaneously.

2.1   Delay Model

The macromodel delay has to be described for each type of logic gate separately. The total delay τ_m of a gate at position m is then approximated by

$$\tau_m = \tau_{in,m} + \tau_{s,m} \quad , \tag{1}$$

with $\tau_{s,m}$ being the step response, which is independent of the signal form, and $\tau_{in,m}$, which is the contribution caused by the finite input signal rise and fall times. The optimization method aims at the minimization of glitches, which necessitates equalizing all path delays. However, the delay τ_m depends on the input transition. For example: The delay of a 2-input AND gate in $0.25\mu m$ technology with a certain load is $0.4ns$ for the input transition $11 \rightarrow 00$ and $0.75ns$ for $00 \rightarrow 11$. Therefore, the different paths can exactly be balanced for one special transition only. Experiments have shown that the worst case delay is a good choice and is easy to formulate in the model. Furthermore, numerous simulations based on this model show, that although the paths cannot be exactly balanced for all transitions, glitching can be eliminated. The delay τ_m turns out to be a function of W and L as optimization parameters. For a 2-input AND gate

$$\tau_m = k_1 L_{m-1} + k_2 L_{m-1}^2 + k_3 \frac{L_{m-1}}{W_{m-1}} L_m W_m +$$

$$k_4 L_m + k_5 L_m^2 + k_6 \frac{L_m}{W_m} L_{m+1} W_{m+1}. \tag{2}$$

The total delay of a path number ν is the sum over all gate delays in this path:

$$\tau_\nu = \sum_{m=1}^{n} \tau_m \quad, \tag{3}$$

where n is the number of gates in the path. For a detailed description of the delay model used here see [7].

2.2   Power Consumption Model

With the objective function (3) only the delay can be considered in an optimization procedure so far. In order to take into account also the transistor size dependency of the short-circuit currents and the total capacitance of a circuit, an objective function for the power consumption of gate number m can be formulated as follows:

$$P_m = P_{m,cap} + P_{m,sc} \quad, \tag{4}$$

where $P_{m,cap}$ denotes the power consumed for charging the gate and drain/source capacitances and $P_{m,sc}$ denotes the short-circuit power consumption of gate m. As described in [7], both terms can also be expressed as functions of W, L and technology constants.

The total transistor size dependent power consumption in path number ν can be formulated as:

$$P_\nu = \sum_{m=1}^{n} P_m \quad, \tag{5}$$

for a path with n gates.

3   Multiobjective Optimization

In order to find a power optimal solution for W and L the designer is confronted with two conflicting design criteria: path balancing by transistor sizing, achieved by enlarging transistors, and low power consumption for charging capacitances which requires small transistors at the same time. This problem usually can be solved with a non-linear programming method. A common approach to find a solution is to keep one of the design criteria within upper and lower bounds and find an optimal solution for the other one under these restrictions. The problem is to determine the upper and lower bounds if they are not previously known. Awkwardly chosen bounds may result in an unsolvable optimization problem. Therefore, not every single criterion is optimized while restricting all the others, but the weighted sum of all the design criteria[3]. In order to equalize all the path delays with respect to the critical path, every path requires individual optimization. Let τ_{crit} denote the critical path delay of the circuit. For every path ν a solution of

$$\min_{W,L} |\tau_\nu - \tau_{crit}| \tag{6}$$

must be calculated to achieve path balancing. The path delay τ_ν is defined in (3). The power consumption according to (5) is minimized by

$$\min_{W,L}(P_\nu = \sum_{m=1}^{n} P_m) \quad . \tag{7}$$

Equations (6) and (7) describe convex optimization problems in W and L. The multiobjective optimization problem is then given by:

$$\min_{W,L}(S_\nu = w \cdot (\tau_\nu - \tau_{crit})^2 + (1 - w) \cdot P_\nu) \quad . \tag{8}$$

The weight factor w varies between 0 and 1, $w \in [0, 1]$. Results of the optimization are highly independent of the choice of w. Only values extremly close to 0 or 1 influence the result. In order to obtain a cost function, which is differentiable everywhere, $|\tau_\nu - \tau_{crit}|$ has been replaced by its square. The upper and lower bounds of the transistor sizes are determined by the minimum feature size of the used technology and the user defined limits for the maximum available area for a single transistor.

Assigning a value to w allows a solution to be chosen depending on which of the design objectives is more desired: low power consumption caused by the total capacitive load or balanced path delays. However, experiments have shown that for many circuits the best low power solution is obtained if $|\tau - \tau_{crit}| = 0$, i.e. for optimally balanced paths. This is usually given when $w = 0.5...1$.

4   Minimizing Gate Capacitances

As mentioned before, the case considered in this paper is the one, when transistors are being made longer. This leads to larger channel resistance of the transistor and increases its gate capacitance. To increase the channel resistance without increasing the gate capacitance one has to be able to change them independently from each other. This is possible if the capacitance and the resistance are no longer part of one common transistor. To achieve that one can split the common transistor into two ("Twin-Transistors"). The resistance can then be assigned to one of them, the capacitance to the other. The gate capacitance of one of the transistors has to be driven by the previous stage and should therefore be minimum size. The other transistor provides the additional delay required for path balancig. Its gate is hard wired to ground (pmos) or vdd! (nmos) and the large capacitance has no influence on the power consumption of the circuit. Introducing "Twin-Transistors" doubles the number of devices in the gate. Even if they can be placed in a area-saving way, together with additional wiring, the area taken is almost doubled. However, it's obvious that, within one block, the transistors, responsible for the increased delay, can be merged together ("Merged Transistors"). This highly compensates for the area increase caused by "Twin-Transistors". A detailed description of the concept of "Merged-" and "Twin-transistors" can be found in [8].

5   Balancing of the Path Delays

The transistor sizing algorithm for the reduction of glitching activity is implemented in the program **GliMATS** (**Gl**itch **M**inimization by **A**utomated **T**ransistor **S**izing). It allows to optimize the circuits automatically. **GliMATS** processes a netlist (e.g. extracted layout) of the circuit. The user can set a value for the weight factor w and specify which of the transistor topologies is to be used - standard, "Twin-Transistors" or "Merged-Transistors". As the output **GliMATS** produces netlist of the glitch minimized circuit. It is assumed that the circuit is already optimized to match eventually given timing constraints, so the critical path must not be manipulated in order to retain the required maximum delay. The input netlist to the path balancing program is given from this previous speed optimization. **GliMATS** has been described in [8].

6   "Zero-Delay" Paths

The simulation results presented below have show, that proposed transistor sizing method can be very efficient in eliminating spurious switching activities. However, in some cases the topology of the circuit does not allow for optimal solution. Consider the following equation:

$$y = (a + b) + c$$

One possible realisation of that equation is to implement an adder, processing the operation on numbers a and b, and another one for adding c to the result of the previous operation (see Figure 1 left).

Obviously path '2' cannot be optimized for the given topology of the circuit. The signals of path '1' will always arrive delayed at the input of the second adder. This effect cannot be compensated by transistor sizing, since no operation is performed on signals of path '2' ("Zero-delay" path). The path consists

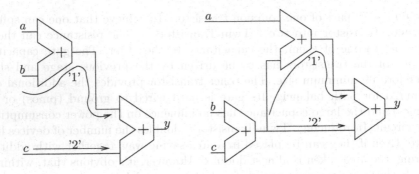

Fig. 1. Paths with "zero-delay" cannot be optimized without introducing additional gates.

only of wiring. Thus no transistors can be sized. In such a situation glitching cannot be avoided without modifying the original topology of the circuit. An obvious solution is to introduce additional gates in path '2' and thus to provide artificial delay. The described transistor sizing algorithm can then be applied to the resulting circuit. The question arises, which gates are best suited as delays for the modified circuit. Two inverters would be the first choice. However, only small delays can be realized with inverters with reasonable transistor lengths. On the other hand, small delays often result in glitches of less significance than large delays. Such glitches may not propagate over many stages of the design. Power consumption of additional inverters may override the savings achieved by optimal path balancing. This has been validated by simulations. Similiar considerations apply to transmission gates. Therefore more complex gates have been sought for, that are best suited for the application. 2-input AND, OR, NOR, NAND, a fulladder as well as different combinations of these cells have been considered. An example with OR gates is shown in Figure 2.

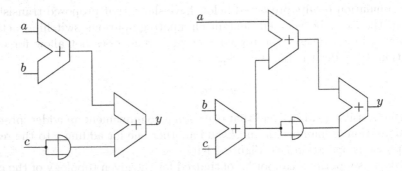

Fig. 2. Paths with "zero-delay" cannot be optimized without introducing additional gates.

Experimental results have shown, that a combination of INV and NAND gates seems to be the most efficient solution.

7   Applications and Experimental Results

The proposed path balancing method has been tested on some example circuits, a few selected are shown here. They include array multipliers and combinational logic blocks (ISCAS'85 Benchmarks). The simulations have been performed with PowerMill before and after transistor optimization for glitch reduction. The different topologies have been tested in the optimization. For simulation 10000 random input vectors have been applied to each circuit. The results are summarized in Table 1. As expected, introducing artificial delay in small circuits (like c17) results in increased power consumption. Also for multipliers no improvement has been achieved. This is due the very special topology of array multipliers (only few "Zero-Delay" paths and most of them with AND as the first gate in the path). However, for large combinatorial blocks, a noticeable improvements (of 18% for c1908) have been observed. Please note, that in that case the simulations have been perfomed with "Twin"-topology.

Table 1. Power consumption in mW for not balanced circuits and percentage savings achieved with transistor sizing performed by GliMATS for different topologies.($0.25\mu m$, $V_{dd} = 2.5V$, PowerMill simulations with 10000 random input vectors).

Circuit	Not balanced	Standard topology	Twin topology	Without Zero-Delay	Merged topology
4 × 4 Mult.	0.157	44%	41%	41%	44%
8 × 8 Mult.	0.822	33%	49%	49%	53%
16 × 16 Mult.	4.000	39%	50%	50%	53%
c17	0.026	12%	30%	25%	40%
c432	0.427	14%	39%	42%	39%
c499	0.997	6%	30%	40%	30%
c880	0.770	22%	45%	49%	44%
c1908	0.935	10%	39%	50%	39%

Note that the percentage of power reduction due to the glitch elimination increases for larger arrays because of the snowball effect that glitches stimulate in these circuits. The CPU-time for the complete optimization of a 8×8 multiplier is about 7 minutes on a Ultra Sparc 2 workstation.

Furthermore, for the 4 × 4, 8 × 8, and 16 × 16 array multipliers the increase of chip area has been estimated by measuring the area increase of a single cell due to transistor sizing and projecting this to the total chip area including the wiring. For the standard topology the expected area increase is between 15% and 31%.

The additional silicon space needed is even greater for the "Twin-Topology", but it's significantly lower for the "Merged-Topology".

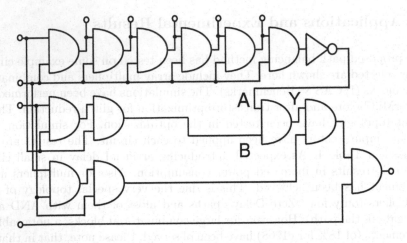

Fig. 3. Logic circuit for demonstration of the path balancing method.

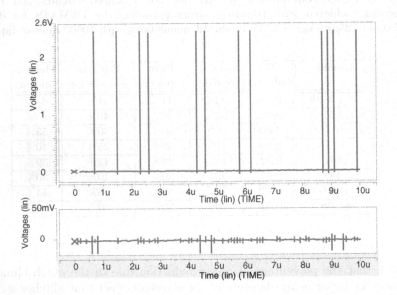

Fig. 4. Signal Y of the circuit in Fig. 3 with unbalanced (top) and balanced path delays (bottom).

To demonstrate the effect of path balancing by transistor sizing a logic circuit shown in Fig. 3 has been designed. If zero delay is assumed for all gates in this circuit, the output is always 0 regardless of the inputs. In a simulation with complex transistor models only glitches due to unbalanced paths are visible at the output.

The signal Y of the unbalanced circuit (all transistor sizes minimal) for random input signals is shown on top of Fig. 4.

The same input signals yield signal Y shown on the bottom of Fig. 4 for the circuit after optimization: glitches are completely eliminated. Note that the voltage axis are scaled differently. It results in power savings of 8%. Fig. 5 shows the slopes of the signals A and B before (top) and after (bottom) path balancing.

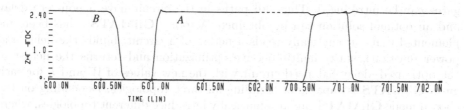

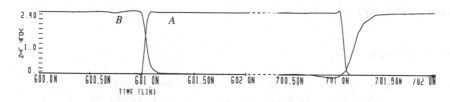

Fig. 5. Example for signals A and B of the circuit in Fig. 3 with unbalanced (top) and balanced path delays (bottom).

This figure demonstrates how signal B is delayed after transistor sizing in order to "wait" on signal A.

The results show significant power savings after **GliMATS** has been applied. However, one must be aware that enlarging of the transistor lengths to increase the delay results in slower signal slopes which may lead to larger short circuit power consumption (this is considered in the results presented) and to an increase of the required chip area. On the other hand, longer transistors result in smaller leakage current which becomes very important in modern technologies. A good application of the method would be in combination with pipelining, where register stages work as glitch barriers. In the combinational logic between two register stages glitching could be eliminated by the presented transistor optimization. In order to reduce design time for the different sized gates, module generators can be applied for automatic scaling of parameterized gate layouts.

8   Conclusion

In this work a method for transistor size optimization to achieve equal path delays in CMOS circuits has been presented. Delay and power consumption of a path can be modeled as functions of the transistor sizes W and L at gate level. With multiobjective optimization the path delay differences and the power consumed for charging capacitances can be minimized simultaneously. The solutions for W and L restricted by upper and lower bounds, given by the minimum feature size and area limitations. By splitting long transistors into two a decrease in gate capacitances has been achieved. In case of "Twin"-topology a significant area increase has to be taken into account. For "Zero-Delay" paths, additional gates can be introduced. Thus, all paths in the circuit have a non-zero delay and an optimal solution can be obtained. A tool – **GliMATS** – has been implemented that automatically reads a netlist of a circuit, builds the delay and power functions, starts multiobjective optimization and returns the netlist of the optimized, delay balanced circuit with the new values of W and L for each gate. **GliMATS** is capable of handling all three topologies. Depending on the chosen mode **GliMATS** can automatically introduce different topologies, where applicable, to achieve best power savings. By applying this method glitching in a circuit can be reduced drastically. Experimental results show significant power savings after optimization.

References

1. M. Borah, R. M. Owens, and M. J. Irwin. Transistor Sizing for Low Power CMOS Circuits. *IEEE Trans. on Computer-Aided Design*, 15(6):665–671, June 1996.
2. J. P. Fishburn and A. E. Dunlop. TILOS: A Posynomial Programming Approach to Transistor Sizing. *Proc. ICCAD*, pages 326–328, 1985.
3. B. Hoppe, G. Neuendorf, D. Schmitt-Landsiedel, and W. Specks. Optimization of High-Speed CMOS Logic Circuits with Analytical Models for Signal Delay, Chip Area, and Dynamic Power Dissipation. *IEEE Trans. on Computer-Aided Design*, 9(3):236–247, March 1990.
4. J.Kim S. Kim and S.-Y.Hwang. Nwe Path Balancing Algorithm for Glitch Power Reduction. *IEE Proc.-Circuit Devicers Syst.*, Vol. 148, June 2001.
5. T. Sakuta, W. Lee, and P. T. Balsara. Delay Balanced Multipliers for Low Power/Low Voltage DSP Core. *IEEE Symp. on Low Power Electronics*, pages 36–37, October 1995.
6. S. Trimberger. Automated Performance Optimization of Custom Integrated Circuits. *Proc. Int. Symp. on Circuits and Systems*, pages 194–197, 1983.
7. A. Wróblewski, C.V. Schimpfle, and J. A. Nossek. Automated Transistor Sizing Algorithm For Minimizing Spurious Switching Activities in CMOS Circuits. *Proc. IEEE Int. Symp. on Circuits and Systems, ISCAS'2000, Geneva 2000 IEEE*.
8. A. Wróblewski, Otto Schumacher, C.V. Schimpfle, and J. A. Nossek. Minimizing Gate Capacitances With Transistor Sizing. *Proc. IEEE Int. Symp. on Circuits and Systems, ISCAS'2001*, Sydney 2001 IEEE.

Modeling Propagation Delay of MUX, XOR, and D-Latch Source-Coupled Logic Gates

M. Alioto[1] and G. Palumbo[2]

[1] DII – Dipartimento di Ingegneria dell'Informazione, Università di Siena,
v. Roma n. 56, I-53100 - Siena (Italy)
malioto@dii.unisi.it

[2] DEES - Dipartimento Elettrico Elettronico e Sistemistico, Universita' di Catania,
viale Andrea Doria 6, I-95125 CATANIA - ITALY
Phone ++39.095.7382313; Fax ++39.095.330793
gpalumbo@dees.unict.it

Abstract. In this paper, an analytical delay model of Source-Coupled Logic (SCL) gates is proposed. In particular, the multiplexer, the XOR and the D-latch gates are considered. The method starts from a linearization of SCL gates, and analysis of the equivalent circuit obtained is simplified by introducing the dominant-pole approximation. The delay expression obtained is quite simple and each term has an evident circuit meaning, hence it is useful to design. The model was validated by extensive comparison with Spectre simulations by using a 0.35-μm CMOS technology. Results show that the predicted delay values agree well with simulated results.

1 Introduction

Among logic circuits based on CMOS technology, Source Coupled Logic (SCL) gates are gaining the market interest in various applications, such as high-resolution high-speed mixed-signal circuits [1]–[4]. However, the SCL logic was demonstrated to be also suitable for the realization of multiplexer circuits in optical-fiber links operating in the range of 10 Gb/s. Indeed, SCL logic gates used in these applications, i.e. the multiplexer (MUX), the XOR and the D-latch gates, have been proved to exhibit a better performance at a lower power dissipation with respect to traditional static logic [5].

Considering that the SCL gates are characterized by a static power dissipation, the power-delay trade-off is crucial in the design of high-performance digital circuits. An analytical model of delay as a function of process and design parameters, such as the bias current that determines the static power dissipation, would be useful to develop a better understanding of the power-delay trade-off.

In this paper, an analytical delay model of MUX, XOR and D-latch SCL gates is proposed. The model is based on a preliminary circuit linearization and the dominant-pole approximation. The expressions obtained can be helpful to the design at the transistor level of SCL gates.

The model proposed was validated by comparison with Spectre simulations in a variety of load and biasing conditions by using a 0.35-μm CMOS process.

B. Hochet et al. (Eds.): PATMOS 2002, LNCS 2451, pp. 429–437, 2002.

2 MUX, XOR, and D-Latch SCL Gates

SCL logic gates are based on the source-coupled pair of NMOS transistors, that work in the saturation or cut-off region and approximate well a voltage-controlled current switch. The bias current, I_B, is steered to one of the two output branches by a network consisting of source-coupled pairs, whose topology depends on the logic function implemented. In order to always guarantee a meaningful logic value at the output, to each input set value there is only one current path connecting the current source and the outputs. Moreover, to ensure the same noise immunity for all inputs, the NMOS transistors are characterized by the same channel width, W_n, and length, L_n [4]. The current steered by the NMOS network is converted to a differential output voltage $v_o = v_{o,1} - v_{o,2}$, through an active load implemented by a PMOS transistors working in the linear region.

The topology of the multiplexer (MUX), XOR and D-Latch Source Coupled gates is reported in Fig. 1, 2 and 3, respectively. In the MUX gate the selection signal, Φ, activates the source-coupled pair associated with input A or B, transferring its input value to the output. Immediate analysis of the circuit in Fig. 2 shows that it implements the XOR function. As far as the D-latch is concerned, the clock signal, CK, selects either the source-coupled pair M3-M4 that transfers input value of D to the output, or the cross-coupled source-coupled pair M5-M6 that stores the previous output value.

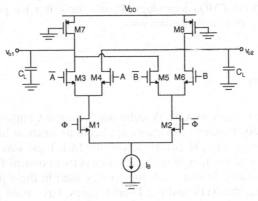

Fig. 1. Topology of an SCL MUX gate.

2.1 Simplified Model of the PMOS Active Load

As demonstrated in [6], where a simple inverter is considered, analysis of circuits in Figs. 1–3 can be simplified by replacing PMOS transistors M7-M8 by an equivalent linear resistance, R_D, in parallel with their parasitic drain-bulk and gate-drain capacitances, that have to be properly linearized since their voltage move over a wide range. More specifically, equivalent resistance R_D, evaluated by using standard BSIM3v3 model [7], is

$$R_D = \frac{R_{int}}{1 - \dfrac{R_{DS}}{R_{int}}} \tag{1}$$

where parameter $R_{DS}=(R_{DSW}*1E-6)/W_p$ models the source/drain parasitic resistance and depends on the empiric model parameter R_{DSW} as well as PMOS transistor effective channel width W_p, and parameter R_{int} is given by

$$R_{int} = \left[\mu_{eff,p} C_{OX} \frac{W_p}{L_p} \left(V_{DD} - |V_{T,p}| \right) \right]^{-1} \tag{2}$$

that represents the contribution to resistance due to the channel of the PMOS transistor in the linear region without accounting for the parasitic drain/source resistance one. In (2), term $\mu_{eff,p}$ represents the effective hole mobility [6], parameter L_p is the PMOS effective channel length, C_{OX} is the oxide capacitance per area, and $V_{T,p}$ is the threshold voltage.

The PMOS drain-bulk junction capacitance, $C_{db,p}$, is linearized by modifying its value in a zero-bias condition via coefficients K_j given by [8]

$$K_j = \frac{\phi^m}{V_2 - V_1} \left[\frac{(\phi - V_1)^{1-m}}{1-m} - \frac{(\phi - V_2)^{1-m}}{1-m} \right] \tag{3}$$

where ϕ is the built-in potential across the junction, m is the grading coefficient of the junction, and V_1 and V_2 are the minimum and maximum direct voltages across the junction, respectively.

Since M7-M8 work in the linear region, their gate-drain capacitance, $C_{gd,p}$, is equal to the sum of overlap contribution and channel contribution evaluated by using the BSIM3v3 model

$$C_{gd,p} = C_{gd0} W_p + \frac{3}{4} A_{bulk,max} W_p L_p C_{OX} \tag{4}$$

where C_{gd0} is the overlap capacitance per unit length, and coefficient $A_{bulk,max}$ is a process-dependent constant slightly greater than unity, equal to [6]

$$A_{bulk} = \max_{L_p} \left\{ \frac{1}{1 + K_{ETA}|V_{SB}|} \left\{ 1 + \frac{K_{1OX}}{2\sqrt{\phi_S - |V_{SB}|}} \left[\frac{A_0 L_p}{L + 2\sqrt{X_J X_{dep}}} \cdot \right. \right. \right. \tag{5}$$

$$\left. \left. \left. \cdot \left(1 - A_{GS} \left(|V_{GS}| - |V_T| \right) \left(\frac{L_p}{L_p + 2\sqrt{X_J X_{dep}}} \right)^2 \right) + \frac{B_0}{W_{p,min} + B_1} \right] \right\} \right\}$$

that in a 0.35-μm CMOS process and $V_{DD}=3.3$ V results 1.34. It is worth noting that, as observed in [6], the channel contribution is significantly different from that obtained for long-channel technologies.

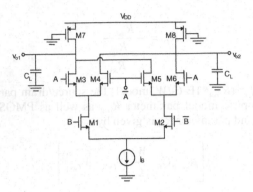

Fig. 2. Topology of an SCL XOR gate.

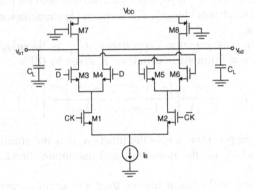

Fig. 3. Topology of an SCL D-latch.

3   Delay Model of MUX, XOR, and D-Latch SCL Gates

In the following, an analytical delay model of MUX and XOR gates is developed and then extended to the D-latch circuit.

3.1   Delay Model of the MUX and XOR SCL Gates

In the SCL MUX shown in Fig. 1, the worst-case delay is obtained by applying the switching input, v_i, to transistors M1-M2, with constant logic values of A and B, that are assumed to be 0 and 1, respectively, without loss of generality. Indeed, when v_i is applied to the transistors M3-M4 (M5-M6), capacitances of transistors M1-M2 do not contribute to the overall delay, since they have already switched.

Under this condition, transistors M3 and M6 work in the saturation region, while M4-M5 are in the cut-off region. In this worst-case condition, the XOR gate delay is equal to that of the MUX, since the former is obtained from the latter simply by setting in it $B = \overline{A}$. Therefore, in the following we will only consider the MUX gate.

The MUX delay can be analytically modeled by considering that transistors M1-M2 of the MUX work in the saturation region for most of the time, and their source voltage is the same for both input logic values, since it is set by the NMOS transistor in the ON state regardless of the input value. Thus, the circuit can be linearized around the logic threshold $v_i=0$, and half-circuit concept applies, due to the topological symmetry and differential input. The resulting linearized half-circuit, that is shown in Fig. 4, transistors M1-M3 (M2-M6) are replaced by their linearized model that includes an equivalent transconductance, $G_{M,n}$, that is evaluated in the following.

In the equivalent circuit in Fig. 4, transistor M7 (M8) is replaced by its equivalent resistance R_D, and input A is driven by an equal logic gate with equivalent load resistance R_D. In Fig. 4, transistors capacitances that must be considered are C_{db} and C_{sb}, that represent the (equal) drain-bulk and source-bulk junction capacitances, C_{gd}, that schematizes the channel and the overlap contribution between gate and drain, and C_{gs}, that is the gate-source capacitance. Moreover, the wiring capacitance and input capacitance of driven logic gates are schematized by an external load capacitance, C_L.

To simplify delay evaluation, let us assume the circuit in Fig. 4 to be well approximated by a first-order network with time constant τ, that can be evaluated by resorting to the time constant method [9]. Thus, the MUX delay for a step input is equal to $0.69\,\tau$, that by inspection of Fig. 4 is given by

$$\tau_{PD,SCL} = 0.69\left[R_D\left(C_{db,3} + 2C_{gd,3} + C_{db,5} + C_{gd,5} + C_{db,7} + C_{gd,7} + C_L\right) + \right. \tag{6}$$

$$\left. + \frac{1}{G_{M,n}}\left(C_{db,1} + C_{gd,1} + C_{sb,3} + C_{gs,3} + C_{gs,4}\right)\right]$$

In relationship (6), parameter $G_{M,n}$ for M1-M3 can be evaluated as the ratio of the total variation of drain current $i_{D,n}$ and total variation of voltage $V_{GS,n}$ during a complete switching. Since the drain current of M1-M3 varies from I_B to zero (or viceversa), and their $V_{GS,n}$ varies from $V_{Tn} + \sqrt{I_B \Big/ \left(\frac{1}{2}\mu_{eff,n}C_{OX}\frac{W}{L}\Big|_3\right)}$ to $V_{T,n}$, the resulting expression of $G_{M,n}$ is

$$G_{M,n} = \sqrt{\frac{1}{2}\mu_{eff,n}C_{OX}\frac{W}{L}\Big|_n I_B} \tag{7}$$

that, compared to the traditional small-signal transconductance, it is reduced by a factor of two (this is due to the wide variation of current and voltage around the bias point). In relationship (7), term $\mu_{eff,n}$ represents the effective electron mobility and can be evaluated as a function of BSIM3v3 parameters as done in [6].

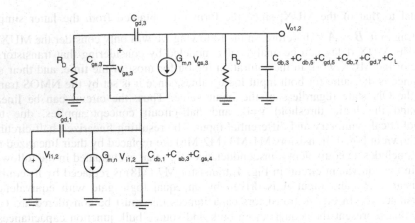

Fig. 4. Equivalent linear circuit of an SCL MUX circuit.

The NMOS capacitances C_{gd1}, C_{gd3} and C_{gd5} (C_{gs4}) are an overlap contribution equal to the product of the overlap capacitance per unit channel width C_{gd0} (C_{gs0}) and their channel width, W_n,

$$C_{gd,1,3,5} = C_{gd0}W_n$$

$$C_{gs,4} = C_{gs0}W_n \tag{8}$$

Moreover, capacitance C_{gs3} of transistor M3 working in the saturation region includes the overlap and the channel contribution

$$C_{gs,3} = C_{gs0}W_n + \frac{2}{3}W_n L_n C_{OX} \tag{9}$$

All drain-bulk and source-bulk NMOS capacitances are linearized by means of (3), as discussed for the PMOS drain-bulk capacitance.

3.2 Input Capacitance Evaluation

When cascaded logic gates are considered, delay of each gate has to be evaluated by accounting for the input capacitance of the driven circuits. To evaluate the input capacitance of an SCL gate, C_{input}, it is useful to remember that the source voltage of a switching source-coupled pair is independent of the input logic value. Thus, assuming the source voltage to be constant, the input capacitance C_{input} seen from the gate of each transistor results equal to its gate-source capacitance evaluated in the saturation region, that is given by

$$C_{input} = \frac{2}{3}W_n L_n C_{OX} \tag{10}$$

3.3 Extension of the Model to the D-Latch Gate

As for the MUX and XOR gates, the worst-case delay of a D-latch occurs when input *CK* switches and other input, *D*, is constant. To evaluate the delay of the D-latch, it is worth noting that it differs from MUX and XOR gate only for the cross-coupled source-coupled pair M5-M6. These transistors load the output nodes by the capacitance seen from their gate terminal, C_{input}, adding to the other contributions, that are equal to those of MUX and XOR gate, by inspection of circuit in Fig. 3. As a consequence, the D-latch delay is again given by relationship (6), where an equivalent load capacitance C_L' equal to

$$C_L' = C_L + C_{input} \tag{11}$$

must be used to account for additive loading effect of transistors M5-M6.

4 Delay Model Validation and Simulation Results

The delay model (6) of the MUX and XOR gates and (6)–(11) of the D-latch were validated by performing extensive simulations by using a 0.35-μm CMOS process.

To explore a wide variety of design and loading conditions, the bias current was varied from 5 μA to 100 μA, and transistors' aspect ratios were sized to obtain a logic swing ranging from 400 mV to 1.6 V, and a voltage gain around the logic threshold ranging from 2 to 7. Moreover, the load capacitance, C_L, was set to 0 F, 50 fF, 200 fF and 1 pF.

Some curves among the results obtained are reported in Figs. 5–6, where delay of the MUX/XOR simulated and predicted is plotted versus bias current, I_B, for a transistor sizing that leads to a logic swing of 700 mV and a voltage gain of 3 (obviously, transistors' aspect ratio are changed with the bias currents according to logic swing and voltage gain requirements). More specifically, Figs. 5-6 refer to the case with load capacitance, C_L, equal to 0 F and 1 pF, respectively. The analytical model (6) agrees well with simulation results, as is shown by Fig. 7, that reports the error with respect to values obtained with Spectre simulations versus I_B for the considered load capacitances. By inspection of Fig. 7, the error is always lower than 20%, and typically lower. Indeed, its average value is 8.9%. Similar results have been obtained for the D-latch gate, for which the maximum and the average error are 22% and 7.4%, respectively. As a result, the model proposed of MUX, XOR and D-latch SCL gates is accurate enough for practical purposes, i.e. for timing analysis of complex digital circuits.

Regarding the input capacitance, comparison with simulation runs shows that relationship (10) differs from the measured value by at most 15%, and typically by less than 10%.

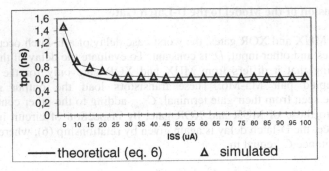

Fig. 5. Delay of MUX gate versus bias current for C_L=0 F.

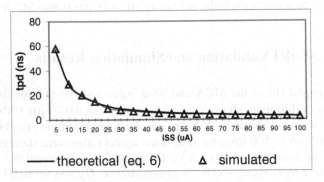

Fig. 6. Delay of MUX gate versus bias current for C_L=1 pF.

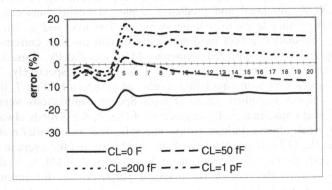

Fig. 7. Error of the model (6) with respect to simulations versus bias current for C_L equal to 0 F, 50 fF, 200 fF and 1 pF.

5   Conclusions

In this paper, an analytical delay model of Source-Coupled Logic (SCL) gates was discussed. The multiplexer, XOR and D-latch gates were analyzed by a first circuit linearization. Successively, the equivalent circuit obtained was simplified by introducing the dominant-pole approximation. The closed-form expression of delay is quite simple and each term has an evident circuit meaning. Hence, delay expression allows a deep understanding of delay dependence on process and design parameters, as well as of dominant contributions. For this reason, delay expression proposed is useful to design. The model was validated by extensive comparison with Spectre simulations by using a 0.35-μm CMOS technology. Analysis shows that the maximum error is about 20%, and its average value is lower than 10%. Therefore, the delay expression proposed is accurate enough for modeling purposes.

References

[1] H. Leopold, G. Winkler, P. O'Leary, K. Ilzer, J. Jernej, "A monolithic CMOS 20-b analog-to-digital converter," *IEEE J. Solid-State Circuits*, vol. 26, pp. 910–916, July 1991.

[2] I. Fujimori *et al.*, "A 5-V single chip delta-sigma audio A/D converter with 111 dB dynamic range," *IEEE J. Of Solid State Circuits*, vol. 32, pp. 329–336, Mar. 1997.

[3] S. Jantzi, K. Martin, A. Sedra, "Quadrature bandpass ΣΔ modulator for digital radio," *IEEE J. Of Solid State Circuits*, vol. 32, pp. 1935–1949, 1997.

[4] S. Kiaei, S. Chee, D. Allstot, "CMOS source-coupled logic for mixed-mode VLSI," *Proc. Int. Symp. Circuits Systems*, pp. 1608–1611, 1990.

[5] A. Tanabe, M. Umetani, I. Fujiwara, T. Ogura, K. Kataoka, M. Okiara, H. Sakuraba, T. Endoh, F. Masuoka, "0.18-μm CMOS 10-Gb/s Multiplexer/Demultiplexer ICs Using Current Mode Logic with Tolerance to Threshold Voltage Fluctuation," *IEEE J. of Solid-State Circuits*, vol. 36, no. 6, June 2001.

[6] M. Alioto - G. Palumbo - S. Pennisi, "Predicting Propagation Delay in SCL Gates", *ECCTD'01*, Espoo (Finland), pp. III/209-212, August 2001.

[7] Y. Cheng, C. Hu, *MOSFET modeling & BSIM3 user's guide*, Kluwer Academic Publishers, Boston, 1999.

[8] J. Rabaey, *Digital Integrated Circuits (A Design Perspective)*, Prentice Hall, 1996.

[9] B. Cochrun, A. Grabel, "A Method for the Determination of the Transfer Function of Electronic Circuits", *IEEE Trans. on Circuit Theory*, vol. CT-20, no. 1, pp. 16–20, Jan. 1973.

Operating Region Modelling and Timing Analysis of CMOS Gates Driving Transmission Lines

Gregorio Cappuccino and Giuseppe Cocorullo

Electronics, Computer Science and Systems Department,
DEIS-University of Calabria
87036 Rende, Italy
{cappuccino, cocorullo}@deis.unical.it

Abstract. The switching behaviour and the operating region of a complementary metal-oxide-semiconductor (CMOS) gate driving a resistance-inductance-capacitance (RLC) transmission line is investigated in this paper. Closed form expressions for the time the transistors operate in the saturation and triode region respectively are proposed. Closed form expressions show predictions within 10% of HSPICE results for a wide range of line and buffer parameters, making them suitable to be applied to the problem of buffer sizing, repeater insertion, short circuit power estimation and generally whenever the accurate knowledge of the operation of CMOS buffers driving a transmission line is required. In the paper useful hints for choosing the most appropriate model for the triode region of the transistors of the inductive-line driver are also given.

1 Introduction

In the last years, the role of interconnects has become one of the most important issues in the design of high performance CMOS circuits. It is well known that capacitive wire parasitic is the main reason for the dynamic power consumption and, in conjunction with the wire resistance, it dominates the overall propagation delay of signals. Moreover the high yield rate and the miniaturisation allowed by progress in technology have led to complex integrated circuits (ICs) reaching 1-2 centimetres edge dimension, and both gate delay and rise time to be lowered to few tens of picoseconds. In these ICs, global scope interconnects such as that encountered in clock distribution networks, control and data buses may traverse up to one chip edge [1]. Thus the electrical length of interconnects can become a significant fraction of a wavelength and, owing to line inductance, lines may exhibit the characteristic behaviour of a transmission line. In this case, the transmission line properties may significantly alter the behaviour of the CMOS buffer, forcing transistors uncommonly to work in linear mode rather than in saturation for a significant percentage of the transition time. The goal of this paper is to provide an accurate estimation of the portions of the switching period in which the transistor operates in saturation and in linear mode respectively. The proposed expressions can be used in the CAD tools

B. Hochet et al. (Eds.): PATMOS 2002, LNCS 2451, pp. 438–447, 2002.

adopted in the critical design phases of nowadays integrated circuits such as in buffer sizing, repeater insertion and power minimization strategies.

The rest of the paper is organised as follows:

a background for the analysis of the commutation transient of a CMOS gate driving a transmission line is provided in section 2. In this section the first phases of the switching transient are analysed and a closed form solution for the time the transistor operates in penthode (saturation) region is reported.

Section 3 provides the description of the remaining part of the switching process, when the transistor operates in triode (linear) mode. In section 3 a linear model for the transistor operating in triode region is also proposed. On the basis of the presented model the closed form solution for the duration of the last transition phase is carried out. Simulation results and accuracy characterization proving the validity of the proposed expressions are reported in section 4. Finally, some conclusions are drawn in section 5.

2 The First Phase of the Switching Transient: The Saturation

In fig.1 two typical CMOS gates, a line driver and a receiver are depicted. The active devices are interconnected through a line with per-unit-length capacitance, resistance and inductance C, R and L, respectively.

In the subsequent discussion we will focus on a $0\text{-}V_{dd}$ transition occurring at the driver output and then involving mainly the PMOS, but it being understood that all results are applicable to the $V_{dd} - 0$ one, once it is recognized that all the voltage polarity in a PMOS transistor are opposite to the corresponding quantities in a NMOS device,

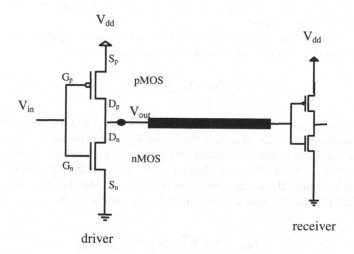

Fig. 1. A CMOS driver and a receiver interconnected through a transmission line

The work is developed assuming that the line resistance being low enough to neglect signal attenuation introduced by the interconnect and that the fall times of the input signal t_{Fin} is lower than the propagation delay across the line $t_{flight}=Z_0C$, where $Z_0=(L/C)^{1/2}$ is the characteristic impedance of the line. These assumptions will be done because either for $t_{Fin} > t_{flight}$ or for high losses the inductive effect of the line parasitics is usually negligible and the transmission line model becomes unnecessary [2]. In fact, in the last cases, the transmission line model can be simply replaced by a lumped or distributed capacitances and resistances. However should be pointing out that in next-generation VLSI circuits, on-chip signal transition times are decreasing, lines are becoming longer and wire resistance is being keeping as low as possible to reduce delay [3]. Therefore, the conditions assumed in the paper will be even more recurrent in future integrated circuits.

During a high-to-low transition at the input of the driver of fig. 1, assuming a full voltage swing (i.e. from the supply voltage V_{dd} to 0V) and a linear relation between the input voltage V_{in} and the time t, V_{in} can be expressed as:

$$
V_{in}(t) = \begin{cases} V_{dd}\left(1-\dfrac{t}{t_{Fin}}\right) & \text{for} \quad 0 \leq t \leq t_{Fin} \\ 0 & \text{for} \quad t \geq t_{Fin} \end{cases} \tag{1}
$$

t_{Fin} being the fall time of the signal.

At $t= t_{ON} = t_{Fin}|V_{Tp}| / V_{dd}$, in accord with (1), $V_{in}(t)$ reaches the value $V_{dd}-|V_{Tp}|$, V_{Tp} being the threshold voltage of the PMOS. As a consequence, the gate-to-source voltage, i.e. $V_{GSp}(t) = V_{in}(t) - V_{dd}$, equals V_{Tp} and the transistor begins to conduct a current $I_{DSp}(t)$. Assuming the dividing line between the triode and saturation regions of PMOS is given by $V_{DSp} = V_{DSSATp}$ where $V_{DSSATp}=V_{GSp}-V_{Tp}$ [4], the transistor is operating in the saturation region.

In fact, by expressing the gate-to-source and drain-to-source voltages in terms of driver input and output voltages, the saturation condition is met if:

$$
V_{in}(t) - V_{out}(t) > V_{T_p} , \tag{2}
$$

and since $V_{in}(t)$ and $V_{out}(t)$ are still equal approximately to V_{dd} and 0V, respectively, it follows that equation (2) holds, at least at the beginning of the transition.

As long as the transistor remains saturated, neglecting the channel length modulation, the output current of the PMOS will depend only on the effective control voltage $V_{GSp}-V_{Tp}$. Then assuming the alpha-power law is used to model the saturation current of the MOSFET [5], the drain-to-source current $I_{DSp}(t)$, for $t > t_{on}$, is

$$
I_{DS_p}(t) = K|V_{in}(t) - V_{dd} - V_{T_p}|^{\alpha} , \tag{3}
$$

where α is the velocity saturation index and K is a coefficient taking into account the geometrical and process parameters of the transistor, respectively [2, 5].

It is evident from (3) that the maximum value of the output current I_{MAX} occurs at $t=t_{Fin}$, when $V_{in}(t)=0$:

$$I_{MAX} = K\left(V_{dd} - |V_{Tp}|\right)^\alpha. \tag{4}$$

Equation (3) can be used to carry out the expression for the output voltage of the buffer; in fact, since during the first transient phase both lossless and lossy transmission lines behave at the input nodes as a resistor equal to the characteristic impedance of the line [6], the output voltage $V_{out}(t)$ is given by the output current of the driver multiplied by Z_0 [2].

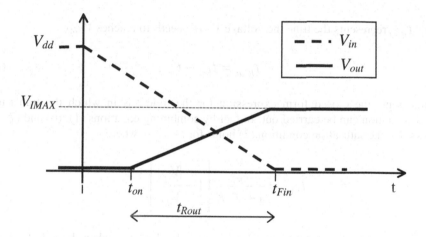

Fig. 2. Schematisation of a Vdd → 0 transition at the input of a CMOS buffer driving a transmission line and corresponding driver response. t_{Fin} and t_{rout} represent the input signal fall-time and the output waveform rise-time, respectively

That is, from (1) and (3), ignoring the effect of the NMOS transistor:

$$V_{out}(t) = Z_0 I_{DS_P}(t) = Z_0 K\left(V_{dd}\frac{t}{t_{Fin}} - |V_{Tp}|\right)^\alpha. \tag{5}$$

This means that at $t=t_{on}$ the voltage at the output of the driver begins to rise towards a relative maximum V_{IMAX} that is reached at $t=t_{Fin}$:

$$V_{IMAX} = V_{out}(t_{Fin}) = Z_0 K\left(V_{dd} - |V_{Tp}|\right)^\alpha, \tag{6}$$

A very realistic hypothesis, without loss of generality, is that $V_{IMAX} \leq V_{dd}/2$, that is the line is not overdriven. In fact, if this is not the case, voltage undershoots and overshoots may occur at both the near and far end of the line. Since the signal may oscillate to voltage levels higher than supply voltage or lower than ground, in practical CMOS this condition is strongly avoided, because it can cause logical errors and severe transistor reliability problems. Moreover, because the saturation index alpha is approximately equal to one for short-channel MOSFET [5], in accord with (5) a linear relation between the output voltage $V_{out}(t)$ and the time t should be also expected for $t_{on} < t < t_{Fin}$, as pictorially depicted in fig. 2. Then $V_{out}(t)$ can be expressed, in this range, as:

$$V_{out}(t) = V_{IMAX}\, \frac{t}{t_{Rout}} \, , \tag{7}$$

where t_{Rout} represents the time the voltage $V_{out}(t)$ needs to reaches V_{IMAX}:

$$t_{Rout} = t_{Fin} - t_{on} \, . \tag{8}$$

At this point, the closed form expression for the time t_{SAT} in which the transistor leaves saturation can be carried out. In fact by combining equations (1), (6) and (7), it follows that the saturation condition (2) holds for $t < t_{SAT}$, where

$$t_{SAT} = \frac{t_{Fin}}{1 + Z_0 K}\left[1 + \frac{|V_{Tp}|}{V_{dd}}\right]. \tag{9}$$

t_{SAT} coincides with the instant when $V_{out}(t)$ equals $|V_{Tp}|$, that is when the gate-to-drain voltage results in inversion charges in the proximity of the drain region, causing transistor to operate in triode region instead of saturation [4].

3   From Saturation to Triode Region

As above mentioned, for $t \geq t_{SAT}$ the transistor enters in triode region and begins to act as a resistor $R_{ON}(t)$ whose value depends on both the gate-to-source and the drain-to-source voltages.
In these conditions the drain-to-source current $I_{DSp}(t)$ is:

$$I_{DS_P}(t) = KV_{DS}(t)\left[V_{GS}(t) - V_{TP} - V_{DS}(t)\right]. \tag{10}$$

The value of resistance $R_{ON}(t)$ seen between the drain and the source of the MOSFET can be carried out by calculating the $V_{DS}(t)/I_{DSp}(t)$ ratio. Taking into account that, in the above mentioned condition, V_{GSp} is equal to $-V_{dd}$ and $V_{in}(t)$ is 0V by now:

$$R_{ON}(t) = \frac{V_{DS}(t)}{I_{DSP\,lin}(t)} = \frac{1}{K\left[\left(V_{dd} - \left|V_{T_p}\right|\right) - V_{DS}(t)\right]}, \quad (11)$$

This denotes that the PMOS can be assumed to be a voltage-dependent resistor whose value ranges between an upper bound R_{ONmax} when $V_{out}(t) = |V_{Tp}|$, i.e. at $t = t_{SAT}$, and a lower bound R_{ONmin} reached at the end of the $0 \rightarrow V_{dd}$ transition, when $V_{out}(t) = V_{dd}$:

$$R_{ON\,min} = \frac{1}{K\left(V_{dd} - \left|V_{T_p}\right|\right)}, \quad (12a)$$

and

$$R_{ON\,max} = \frac{2}{K\left(V_{dd} - \left|V_{T_p}\right|\right)}, \quad (12b)$$

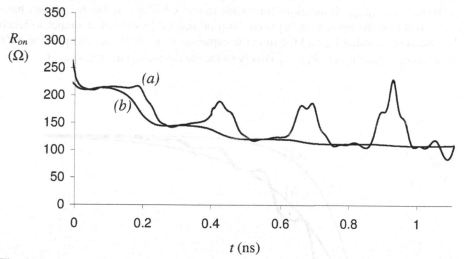

Fig. 3. Actual drain-to-source voltage/current ratio of a PMOS driving an RLC transmission line (a) and corresponding estimated value (b). As noticeable, the channel resistance as the transistor starts to operate in triode region, $R_{ONmax} \approx 215\Omega$, is twice the value it takes at the end of the switching transient, $R_{ONmin} \approx 107\Omega$.

Fig. 3 compares the HSPICE plot of the $V_{DSp}(t) / I_{DSp}(t)$ ratio of a buffer driving a transmission line and the value of R_{ON} given by (11). The active device is a 0.35μ PMOS supplied at $V_{dd} = 3.3\text{V}$, with a channel width of 60μ, $K = 3.5 \cdot 10^{-3}\text{A/V}^2$ and a threshold $V_{Tp} = -0.59\text{V}$. The line has a characteristic impedance of 58Ω. As shown, despite of its straightforwardness, the (11) captures with adequate accuracy the development of the channel resistance of the transistor during the signal round-trips needed to complete the transition. The visible points of disagreement between the measured and the estimated values occur only during the arrival of the signal reflections from the far end of the line.

Moreover, the above mentioned results allow important remarks to be done: the drain-to-source resistance of the MOSFET (namely the output resistance of the buffer) as the transistor starts to operate in triode region, R_{ONmax}, is twice the value it takes at the end of the switching transient, R_{ONmin}. Moreover, as the major part of the energy exchange between the supply and the line occurs just during the first round trips of the signal, the above discussion suggests that for $t \geq t_{SAT}$ the MOSFET can be replaced with a linear resistor equal to R_{ONmax} for a wide range of analysis purposes. Fig. 4 reports the output voltage of a 0.35μm CMOS buffer with a channel length and width of 0.3μm and 60μm, respectively. It drives a transmission line with a capacitance of 1.8pF, a resistance of 15Ω and an inductance of 6nH. In the same figure two further plots of the voltage across a capacitor of capacitance C =1.8pF charged from 0 to V_{dd} are reported. The two latter plots refer to a charge process through a resistor of value R_{ONmax} = 215Ω and through a resistor of value R_{ONmin}= 107Ω, respectively. By comparing these three waveforms, is evident the correlation between the output waveform of the actual buffer and the voltage across the capacitor charged by means of the resistor R_{ONmax}. Simulations for a wide range of buffer and line parameters have confirmed a strong agreement between the time needed by the buffer output to reach the steady-state value V_{dd} and the time the capacitor C needs to charge up to the 86% of the same value through R_{ONmax}, namely twice the time-constant $\tau = R_{ONmax}C$.

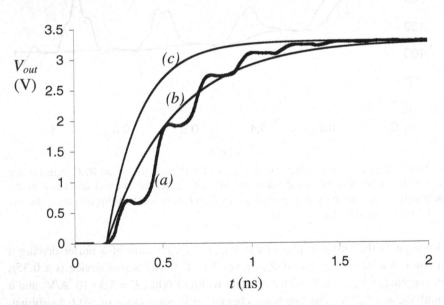

Fig. 4. Comparison among the output waveform of a 0.35μ CMOS buffer driving a transmission line with C=1.8pF, R=15Ω and L=6nH(a) and the charging voltage of a 1.8pF capacitor charged from 0 to 3.3V through: a resistor R_{ONmax} = 215Ω (b), and a resistor R_{ONmin} = 107Ω (c)

On the basis of the experimentally evidenced correlation, the time the transistor spends in the last phase of the $0 \rightarrow V_{dd}$ transition, that is the time the transistor operates in linear mode, can be finally calculated:

in fact, under the assumption that at $t_{end} = 2\tau$ the $0 \rightarrow V_{dd}$ transition is completed, the interval the transistor works in triode region is:

$$t_{end} - t_{SAT} = 2\tau - t_{SAT} . \tag{13a}$$

Moreover, because the line capacitance C can be expressed as T_{flight} / Z_0 [7], the (13a) becomes:

$$2T_{flight} \frac{R_{ON\,max}}{Z_0} - t_{SAT} \tag{13b}$$

or, from (9):

$$2T_{flight} \frac{R_{ON\,max}}{Z_0} - \frac{t_{Fin}}{1 + Z_0 K} \left(1 + \frac{|V_{T_p}|}{V_{dd}} \right). \tag{13c}$$

4   Simulation Results

In order to demonstrate the validity of the proposed expressions, HSPICE simulations have been carried out for a wide range of line and transistor parameters.

Table 1 compares the estimated and the HSPICE values of the PMOS channel resistance R_{ONmax} when the transistor leaves the saturation, for two different transistors. Input signal fall times of 75 and 175 picoseconds have been used. Reported results refer to typical lines with a characteristic impedance of 58Ω and 65Ω, respectively. As shown, model predictions are within 8% of HSPICE results.

Table 1. Measured and estimated values for the PMOS channel resistance R_{ONmax} of a CMOS buffer driving a transmission line.

t_{Fin} (ps)	Z_0 (Ω)	K (A/V^2)	R_{ONmax}(Ω)		
			measured	estimated	error %
75	58	$2.3 \cdot 10^{-3}$	337	316	−6
		$3.5 \cdot 10^{-3}$	205	211	3
	65	$2.3 \cdot 10^{-3}$	332	316	−5
		$3.5 \cdot 10^{-3}$	210	211	0
175	58	$2.3 \cdot 10^{-3}$	343	316	−8
		$3.5 \cdot 10^{-3}$	214	211	−1
	65	$2.3 \cdot 10^{-3}$	338	316	−6
		$3.5 \cdot 10^{-3}$	210	211	0

Table 2 reports the values of the instant in which the transistor finishes to operate in the penthode region and begins to work in the triode region, for the same set of transistor/line parameters of table 1. The results show again a good accordance between HSPICE and estimated values.

Table 2. Comparison among the measured and estimated value of t_{SAT}

tfin (ps)	Z_0 (Ω)	K (A/V^2)	t_{SAT} (ps)		
			measured	estimated	error %
75	58	$2.3 \cdot 10^{-3}$	71	78	10
		$3.5 \cdot 10^{-3}$	69	74	7
	65	$2.3 \cdot 10^{-3}$	70	77	10
		$3.5 \cdot 10^{-3}$	70	72	3
175	58	$2.3 \cdot 10^{-3}$	165	182	10
		$3.5 \cdot 10^{-3}$	158	172	9
	65	$2.3 \cdot 10^{-3}$	164	179	9
		$3.5 \cdot 10^{-3}$	153	168	10

Finally, table 3 compares the measured and estimated values of the duration of the switching transient t_{end} used in the (13). Simulation results for three different line lengths of 9, 12 and 15 mm are shown. As noticed, the proposed model allows accurate evaluating (less than 10 % error in the examined parameters range) of the time the CMOS buffer requires to complete a full-swing transition, and consequently, by means of the (13), of the transient period the transistor works in triode region.

Table 3. The HSPICE and estimated values of the duration of a $0 \rightarrow V_{dd}$ transition at the output of a CMOS buffer as response to a $V_{dd} \rightarrow 0$ input transition with a fall time of 75 ps

line length (mm)	Z_0 (Ω)	K (A/V^2)	t_{end} (ps)		
			measured	estimated	error %
9	58	$2.3 \cdot 10^{-3}$	715	684	−5
		$3.5 \cdot 10^{-3}$	498	456	−9
	65	$2.3 \cdot 10^{-3}$	707	684	−3
		$3.5 \cdot 10^{-3}$	436	456	4
12	58	$2.3 \cdot 10^{-3}$	897	911	2
		$3.5 \cdot 10^{-3}$	632	608	−4
	65	$2.3 \cdot 10^{-3}$	954	911	−5
		$3.5 \cdot 10^{-3}$	611	608	−1
15	58	$2.3 \cdot 10^{-3}$	992	1081	5
		$3.5 \cdot 10^{-3}$	755	760	1
	65	$2.3 \cdot 10^{-3}$	1142	1139	0
		$3.5 \cdot 10^{-3}$	705	760	7

5 Conclusions

Line inductance may play an important role in the switching behaviour of a CMOS buffer. Line inductance, in fact, causes the driver transistors to work mainly in linear region instead of saturation, despite of what is generally assumed. An accurate estimation of the periods in which the switching transistor operates in saturation and linear mode may be of great interest for the development of high performance ICs design strategies. However, in the VLSI application, an analytical approach to the problem is practically unfeasible. In the paper closed expressions for rapid estimation of the portion of the transient time actually spent by the active devices both in saturation and triode region has been presented. The paper proposes also a model for the RLC-line-driver transistor in the triode region. The model agrees well with SPICE simulations and is particularly suitable to be implemented in CAD tools.

References

1. Deutsch A., Koposay G. V., Surovic C. W., Rubin B. J., Terman L. M., Dunne R. P. Jr., Gallo T. A., Dennard R. H.: Modeling and Characterization of Long on Chip Interconnections for High Performance Microprocessors. IBM Journal Res. Dev., Vol 39, n. 5, (1995) 547–567
2. Ismail Y. I., Friedman E. G., Neves J. L.: Dynamic and Short-Circuit Power of CMOS Gates Driving Lossless Transmission Lines. IEEE trans. On Circuit and Systems-I: Fund. Theory and Applications, Vol. 46, n. 8 (1999) 950–961
3. Restle, P. J., Jenkins, K. A., Deutsch, A., Cook P. W.: Measurement and Modelling of On-Chip Transmission Line Effects in a 400 MHz Microprocessor. IEEE Journal of Solid-State Circuits, Vol. 33, n. 4 (1998) 662–665
4. Millman, J., Grabel, A.: Microelectronics. 2nd edn. McGraw-Hill, New York (1987)
5. Sakurai, T., Newton A. R.: Alpha-Power Law MOSFET Model and its Applications to CMOS Inverter Delay and Other Formulas. IEEE Journal of Solid-State Circuits, Vol. 25, n. 2 (1990) 584–593
6. Cappuccino, G., Cocorullo, G.: Time-domain Macromodel for Lossy VLSI Interconnect. IEE Electronics Letters, vol. 36, N. 14 (2000) 1207–1208
7. Bakoglu, H.B.: Circuits, Interconnections and Packaging for VLSI. Addison-Wesley, Reading, Massachusetts (1990)

Selective Clock-Gating for Low Power/Low Noise Synchronous Counters [1]

Pilar Parra, Antonio Acosta, and Manuel Valencia

Instituto de Microelectrónica de Sevilla-CNM / Universidad de Sevilla
Avda. Reina Mercedes s/n, 41012-Sevilla, SPAIN
Phone: +34-95-505-66-66; Fax: +34-95-505-66-86; acojim@imse.cnm.es

Abstract. The objective of this paper is to explore the applicability of clock gating techniques to binary counters in order to reduce the power consumption as well as the switching noise generation. A measurement methodology to establish right comparisons between different implementations of gate-clocked counters is presented. Basically two ways of applying clock gating are considered: clock gating on independent bits and clock gating on groups of bits. The right selection of bits where clock gating must be applied and the suited composition of groups of bits is essential when applying this technique. We have found groupment of bits is the best option when applying clock gating to reduce power consumption and specially to reduce noise generation.

1 Introduction

In sequential circuits, it can be often found idle parts where no computation is performed. We can selectively stop the clock of these portions of the circuit using a signal to stop the global clock. This technique is called clock-gating and has been traditionally used to reduce the power consumption due to the system clock transitions [1-3]. For those cycles in which the partial or total state of the system does not change, it is not necessary to let the clock to have a transition. The same reasoning can be made in terms of switching noise. Clock transitions taking place in idle cycles are source of non-productive noise.

In this paper we have centered on the study of binary counters. Such circuits can be considered as a good example of non-productive noise generation because, most of the cycles, the switching bits are only a few. In this sense, most of the significant bits change at a lower rate than the least significant bits.

Synchronous counters can be considered as key subsystems in modern VLSI circuits because of their wide use and importance. Unlike ripple counters, they do not present spurious states, although the excess in hardware and clock synchronized transitions can make them unsuited for low power/low noise applications [4].

Usually, when applying clock gating techniques, specific conditions should be met that stop the system clock totally or in a big part of the system. So, the excess in hard-

---

1. This work has been sponsored by the Spanish MCYT TIC2000-1350 MODEL and TIC2001-2283 VERDI Projects

B. Hochet et al. (Eds.): PATMOS 2002, pp. 448-457

ware introduced by the clock gating can be compensated by the number of elements that will save power and noise. Only in those cases where a significant part of the clocked system can be disabled, the clock gating technique can bring some advantages in terms of power/noise reduction [5].

For counters, there are no idle cycles (except in the case of "no operation" function if this exists), on the contrary we have that, in every clock cycle, there is at least one state variable switching. For this reason, it is not possible to globally stop the clock and if we want to apply clock gating techniques, we will have to do it on individual bits or on groups of bits. This feature gives a special sense to the study of clock gating in counters.

In this work we will establish the way of stopping the clock in counters separately in some of the bits or, in the case that we select a group strategy, we will establish the suited size of the groups. We will also quantify what is the benefit obtained when applying this technique. In order to do this, a comparison methodology based on switch level simulation has been defined. We will also determine the timing implications of all these techniques.

The structure of the paper is as follows. In the next section we will present the counter that has been used to illustrate the power/noise reduction technique and the selected measurements made to quantify the savings. In section 3 the different options to gate the clock for sequential circuits are summarized and the timing implications of the clock gating strategies are analysed. In section 4 simulation results are presented and finally, we will draw our conclusions.

2   The Counter and the Measurement Methodology

We have selected as demonstrator a 16 bit binary counter with four operations: count up, count down, parallel load and inhibition. It also has an asynchronous reset. In Fig 1a the counter functional and structural descriptions are shown. To implement it, a modular design has been developed based on the cell in Fig 1b. The circuit has been designed using the AMS standard 0.35 μm CMOS library. Although power consumption and noise generation strongly depends on the kind of flip-flops used [6,7] our goal is to explore what improvements can be done in a counter with a given flip-flop.

Concerning the measurement methodology, a simulation-based method has been selected to estimate the power consumption and the noise generated by the counter and by its gate-clocked versions. Focusing on the count up operation, 500 cycles have been simulated at switch level (Mach PA from Mentor Graphics) and the peak to peak deviation (Vpp) in the power supply (VDD) and the RMS noise (PRMS) have been considered as an indirect measurement of noise. For measuring the power consumption the average value of the supply current (AVG(I_{VDD})) was computed. In the simulations, we have considered 1nH parasitics coupling inductances between the circuit and power and between the circuit and ground lines [8] (Fig 2).

We first applied the measurement methodology to the 16 bit binary counter, before applying clock gating, in order to have a reference value to establish comparisons with the gate-clocked solutions. The obtained results are shown in Fig 3, where the varia-

tions in power supply and in supply current versus the time of simulation have been
plotted in order to show the counter behavior in a qualitative way.

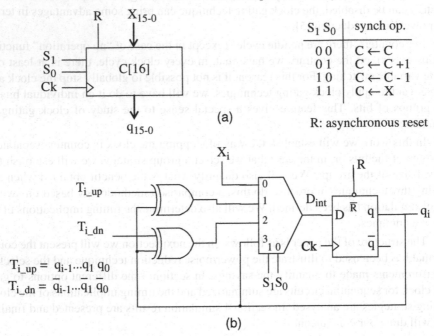

$S_1 S_0$	synch op.
0 0	$C \leftarrow C$
0 1	$C \leftarrow C + 1$
1 0	$C \leftarrow C - 1$
1 1	$C \leftarrow X$

R: asynchronous reset

$$T_{i\_up} = \overline{q_{i-1} \cdots q_1 \, q_0}$$
$$T_{i\_dn} = \overline{\overline{q_{i-1}} \cdots \overline{q_1} \, \overline{q_0}}$$

Fig. 1. (a) Structural and functional descriptions of the implemented
counter (b) Basic cell for the modular design

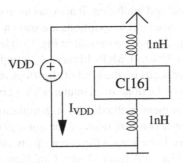

Fig. 2. Environment for simulating realistic conditions

There is a remarkable feature on I_{VDD} vs time graphic: some big-sized peaks of
current appear periodically. The explanation for this fact can be given attending on the
moments where these peaks are produced. They correspond to the times when more
than five or six cells of the counter change simultaneously of state. For example, the
biggest peak in the graphic is produced when the counter goes from 255 to 256
(011111111 -> 100000000 in radix 2). so there are 9 flip-flops switching at a time. The
following peaks in size are produced when the counter goes from 127 to 128 (8 flip-

flops switching), from 63 to 64 (7 flip-flops switching), 31 to 32 (6 flip-flops switching) and others like these.

The values of Vpp (mean and standard deviation) and PRMS are shown in Fig 3 too. We have also computed the mean of Vpp when Ck = 1 (Vppr) and when Ck = 0 (Vppf). It can be noticed that the noise is higher for Ck = 1 (just after the active edge). This effect is due to the particular configuration of the flip-flops. The value of the average current supply consumption (AVG(I)) is shown. With all these quantities we have a reference to estimate how the clock gating technique improves the counter performance in terms of power consumption and noise generation.

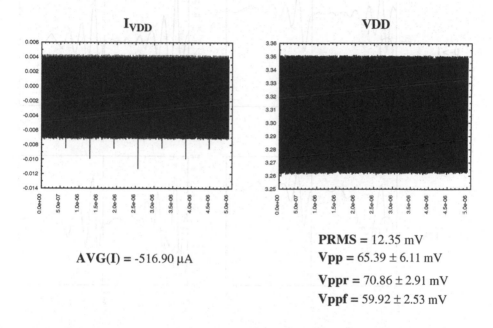

$$AVG(I) = -516.90\ \mu A$$

PRMS = 12.35 mV
Vpp = 65.39 ± 6.11 mV
Vppr = 70.86 ± 2.91 mV
Vppf = 59.92 ± 2.53 mV

Fig. 3. Power supply and supply current versus time of simulation

In Fig 4 we have plotted VDD and I_{VDD} graphics in detail for some cycles of interest: when the counter goes from 0 to 1 (there is only one flip-flop changing); when the counter goes from 127 to 128 (there are 8 flip-flops switching at the same time) and when the counter goes from 255 to 256 (9 flip-flops switching simultaneously). Each graphic is divided in two parts by a vertical line. The first part corresponds to the semicycle where the active edge occurs and the second part corresponds to the other semicycle.

We have marked two peaks in each graphic. The first one (marked with continuous line) is the same in the three cases (0->1, 127->128, 255->256), and is due to the noise produced by the clock signal arriving to the flip-flops. The second one, grows with the number of flip-flops that switch simultaneously, being for example I_{VDD} under 0.005

in the case 0->1, over 0.010 in the case 127->128 and near 0.012 in the case 255->256. We have already pointed up these peaks when looking at Fig 3.

From this analysis we can conclude that there are two sources of noise and it seems clear that clock gating techniques are directed to reduce the first type of noise that it is not needed if there are non-switching bistables, but the second type of noise will need some different considerations.

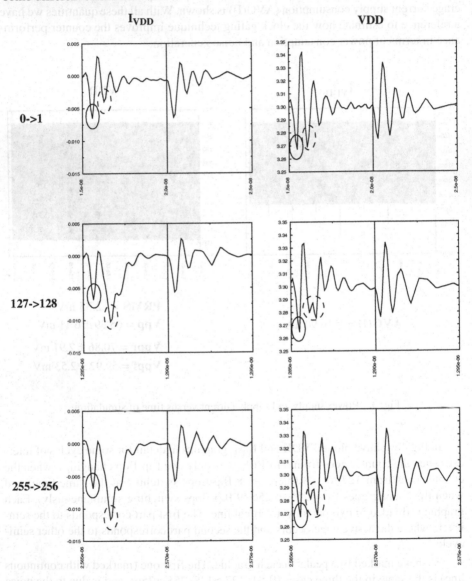

Fig. 4. I$_{VDD}$ and VDD versus time of simulation

3 Selective Clock Gating

3.1 Clock-Gating Circuits

Here, the different approaches we have used to gate the clock are presented. Basically, two cases have been considered: the based on latches solution and the latch-free solution [5]. In the latch-free option we use only one gate (AND or OR) but we need to restrict the changes in the signal that stops the clock (INH). In Fig 5a the latch-free gated clock generator with AND gate is shown. Below the circuit a timing diagram where it can be seen how a positive edge in INH when Ck = 1 provokes an erroneous (no synchronised) edge in the signal gated_ck. The same situation can be found for the OR gate at Ck = 0 (Fig 5b). This effect can be avoided by adding a latch to filter the glitches in INH (Fig 5c and Fig 5d), yielding the based on latches solution.

Let us analyse what circuits from those in Fig 5 are suited for our purposes. In the counter case, the signal INH comes from the comparison between D_{int} (excitation signal of flip-flop i) and q (state of the cell i). Only when both are different there will be a change in the state. Having into account that in our circuit the active edge is the positive one, the transitions in D_{int} and q and hence in INH will probably occur when Ck = 1.

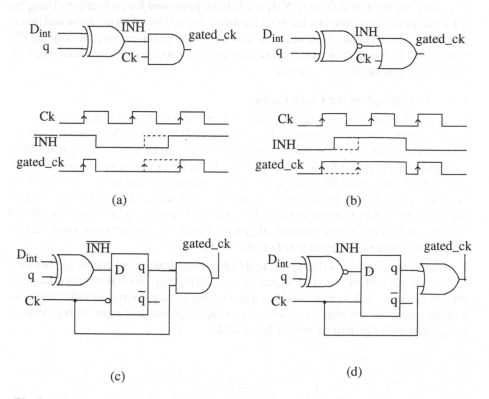

Fig. 5. a), b) free of latches gated clock generation c), d) based on latches
gated clock generation

For this reason we must not consider the case of the Fig 5a that would lead erroneous edges in gated_ck. However, we could consider the case of Fig 5b if we guarantee that the operating frequency in the counter is

$$T = (1/f) < Tp(INH) = Tp\ (q) + Tp(AND) + Tp\ (XNOR).$$

Finally, we can choose whatever between the cases in Fig 5c and Fig 5d, the based on latches solutions, because they filter the glitches and do not impose any restrictions on frequency.

As it was previously said there are two possibilities to apply clock gating techniques to counters: doing it over some of the bits or over groups of bits.

In the first case, the cell i is substituted by a clock gating cell. If based on latches clock gating is selected, a latch and two gates are the overhead for each substituted cell. With this, the extra hardware will produce noise and power consumption comparable with the saving we would get. If clock gating without latches is selected the overhead is smaller but the operating frequency will be reduced.

The second option is to consider groups of bits sharing the same gated clock. This way, only one inhibition signal (INH) needs to be generated for each group. Then, the noise and power introduced by the extra hardware is much less than the noise and power we can save. Some efforts have been made to speed long counters using partitioning [9]. Our interest here is not this but to study how noise and power can be reduced when we partition the counter in several parts.

3.2   Grouping Bits for Clock Gating

To make groups of bits sharing the same clock we must select bits that are consecutive in significance. The least significant bit in the group is the one with the biggest switching frequency and it will impose the frequency of the group clock. In Fig 6 we show three groupment options for the 16 bit counter. In the first we have two blocks of 8 bits, one of them works with the initial clock of the counter (ck) and for the other we have generated a gated clock for the 8th bit so that this cell and the following ones are synchronised. In the second groupment we have generated gated clocks for the 4th, 8th and 12th bits so we have four synchronised groups. Finally the last option implies the biggest cost because we generate gated clocks for 7 groups.

It must be pointed out that applying this grouping strategy the counter do not have full synchronisation between its stages because each group clock is obtained from the previous groups clock signals. This has a good consequence in terms of switching noise reduction because simultaneous switching in flip-flops was the cause of big peaks in I_{VDD} (section 2) that will be indirectly reduced.

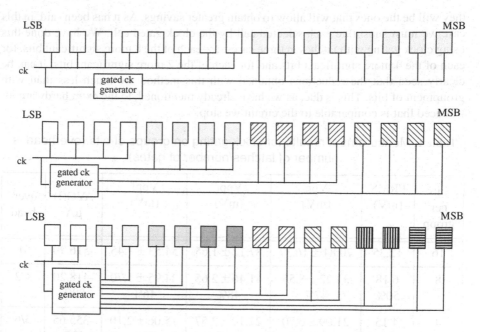

Fig. 6. Three approximations four grouping bits gate clock technique

4   Simulation Results

In this section the obtained results for the implemented clock gated counters are shown. In first place, we will centre on the grouping strategy and later we will compare this with the bit strategy.

In Table 1 are the results for the grouping strategy being applied with blocks of 8, 4 and 2 bits. At the first row the reference values of 16 bit binary counter are repeated to easily establish comparisons. In each of the other rows, the absolute values for the three gated clock cases and also the percentages of reduction are collected. An overhead reference has been introduced in the table. For each gate-clocked counter we show two values separated by a slash. The first one corresponds to the number of latches and the second one to the number of gates. As we can see the 8 bit groupment allows a 50% reduction in noise with a minimum extra cost (only two gates and a latch) and minimum desynchronisation. However the power consumption is only reduced in a 20%. The 4 bit groupment allows to low the generated noise until 35% and power consumption until 70%. The 2 bit groupment improves the behavior only in 10% but with much more extra hardware (7 latches and 14 gates). As we can see this technique is very suited for reducing noise.

In Table 2 we show the results when we apply clock gating on independent bits. In this case we focus on bits that switch at a lower rate (the more significant bits), because they will be the ones that will allow to obtain greater savings. As it has been said, in this case we must substitute each selected cell by the clock gated cell. We have done this using clock gating with latches in three cases: for each of the 8 more significant bits, for each of the 4 more significant bits and for each of the 2 more significant bits. It can be clearly seen that the reductions obtained with this method are much less than with groupment of bits. This is due, as we have already mentioned, to the extra hardware introduced that is comparable to the circuit we stop.

Table 1. Measurement results for clock gating on groups of bits (overhead is number of latches/number of gates).

bits per group	PRMS (mV)	Vpp (mV)	Vppr (mV)	Vppf (mV)	AVG(I) μA	over-head
16	12.35	70.84 ± 16.37	87.12 ± 1.65	54.53 ± 1.45	-516.90	0
8	6.18 50%	33.22 ± 8.55 47%	41.46 ± 2.65 48%	24.95 ± 1.70 46%	-418.20 80%	1/2
4	4.15 34%	21.09 ± 6.70 30%	27.10 ± 3.57 31%	15.06 ± 2.10 28%	-355.65 69%	3/6
2	3.13 25%	13.63 ± 5.52 19%	17.02 ± 5.59 20%	10.23 ± 2.57 19%	-326.41 63%	7/14

Table 2. Measurement results for clock gating on independent bits

bits	PRMS (mV)	Vpp (mV)	Vppr (mV)	Vppf (mV)	AVG(I) μA	over-head
none	12.35	70.84 ± 16.37	87.12 ± 1.65	54.53 ± 1.45	-516.90	0
all	10.86	51.81 ± 19.87 73%	71.51 ± 2.75 82%	32.08 ± 2.08 60%	-683.02 132%	16/32
8 MSB	10.75 87%	52.86 ± 14.39 74%	67.04 ± 2.24 77%	38.64 ± 2.43 71%	-513.63 99.4%	8/8
4 MSB	11.39 92%	59.83 ± 17.53 84%	77.24 ± 1.74 89%	42.38 ± 1.87 78%	-514.99 99.6%	4/4
2 MSB	11.81 96%	64.96 ± 18.59 92%	83.44 ± 1.50 96%	46.43 ± 1.78 85%	-515.65 99.8%	2/2

5   Conclusions

Nowadays it is widely accepted that power consumption and switching noise are crucial factors that must be reduced when designing high performance circuits. This communication analyses the applicability of the clock gating technique to binary counters. Although this technique has been usually used to reduce power consumption, in this work, its influence in switching noise generation has been analysed. Two different clock gating strategies have been considered and compared: clock gating on independent bits and clock gating on groups of bits. We have found how clock gating techniques when are conveniently applied on counters are a very good option to reduce power and, specially, switching noise. Particularly, selecting 2 bit groups for clock-gating yields in a reduced 25% of generation noise and 63% of power consumption when comparing to the original counter. Timing implications in each solution have also been considered and we have conclude that there is a small desynchronization that implies a benefit on switching noise reduction.

References

1. Benini, L., De Micheli, G., "Automatic Synthesis of Low Power Gated-Clock Finite-State Machines", *IEEE Transactions on Computer-Aided Design of Integrated Circuits and Systems,* Vol.15, no. 6, June 1996., pp. 630–643

2. Benini, L., DeMicheli, G., Macii, E., Poncino, M. y Scarsi, R. "Symbolic Synthesis of Clock-Gating Logic for Power Optimization of Control-Oriented Synchronous Networks". *European Design & Test Conference (EDTC-97),* pp. 514–520.

3. Piguet, C., "Low-Power Design of Finite State Machines", in PATMOS, pp. 25-34, Bologna, Sept. 1996.

4. A.J. Acosta, R. Jiménez, J., M. J. Juan, Bellido, and M. Valencia, "Influence of clocking strategies on the design of low switching-noise digital and mixed-signal VLSI circuits", in 10th PATMOS, pp. 316–326, Göttingen, Sept. 2000.

5. Emnett, F., Biegel, M., "Power Reduction through RTL Clock Gating". SNUG San Jose 2000.

6. Stojanovic, V., OKlobdzija, V.G., "Comparative Analysis of Master-Slave LAtches and Flip-Flops for High-Performance and Low-Power Sistems", IEEE Journal of Solid State Circuits, Vol. 34, No. 4, April 1999.

7. Jiménez, R., Parra, P., Sanmartín, P. and Acosta, A.J.: "Analysis of high-performance flip-flops for submicron mixed-signal applications", International Journal of Analog Integrated Circuits and Signal Processing. Kluwer Academic Publishers. (accepted)

8. X. Aragonès, J. L. González and A. Rubio, "Analysis and Solutions for Switching Noise Coupling in Mixed-Signal ICs". Kluwer Academic Publishers, 1999.

9. Stan,M.,Tenca,A. and Ercegovac, M., "Long and Fast Up/Down Counters". IEEE Transactions on Computers, Vol. 47, No. 7, July 1998.

Probabilistic Power Estimation for Digital Signal Processing Architectures*

Achim Freimann

Universität Hannover
Institut für Mikroelektronische Schaltungen und Systeme
freimann@ims.uni-hannover.de
http://www.ims.uni-hannover.de/~freimann

Abstract. A method for estimating the power on architecture-level is described. Originally based on simulations with data sequences, the method is extended by an simulation-free approach. The statistical properties required for the underlying Dual-Bit-Type model are propagated through the circuit. The necessary computation formulas are presented. For both approaches, the model accuracy for base modules as for signal processing applications is comparably low.

1 Introduction

Driven by mobile applications, power dissipation has become a key issue in the design of integrated circuits. Additionally, the time available for designing a circuit is decreased by a shorter cycle time of the product.

To reduce the power consumption of a product without increasing the so-called *time-to-market*, different implementations have to be compared with respect to their power consumption in an early design stage. For this comparison, the architecture-level is best suited. In case the comparison isn't performed manually but instead an architecture synthesis stage shall be used, a fast approach for determining the architecture's power consumption is required.

Several different approaches for power estimation are known from literature [3], [4], which normally rely on simulating an architecture with data sequences. Based on a known procedure which utilizes the *Dual-Bit-Type* (DBT) method [4] in a design-flow employing a circuit synthesis step [2], hereinafter an approach is presented which omits the simulation step.

2 Power Modelling

For modelling the power consumption on architecture-level, often a macro model approach is used. It models each base module like adders, multipliers, registers, e.g., as separate instance with respect to its power consumption. The total power consumption of an architecture is computed as the sum of each module's power.

* The work presented is supported by the German Research Foundation, Deutsche Forschungsgemeinschaft (DFG), within the research initiative "VIVA" under contract number PI 169/14.

B. Hochet et al. (Eds.): PATMOS 2002, LNCS 2451, pp. 458–467, 2002.
© Springer-Verlag Berlin Heidelberg 2002

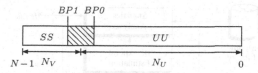

Fig. 1. Dual-Bit-Type data model

2.1 Modelling Approach

The model described in [4] is based on the observation that signals in two's complement number representation show two regions with completely different statistical behavior. In Fig. 1, the decomposition of a data word into a sign region SS, a region behaving like uniform white noise (UWN) UU, and a intermediate region is shown.

To estimate the power of a base module, power coefficients for both types of region in the data word have to be determined once in a characterization phase. Thereafter, these coefficients can be used in the computation of the module's power in an estimation phase.

2.2 Characterization Phase

During the characterization phase, base modules are mapped from a functional description to a gate-level netlist. This technology mapping step enables the use of power estimation in a synthesis-based design flow [2]. Using characteristic stimuli data, a power simulation of the gate-level netlist is performed to obtain power coefficients.

For modules with one input a coefficient P_{UU} representing the UWN region has to be determined. Additionally, four coefficients P_{++}, P_{+-}, P_{-+}, and P_{--}, in short P_{SS}, are required for the sign region, where e.g. $+-$ denotes a sign transition form plus to minus. Modules with two inputs require 64 coefficients $P_{SS/SS/SS}$ for the sign region, as the two inputs and the output can not be considered separately. Again, one coefficient $P_{UU/UU}$ for the UWN region of both inputs has to be determined. The resulting power coefficients are stored in a power database for later use in the estimation phase.

Compared to [4], module characterization is much faster because simulations are performed on gate- instead on circuit-level. Additionally, a parametric model depending on the word sizes is omitted, because it requires the knowledge of the internal module's implementation thus permitting the use of a circuit synthesis. For the approach given in [3], characterization times in the range of hours and days are given, whereas the chosen method performs the module's characterization in just a few seconds.

2.3 Estimation Phase

The estimation phase depicted in Fig. 2 consists of the steps *statistics analysis, breakpoint estimation*, and *power computation*.

In the statistics analysis a functional simulation of an architecture consisting of several base modules is performed using data signals. During the simulation the statistic measures mean value μ, standard deviation σ, and temporal correlation ρ are computed on all module inputs. For modules with one input the number of sign transitions at

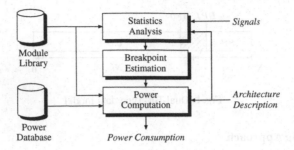

Fig. 2. Details of the estimation phase

the input is counted, whereas for modules with two inputs the combined sign transitions of both inputs and the output is determined. The transition probabilities P_{Ass}, respectively $P_{Ass,Bss,Xss}$ for the combined inputs A and B as well as output X, with $SS \in \{++, +-, -+, --\}$, are computed from the number of their corresponding sign transitions divided by the total number of transitions.

The breakpoints $BP1$ and $BP0$ shown in Fig. 1 are computed in the breakpoint estimation using statistical properties determined in the statistics analysis:

$$\begin{aligned} BP0 &= \log_2 \left(2\sigma \left(1 - \rho \right) \right) \\ BP1 &= \log_2 \left(6\sigma \sqrt{1 - \rho} \right) \end{aligned} \tag{1}$$

Using linear interpolation of the intermediate region, the number of bits in the sign region N_S and the UWN region N_U is computed from the breakpoints:

$$\begin{aligned} N_S &= (N - BP1) + \frac{BP1 - BP0 - 1}{2} \\ N_U &= (BP0 + 1) + \frac{BP1 - BP0 - 1}{2} \end{aligned} \tag{2}$$

The power consumption of a module and the total power consumption of an architecture is computed using the power coefficients which have been determined in the characterization phase. E.g., for a module with one input, the module's power consumption is:

$$P_{Module} = \frac{N_U}{N} P_{UU} + \frac{N_S}{N} \sum_{\forall SS} P_{Ass} P_{SS} \tag{3}$$

3 Probabilistic Statistics Analysis

To compute the power consumption of an architecture, a functional simulation with real data streams is required in the statistics analysis step of the power estimation phase. An approach is proposed here, called probabilistic statistics analysis, that avoids the partly time-consuming simulations. Instead, the statistical properties are propagated through the circuit and all information required in the breakpoint estimation and the power computation are derived from the statistics.

3.1 Requirements

According to (1) - (3), the statistic properties σ and ρ as well as the number of sign transitions, respectively the sign transition probabilities, at the inputs and if necessary at the output of a module are required. This leads to the requirement to provide the statistical properties at the module's output for successive modules.

As the circuit isn't simulated with real data stimuli, a data model is necessary that models the typical data closely. The DBT-model implicitly bases on the assumption of data with gaussian distribution. This can be modelled for the probabilistic approach using an first-order autoregressive noise process, short AR(1)-process. The properties of this AR(1)-process are implicitly exploited in the following derivations.

3.2 Computation of Sign Transition

The major problem in the probabilistic statistics analysis is the computation of the sign transition probability $P_{A_{ss}}$, respectively $P_{A_{ss},B_{ss},X_{ss}}$.

Single Input Modules. The four sign transition probabilities $++$, $+-$, $-+$, and $--$ of a module with one input can generally be computed using the statistical properties of the input signal. In a first step, the *sign probability* P_{A_+} of a positive value on input A, respectively P_{A_-} of a negative value, are computed using the density function $\Phi(x)$ of a gaussian-distribution:

$$
\begin{aligned}
P_{A_+} &= P(A_t \geq 0) = \Phi\left(\frac{-\mu_A}{\sigma_A}\right) \\
P_{A_-} &= P(A_t < 0) = 1 - P_{A_+}
\end{aligned}
\tag{4}
$$

Where A_t is the random variable of input A at an imaginary time t. The transition probabilities can be expressed as conditional probabilities at time t and $t-1$. E.g., $P_{A_{+-}}$ is the probability of a positive to negative sign transition at A and computes to:

$$
P_{A_{+-}} = P(A_t < 0 \mid A_{t-1} \geq 0)
\tag{5}
$$

For $P_{A_{++}}$, $P_{A_{-+}}$, and $P_{A_{--}}$ similar relationships with respect to $P(A_t \mid A_{t-1})$ hold. These probabilities $P_{A_{ss}}$ can't be computed generally from (4), as they depend on the temporal correlation $\rho_A = \mathrm{Corr}(A_{t-1}, A_t)$. The dependency of ρ_A and $P_{A_{+-}}$, respectively $P_{A_{++}}$, is shown in Fig. 3 and was obtained from simulations. The probabilities $P_{A_{ss}}$ can be derived to:

$$
\begin{aligned}
P_{A_{+-}} &= \begin{cases} P_{A_+} P_{A_-} \sqrt{1 - \rho_A} & \rho_A \geq 0 \\ P_{A_+} \left(1 - P_{A_-} \sqrt{1 + \rho_A}\right) & \rho_A < 0 \end{cases} \\
P_{A_{++}} &= P_{A_+} - P_{A_{+-}} \\
P_{A_{-+}} &= \begin{cases} P_{A_-} P_{A_+} \sqrt{1 - \rho_A} & \rho_A \geq 0 \\ P_{A_-} \left(1 - P_{A_+} \sqrt{1 + \rho_A}\right) & \rho_A < 0 \end{cases} \\
P_{A_{--}} &= P_{A_-} - P_{A_{-+}}
\end{aligned}
\tag{6}
$$

With (4), (6), the sign transition probabilities for the power estimation of modules with one input can be computed from statistics μ_A, σ_A, and ρ_A of input A.

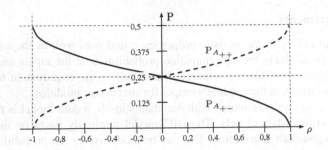

Fig. 3. Sign transition probabilities P_{A++} and P_{A+-} for AR(1)-process ($\mu_A = 0$)

Two Input Modules. For modules with two inputs, the combined sign transition probabilities $P_{Ass,Bss,Xss}$ for both inputs and the output are required. Two out of 64 of them are given here as an example:

$$
\begin{aligned}
P_{A++,B++,X++} &= P_{B++|A++} \, P_{X+|A+,B+} \, P_{X+|A+,B+} \\
P_{A++,B++,X+-} &= P_{B++|A++} \, P_{X+|A+,B+} \, P_{X-|A+,B+}
\end{aligned}
\tag{7}
$$

For (7) the conditional probabilities $P_{Bss|Ass}$ are required. E.g. $P_{Bss|A++}$ computes with $\rho_{AB} = \mathrm{Corr}(A_t, B_t)$ the spatial correlations of A and B to:

$$
P_{B++|A++} = \begin{cases} \left. \begin{array}{c} P_{A++} - P_{B+-|A++} \\ -P_{B-+|A++} - P_{B--|A++} \end{array} \right\} & \rho_{AB} \geq 0 \\ P_{A++} P_{B++} \sqrt{1 + \rho_{AB}} & \rho_{AB} < 0 \end{cases}
$$

$$
P_{B+-|A++} = P_{A++} P_{B+-} \sqrt{1 - |\rho_{AB}|}
$$

$$
P_{B-+|A++} = P_{A++} P_{B-+} \sqrt{1 - |\rho_{AB}|}
\tag{8}
$$

$$
P_{B--|A++} = \begin{cases} P_{A++} P_{B--} \sqrt{1 - \rho_{AB}} & \rho_{AB} \geq 0 \\ \left. \begin{array}{c} P_{A++} - P_{B++|A++} \\ -P_{B+-|A++} - P_{B-+|A++} \end{array} \right\} & \rho_{AB} < 0 \end{cases}
$$

The conditional probabilities $P_{Xs|Bs,As}$ of output X are required in the computation of (7), but can't be given independently of the module's function.

3.3   Computation of Statistical Properties

As an example, the formulae for computation of statistical properties and sign transition probabilities at the module's output are given for adders and multipliers. Accordingly, similar calculations can be given for other modules provided the assumptions given by the AR(1)-process hold. E.g. subtractor, shifter, register, and selector are modules which have been successfully modeled. With this set of modules, many signal processing architectures can be investigated.

Adder. The statistics at the output of the adder $X = A + B$ using the spatial temporal correlations $\rho_{A'B} = \mathrm{Corr}(A_{t-1}, B_t)$ and $\rho_{AB'} = \mathrm{Corr}(A_t, B_{t-1})$ are:

$$\mu_X = \mu_A + \mu_B$$
$$\sigma_X^2 = \sigma_A^2 + \sigma_B^2 + 2\rho_{AB}\sqrt{\sigma_A\sigma_B} \tag{9}$$
$$\rho_X = \frac{\rho_A\sigma_A^2 + (\rho_{A'B} + \rho_{AB'})\sigma_A\sigma_B + \rho_B\sigma_B^2}{\sigma_X^2}$$

The sign probabilities of X for the computation of (7) are:

$$P_{X_+|A_+,B_+} = 1$$

$$P_{X_+|A_+,B_-} = \begin{cases} \Phi\left(\frac{-\mu_X}{\sigma_X}\right) \cdot \sqrt{1 - \rho_{AB}} & \rho_{AB} \geq 0 \\ 1 - P_{X_-|A_+,B_-} & \rho_{AB} < 0 \end{cases}$$

$$P_{X_+|A_-,B_+} = P_{X_+|A_+,B_-}$$

$$P_{X_+|A_-,B_-} = 0$$

$$P_{X_-|A_+,B_+} = 0 \tag{10}$$

$$P_{X_-|A_+,B_-} = \begin{cases} 1 - P_{X_+|A_+,B_-} & \rho_{AB} \geq 0 \\ \left(1 - \Phi\left(\frac{-\mu_X}{\sigma_X}\right)\right) \cdot \sqrt{1 + \rho_{AB}} & \rho_{AB} < 0 \end{cases}$$

$$P_{X_-|A_-,B_+} = P_{X_-|A_+,B_-}$$

$$P_{X_-|A_-,B_-} = 1$$

Multiplier. In case of the multiplier, there is the problem that it is not possible to give the expected value $E(X)$ in case of correlated input signals A and B. Therefore, exclusively the constant multiplication $X = c \cdot A$ is discussed here. For typical signal processing applications, this imposes no limitations on the applicability of the probabilistic approach. The statistics at the output X result in:

$$\mu_X = c \cdot \mu_A$$
$$\sigma_X^2 = c^2 \cdot \sigma_A^2 \tag{11}$$
$$\rho_X = \rho_A$$

A constant multiplier can be implemented as a dedicated multiplier and corresponds to a module with one input, thus no additional information are required for the module's power estimation in (3).

3.4   Propagation of Statistical Properties

To be able to replace the statistics analysis by a simulation-free approach, all statistics have to be propagated through the whole circuit consisting of various modules. Besides the statistics which can be computed at the module's output independent of other modules, spatial temporal correlations ρ_{AB}, $\rho_{A'B}$, and $\rho_{AB'}$ are required in the computation

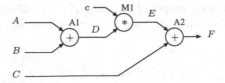

Fig. 4. Signal flow graph von $F = c \cdot (A - B) + C$

of (8)-(10), which have to be determined using statistics applied to the primary inputs of the circuit. They are impacted by the function of all modules which exist on the path between primary inputs and the inputs of the module under investigation.

E.g., cross-correlations of the inputs of Adder A2 shown in the signal flow graph in Fig. 4 can't be computed from Signals E and C directly. Including multiplier M1 and adder A1 on the path from E to A and B, the spatial correlation is:

$$\rho_{EC} = \frac{\sigma_A \cdot \sigma_C \cdot \rho_{AC} + \sigma_B \cdot \sigma_C \cdot \rho_{BC}}{\sigma_E \cdot \sigma_C} \tag{12}$$

For $\rho_{E'C}$ and $\rho_{EC'}$ similar formulae exist. Using (9), (11), σ_E in (12) computes to:

$$\sigma_E = c \cdot \sigma_D = c \cdot \sqrt{\sigma_A^2 + \sigma_B^2 + 2\rho_{AB}\sqrt{\sigma_A \sigma_B}} \tag{13}$$

An efficient algorithm exists for the propagation and computation of all necessary statistics, including spatial temporal correlations, at the modules' inputs and outputs.

4   Model Accuracy

The investigation of the model accuracy provides information about the difference of the module's power estimated on architecture-level compared to a reference computed on gate-level. As an example, results for adders and multipliers are given which have been obtained by performing a large number of experiments.

For the assessment of the model, the relative mean error $\bar{\epsilon}$ and the square root $\tilde{\epsilon}$ of the relative mean square error have been used:

$$\bar{\epsilon} = \frac{1}{L} \sum_{i=1}^{L} \frac{P_{k,i} - P_{l,i}}{P_{k,i}}, \qquad \tilde{\epsilon} = \sqrt{\frac{1}{L} \sum_{i=1}^{L} \left(\frac{P_{k,i} - P_{l,i}}{P_{k,i}} \right)^2} \tag{14}$$

with $P_k \in \{P_{Ref}, P_{Sim}\}, \; P_l \in \{P_{Sim}, P_{Prob}\}$

In (14), P_{Ref} and P_{Sim} are the module's power consumptions obtained simulating on gate- respectively architecture-level. P_{Prob} is the estimated power using the probabilistic approach. L is the number of experiments.

For the determination of the model accuracy of base modules, a noise generator, whose statistical properties μ, σ, and ρ were varied, was used for the experiments. For modules with two inputs, two independent noise generators were used. As their parameters were changed independently, more than 2000 different experiments have been performed per module and given word size of the inputs.

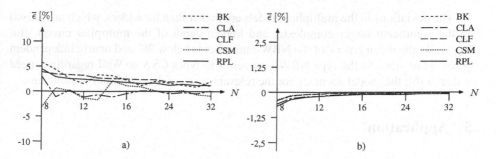

Fig. 5. Error $\bar{\epsilon}$ for different adder implementations: a) P_{Sim} to P_{Ref}, b) P_{Prob} to P_{Sim}

4.1 Adders

The module library contains several different implementations of the adder, like Brent-Kung (BK), Carry Look-Ahead (CLA), Fast Carry Look-Ahead (CLF), Conditional-Sum (CSM), and Ripple-Carry (RPL). The relative mean error $\bar{\epsilon}$ of P_{Sim} to P_{Ref} for these five types of adders depending on the word size N is depicted in Fig. 5 a). In general, the error is noticeable below 5%. The error $\tilde{\epsilon}$, which is not shown here, has a maximum of 7.5% but is for the most part clearly below 3%.

The power obtained by the probabilistic approach P_{Prob} compared to the one computed by a gate-level simulation P_{Ref} results in a nearly identical chart. Therefore, the error $\bar{\epsilon}$ of P_{Prob} to P_{Sim} is given in Fig. 5 b) to emphasize the marginal deviation of the two methods on architecture-level. Mostly, the error is clearly below 1%. The deviation is mainly caused by the different way the number of sign transitions is determined in both approaches.

4.2 Multipliers

Three implementation alternatives of the multipliers were investigated: Carry Save Array (CSA), Non-Booth-coded Wallace Tree (NBW), and Booth-coded Wallace Tree (Wall). The comparison of the power consumption obtained by simulation on gate- and architecture-level are depicted in Fig. 6 for $\bar{\epsilon}$ and $\tilde{\epsilon}$, respectively.

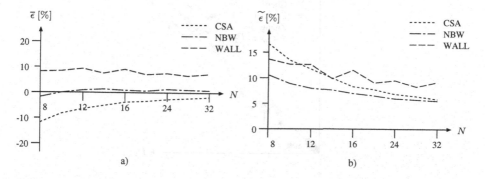

Fig. 6. Errors $\bar{\epsilon}$ and $\tilde{\epsilon}$ for comparison of P_{Sim} to P_{Ref} for different multiplier implementations

The deviations in the multiplier models are higher than for adders, which are caused by the significant larger complexity and logical depth of the multiplier circuit. But especially the mean error $\bar{\epsilon}$ of the NBW-multiplier is below 2% and nearly independent of the word size. As the type NBW is superior to types CSA or Wall regarding speed and area [6], the model accuracy for the relevant cases of the multiplier is sufficient.

5   Application

As an example for a signal processing application, the discrete cosine transform (DCT) was chosen, which is an integral part of many picture and video coding standards. Many different implementations of the DCT are known from literature which were mainly designed with the optimization criteria speed and area in mind, but not power consumption. Following, three different architectures of the 8×1-DCT corresponding to Chen [1], Zhang [7], and Loeffler [5] are investigated. Additionally, an assessment of the power estimation on architecture-level and its suitability for design decisions in an early design stage is given.

Figure 7 shows the model accuracy of each module's power for the DCT implementations. The graph depicts P_{Sim} over P_{Ref} and shows a good correspondence. Comparing the estimated power consumption P_{Sim} and P_{Prob} for the whole architecture, the one for the Chen- and Zhang-DCT is about 10% above the one obtained by simulating the gate-level netlist, whereas for the Loeffler-DCT the power estimation on architecture-level is about 10% too low. Thus, the accuracy for the architectural power estimation is in the range of $\pm 10\%$.

This error range $P_{Sim} \pm 10\%$ is shown for a final assessment in Fig. 8 together with the gate-level power P_{Ref}. Although the sign of the deviation of P_{Sim} compared to P_{Ref} is different for the Zhang- and Loeffler-DCT, a decision based on the architectural power estimation P_{Sim} or P_{Prob} leads correctly to the power efficient Loeffler-DCT.

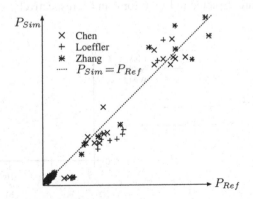

Fig. 7. P_{Sim} compared to P_{Ref} for each module of the DCT for sequence "Coastguard"

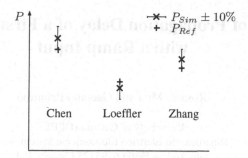

Fig. 8. Comparison of power P_{Sim} including error range $\pm 10\%$ to P_{Ref}

6 Conclusion

In this paper, an extension to an existing approach for architectural power estimation of signal processing architectures is proposed, which omits the usually employed functional simulation. Instead, the statistical properties required for the power estimation of an architecture are applied to the primary circuit inputs. These statistics are propagated through the whole circuit. The problem of computing the sign transitions required at the input of base modules for the employed Dual-Bit-Type data model was solved.

The investigation of model accuracy for base modules shows a good correspondence between estimation on architecture-level and simulation on gate-level. For the application of a discrete cosine transform, the deviation between gate- and architecture-level power estimation is less than 10% and allows for the selection of the power efficient implementation in an early design stage. Finally, the proposed probabilistic power estimation imposes no additional decrease in accuracy compared to a method employing a functional simulation.

References

1. W.-H. Chen, H. Smith, and S. C. Fralick. A fast computational algorithm for the discrete cosine transform. *IEEE Transactions on Communications*, 25(9):1004–1009, Sept. 1977.
2. A. Freimann. Framework for high-level power estimation. In D. Soudris, P. Pirsch, and E. Barke, editors, *Integrated Circuit Design: Power and Timing Modeling, Optimization and Simulation (PATMOS), 10th International Workshop*, volume 1918 of *Lecture Notes in Computer Science*, pages 56–65, Heidelberg, Sept. 2000. Springer Verlag.
3. S. Gupta and F. N. Najm. Power modeling for high level power estimation. In *ACM/IEEE Design Automation Conference*, pages 365–370, 1997.
4. P. E. Landman and J. M. Rabaey. Architectural power analysis: The dual bit type method. *IEEE Transactions on Very Large Scale Integration (VLSI) Systems*, 3(2):173–187, June 1995.
5. C. Loeffler, A. Ligtenberg, and G. S. Moschytz. Practical fast 1-d dct algorithms with 11 multiplications. In *IEEE International Conference on Acoustics, Speech, and Signal Processing*, volume II, pages 988–991, 1989.
6. Synopsys, Inc. *DesignWare Foundation Quick Reference Guide, V2000.11*, 2000.
7. J. Zhang and N. W. Bergmann. A new 8×8 fast dct algorithm for image compression. In *IEEE Workshop on Visual Signal Processing and Communications*, pages 57–60, Sept. 1993.

Modeling of Propagation Delay of a First Order Circuit with a Ramp Input

Rosario Mita and Gaetano Palumbo

University of Catania (DEES)
Dipartimento Elettrico Elettronico e Sistemistico
viale Andrea Doria 6, 95125 Catania, Italy
{rmita, gpalumbo}@dees.unict.it
http://graymalkin.dees.unict.it

Abstract. In this paper a simple analytical model which evaluate the propagation delay of a first-order circuit with a linear input is presented. The model can be used to estimate the propagation delay both of current mode logic (CML) and source coupled logic (SCL) circuits and wires in a VLSI process. The approximation gives an error lower than 6%, and it is a continuous function. The model compared with an ideal RC circuit is successively adopted in a real case such as a CML gate. In particular, this gate was designed with a 6-GHz technology and Spice simulations are performed showing an error lower than 5% in excellent agreement with the theoretical results.

1 Introduction

In the analysis of digital systems it is very important to estimate the maximum operative frequency. A measure to evaluate it is the propagation delay defined as the time between when the output and the input signal reach 50% of their final value. In order to evaluate the propagation delay, some textbooks suggest modeling digital gates with a first order RC circuit [1]–[2]. These models are particularly suitable in CML circuits where the transistors can be assumed to be working in an almost linear region, hence allowing the development of an equivalent small signal model [3] – [4]. Others areas where the first order RC model is increasingly applied are the modeling of gates loaded with an equivalent RC component [5] – [7], and the modeling of wire which is becoming a hot topic in the design of digital systems [8] – [10]. In order to get simple results the particular case of first order RC response with step input signal is often considered. As well-known the resulting propagation delay is

$$\tau_{PD} = 0.69RC \tag{1}$$

However, more realistically cases need to consider at least one linear input signal [11]-[12]. Consider for example two gates in series, the latter is driven by the output of the former that in general cannot be considered a step.

B. Hochet et al. (Eds.): PATMOS 2002, LNCS 2451, pp. 468–476, 2002.

In contrast with the case of a step input, analysis and modeling of the propagation delay in a first order RC circuit driven by a linear ramp is not simple.
In this paper we propose a simple approximation which allows us to simply account for the input slope in the propagation delay. The model is useful for pencil and paper estimation. Model validation is carried out by comparing the propagation delay both of an ideal RC circuit and a CML inverter under many input slope conditions.

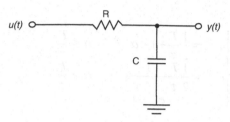

Fig. 1. RC-circuit

2 Time Response of a *RC*-Circuit with a Ramp Input

Let us consider the first order *RC*-circuit in Fig. 1 whose linear ramp input has a rise time T and an amplitude equal to unity according to the following relationship

$$u_{RAMP}(t) = \begin{cases} \dfrac{t}{T} & t \leq T \\[3mm] 1 & t > T \end{cases} \tag{2}$$

The output response in the Laplace domain of the circuit in Fig. 1 with the input given by (2) is

$$Y_{RAMP}(s) = H(s) \cdot U_{RAMP}(s) \tag{3}$$

where $H(s)$ is the transfer function of the RC-circuit and $U_{RAMP}(s)$ is the Laplace transform of the input ramp (2). After some algebraic manipulations, it is easy to obtain the ramp response, in the time domain, as

$$y_{RAMP}(t) = \begin{cases} \dfrac{1}{T}\left(t - \tau + \tau \cdot e^{-\frac{t}{\tau}} \right) & t \leq T \\[4mm] 1 + \dfrac{\tau}{T} \cdot e^{-\frac{t}{\tau}} \cdot \left(1 - e^{\frac{T}{\tau}} \right) & t \geq T \end{cases} \tag{4}$$

The propagation delay is the difference, in time, between when the output and the input signal reach 50% of their final value. The two signals are named τ_{PDout} and τ_{PDin}, respectively. Hence, from definition, the normalized τ_{PDin} is given by

$$\frac{\tau_{PDin}}{\tau} = \frac{1}{2} \cdot \frac{T}{\tau} \tag{5}$$

To evaluate the output propagation delay, we have to solve equation $y_{RAMP}(\tau_{PDout})=0.5$ and, from (4), we get

$$
\begin{cases}
e^{-\alpha} = \dfrac{1}{2}\dfrac{T}{\tau} + 1 - \alpha & \alpha \le \dfrac{T}{\tau} \tag{6a}\\[3mm]
e^{-\alpha} = \dfrac{1}{2}\dfrac{T}{\tau}\dfrac{1}{e^{\frac{T}{\tau}}-1} & \alpha \ge \dfrac{T}{\tau} \tag{6b}
\end{cases}
$$

where parameter α is

$$\alpha = \frac{\tau_{PDout}}{\tau} \tag{7}$$

Relationships (6) are shown in Fig. 2, where, for both (6a) and (6b), parameter α is plot versus T/τ in the range from 0 to 10. Equating (6a) with (6b) and assuming $\alpha=T/\tau$ we found the intersection point between the two curves which is equal to $\alpha=T/\tau=1.6$, and defines the limit of validity for the two relationships.

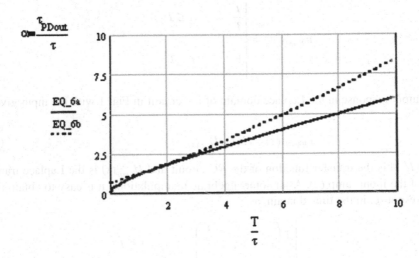

Fig. 2. Equations (6a) and (6b)

Hence, for T/τ greater than 1.6, the propagation delay can be evaluated using (6a) while for T/τ lower than 1.6 we have to use equation (6b). It can be verified that point $T/\tau=1.6$ represents a flex point for function (6).

Solving numerically this equation in the two ranges and using (5), we can plot the normalized propagation delay τ_{PD}/τ versus the normalized input ramp slop T/τ as shown in Fig. 3.

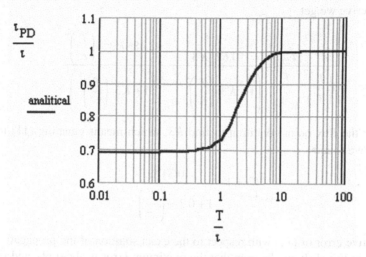

$\dfrac{t_{PD}}{t}$

analitical

$\dfrac{T}{t}$

Fig. 3. Normalized analytical propagation delay of an RC circuit in logarithm scale

3 The Proposed Model

Since an analytical solution of (6) is difficult to reach, and we will use an approximation to solve it. By inspection of Fig. 3, we see that with an input step (i.e., $T=0$) the normalized propagation delay, τ_{PD}/τ, tends to 0.69, while for a very low input slope (i.e., T tending to infinity) the normalized propagation delay tends to be unitary [12]. Hence, in both cases we have a limited behavior. Moreover the curve has the flex point at $T/\tau=1.6$. Thus, the curve on Fig. 3 can be represented by an equation with an upper and lower limit and a definite second order derive. On the basis of these considerations a simple expression that satisfies them is

$$\frac{\tau_{PD}}{\tau} = K1 - \frac{K2}{1 + K3 \cdot \left(\dfrac{T}{\tau}\right)^2} \tag{8}$$

where coefficient $K1$, $K2$, and $K3$ have to be defined to satisfy the previous conditions. In particular, from the upper and lower boundary, we get

$$\lim_{\frac{T}{\tau} \to +\infty} \frac{\tau_{PD}}{\tau} = K1 = 1 \tag{9}$$

$$\lim_{\frac{T}{\tau} \to 0} \frac{\tau_{PD}}{\tau} = K1 - K2 = 0.69 \tag{10}$$

hence, $K1=1$ and $K2=0.31$. Substituting the values found in (12) and evaluating the second derive we get

$$\frac{d^2}{d\left(\frac{T}{\tau}\right)^2} \frac{\tau_{PD}}{\tau} = \frac{0.62 \cdot K3}{\left(1 + K3 \cdot \left(\frac{T}{\tau}\right)^2\right)^2} - \frac{2.48 \cdot K3^2 \cdot \left(\frac{T}{\tau}\right)^2}{\left(1 + K3 \cdot \left(\frac{T}{\tau}\right)^2\right)^3} \tag{11}$$

By using the flex point condition to find $K3$, which means equating (11) to zero for $T/\tau=1.6$, we found

$$\frac{\tau_{PD}}{\tau} = 1 - \frac{0.31}{1 + 0.13 \cdot \left(\frac{T}{\tau}\right)^2} \tag{12}$$

The relative error of (12) with respect to the exact solution of the propagation delay is reported in Fig. 4. It can be seen that the maximum error is about 6% and is less than 2% outside the range of T/τ from 1 to 10.

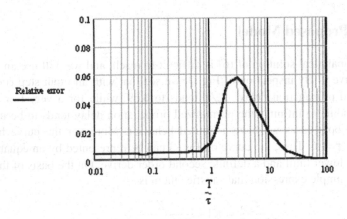

Fig. 4. Relative error of the proposed model

4 Simulation Results

To evaluate the accuracy of the model in a more realistic case, we used it to predict the propagation delay of the CML inverter depicted in Fig. 5.

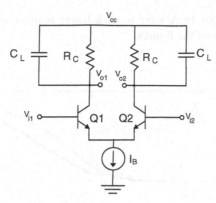

Fig. 5. Simplified schematic of a CML inverter

Following the approach proposed in [3], we can approximate the inverter with a first order circuit having the time constant

$$\tau = \frac{r_e + r_b}{1 + g_m r_e} C_{be} + r_b C_{bci}\left[1 + \frac{g_m (r_c + R_c)}{1 + g_m r_e}\right] + $$
$$+ (r_c + R_c)(C_{bci} + C_{bcx} + C_{cs}) + R_c C_L$$

(13)

where r_e, r_c and r_b are the emitter, collector and base parasitic resistances respectively, g_m is the small-signal transconductance, C_{be} is the base-emitter capacitance whose contribution is due to the junction capacitance C_{je} and the diffusion capacitance C_D, C_{bci} is the base-collector junction capacitance, C_{bcx} is the junction capacitance between collector and base contact and, finally, C_{cs} is the collector-substrate junction capacitance. To take into account the rapid movement and the wide range of the voltage, the junction capacitances were modified, from their value in a zero-bias condition, via coefficients K_j given by [12]

$$K_j = \frac{\phi^m}{V_H - V_L}\left[\frac{(\phi - V_L)^{1-m}}{1 - m} - \frac{(\phi - V_H)^{1-m}}{1 - m}\right]$$

(14)

where ϕ is the built-in potential across the junction, m is the grading coefficient of the junction, VH and VL are the minimum and maximum direct voltages across the junction.

SPICE simulations using a 6 GHz BiCMOS process tested the proposed model.

The CML inverter, was designed for three different load capacitance (0.1pF, 1pF and 10pF) and five bias current (from 100µA to 1mA) conditions. Since the power supply was 5V, to obtain a logic swing equal to 250mV we calculated the load R_c for each bias current.

Figures 6, 7 and 8 give the simulation results of the CML gate when an input ramp with variable slope is applied. Each curve was plotted for the different conditions summarized in Table I. Although for each load condition five different bias current in the range from 0.1mA to 1mA were used, a lower number of curves are shown to allow greater readability of the figures.

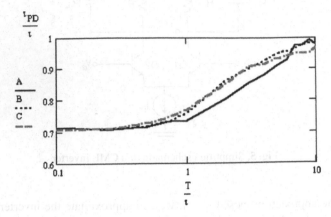

Fig. 6. Propagation delay of the CML inverter

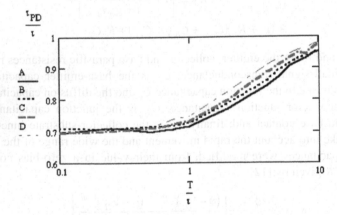

Fig. 7. Propagation delay of the CML inverter

Comparing these curves with respect to the results from the model, we found that the maximum error was represented by simulation with C_L=10pF, I_{SS}=100µA (with R_C=1250Ω and τ=12.5ns). The maximum error for the model is plotted in Fig. 9, which is slightly higher than 5% only in a narrow range. Outside is always lower than this value.

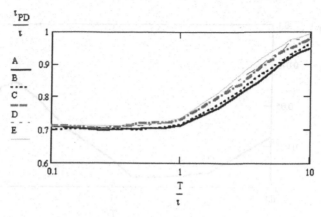

Fig. 8. Propagation delay of the CML inverter

Table 1. Different conditions considered for the simulations

Curve	C_L	I_{ss}	R_c	τ
A – Fig. 6	0.1 pF	100 μA	1250 Ω	155.6 ps
B – Fig. 6	0.1 pF	300 μA	417 Ω	69 ps
C – Fig. 6	0.1 pF	1 mA	125 Ω	57.3 ps
A – Fig. 7	1 pF	100 μA	1250 Ω	1.28 ns
B – Fig. 7	1 pF	300 μA	417 Ω	444.3 ps
C – Fig. 7	1 pF	500 μA	250 Ω	282.14 ps
D – Fig. 7	1 pF	700 μA	179 Ω	215.9 ps
A – Fig. 8	10 pF	100 μA	1250 Ω	12.5 ns
B – Fig. 8	10 pF	300 μA	417 Ω	4.2 ns
C – Fig. 8	10 pF	500 μA	250 Ω	2.5 ns
D – Fig. 8	10 pF	700 μA	179 Ω	1.82 ns
E – Fig. 8	10 pF	1 mA	125 Ω	1.29 ns

5 Conclusion

In this paper a new simple model to evaluate the propagation delay of a first order circuit with a ramp input has been presented. The proposed model is very accurate giving a maximum error lower than 6%. In order to validate the accuracy of the model in a practical case, has been considered a CML inverter which have a first order behavioral.

SPICE simulations in different load conditions and current supply were been performed confirming the accuracy of the proposed model, giving a maximum error about 5%.

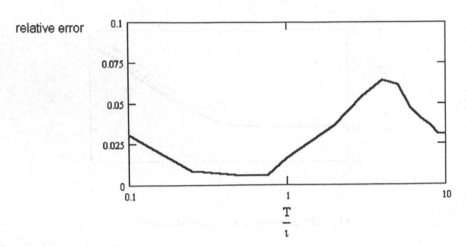

Fig. 9. Maximum error of the proposed model with respect to a CML simulated circuit

References

1. S. Kang, Y. Leblebici: CMOS Digital Integrated Circuits 2° ed. McGraw-Hill (1999)
2. K. Martin: Digital Integrated Circuit Design. Oxford University Press (2000)
3. M. Alioto, G. Palumbo: Modeling and Optimized Design of Current Mode MUX/XOR and D Flip-Flop. IEEE Trans. on CAS part II, Vol. 47, No. 5, pp. 452-461, May 2000
4. M. Alioto, G. Palumbo. Highly Accurate and Simple Models for CML and ECL Gates. IEEE Trans. on CAD, Vol. 18, No. 9, pp. 1369-1375, Sept. 1999
5. F. Dartu, N. Menezes, L. Pileggi: Performance Computation for Pre-characterized CMOS Gates with RC Loads. IEEE Trans. on CAD of Integrated Circuits and Systems, Vol. 15, No. 5, pp. 544-553, May 1996
6. V. Adler, E. Friedman: Delay and Power Expressions for a CMOS Inverter Driving a Resistive-Capacitive Load. Analog Integrated Circuits and Signal Processing, vol. 14, no. 1/2, pp. 29-39, Sept. 1997
7. M. Hafed, M. Oulmane, N. Rumin: Delay and Current Estimation in a CMOS Inverter with an RC Load. IEEE Trans. on CAD of Integrated Circuits and Systems, Vol. 20, No. 1, pp. 80-89, January 2001
8. D. Sylvester, K. Keutzer: Impact of Small Process Geometries on Microarchitectures in Systems on a Chip. Proceedings of the IEEE, Vol. 89, No. 4, pp. 467-489, April 2001
9. R. Ho, K. Mai, M. Horowitz: The Future of Wire. Proceedings of the IEEE, Vol. 89, No. 4, pp. 490-504, April 2001
10. F. Caignet, S. Delmas-Bendhia, E. Sicard: The Challenge of Signal Integrity in Deep-Submicrometer CMOS Technology. Proceedings of the IEEE, Vol. 89, No. 4, pp. 556-573, April 2001
11. E. G. Friedman, J. H. Mulligan, Jr.: Ramp input response of RC tree networks. Analog Integrated Circuits and Signal Processing, vol. 14, no. 1/2, pp. 53-58, Sept. 1997
12. M. Alioto, G. Palumbo: Oscillation Frequency in CML and ESCL Ring Oscillators. IEEE Trans. on CAS part I, Vol. 48, No. 2, pp. 210-214, February 2001.

Characterization of Normal Propagation Delay for Delay Degradation Model (DDM)*

Alejandro Millán, Jorge Juan, Manuel J. Bellido, Paulino Ruiz-de-Clavijo, and David Guerrero

Instituto de Microelectronica de Sevilla – Centro Nacional de Microelectronica
Av. Reina Mercedes, s/n (Edificio CICA) - 41012 Sevilla (Spain)
Tel.: +34 955056666 - Fax: +34 955056686
http://www.imse.cnm.es
Departamento de Tecnologia Electronica – Universidad de Sevilla
Av. Reina Mercedes, s/n (E. T. S. Ingenieria Informatica) - 41012 Sevilla (Spain)
Tel.: +34 954550974 - Fax: +34 954552764
http://www.dte.us.es
{amillan, jjchico, bellido, paulino, guerre}@imse.cnm.es

Abstract. In previous papers we have presented a very accurate model that handles the generation and propagation of glitches, which makes an important headway in logic timing simulation. This model is called *Delay Degradation Model* (DDM). Characterizing DDM completely also implies the characterization of the *normal propagation delay*. In this paper, we propose a simple heuristic model that includes its dependence on the *output load* and the *input transition time*. We have tested this model and found a mean deviation lower than 4%. Also, we present a characterization process for this model that is fully integrated into AUTODDM without affecting the total simulation time needed to characterize a standard cell.

1 Introduction

In the field of logic simulation of digital CMOS circuits, delay models exist that take into account most issues affecting accuracy [1,2,3,4]: low voltage, submicron and deep submicron devices, transition waveform, etc. There are also dynamic effects, the most important being the so-called *input collisions* [5], which happens when two or more input signals change almost simultaneously. The type of input collision that more notably affects the behaviour of digital circuits are the *glitch collisions*, or those that may cause narrow pulses or glitches. In previous papers [6,7,8] we have presented a very accurate model that handles the generation and propagation of glitches, which makes an important headway in logic-timing simulation. This model is called *Delay Degradation Model* (DDM).

One important point in any delay model (including the DDM) is the definition of the model parameters and the set up of a useful characterization process

---

* This work has been partially supported by the MCYT MODEL project TIC 2000-1350 and MCYT VERDI project TIC 2002-2283 of the Spanish Government.

B. Hochet et al. (Eds.): PATMOS 2002, LNCS 2451, pp. 477–486, 2002.

that describes how the model parameter values are obtained. This information is necessary to be able to reproduce simulation results by others and also to check the viability of the approach: a model that is very hard or expensive to characterize may be useless. In previous papers [8,9] we have described the characterization process of the degradation parameters of DDM and we have presented a tool that automates the process, called AUTODDM.

The mentioned DDM is compatible with any model for the *normal propagation delay*, where "normal" means the conventional delay considered by most logic-timing simulators when degradation effect is not taken into account. In the specialized literature there are different papers [1,2] where authors present accurate normal delay models and it would be possible to select one of these models to provide a normal propagation delay model for the DDM, though, since they are models focused on the geometric level, they are not suited to our aims.

At this time, the DDM focuses on circuits described at the gate-level, and is being implemented in a logic-timing simulator based on standard cells, called HALOTIS [10]. From this perspective, an appropriate normal propagation delay model that complements the DDM should be described at the same level. It should be simple enough to be fast and easy to implement without significant loss of accuracy, and must be also easy to characterize, possibly using the same data extracted from the DDM characterization.

In this work, we have obtained such a model for the normal propagation delay, suited for the DDM, that includes its dependence on the *output load* and the *input transition time* at the gate-level. We have also developed a characterization process and included it in the previously developed tool AUTODDM [9]. The analysis is carried out in a 0.35 μm CMOS technology using the standard cell library provided by The Foundry.

The organization of the paper is as follows: in Sect. 2 the characterization process of the degradation parameters is presented; in Sect. 3 we present the results of the normal propagation delay evaluation and we propose a simple model that fits the real behaviour very well; Sect. 4 presents the characterization process for the proposed model; finally we will finish with the main conclusions of this work.

2   Characterization Process of the Degradation Parameters

The equation to evaluate the propagation delay according to the DDM is:

$$t_p = t_{p0} \left[1 - \exp\left(-\frac{T - T_0}{\tau} \right) \right] \tag{1}$$

where T is the time elapsed since the last output transition, t_{p0} is the normal propagation delay and T_0 and τ are the degradation parameters.

For each gate, τ and T_o depend on the output load (C_L), the supply voltage (V_{DD}), the input transition time (τ_{in}) and the position of the input that is changing state (i). It has been obtained [8] that this dependence can be expressed as:

$$\tau_x V_{DD} = A_{xi} + B_{xi} C_L \tag{2}$$

$$T_{0x} = \left(\frac{1}{2} - \frac{C_{xi}}{V_{DD}} \right) \tau_{in} \tag{3}$$

where x stands for r or f depending on the sense of the output transition (rise or fall respectively).

A CMOS gate is fully characterized with respect to the degradation effect when the set $\{A_{xi}, B_{xi}, C_{xi}\}$ is obtained for each gate input. So, the objective of the characterization process is to obtain the values of the set of degradation parameters of (2) and (3) for a particular gate, i.e.:

$$\{A_{xi}, B_{xi}, C_{xi}\} \qquad x = r, f \qquad i = 1...n \tag{4}$$

The characterization process is composed of three main tasks [9]: (a) obtain t_p vs. T curves corresponding to (1); (b) obtain τ vs. C_L curves corresponding to (2); and (c) obtain T_0 vs. τ_{in} curves corresponding to (3).

The main idea in this process is to establish the adequate variation ranges of C_L and τ_{in} in order to obtain accurate values of A, B, and C. With respect to the variation of C_L, the range depends on the gate's input capacitance (C_{in}) varying between $2C_{in}$ and $10C_{in}$, while the range of τ_{in} is calculated as a function of the normal propagation delay when the input transition time is zero (this parameter is called t_{ps}). So, an adequate range for τ_{in} varies between $0.1t_{ps}$ (corresponding to sufficiently fast transitions) and $10t_{ps}$ (corresponding to sufficiently slow transitions).

3   Normal Propagation Delay Analysis and Modeling for DDM

Actually, characterizing the DDM completely also implies the characterization of the normal propagation delay (t_{p0}), and the value of t_{p0} depends on both C_L and τ_{in} [1,4]. Our main objective is to analyse the behaviour of t_{p0} in order to implement it, as part of DDM, in a logic timing simulator (HALOTIS) focused on the simulation of circuits based on standard cell libraries. The model for t_{p0} should be simple and fast in terms of computation time, though it must be accurate enough inside the C_L and τ_{in} variation ranges exposed in the previous section. This model should also be developed at the same level than the DDM (at the gate-level) providing a set of characteristic gate parameters.

We have studied the value of t_{p0} with respect to C_L and τ_{in} for three different gates: an inverter (INV), a two-inputs NAND gate (NAND2), and a two-inputs NOR gate (NOR2). For these gates we have measured the delay from each input to the output of the gate for both falling and raising output transitions. We will note each case as GATE i-R/F, where GATE is INV (inverter), NAND2 (two-inputs NAND gate), or NOR2 (two-inputs NOR gate); i is the number of the input changing; R means a rising output transition; and F means a falling one.

Figure 1.a presents the three-dimensional representation of t_{p0} with respect to C_L and τ_{in} obtained by electric simulation with HSPICE [11] (in subsequent paragraphs, we will refer to these data as a *HSPICE-grid*) for the case of INV 1-R (inverter, input 1 changing with raising output). As we can observe, the grid surface conforms practically to a plane. Figures 1.b and 1.c show the corresponding grid obtained for the cases of NAND2 2-R and NOR2 1-F respectively. In these last two figures it can be seen the same behaviour as in the first one.

Due to these results, we propose the next simple heuristic model in order to fit the normal propagation delay:

$$t_{p0} = D_{xi}C_L + E_{xi}\tau_{in} + F_{xi} \tag{5}$$

where D_{xi}, E_{xi}, and F_{xi} are the model parameters. An individual parameter value is obtained for each type of output transition (r or f, noted by x) and each input of the gate (noted by i).

This model relies on the mentioned set of parameters $\{D_{xi}, E_{xi}, F_{xi}\}$ which have to be characterized for each gate and transition type. In order to verify that this simple model correctly adjusts the gates behaviour, we have fitted these parameters using multiple linear regression over the HSPICE-grid. Figure 2 shows the same representations of Fig. 1's but, in this case, t_{p0} is calculated applying (5). It is clear that the behaviour of this simple model correctly adjusts the HSPICE-grid.

The mentioned result has been obtained for the whole set of studied cases. In table 1 we can see, for each case: the value of the parameters (D, E, and F), the mean absolute error ($\overline{err}$) in ps, and the mean deviation ($\overline{dev}$) expressed into percentages. This error measures are calculated contrasting the value of t_{p0} in the HSPICE-grid with the value obtained from the proposed model (5). It shows clearly that the approximation is adequate, since the mean deviation is always lower than 4%.

4   Characterization Process of Normal Propagation Delay

Once we have established a linear model for the value of t_{p0}, we have to develop a characterization process to be included in AUTODDM. Our intention is also to reduce the impact on the total characterization time as much as possible.

Actually, it is possible to perform an adequate characterization of the D, E, and F parameters using the same data reported by AUTODDM. This tool performs two groups of simulations: one for a set of C_L values and a fixed typical τ_{in} and the other for a set of τ_{in} values and a fixed typical C_L. So, data reported by AUTODDM provide two lines in the HSPICE-grid (Fig. 3).

Figure 4 shows the approximation obtained starting from AUTODDM data. As we can see, these values are practically the ones obtained for Fig. 2. In table 2 we present the characterization data obtained from AUTODDM results for the

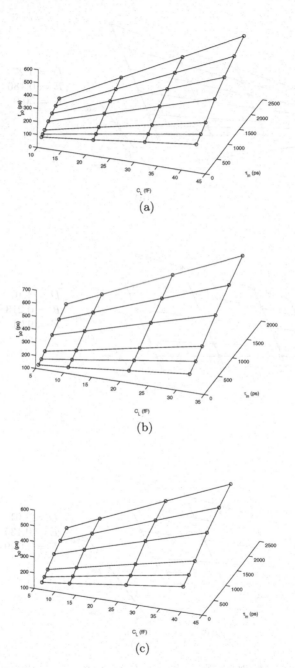

Fig. 1. HSPICE-grids for: (a) INV 1-R, (b) NAND2 2-R, and (c) NOR2 1-F

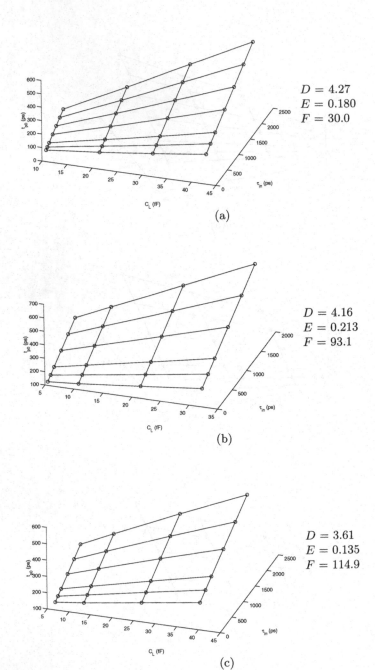

Fig. 2. Grids obtained with (5) applying multiple linear regression to HSPICE data for: (a) INV 1-R, (b) NAND2 2-R, and (c) NOR2 1-F

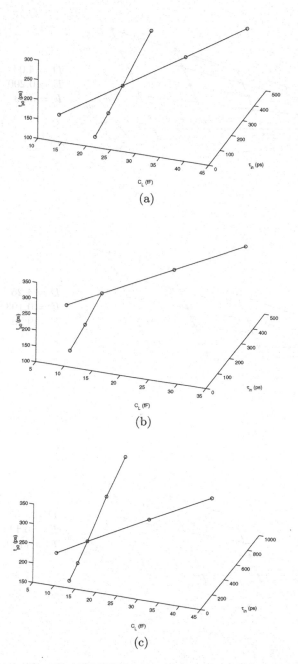

Fig. 3. Points obtained with AUTODDM for: (a) INV 1-R, (b) NAND2 2-R, and (c) NOR2 1-F

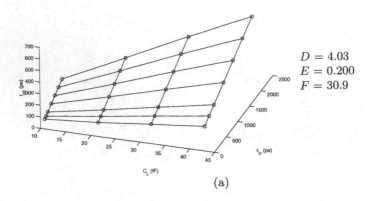

$$D = 4.03$$
$$E = 0.200$$
$$F = 30.9$$

(a)

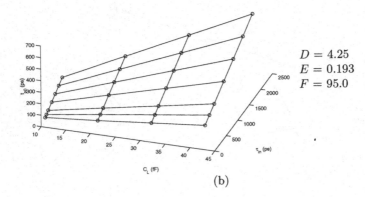

$$D = 4.25$$
$$E = 0.193$$
$$F = 95.0$$

(b)

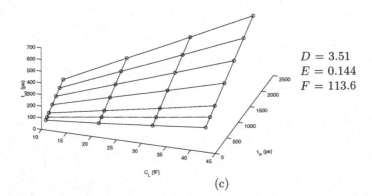

$$D = 3.51$$
$$E = 0.144$$
$$F = 113.6$$

(c)

Fig. 4. Grids obtained with (5) applying multiple linear regression to AUTODDM data for: (a) INV 1-R, (b) NAND2 2-R, and (c) NOR2 1-F

Table 1. Characterization of normal propagation delay parameters using HSPICE data

Gate	Case	D (ps/fF)	E	F (ps)	$\overline{err}$ (ps)	$\overline{dev}$ (%)
INV	1-R	4.27	0.180	30.0	4.6	1.74
INV	1-F	3.57	0.100	31.5	6.8	3.75
NAND2	1-R	4.21	0.202	55.4	2.2	1.03
NAND2	2-R	4.16	0.213	93.1	3.0	1.20
NAND2	1-F	2.77	0.083	42.4	3.9	3.27
NAND2	2-F	2.75	0.019	61.1	3.8	3.34
NOR2	1-R	4.07	0.087	99.7	3.8	1.63
NOR2	2-R	3.95	0.160	43.9	2.6	1.24
NOR2	1-F	3.61	0.135	114.9	4.8	1.58
NOR2	2-F	3.47	0.125	59.7	4.6	2.29

same cases contemplated in table 1. The error values shown in this second table have been calculated in reference to the HSPICE-grid.

So, on the one hand, the use of this simple heuristic model allows us to include the whole calculus into AUTODDM without affecting the total simulation time needed to characterize a standard cell. On the other hand, these data allow us to obtain practically the same values for the parameters D, E, and F, and to keep the mean deviation under 4%.

Table 2. Characterization of normal propagation parameters using AUTODDM data

Gate	Case	D (ps/fF)	E	F (ps)	$\overline{err}$ (ps)	$\overline{dev}$ (%)
INV	1-R	4.03	0.200	30.9	8.0	2.47
INV	1-F	3.53	0.109	31.3	7.4	3.89
NAND2	1-R	4.22	0.188	57.0	5.2	1.90
NAND2	2-R	4.25	0.193	95.0	9.8	2.72
NAND2	1-F	2.88	0.078	42.7	4.1	3.57
NAND2	2-F	2.86	0.014	62.1	3.9	3.47
NOR2	1-R	3.97	0.103	99.1	7.8	2.87
NOR2	2-R	3.89	0.167	43.6	2.7	1.15
NOR2	1-F	3.51	0.144	113.6	5.2	1.60
NOR2	2-F	3.53	0.120	61.1	4.9	2.59

5 Conclusions

Characterizing the DDM completely also implies the characterization of the normal propagation delay (t_{p0}), and the value of t_{p0} depends on both C_L and τ_{in}. This paper presents the analysis we have carried out about the value of t_{p0} in a 0.35 μm CMOS technology. In this way, we have proposed a simple heuristic model for t_{p0} that includes its dependence on the output load (C_L) and the input transition time (τ_{in}). Despite its simplicity, the model is accurate enough in the range of interest. Also, we have presented a characterization process for this model that is fully integrated into AUTODDM without affecting the total simulation time needed to characterize a standard cell. Finally, we have tested this model and found that its mean deviation is, for all cases, lower than 4%.

References

1. Bisdounis, L., Nikolaidis, S., Koufopavlou, O.: Analytical transient response and propagation delay evaluation of the CMOS inverter for short-channel devices. IEEE Journal of Solid-State Circuits **33(2)** (February 1998) 302–306
2. Daga, J. M., Auvergne, D.: A comprehensive delay macro modeling for submicrometer CMOS logics. IEEE Journal of Solid-State Circuits **34(1)** (January 1999)
3. Kayssi, A. I., Sakallah, K. A., Mudge, T. N.: The impact of signal transition time on path delay computation. IEEE Transactions on Circuits and Systems II: Analog and Digital Signal Processing **40(5)** (May 1993) 302–309
4. Auvergne, D., Azemard, N., Deschacht, D., Robert, M.: Input waveform slope effects in CMOS delays. IEEE Journal of Solid-State Circuits **25(6)** (December 1990) 1588–1590
5. Melcher, E., Rothig, W., Dana, M.: Multiple input transitions in CMOS gates. Microprocessing and Microprogramming **35**, North Holland, (1992) 683–690
6. Bellido-Diaz, M. J., Juan-Chico, J., Acosta, A. J., Valencia, M., Huertas, J. L.: Logical modelling of delay degradation effect in static CMOS gates. IEE Proc. Circuits Devices and Systems **147(2)** (April 2000) 107–117
7. Juan-Chico, J., Ruiz-de-Clavijo, P., Bellido, M. J., Acosta, A. J., Valencia, M.: Inertial and degradation delay model for CMOS logic gates. In Proc. IEEE International Symposium on Circuits and Systems (ISCAS), Geneva, (May 2000) I-459-462
8. Juan-Chico, J., Ruiz-de-Clavijo, P., Bellido, M. J., Acosta, A. J., Valencia, M.: Degradation delay model extension to CMOS gates. In Proc. Power and Timing Modelling, Optimization and Simulation (PATMOS) (September 2000) 149–158
9. Juan-Chico, J., Bellido, M. J., Ruiz-de-Clavijo, P., Baena, C., Valencia, M.: AUTODDM: AUTOmatic characterization tool for the Delay Degradation Model. In Proc. 8th IEEE International Conference on Electronics, Circuits and Systems (ICECS), Malta, (September 2001) 1631–1634
10. Ruiz-de-Clavijo, P., Juan, J., Bellido, M. J., Acosta, A. J., Valencia, M.: HALOTIS: High Accuracy Logic Timing Simulator with Inertial and Degradation Delay Model. Design, Automation and Test in Europe (DATE) Conference and Exhibition, Munich (Germany), (March 2001)
11. HSPICE User's Manual. Meta-Software (1999)

Automated Design Methodology for CMOS Analog Circuit Blocks in Complex Systems

Razvan Ionita[1], Andrei Vladimirescu[2], and Paul Jespers[3]

[1] Polytechnic Institute, Bucharest, Romania, and,
Institut Superieur d'Electronique de Paris, ISEP,
21 rue d'Assas, 75006 Paris, France,
[2] University of California, Berkeley,
BWRC, 2208 Allston Way, Berkeley, CA 94704 , and,
Insitut Superieur d'Electronique de Paris, ISEP, Paris, France,
[3] Laboratoire de Microelectronique, Universite Catholique de Louvain,
B-1348 Louvain-la-Neuve,Belgium

Abstract. A design methodology for analog circuit blocks is proposed which combines circuit knowledge for predefined topologies with CAD simulation tools. In a more complex analog or mixed-signal system design metrics for each block are first established during an exploration phase based on analytical descriptions. In the proposed approach behavioral equations for each circuit are derived using a basic EKV model. Selection of a performance point in the design space results in a first sizing of the transistors in each block. This is followed by SPICE verification using foundry-provided process data. Additional numerical optimization is applied in a second iteration for closing the gap with the requirement specification.

1 Introduction

Analog circuit blocks continue to play an important role in mixed-signal systems built today which interface to the outside world through analog front-ends. While complexity of circuits continues to grow design cycles shrink. This trend has revived an older interest in analog synthesis and some products are already available on the market [1], [2].

This paper presents an approach to automating the design of analog CMOS blocks based solely on circuit knowledge and existing simulation tools. The most important step in this methodology is to capture the input to output behavior of the circuit blocks in straightforward mathematical expressions. Matlab [3] is one possible choice for capturing the equations and performing the design exploration phase. Although this step is time consuming it is necessary to understanding circuit operation and generating a robust design. Once the analytical performance description has been generated

B. Hochet et al. (Eds.): PATMOS 2002, LNCS 2451, pp. 487–493, 2002.
© Springer-Verlag Berlin Heidelberg 2002

re-targeting the same topology for another specification or a finer technology becomes trivial.

Design exploration is an essential step for tradeoffs among the different requirements such as gain, bandwidth, power and area. Visualization of the design space is an essential productivity tool and is the step which determines the selection of the appropriate circuit topology among those that have been pre-characterized for meeting all design constraints. Sec. 2 presents the key elements for deriving the performance equations for a two-stage Miller CMOS opamp and the design space information which can be obtained at this stage. Once a performance point has been selected the widths and lengths of the transistors can be generated.

One challenge in developing behavioral representations of basic analog blocks such as operational transconductance amplifiers (OTAs), opamps, low-noise amplifiers (LNA) or mixers was the perceived absence of a simple enough yet sufficiently accurate MOSFET model for submicron devices for hand calculations. The EKV model [4] describes operation from subthreshold to strong inversion and can be stripped down to a format which is only slightly more complex than the Shichman-Hodges (Level=1) quadratic model. As described in Sec. 3 the model used in deriving the performance equations does not need to match the complexity of device models used in simulations such as BSIM3 [5] or MOS9 [6] as long as the appropriate current- and charge-level are predicted.

The next step in the design is performing the SPICE verification with actual process data. Inevitably there will be some discrepancy between predicted and simulated performance. As discussed in Sec. 3 a very important value is the output conductance of the transistor; this value depends very strongly on the operating point as well as some second-order corrections which are impractical for hand calculations. The values of the output conductance which depend strongly on geometry and operating point proves to be principal source of error between exploration and accurate simulation. Another iteration can be done at the Matlab level with refined parameters based on the SPICE simulation.

The last step in the electrical design is to apply the optimization capability of simulators like Eldo or Hspice to minimize or maximize a goal function subject to a number of constraints. This may or may not lead to a better design.

The electrical design has to be followed by the physical implementation; this step should be automated using a capability such as NeoCell [2] which can generate a first layout starting from the SPICE netlist and predefined transistor cells.

A final verification is mandatory which includes the parasitic elements obtained from the layout extraction. Unpleasant surprises can be minimized at this step if the geometries of the predefined layout cells with all parasitics are already included in the SPICE simulation phase.

2 Design Space Exploration for Miller CMOS Opamp

The g_m/I_D methodology proposed by Silveira et al. [7], [8], is used to derive the performance characteristics for the desired circuit. The g_m/I_D function of the normalized current $(I_D/(W/L))$ relates transconductance to current level (power) and therefore provides an excellent tool for deriving the geometry of transistors for given gain and power. Furthermore, an equation relating the current to the transconductance and the inversion level a can be established using the EKV formulation:

$$I_D = nV_{th}g_m \frac{\sqrt{a}}{1-\exp(-\sqrt{a})} \tag{1}$$

where n is the subthreshold slope and a is the ratio between I_D and the specific current I_S:

$$I_s = 2nV_{th}^2\mu C_{ox} \tag{2}$$

with V_{th} the thermal voltage.

The design equations for a simple two-stage CMOS Miller opamp as shown in Fig. 1 can be derived following the criteria below.

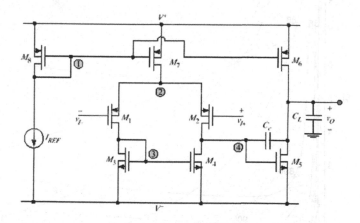

Fig. 1. Two-stage Miller Opamp

Transistors M_1 and M_2 are sized based on the desired bandwidth with

$$g_{m1} = \omega_T C_m \tag{3}$$

For stability reasons the non-dominant second pole and the zero need to be placed to satisfy the phase margin requirement, e.g., for a phase margin of 60° the two frequencies can be

$$\omega_{NDP} = 2.2\omega_T\,; \omega_Z = 10\omega_T \qquad (4)$$

The position of the right-hand-plane zero can be assured by setting $g_{m5} = 10g_{m1}$. Transistors M_3 and M_4 are sized identically for minimizing offset by setting $V_{G3} = V_{G4}$; the two transistors also need to respect the equality of current with M_1 and M_2. The remaining transistors M_6, M_7 and M_8 are sized in a way to provide the desired current levels in the differential pair and second stage. The widths and lengths of the transistors can be computed from the following relation:

$$\frac{W}{L} = \frac{I_D}{2nV_{th}^2\mu C_{ox}}\frac{1}{a} \qquad (5)$$

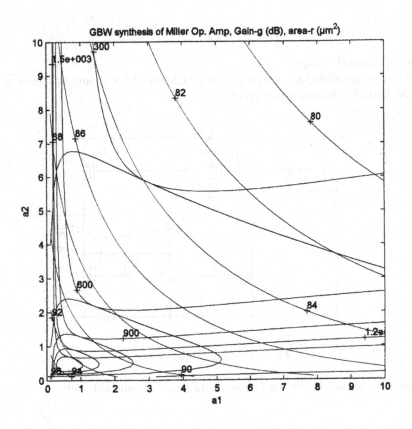

Fig. 2. Design Space showing equal area, gain and current supply curves with $a1$ and $a2$ as parameters;

The inversion levels a of transistors M_1 and M_2 on one hand, and, M_5 on the other hand, can be taken as parameters. A graphical representation of the design space is shown in Fig. 2 for an $f_T = 5$ MHz; the inversion level $a1$ for M_1 and M_2 is on the x-axis and $a2$ for M_5 on the y-axis. The graph contains curves of equal gain superposed with five curves of constant total current and area. Based on the primary design objective a point can be chosen in this design space resulting in the bias level for the differential pair and second stage for a desired gain and area.

The main challenges for a very accurate estimation are the output conductances of devices M_1 to M_6, and, the capacitances at nodes 3, 4 and the output. The principal discrepancies in the SPICE verification phase are caused by the above.

With the unity-gain frequency and phase margins set by design we choose as performance point the minimum area design which achieves best performance in gain and power at the intersection with equi-gain and equi-power curves. This design point corresponds to both stages operating in moderate inversion with $a_1=2.9$ and $a_2=6$. The resulting W and Ls for this design point in a 0.25μ technology are listed in Table 1.

Table 1. Transistor Sizes estimated for the chosen design point

Transistor	W(μm)	L(μm)
M1-M2	10.8	0.75
M3-M4	13.6	1.75
M5	15	0.25
M6	15	0.75
M7	10	2.5
M8	10	2.5

3 Circuit Performance Verification

The remaining step is to verify the design with SPICE. As mentioned in the introduction the main challenge for obtaining an accurate design the first-time around is to match the transistor characteristics of the hand-evaluation model used in the analytical equations with those described by the process files used in SPICE.

There are several ways of calibrating the hand model to the simulation model. The best approach is to derive the transconductance factor k' from simulated $I_D = f(V_{DS}, V_{GS})$ curves. The threshold voltage can be taken directly from the SPICE process file. The capacitance contributions to nodes 3, 4 and *out* also require the thin-oxide capacitance and junction, area and perimeter, information provided by SPICE in the *Operating Point Information* section.

The gain and phase plot produced by SPICE for the Miller opamp is shown in Fig. 3. It can be seen that the two leading design goals, bandwidth and phase margin are met; the simulated results are $f_T = 8.79$ MHz and phase margin of 69°. The gain of 83 dB also matches very well the prediction of 84 dB of the performance point in Fig. 2.

The value of the gain is very sensitive to the operating point of the transistors and the Early voltage parameter in the estimation model. The good fit obtained is after fine tuning the Early voltages for each transistor according to the channel lengths selected and the operating points which result after a first simulation. The actual current consumption also results higher by 10-15% than the design curves due to geometry effects and threshold voltage corrections not available in the simple prediction model. The graph also shows the gain and phase at the output node of the differential pair.

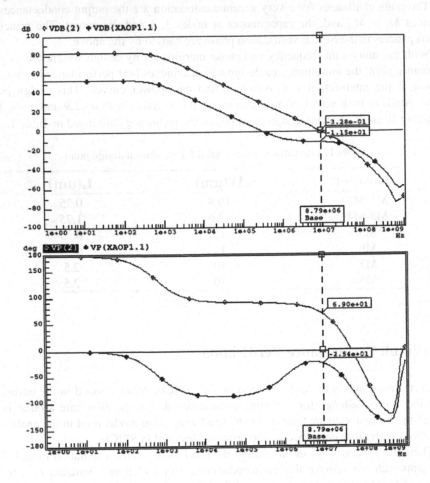

Fig. 3. Simulated Gain in dB and Phase for the opamp output VDB(2), VP(2), and, first stage

The charge effects can be incorporated accurately in the exploration model; this can be noticed in the accurate positioning of the poles and the zero and the resulting bandwidth and phase margin.

The merit of this approach is that once the transistor model parameters used in Matlab are fine tuned to the process parameters all transistors are sized automatically to the desired performance goal and the resulting design matches the simulated circuit within less than 10%.

Further gains in performance can be envisaged by applying numeric optimization available in HSPICE or Eldo. First results show that optimization works very well at the linear circuit level but greatly affects the large-signal bias point during the optimization process. This results in taking some transistors out of the saturation region leading to wrong solutions and failure of the optimization process. The desired approach is to perform small-signal gain-bandwidth optimization with large-signal DC operating point constraints placed on individual transistors.

4 Conclusion

An automated design methodology for analog circuits has been presented based on good circuit knowledge and available simulation tools. This approach applied to the design of complex mixed-signal systems can considerably reduce design time and allows for easy technology and application re-targeting in conjunction with design reuse.

The major up front investment is in developing the analytical description for the circuits of interest. Once this is done the library of behavioral descriptions can be reused for the entire specification spectrum which applies to a given circuit topology. The same analytical formulations can be used whether the circuit is implemented in CMOS bulk or SOI.

References

1. Prado Synthesis Platform, Barcelona Design, Sunnyvale, CA, 2002.
2. NeoCircuit and NeoCell, Neolinear Inc., Pittsburgh, PA, 2002.
3. The Mathworks Inc., Natick, MA, 2002.
4. C. Enz, F. Krummenacher, and, E.A. Vittoz, The EKV MOSFET Model, Technical Report, Swiss Federal Institute of Technology (EPFL), Lausanne, Switzerland, 1997.
5. Y. Cheng and C. Hu, *MOSFET Modeling & BSIM3 User's Guide,* Kluwer Academic Publishers, Boston, 1999.
6. R. Velghe, D. Klaasen and F. Klaasen, Compact MOS modeling for analog circuit simulation, *IEDM Technical Digest,* 1993, pp.485–488.
7. F. Silveira, D. Flandre, and P. Jespers, A g_m/I_D Based Methodology for the Design of CMOS Analog Circuits and its Application to the Synthesis of a Silicon-on-Insulator Micropower OTA, *IEEE JSSC,* vol. 31, No 9, Sept 1996, pp. 1314–1319.
8. P. Jespers, CMOS Analog Circuits, *Lecture Notes,* ISEP, 2001.

Author Index

Lecture Notes in Computer Science

For information about Vols. 1–2358
please contact your bookseller or Springer-Verlag

Vol. 2398: K. Miesenberger, J. Klaus, W. Zagler (Eds.), Computers Helping People with Special Needs. Proceedings, 2002. XXII, 794 pages. 2002.

Vol. 2399: H. Hermanns, R. Segala (Eds.), Process Algebra and Probabilistic Methods. Proceedings, 2002. X, 215 pages. 2002.

Vol. 2400: B. Monien, R. Feldmann (Eds.), Euro-Par 2002 – Parallel Processing. Proceedings, 2002. XXIX, 993 pages. 2002.

Vol. 2401: P.J. Stuckey (Ed.), Logic Programming. Proceedings, 2002. XI, 486 pages. 2002.

Vol. 2402: W. Chang (Ed.), Advanced Internet Services and Applications. Proceedings, 2002. XI, 307 pages. 2002.

Vol. 2403: Mark d'Inverno, M. Luck, M. Fisher, C. Preist (Eds.), Foundations and Applications of Multi-Agent Systems. Proceedings, 1996-2000. X, 261 pages. 2002. (Subseries LNAI).

Vol. 2404: E. Brinksma, K.G. Larsen (Eds.), Computer Aided Verification. Proceedings, 2002. XIII, 626 pages. 2002.

Vol. 2405: B. Eaglestone, S. North, A. Poulovassilis (Eds.), Advances in Databases. Proceedings, 2002. XII, 199 pages. 2002.

Vol. 2406: C. Peters, M. Braschler, J. Gonzalo, M. Kluck (Eds.), Evaluation of Cross-Language Information Retrieval Systems. Proceedings, 2001. X, 601 pages. 2002.

Vol. 2407: A.C. Kakas, F. Sadri (Eds.), Computational Logic: Logic Programming and Beyond. Part I. XII, 678 pages. 2002. (Subseries LNAI).

Vol. 2408: A.C. Kakas, F. Sadri (Eds.), Computational Logic: Logic Programming and Beyond. Part II. XII, 628 pages. 2002. (Subseries LNAI).

Vol. 2409: D.M. Mount, C. Stein (Eds.), Algorithm Engineering and Experiments. Proceedings, 2002. VIII, 207 pages. 2002.

Vol. 2410: V.A. Carreño, C.A. Muñoz, S. Tahar (Eds.), Theorem Proving in Higher Order Logics. Proceedings, 2002. X, 349 pages. 2002.

Vol. 2412: H. Yin, N. Allinson, R. Freeman, J. Keane, S. Hubbard (Eds.), Intelligent Data Engineering and Automated Learning – IDEAL 2002. Proceedings, 2002. XV, 597 pages. 2002.

Vol. 2413: K. Kuwabara, J. Lee (Eds.), Intelligent Agents and Multi-Agent Systems. Proceedings, 2002. X, 221 pages. 2002. (Subseries LNAI).

Vol. 2414: F. Mattern, M. Naghshineh (Eds.), Pervasive Computing. Proceedings, 2002. XI, 298 pages. 2002.

Vol. 2415: J.R. Dorronsoro (Ed.), Artificial Neural Networks – ICANN 2002. Proceedings, 2002. XXVIII, 1382 pages. 2002.

Vol. 2417: M. Ishizuka, A. Sattar (Eds.), PRICAI 2002: Trends in Artificial Intelligence. Proceedings, 2002. XX, 623 pages. 2002. (Subseries LNAI).

Vol. 2418: D. Wells, L. Williams (Eds.), Extreme Programming and Agile Methods – XP/Agile Universe 2002. Proceedings, 2002. XII, 292 pages. 2002.

Vol. 2419: X. Meng, J. Su, Y. Wang (Eds.), Advances in Web-Age Information Management. Proceedings, 2002. XV, 446 pages. 2002.

Vol. 2420: K. Diks, W. Rytter (Eds.), Mathematical Foundations of Computer Science 2002. Proceedings, 2002. XII, 652 pages. 2002.

Vol. 2421: L. Brim, P. Jančar, M. Křetínský, A. Kučera (Eds.), CONCUR 2002 – Concurrency Theory. Proceedings, 2002. XII, 611 pages. 2002.

Vol. 2423: D. Lopresti, J. Hu, R. Kashi (Eds.), Document Analysis Systems V. Proceedings, 2002. XIII, 570 pages. 2002.

Vol. 2425: Z. Bellahsene, D. Patel, C. Rolland (Eds.), Object-Oriented Information Systems. Proceedings, 2002. XIII, 550 pages. 2002.

Vol. 2426: J.-M. Bruel, Z. Bellahsene (Eds.), Advances in Object-Oriented Information Systems. Proceedings, 2002. IX, 314 pages. 2002.

Vol. 2430: T. Elomaa, H. Mannila, H. Toivonen (Eds.), Machine Learning: ECML 2002. Proceedings, 2002. XIII, 532 pages. 2002. (Subseries LNAI).

Vol. 2431: T. Elomaa, H. Mannila, H. Toivonen (Eds.), Principles of Data Mining and Knowledge Discovery. Proceedings, 2002. XIV, 514 pages. 2002. (Subseries LNAI).

Vol. 2435: Y. Manolopoulos, P. Návrat (Eds.), Advances in Databases and Information Systems. Proceedings, 2002. XIII, 415 pages. 2002.

Vol. 2436: J. Fong, C.T. Cheung, H.V. Leong, Q. Li (Eds.), Advances in Web-Based Learning. Proceedings, 2002. XIII, 434 pages. 2002.

Vol. 2438: M. Glesner, P. Zipf, M. Renovell (Eds.), Field-Programmable Logic and Applications. Proceedings, 2002. XXII, 1187 pages. 2002.

Vol. 2440: J.M. Haake, J.A. Pino (Eds.), Groupware: Design, Implementation and Use. Proceedings, 2002. XII, 285 pages. 2002.

Vol. 2442: M. Yung (Ed.), Advances in Cryptology – CRYPTO 2002. Proceedings, 2002. XIV, 627 pages. 2002.

Vol. 2443: D. Scott (Ed.), Artificial Intelligence: Methodology, Systems, and Applications. Proceedings, 2002. X, 279 pages. 2002. (Subseries LNAI).

Vol. 2444: A. Buchmann, F. Casati, L. Fiege, M.-C. Hsu, M.-C. Shan (Eds.), Technologies for E-Services. Proceedings, 2002. X, 171 pages. 2002.

Vol. 2445: C. Anagnostopoulou, M. Ferrand, A. Smaill (Eds.), Music and Artificial Intelligence. Proceedings, 2002. VIII, 207 pages. 2002. (Subseries LNAI).

Vol. 2447: D.J. Hand, N.M. Adams, R.J. Bolton (Eds.), Pattern Detection and Discovery. Proceedings, 2002. XII, 227 pages. 2002. (Subseries LNAI).

Vol. 2451: B. Hochet, A.J. Acosta, M.J. Bellido (Eds.), Integrated Circuit Design. Proceedings, 2002. XVI, 496 pages. 2002.

Vol. 2453: A. Hameurlain, R. Cicchetti, R. Traunmüller (Eds.), Database and Expert Systems Applications. Proceedings, 2002. XVIII, 951 pages. 2002.

Vol. 2454: Y. Kambayashi, W. Winiwarter, M. Arikawa (Eds.), Data Warehousing and Knowledge Discovery. Proceedings, 2002. XIII, 339 pages. 2002.

Vol. 2455: K. Bauknecht, A M. Tjoa, G. Quirchmayr (Eds.), E-Commerce and Web Technologies. Proceedings, 2002. XIV, 414 pages. 2002.